ALL ■ IN ■ ONE

Fundamentals of Engineering FE Civil

EXAM GUIDE

ABOUT THE AUTHOR

Indranil Goswami, Ph.D., P.E., is a registered Professional Engineer in Maryland and has been a professor at Morgan State University in Baltimore. He is currently a senior consulting engineer. He is the author of *Civil Engineering PE All-in-One Exam Guide* and was the recipient of the 2016 Engineering Education Excellence Award from the National Society of Professional Engineers.

ALL ■ IN ■ ONE

Fundamentals of Engineering FE Civil

EXAM GUIDE

Indranil Goswami, Ph.D., P.E.

McGraw Hill Education

New York • Chicago • San Francisco • Athens
London • Madrid • Mexico City • Milan
New Delhi • Singapore • Sydney • Toronto

Library of Congress Control Number: 2018931205

Fundamentals of Engineering FE Civil All-in-One Exam Guide

1 2 3 4 5 6 7 8 9 LCR 23 22 21 20 19 18

ISBN 978-1-260-01134-0
MHID 1-260-01134-8

Sponsoring Editor
Lauren Poplawski

Editing Supervisor
Stephen M. Smith

Production Supervisor
Lynn M. Messina

Acquisitions Coordinator
Elizabeth M. Houde

Project Manager
Jyotsna Ojha, Cenveo® Publisher Services

Copy Editor
Surendra Shivam, Cenveo Publisher Services

Proofreader
Manish Tiwari, Cenveo Publisher Services

Art Director, Cover
Jeff Weeks

Composition
Cenveo Publisher Services

This book is dedicated to the memory of my mother and my uncle Kakamoni, both of whom were a source of endless support and inspiration for me but could not see it published.

CONTENTS

PREFACE

This book was written to assist the student preparing for the Fundamentals of Engineering (FE) Civil examination. It consists of condensed review material and approximately 440 solved examples of difficulty level appropriate for this exam. As of January 2014, the FE Civil exam has consisted of approximately 110 questions in the following subject areas:

Mathematics	7–11 problems
Probability and Statistics	4–6 problems
Computational Tools	4–6 problems
Ethics and Professional Practice	4–6 problems
Engineering Economics	4–6 problems
Statics	7–11 problems
Dynamics	4–6 problems
Mechanics of Materials	7–11 problems
Materials	4–6 problems
Fluid Mechanics	4–6 problems
Hydraulics and Hydrologic Systems	8–12 problems
Structural Analysis	6–9 problems
Structural Design	6–9 problems
Geotechnical Engineering	9–14 problems
Transportation	8–12 problems
Environmental Engineering	6–9 problems
Construction	4–6 problems
Surveying	4–6 problems

In this book, the content is organized in the same order as the official (NCEES) syllabus. Whenever appropriate, solutions include specific suggestions and relevant strategic advice.

When attempting the practice problems, students are advised to use a digital (PDF) copy of the *FE Reference Handbook* (currently in version 9.4), so they can get accustomed to navigating it in electronic form, rather than as a paper document. Make sure that you use the most current version because there have been significant changes in organization and content from the eighth to the ninth edition.

On the actual exam, the examinee will have 5 hours and 20 minutes to answer 110 questions. That is an average pace of about 3 minutes per problem. The first 55 questions will represent the first half of the exam (covered in Chaps. 1 to 10 in this book). A break will be offered after the first half is completed, no matter how long this takes. After the break, the examinee will use the remaining time to complete the second half of the exam (covered in Chaps. 11 to 18). Thus, an examinee may take only an hour to complete the first 55 questions and, after the break, will have more than 4 hours available to answer the remaining 55 questions. Another examinee may do exactly the opposite.

In these practice problems, I have tried to keep the difficulty level reasonable. Some are easier than others. However, some of the questions, in spite of being more difficult than the "3-minute" variety, have been included because they reinforce a critical area of learning. On the actual test, there will invariably be a mix of questions of various difficulty levels, and a well-prepared examinee will be able to recognize quickly whether a particular problem is easy enough to attempt right away or one that should be put aside and come back to later. This is a very important strategy for an exam as fast-paced and as diverse as the FE Civil exam.

Indranil Goswami, Ph.D., P.E.

UNITS AND UNIT CONVERSIONS

Angle

1 radian = 57.29578 degrees

1 gradian = 0.9 degree = 0.015708 radian

Length

1 ft = 0.3048 m = 0.3333 yd

1 m = 3.281 ft = 39.37 in = 1.0936 yd

1 micron = 1 μm = 1.0×10^{-6} m

1 angstrom = 1.0×10^{-10} m

1 mil = 0.001 in

1 mile = 1760 yd = 5280 ft = 1.6093 km

1 yd = 3 ft = 36 in = 0.9144 m

Area

1 acre = 43,560 ft^2 = 1/640 sq. mile = 0.40469 hectare (ha)

1 darcy = 9.86923×10^{-13} m^2 (approx. the area of 1 micron × 1 micron)

1 ha = 10,000 m^2 = 2.471 acres = 107,639 ft^2 = 0.00386 sq. mile

1 sq. km = 247.105 acres = 100 ha = 1,195,990.05 yd^2

1 sq. mile = 640 acres = 259 ha

Volume

1 acre-ft = 43,560 ft^3 = 325,851.43 gal (U.S.)

1 fl. oz. (U.S.) = 1/128 gal (U.S.) = 29.574 mL = 1.80469 in^3

1 ft^3 = 7.4805 gal (U.S.) = 28.3168 L = 957.5 fl. oz.

1 gal (Imperial) = 4.5461 L

1 gal (U.S.) = 0.13368 ft^3 = 3.7854 L = 4 quarts = 8 pints = 16 cups = 128 fl. oz.

1 L = 0.2642 gal (U.S.) = 0.03531 ft^3

1 gal (water) = 8.34 lb

1 ft^3 (water) = 62.45 lb

1 lb (water) = 0.11983 gal = 0.016 ft^3

Velocity

1 mph = 1.46667 ft/s = 0.447 m/s = 1.609 kmph

1 m/s = 3.281 ft/s = 3.6 kmph = 2.237 mph

Mach 1 (speed of sound at standard conditions) = 1225.1 kmph = 761.2 mph

Volumetric Flow Rate

1 in^3/s = 1.63871 × 10^{-5} m^3/s = 5.787 × 10^{-4} ft^3/s

1 ft^3/s (cusec) = 0.02832 m^3/s = 448.83 gal/min = 0.6463 mgd

1 m^3/s (cumec) = 35.3147 ft^3/s = 15,850 gal/min = 22.8245 mgd

1 gal/min = 0.002228 ft^3/s = 0.00144 mgd

1 mgd = 1.5472 ft^3/s = 0.0438 m^3/s = 694.44 gal/min

1 acre-in/h = 1.00833 ft^3/s

Permeability

1 cm/s = 9.225 × 10^8 gal/acre/day

1 in/h = 2 ft/day = 6.517 × 10^5 gal/acre/day

Mass

1 kg = 2.204 lb-m = 1 Ns^2/m

1 slug = 1 lb-s^2/ft = 14.59 kg

1 tonne (metric ton) = 1000 kg

Density

1 g/cm^3 = 1000 kg/m^3 = 62.45 lb-m/ft^3

Runoff or Precipitation Depth

1 acre-ft/sq. mile = 0.01875 in

1 in = 53.333 acre-ft/sq. mile

Pressure/Stress

1 std atm = 1.01325 bar = 1.013 × 10^5 Pa = 101.325 kPa = 14.696 lb/in^2 = 2116.2 lb/ft^2

1 std atm = 10.33 m water = 33.9 ft water

1 std atm = 29.92 in mercury (Hg) = 760 mm Hg

1 bar = 1.0 × 10^5 Pa = 100 kPa

1 kip/in^2 (ksi) = 6.895 MPa = 6.895 N/mm^2

1 MPa = 0.14504 ksi = 145.04 psi

1 ton/ft^2 = 13.889 lb/in^2 = 95.76 kPa = 0.945 std atm

Force

1 lb = 4.448 N

1 kip = 4448 N = 4.448 kN = 0.5 ton (U.S.)

1 N = 1 kg-m/s^2 = 0.22481 lb = 1.0 × 10^5 dynes

1 dyne = 1 g-cm/s^2 = 2.2481 × 10^{-6} lb

1 ton (U.S.) = 2000 lb = 2 kips = 8896 N = 8.896 kips

Energy

1 lb-ft = 1.3558 J = 0.001285 Btu = 0.32383 cal (calorie)

1 Btu = 778.169 lb-ft = 1055.056 J = 2.928 × 10^{-4} kW-h

1 cal = 4.1868 J = 3.088 lb-ft

Power

1 hp = 550 lb-ft/s = 745.7 W = 0.746 kW = 42.41 Btu/min = 0.707 Btu/s

1 kW = 56.87 Btu/min = 0.948 Btu/s = 737.6 lb-ft/s = 1.341 hp

Concentration

1 mg/L = 8.3454 lb-m/Mgal

1 mg/L = 1 ppm (if solvent is water, since 1 L of water = 1 kg)

Earth Parameters

Earth radius = 6371 km = 3960 miles (approx.)

Earth mass = 5.9736 × 10^{24} kg

Universal Constants

Acceleration due to gravity (g) = 9.80665 m/s^2 = 32.174 ft/s^2

Atomic mass unit (m_u = 1 u) = 1.66054 × 10^{-27} kg

Avogadro number (N_A) = 6.022141 × 10^{23}/mol

Boltzmann constant (k) = 1.38065 × 10^{-23} J/K

Electric permittivity of free space (ε_o) = 8.854187… × 10^{-12} F/m

Electron charge (e) = 1.602176 × 10^{-19} C

Electron rest mass (m_e) = 9.109382×10^{-31} kg

Electron volt (eV) = 1.602176×10^{-19} J

Euler's constant (γ) = 0.577215664901532...

Faraday's constant (F) = 96485.34 C/mol

Gravitational constant (G) = 6.6738×10^{-11} N-m²/kg²

Magnetic permeability of free space (μ_o) = $4\pi \times 10^{-7} = 1.25664 \times 10^{-6}$ H/m

Molar volume of an ideal gas (at 0°C) = 22.414 L

Natural (Naperian) logarithm base (e) = 2.718281828459045...

Pi (π) = 3.14159265358979...

Planck constant (h) = 6.62607×10^{-34} J-s

Speed of light in vacuum (c) = 2.99792458×10^{8} m/s

Stefan-Boltzmann constant (σ) = 5.6704×10^{-8} W/m^2-K^4

Universal gas constant (R) = 8.314462 J/mol-K

CHAPTER 1

Mathematics

Analytic Geometry

Straight Line

The equation of a straight line in the x-y plane can be described in various ways.

Slope-Intercept Form

The line is represented as

$$y = mx + b \tag{1.1}$$

where m is the slope ("rise over run," or Δy over Δx) and b is the y intercept (the y coordinate of the point where the line crosses the y axis).

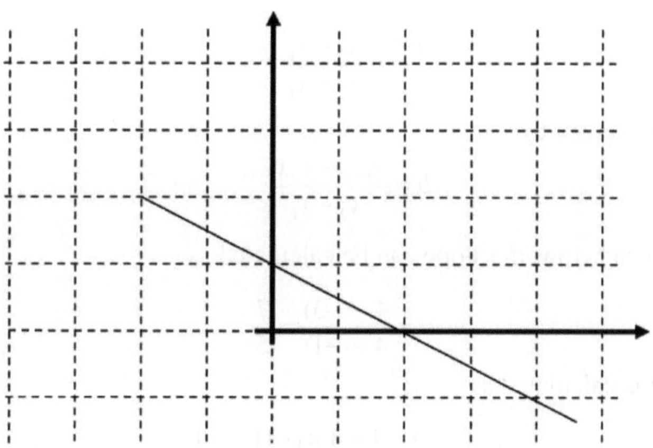

In the example shown above, the "rise-over-run" is $(-1)/(+2) = -0.5$. The line crosses the y axis at (0, 1). Therefore, the y intercept is +1.

The equation of the line is

$$y = -0.5x + 1 \Rightarrow x + 2y = 2$$

Double-Intercept Form

The straight line equation can also be expressed in the double-intercept form

$$\frac{x}{a}+\frac{y}{b}=1 \tag{1.2}$$

where a and b are the x and y intercepts, respectively

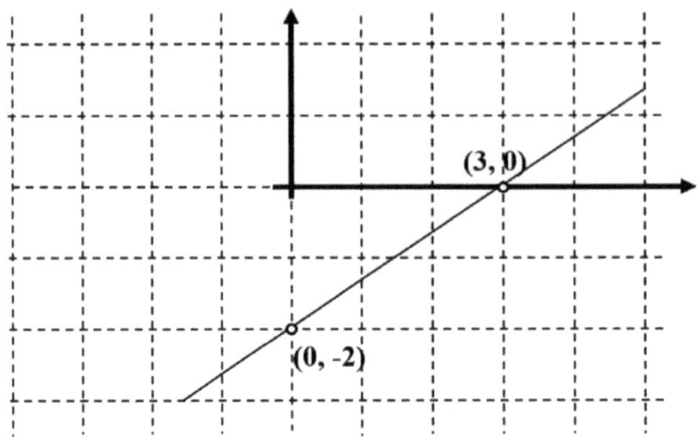

The equation of the line shown above (x intercept = +3, y intercept = −2) is

$$\frac{x}{3}+\frac{y}{-2}=1 \Rightarrow 2x-3y=6$$

Two-Point Form

A straight line may be defined using two points (x_1, y_1) and (x_2, y_2). The slope is given by

$$m=\frac{y_2-y_1}{x_2-x_1} \tag{1.3}$$

and the y intercept is given by

$$b=\frac{y_1x_2-y_2x_1}{x_2-x_1} \tag{1.4}$$

For the line shown below, the slope can be calculated as

$$m=\frac{4-(-3)}{4-(-1)}=\frac{7}{5}$$

and the y intercept is calculated as

$$b=\frac{(-3)\times 4-4\times(-1)}{4-(-1)}=-\frac{8}{5}$$

The equation of the line is

$$y=\frac{7}{5}x-\frac{8}{5} \Rightarrow 7x-5y=8$$

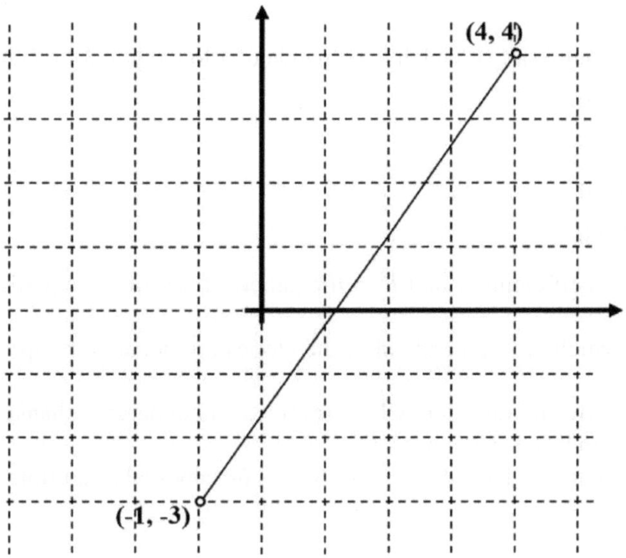

Slope and One Point Form

When the slope m and one point (x_1, y_1) are known, the equation of the line is

$$y - y_1 = m(x - x_1) \tag{1.5}$$

For a line with slope = –0.8 and passing through the point (5, 3), the equation is

$$y - 3 = -0.8(x - 5) = -0.8x + 4 \Rightarrow 0.8x + y = 7 \Rightarrow 4x + 5y = 35$$

Parallel and Perpendicular Lines

Two lines which are parallel have the same slope:

$$m_1 = m_2$$

Two lines that are perpendicular have slopes m_1 and m_2 such that

$$m_1 m_2 = -1$$

$$m_1 = -\frac{1}{m_2}$$

Angle between Two Straight Lines

Angle α between two lines with slope m_1 and m_2 is calculated as

$$\alpha = \tan^{-1}\left(\frac{m_1 - m_2}{1 + m_1 m_2}\right) \tag{1.6}$$

Example 1.1

Which of the following is a line which is perpendicular to the line $3x + 7y + 2 = 0$?

A. $7x + 3y + 8 = 0$

B. $3x - 7y + 1 = 0$

C. $7x - 3y + 5 = 0$

D. $3x + 7y + 1 = 0$

Solution The coefficients A and B in the general form of the equation yield the slope $m = -A/B$.

For two lines which are perpendicular, the slopes are negative reciprocal of each other ($m_1 = -1/m_2$).
This means that the A:B relationship will be reversed and undergo a change in sign.
The original equation has $A = 3$ and $B = 7$.
The perpendicular line will have $A = +7$ and $B = -3$ (or any scalar multiple).

Answer is C.

Example 1.2

What is the equation of the line shown in the figure?

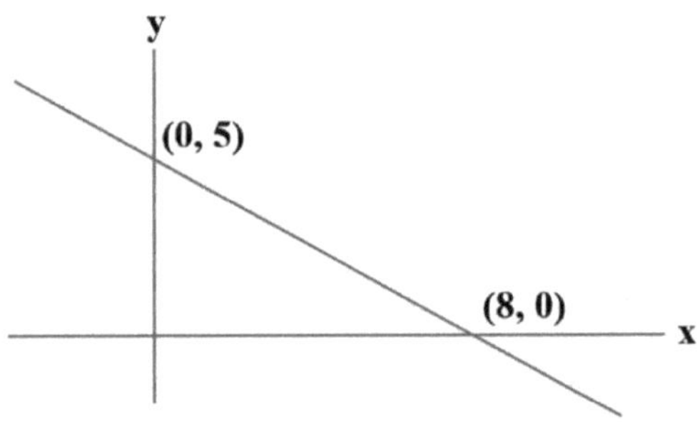

A. $8x + 5y - 25 = 0$

B. $8x - 5y = 0$

C. $5x + 8y + 25 = 0$

D. $5x + 8y - 40 = 0$

Solution The double-intercept form of a straight line can be used here:

$$\frac{x}{8} + \frac{y}{5} = 1 \Rightarrow 5x + 8y = 40$$

The slope-intercept form of the straight line can also be used. The y intercept is 5 and the slope can be computed using $m = \Delta y/\Delta x = -5/8$.

Therefore, the equation is $y = mx + b = -\frac{5}{8}x + 5 \Rightarrow 8y = -5x + 40 \Rightarrow 5x + 8y = 40$

Answer is D.

Distance Formula

In the two-dimensional x-y plane, the distance between two points (x_1, y_1) and (x_2, y_2) is given by

$$d = \sqrt{(x_2 - x_1)^2 + (y_2 - y_1)^2} \tag{1.7}$$

In three dimensions, the analogous formula is

$$d = \sqrt{(x_2 - x_1)^2 + (y_2 - y_1)^2 + (z_2 - z_1)^2} \tag{1.8}$$

Quadratic Equation

The quadratic equation, if expressed in the standard form $ax^2 + bx + c = 0$ has the roots

$$x_1, x_2 = \frac{-b \pm \sqrt{b^2 - 4ac}}{2a} \tag{1.9}$$

When $b^2 - 4ac = 0$, this results in two identical real roots.

When $b^2 - 4ac < 0$, this results in two complex roots, which are conjugates. For the special case when $b = 0$, the two roots are pure imaginary.

When $b^2 - 4ac > 0$, this results in two distinct real roots.

Trigonometry

The basic definitions of trigonometric functions are based on the right triangle shown in Fig. 1.1. The functions $\sin(\theta)$, $\cos(\theta)$, and $\tan(\theta)$ are defined as

$$\sin\theta = \frac{\text{opposite}}{\text{hypotenuse}} = \frac{AB}{AC} \tag{1.10a}$$

$$\cos\theta = \frac{\text{adjacent}}{\text{hypotenuse}} = \frac{BC}{AC} \tag{1.10b}$$

$$\tan\theta = \frac{\text{opposite}}{\text{adjacent}} = \frac{AB}{BC} \tag{1.10c}$$

Figure 1.1 Right triangle.

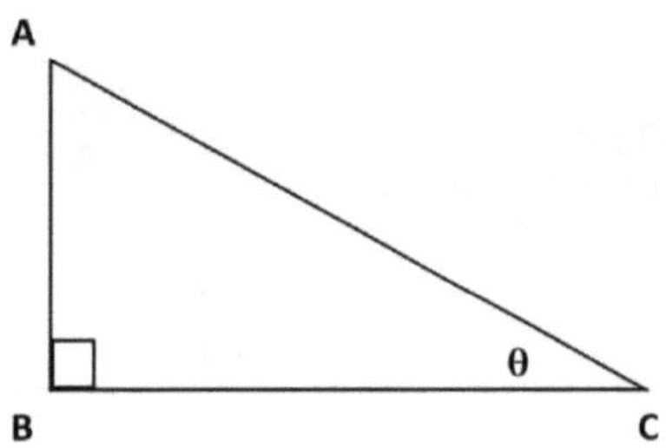

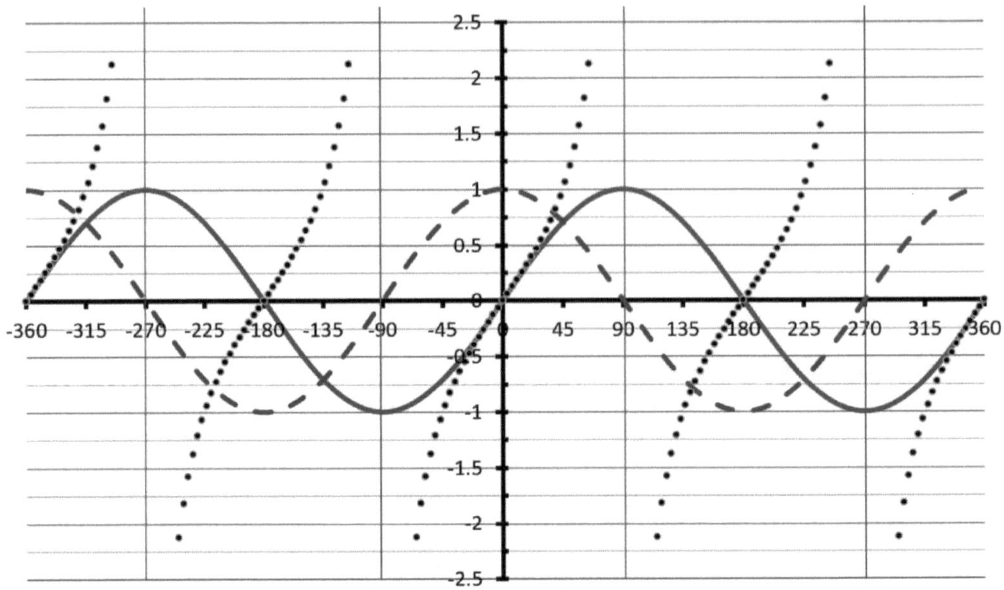

Figure 1.2 Plots of sin(θ), cos(θ), and tan(θ) functions within the domain $-2\pi < \theta < 2\pi$.

The sine and cosine functions have a period of 360°, or 2π radians, while the tangent function has a period of 180° (π radians). The plots of these functions over the range -2π to 2π are shown in Fig. 1.2. As may be seen, the sine (solid line) and cosine (broken line) functions are identical except for a "phase shift" of 90°. These functions are bounded between the extreme values of −1 and +1. On the other hand, the tangent function (dotted line) approaches $+\infty$ or $-\infty$ for angles which are odd multiples of 90°. In Fig. 1.2, these vertical asymptotes are shown at $\theta = -270°, -90°, +90°$, and $+270°$.

Law of Sines and Law of Cosines

For any triangle (not necessarily a right triangle), as shown in Fig. 1.3, the law of sines states:

$$\frac{a}{\sin A} = \frac{b}{\sin B} = \frac{c}{\sin C} \tag{1.11}$$

Figure 1.3
Triangle geometry.

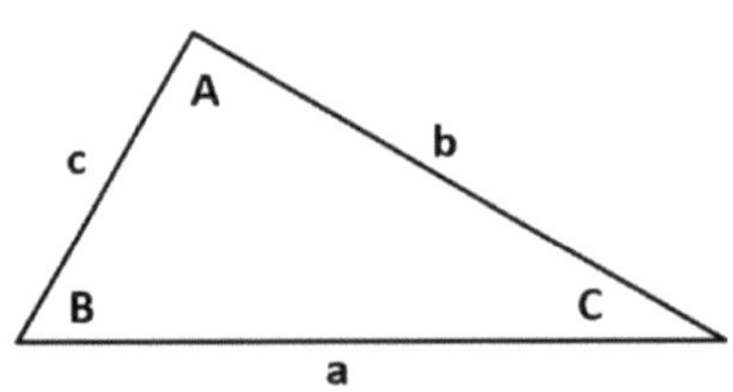

and the law of cosines can be written

$$a^2 = b^2 + c^2 - 2bc\cos A \tag{1.12a}$$

$$b^2 = a^2 + c^2 - 2ac\cos B \tag{1.12b}$$

$$c^2 = a^2 + b^2 - 2ab\cos C \tag{1.12c}$$

Example 1.3

For the triangle shown below, what is the angle A?

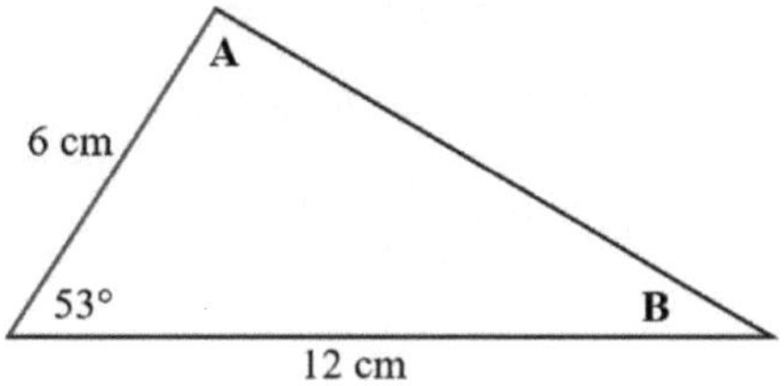

A. 81.3°

B. 82.1°

C. 82.8°

D. 83.5°

Solution Designating the side opposite the angle of 53° as "x," we can use the law of cosines to find side length "x"

$$x^2 = 6^2 + 12^2 - 2\times 6\times 12\times \cos 53 = 93.34$$
$$x = 9.66 \text{ cm}$$

Now, using law of sines,

$$\frac{x}{\sin 53} = \frac{12}{\sin A} \Rightarrow A = \sin^{-1}\left(\frac{12\sin 53}{9.66}\right) = 82.8°$$

Answer is C.

Trigonometric Identities

The secondary trigonometric functions are cosecant (csc), secant (sec), and cotangent (cot), which are the reciprocals of the sine, cosine, and tangent functions, respectively.

$$\csc\theta = \frac{1}{\sin\theta};\ \sec\theta = \frac{1}{\cos\theta};\ \cot\theta = \frac{1}{\tan\theta}$$

Based on these basic relationships, several trigonometric identities can be derived.

$$\sin^2\theta + \cos^2\theta = 1 \tag{1.13a}$$

$$\sec^2\theta - \tan^2\theta = 1 \tag{1.13b}$$

$$\csc^2\theta - \cot^2\theta = 1 \tag{1.13c}$$

Sum of Angles

$$\sin(A \pm B) = \sin A \cos B \pm \cos A \sin B \tag{1.14a}$$

$$\cos(A \pm B) = \cos A \cos B \mp \sin A \sin B \tag{1.14b}$$

$$\tan(A \pm B) = \frac{\tan A \pm \tan B}{1 \mp \tan A \tan B} \tag{1.14c}$$

Double Angles

$$\sin(2A) = 2\sin A \cos A \tag{1.15a}$$

$$\cos(2A) = \cos^2 A - \sin^2 A \tag{1.15b}$$

$$\tan(2A) = \frac{2\tan A}{1 - \tan^2 A} \tag{1.15c}$$

Products

$$\sin A \cos B = \frac{1}{2}\{\sin(A+B) + \sin(A-B)\} \tag{1.16a}$$

$$\sin A \sin B = \frac{1}{2}\{\cos(A-B) - \cos(A+B)\} \tag{1.16b}$$

$$\cos A \cos B = \frac{1}{2}\{\cos(A-B) + \cos(A+B)\} \tag{1.16c}$$

Sums

$$\sin A + \sin B = 2\sin\left(\frac{A+B}{2}\right)\cos\left(\frac{A-B}{2}\right) \tag{1.17a}$$

$$\sin A - \sin B = 2\cos\left(\frac{A+B}{2}\right)\sin\left(\frac{A-B}{2}\right) \tag{1.17b}$$

$$\cos A + \cos B = 2\cos\left(\frac{A+B}{2}\right)\cos\left(\frac{A-B}{2}\right) \tag{1.17c}$$

$$\cos A - \cos B = -2\sin\left(\frac{A+B}{2}\right)\sin\left(\frac{A-B}{2}\right) \tag{1.17d}$$

Half Angles

$$\sin\left(\frac{A}{2}\right) = \pm\sqrt{\frac{1-\cos A}{2}} \tag{1.18a}$$

$$\cos\left(\frac{A}{2}\right) = \pm\sqrt{\frac{1+\cos A}{2}} \tag{1.18b}$$

$$\tan\left(\frac{A}{2}\right) = \pm\sqrt{\frac{1-\cos A}{1+\cos A}} \tag{1.18c}$$

Example 1.4

Which of the following is equal to $4\sin(50)\sin(120)$?

A. 2 sin 170 – 2 cos 70

B. 2 cos 70 – 2 cos 170

C. 2 cos 170 – 2 cos 70

D. 2 sin 170 – 2 sin 70

Solution According to Eq. (1.17d), $\cos A - \cos B = -2\sin\left(\frac{A+B}{2}\right)\sin\left(\frac{A-B}{2}\right)$. In this example, if $A = 70$ and $B = 170$, the right-hand side of the identity becomes –2 sin(120) sin(–50) = $2\sin(120)\sin(50) = \cos(70) - \cos(170)$. Therefore, $4\sin(50)\sin(120) = 2\cos 70 - 2\cos 170$.

Answer is B.

Conic Sections

Conic sections are four different curves that are generated when a plane intersects a double cone, as shown in Fig. 1.4.

If the intersecting plane makes an angle θ with the axis of the cone, then we see from Fig. 1.4 that

- When the angle $\theta = 90°$, the intersection of the plane and the cone is a circle, which is a closed curve.
- When the angle $\phi < \theta < 90°$, the intersection of the plane and the cone is an ellipse, which is also a closed curve.

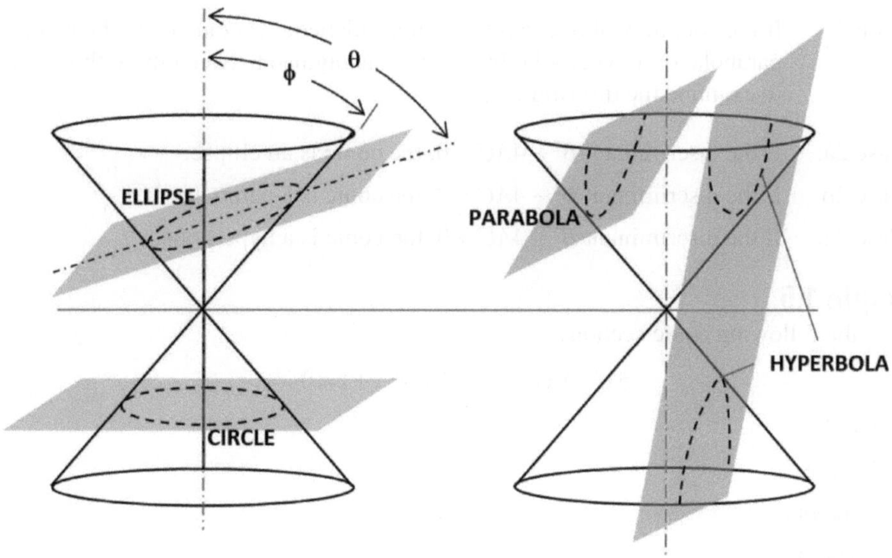

Figure 1.4 Conic sections.

- When the angle $\theta = \phi$, the intersection of the plane and the cone is a parabola, which is an open curve.
- When the angle $0° < \theta < \phi$, the intersection of the plane and the cone is a hyperbola, which is an open curve composed of two halves.

Eccentricity

The eccentricity of a conic section is defined as

$$e = \frac{\cos\theta}{\cos\phi} \tag{1.19}$$

where θ is the angle between the axis of the cone and the cutting plane, and ϕ is the cone angle.

Thus, for a circle, eccentricity $e = 0$. For an ellipse, $0 < e < 1$. A parabola has $e = 1$ and for a hyperbola, eccentricity $e > 1$.

Generalized Equation of Conic Sections

All of the four curves mentioned (circle, ellipse, parabola, and hyperbola) are plane curves of second order, which means that they can be fully described using an equation involving no higher than second-order terms in two variables x and y. Let's say that the general equation of these conic sections is

$$Ax^2 + Bxy + Cy^2 + Dx + Ey + F = 0 \tag{1.20}$$

Obviously, if all three second-order terms were missing ($A = B = C = 0$), the equation would describe a (first-order) straight line $Dx + Ey + F = 0$. Thus, in order to qualify as a conic section, we must have *at least one* nonzero coefficient A, B, or C.

Case 1: If $B = 0$ and $A = C$, the conic is a *circle.*

Case 2: If the condition above (case 1) is not satisfied, the conic could be an ellipse, parabola, or a hyperbola. In order to distinguish between these three, we need to examine the discriminant $B^2 - 4AC$.

Case 2a: If the discriminant $B^2 - 4AC < 0$, the conic is an ellipse.

Case 2b: If the discriminant $B^2 - 4AC = 0$, the conic is a parabola.

Case 2c: If the discriminant $B^2 - 4AC > 0$, the conic is a hyperbola.

Example 1.5

What is the following conic section?

$$5x^2 + 8xy + 3y^2 + 7x - 8y + 1 = 0$$

A. Circle

B. Ellipse

C. Parabola

D. Hyperbola

Solution The necessary conditions for a *circle* are $A = C \neq 0$ and $B = 0$. These conditions are not met. Therefore, it is either an ellipse, parabola, or hyperbola. The discriminant is calculated as

$$B^2 - 4AC = 8^2 - 4 \times 5 \times 3 = 4 > 0$$

Since the discriminant is greater than zero, the equation describes a hyperbola.

Answer is D.

Circle

The equation of a circle (see Fig. 1.5) with center at (h, k) and radius $= r$ is given by

$$(x-h)^2 + (y-k)^2 = r^2 \tag{1.21}$$

An important skill in determining the parameters of a circle is the method of completion of squares.

Example 1.6

The equation of a circle is given as

$$2x^2 + 8x + 2y^2 + 12y - 6 = 0$$

Determine the center and radius of the circle.

Solution The coefficient (B) of the xy term is zero and the coefficients of the x^2 and y^2 terms (A and C, respectively) are equal. So, this is a circle.

$$2x^2 + 8x + 2y^2 + 12y - 6 = 0$$
$$x^2 + 4x + y^2 + 6y = 3$$

To complete the x and y quadratics, we need to add $(2)^2$ and $(3)^2$ on the left-hand side

$$x^2 + 4x + 2^2 + y^2 + 6y + 3^2 = 3 + 4 + 9 = 16$$
$$(x+2)^2 + (y+3)^2 = 4^2$$

Therefore, the center of the circle is at (–2, –3) and radius is 4.

Figure 1.5 Parameters of the circle.

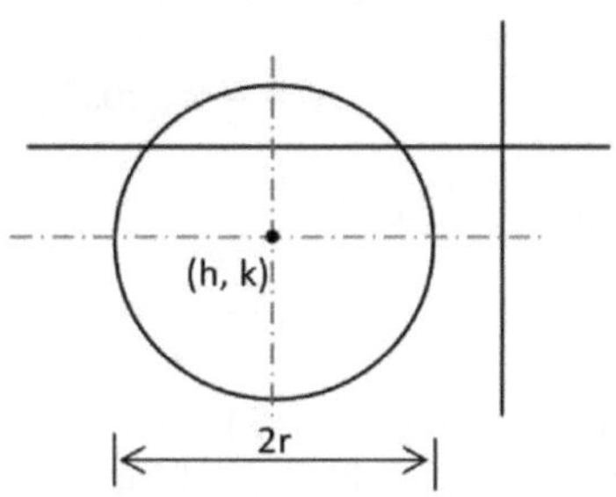

Example 1.7

A tangent is drawn from the point (10, 1) to the circle with center at (4, 6) and radius 3. What is the angle α between the tangent and the radial line?

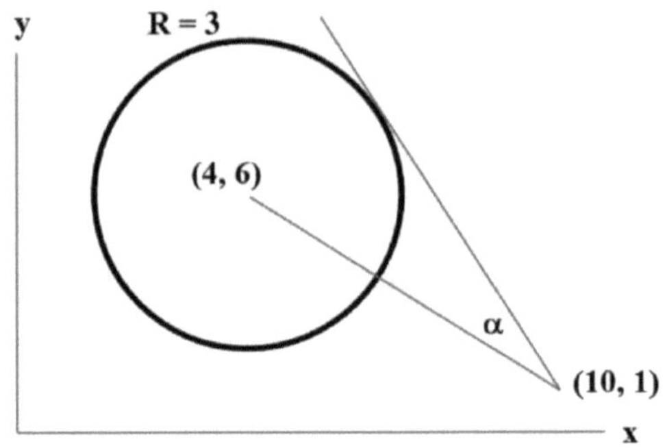

A. 22.6°

B. 24.6°

C. 36.8°

D. 65.4°

Solution The general form of the circle shown is

$$(x-4)^2+(y-6)^2=3^2$$

The length of the tangent (to the point of tangency) is calculated as

$$\sqrt{(x'-h)^2+(y'-k)^2-r^2}=\sqrt{(10-4)^2+(1-6)^2-3^2}=7.21$$

$\tan\alpha = 3/7.21 = 0.416$

$\alpha = 22.6°$

Answer is A.

Mensuration (Segment and Sector of a Circle)

The segment of a circle (shaded) defined by a central angle ϕ is shown in Fig. 1.6. The angle ϕ (radians) is related to the arc length s according to

$$\phi=\frac{s}{r}=2\cos^{-1}\left(\frac{r-d}{r}\right) \tag{1.22}$$

The area of the segment is given by

$$A=\frac{r^2}{2}(\phi-\sin\phi) \tag{1.23}$$

In this expression, the first term ($r^2\phi/2$) is the area of the circular sector and the second term subtracted from it ($r^2\sin\phi/2$) is the area of the triangle.

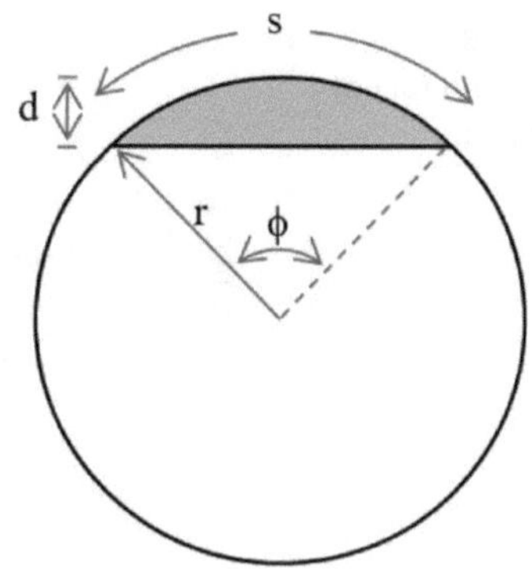

Figure 1.6 Segment of a circle.

Ellipse

An ellipse is the locus of a point such that the sum of its distances from two fixed points (the foci) is constant. The ellipse could be described as an oval or a distorted circle.

Ellipse with Axes Parallel to Cartesian Axes

When the axes of the ellipse are parallel to the Cartesian axes (see Fig. 1.7) and the center of the ellipse is located at coordinates (h, k), the standard form of the ellipse is

$$\left(\frac{x-h}{a}\right)^2+\left(\frac{y-k}{b}\right)^2=1 \tag{1.24}$$

For this shape, the coefficients A and C are unequal but of the same sign and the coefficient $B = 0$. The method of completion of squares is useful again to convert the equation of an ellipse to standard form.

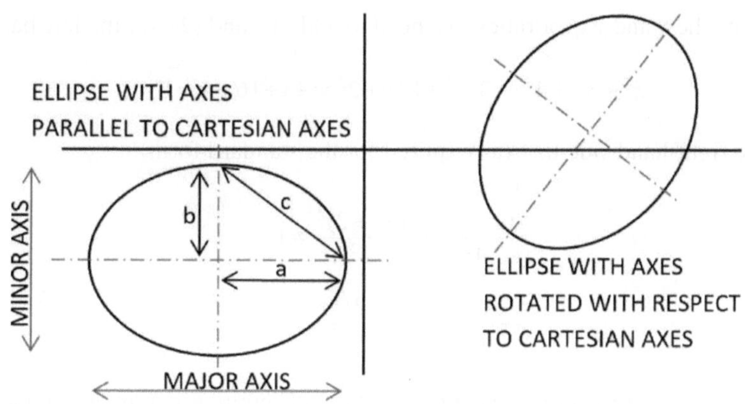

Figure 1.7 Ellipse.

Figure 1.8
Center and foci of an ellipse.

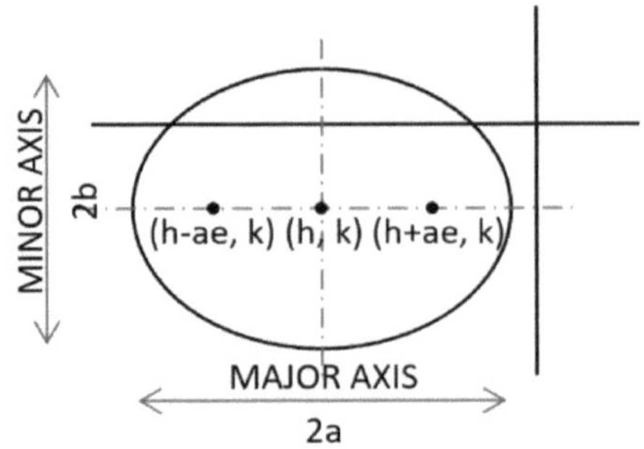

The eccentricity of an ellipse is given by

$$e = \sqrt{1 - \frac{b^2}{a^2}} \tag{1.25}$$

The distance from the center of the ellipse to either focus is equal to the product of the semimajor axis length (a) and the eccentricity (e). See Fig. 1.8.

Example 1.8

The equation of an ellipse is given as

$$x^2 + 8x + 9y^2 + 12y + 4 = 0$$

Determine the parameters (center and length of axes) of the ellipse.

Solution The coefficient (B) of the xy term is zero and the coefficients of the x^2 and y^2 terms (A and C, respectively) are *unequal but of the same sign*. So, this is an ellipse with axes parallel to the Cartesian axes.

$$x^2 + 8x + 9y^2 + 12y + 4 = 0$$

To complete the x and y quadratics, we need to add $(4)^2$ and $(2)^2$ on the left-hand side

$$x^2 + 8x + 4^2 + (3y)^2 + 12y + 2^2 = -4 + 16 + 4 = 16$$

To reduce the right-hand side to 1 (as required for the standard form)

$$\frac{(x+4)^2}{4^2} + \frac{(3y+2)^2}{4^2} = 1$$

$$\frac{(x+4)^2}{4^2} + \frac{(y+2/3)^2}{(4/3)^2} = 1$$

Therefore, the center of the ellipse is at $(-4, -2/3)$ and semimajor axis, $a = 4$ and semiminor axis, $b = 4/3$.

Example 1.9

What is the eccentricity of the ellipse whose equation is $4x^2 + y^2 - 3x + 7y - 2 = 0$?

A. 0.87

B. 0.75

C. 0.50

D. 0.25

Solution The *FE Reference Handbook*, version 9.4, shows an ellipse with axes that are horizontal and vertical. Such an ellipse would have no mixed (xy) terms. In order to extract the a and b coefficients (all that's needed for the calculation of eccentricity), one needs to simply see that the standard form will be like

$$\frac{(x-h)^2}{1^2} + \frac{(y-k)^2}{2^2} = 1 \rightarrow a = 1; b = 2$$

Calculation of the rest is not necessary for this question. Therefore, the ratio of minor axis to major axis = $a/b = 1/2 = 0.5$.

Therefore, eccentricity, $e = \sqrt{1 - 0.5^2} = 0.866$.

Answer is A.

NOTE The *Handbook* gives the eccentricity equation for the special case "when $h = k = 0$." This expression for the eccentricity is dependent only on the length of the minor and the major axes and therefore is true for the more general case when h and k are not zero as well.

Mensuration (Perimeter and Enclosed Area of an Ellipse)

For an ellipse whose semimajor axis length is a and semiminor axis length is b (see Fig. 1.7), the area enclosed by an ellipse is given by $A = \pi ab$

The perimeter is given by

$$P = \pi(a+b)\left[1 + \frac{\lambda^2}{4} + \frac{\lambda^4}{64} + \cdots\right] \tag{1.26}$$

where $\lambda = \dfrac{a-b}{a+b}$.

The perimeter P can be approximated by

$$P \approx 2\pi\sqrt{\left(\frac{a^2+b^2}{2}\right)} \approx \sqrt{2}\pi c \tag{1.27}$$

Example 1.10

What is the area of the ellipse whose equation is $4x^2+y^2-3x+7y-2=0$?

A. 541

B. 172

C. 93

D. 23

Solution In this case, it is not enough to simply calculate the ratio b/a, but the actual lengths "a" and "b," because the area is asked for.

The given equation can be adapted into the standard form by the completion of squares:

Step 1: Complete the squares

$$4\left[x^2-2\left(\frac{3}{8}\right)x+\left(\frac{3}{8}\right)^2\right]+\left[y^2+2\left(\frac{7}{2}\right)y+\left(\frac{7}{2}\right)^2\right]-2=4\left(\frac{3}{8}\right)^2+\left(\frac{7}{2}\right)^2=\frac{205}{16}$$

$$4\left[x-\frac{3}{8}\right]^2+\left[y+\frac{7}{2}\right]^2=\frac{237}{16}$$

$$\frac{\left(x-\frac{3}{8}\right)^2}{\frac{237}{64}}+\frac{\left(y+\frac{7}{2}\right)^2}{\frac{237}{16}}=1$$

Therefore, semimajor axis length, $a=\sqrt{(237/16)}$ and semiminor axis length, $b=\sqrt{(237/64)}$. Therefore, area, $A=\pi ab=23.27$.

Answer is D.

Hyperbola

Hyperbola with Axes Parallel to Cartesian Axes

When the axes of a hyperbola are parallel to the Cartesian axes and the center of the hyperbola is located at coordinates (h, k), the standard form of the hyperbola is

$$\left(\frac{x-h}{a}\right)^2-\left(\frac{y-k}{b}\right)^2=1 \tag{1.28}$$

For this shape, the coefficients A and C are of opposite sign and the coefficient $B=0$. The hyperbola consists of two halves which are mirror images of each other and are "contained" by two asymptotes, as shown in Fig. 1.9. The two asymptotes have slope $\pm b/a$.

The eccentricity of a hyperbola is given by

$$e=\sqrt{1+\frac{b^2}{a^2}} \tag{1.29}$$

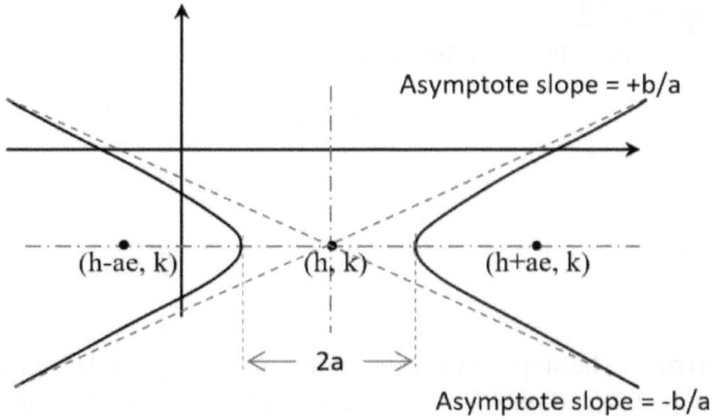

Figure 1.9 Hyperbola.

The distance from the center of the hyperbola to either focus is equal to the product of the semimajor axis length a and the eccentricity e.

Example 1.11

The equation of a hyperbola is given as

$$x^2 + 8x - 4y^2 + 12y - 13 = 0$$

Determine the parameters of the hyperbola.

Solution The coefficient B of the xy term is zero and the coefficients of the x^2 and y^2 terms (A and C, respectively) are of *opposite sign*. So, this is a hyperbola with axes parallel to the Cartesian axes.

$$x^2 + 8x - 4y^2 + 12y - 13 = 0$$

The method of completion of squares is used again, as in Example 1.10.

To complete the x and y quadratics, we need to add $(4)^2$ and subtract $(3)^2$ on the left-hand side

$$x^2 + 8x + 4^2 - (2y)^2 + 12y3^2 = 13 + 16 - 9 = 20$$

To reduce the right-hand side to 1 (as required for the standard form)

$$\frac{(x+4)^2}{20} - \frac{(2y-3)^2}{20} = 1$$

$$\frac{(x+4)^2}{20} - \frac{(y-1.5)^2}{5} = 1$$

Therefore, the center of the hyperbola is at (–4, 1.5) and semimajor axis, $a = \sqrt{20} = 2\sqrt{5}$ and semiminor axis, $b = \sqrt{5}$.

The asymptotes have slope $m_1, m_2 = \pm b/a = \pm 1/2$.

Example 1.12

What is the angle between the asymptotes of the hyperbola whose equation is

$$9(x-2^2)-16(y-5)^2=144$$

A. 37°

B. 74°

C. 126°

D. 143°

Solution The slope of the asymptotes of any hyperbola is given by $\pm b/a$. The given equation can be adapted into the standard form by reducing the right side to 1:

$$\frac{(x-2)^2}{16}-\frac{(y-5)^2}{9}=1$$

Therefore, $a=4$ and $b=3$.

$$\tan\theta=\pm\frac{3}{4}\Rightarrow\theta=\pm 36.9°$$

Therefore, the acute angle between the asymptotes $=2\theta=73.8°$ and the obtuse angle $=106.2°$.

Answer is B.

Rectangular Hyperbola

A special form of the hyperbola is the rectangular hyperbola which has asymptotes perpendicular to each other. When these asymptotes are the Cartesian axes themselves, the equation of the hyperbola is of the form

$$xy=c \tag{1.30}$$

When the constant c is positive, the hyperbola consists of two curves in the 1st and 3rd quadrants (solid lines in Fig. 1.10). When c is negative, the curves are in the 2nd and 4th quadrants (broken lines).

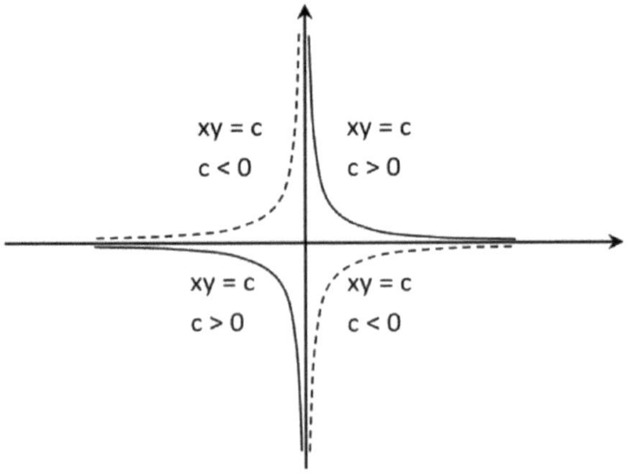

Figure 1.10 Rectangular hyperbola with Cartesian axes as asymptotes.

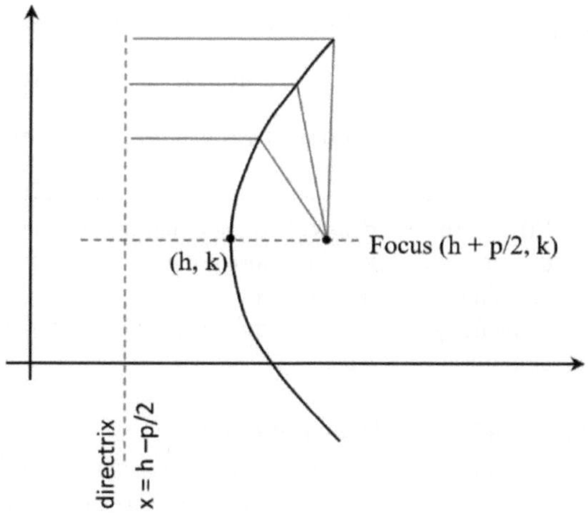

Figure 1.11
Parabola.

Parabola

A parabola is the locus of a point which is equidistant from a point (the focus) and a line (the directrix).

Figure 1.11 shows a parabola which is linear in the variable x and quadratic in the variable y [Eq. (1.31)]. Such a parabola has a horizontal line as its axis of symmetry. The positive sign of the variable p makes the parabola open to the right as shown in Fig. 1.11. The equation of the parabola shown is

$$(y-k)^2 = 2p(x-h) \tag{1.31}$$

The directrix of the parabola shown is the line $x = h - p/2$.

On the other hand, if the equation had been linear in y and quadratic in x, then it would open upward ($p > 0$) or downward ($p < 0$) and its equation would be

$$(x-h)^2 = 2p(y-k) \tag{1.32}$$

The latter one is not shown in the *Handbook*. Its directrix would have the equation $y = k - p/2$.

Example 1.13

The equation of a parabola is given as

$$x^2 + 8x - 12y + 4 = 0$$

Determine (a) the location of the focus and (b) the equation of the directrix.

Solution The given equation is linear in the variable y and quadratic in the variable x. This is a parabola with an axis of symmetry that is parallel to the x axis (a horizontal line).

Completing the square on the x variable

$$x^2 + 8x = 12y - 4$$
$$x^2 + 8x + 16 = 12y + 12$$
$$(x+4)^2 = 12(y+1)$$

Comparing to the standard form, $h = -4$, $k = -1$, $p = 6$.

(a) The focus is located at $(h, k + p/2)$, which is $(-4, 2)$.

(b) The equation of the directrix is $y = k - p/2 = -4$. This can be also written as $y + 4 = 0$.

NOTE The *Handbook* gives the directrix equation for the special case "when $h = k = 0$, directrix is $x = -p/2$." The more general case when h and k are not zero would be the equation of the directrix is $x = h - p/2$.

Similarly, when the parabola opens up/down instead of right/left, the equation of the directrix is $y = k - p/2$.

Mensuration: Area Bounded by a Parabola

If a rectangle of width b and height h is drawn as shown in Fig. 1.12, then of all the parabolas that can be drawn between opposite corners, two special parabolas are those that have zero slope at either corner. These two parabolas divide the rectangle into equal thirds, as shown below.

Mensuration: Regular Polygon

Figure 1.13 shows a regular polygon circumscribed on a circle of radius r.

A regular polygon with $n \geq 3$ sides has the following characteristics:

Interior angle subtended (at center) by a side: $\phi = \dfrac{2\pi}{n}$ (1.33)

Angle between adjacent sides: $\theta = \pi\left(1 - \dfrac{2}{n}\right)$ (1.34)

Length of each side: $s = 2r\tan\dfrac{\phi}{2}$ (1.35)

Perimeter of polygon: $P = ns$ (1.36)

Area of polygon: $A = \dfrac{nsr}{2}$ (1.37)

Figure 1.12 Area bounded by a parabola.

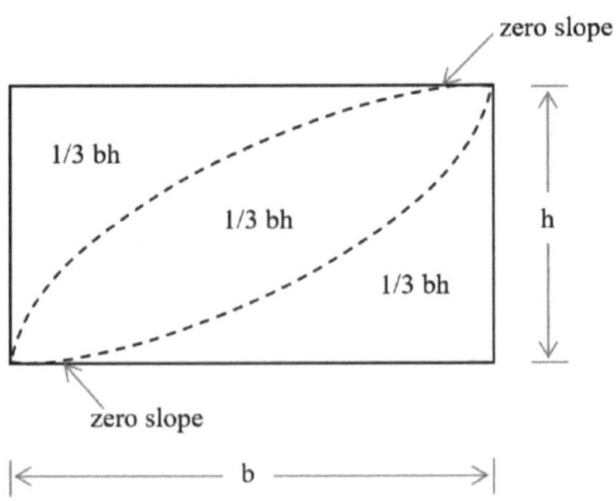

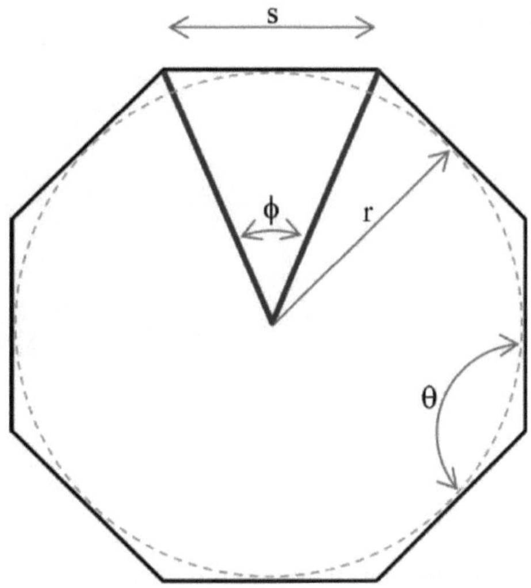

Figure 1.13 Regular polygon circumscribed on a circle.

Mensuration

Mensuration: Area of a Parallelogram

Page 25 of the *Handbook* has a parallelogram with pairs of opposite sides labeled "a" and "b" with the interior acute angle between the two sides labeled "ϕ." The diagonals d_1 and d_2 are given by

$$d_1 = \sqrt{a^2 + b^2 - 2ab\cos\phi} \tag{1.38}$$

$$d_1 = \sqrt{a^2 + b^2 + 2ab\cos\phi} \tag{1.39}$$

and the area of the parallelogram is given by Eq. 1.40, which is equal to the magnitude of the cross product between the vectors representing sides a and b

$$A = ab\sin\phi \tag{1.40}$$

Volume of Solids

Figure 1.14 shows a prismoid with two parallel surfaces A_1 and A_2 with a perpendicular distance h separating them. The cross section (parallel to both A_1 and A_2) at midheight is A. The volume of the prismoid is given by

$$V = \frac{h}{6}(A_1 + A_2 + 4A) \tag{1.41}$$

Pages 25 and 26 of the *Handbook* also have surface area and volume of the following solids: sphere, right circular cone, right circular cylinder, and a paraboloid of revolution.

Figure 1.14
Prismoid.

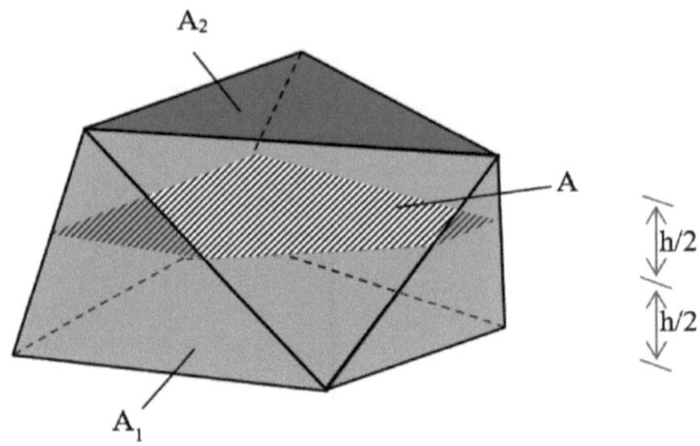

Vectors

A vector is a quantity that has both magnitude and direction, for example, forces and moments. As an illustration, a force is shown in Fig. 1.15. The orientation of the force is along the line directed from the point *A* (−3, 1, 5) to the point *B* (5, 3, −2).

The Cartesian components of the line are given by

$$\mathbf{AB} = [5-(-3)]\mathbf{i} + [3-1]\mathbf{j} + [-2-5]\mathbf{k} = 8\mathbf{i} + 2\mathbf{j} - 7\mathbf{k}$$

The length of the line AB is given by $|AB| = \sqrt{8^2+2^2+7^2} = 10.817$

The unit vector (dimensionless) along the line AB is given by

$$\mathbf{u}_{AB} = \frac{8\mathbf{i}+2\mathbf{j}-7\mathbf{k}}{10.817} = 0.740\mathbf{i} + 0.185\mathbf{j} - 0.647\mathbf{k}$$

If the magnitude of the force is 250 lb, then the force **F** can then be expressed as

$$\mathbf{F} = 250(0.740\mathbf{i} + 0.185\mathbf{j} - 0.647\mathbf{k}) = 185\mathbf{i} + 46.3\mathbf{j} - 161.8\mathbf{k}$$

Figure 1.15
Vector in Cartesian coordinates.

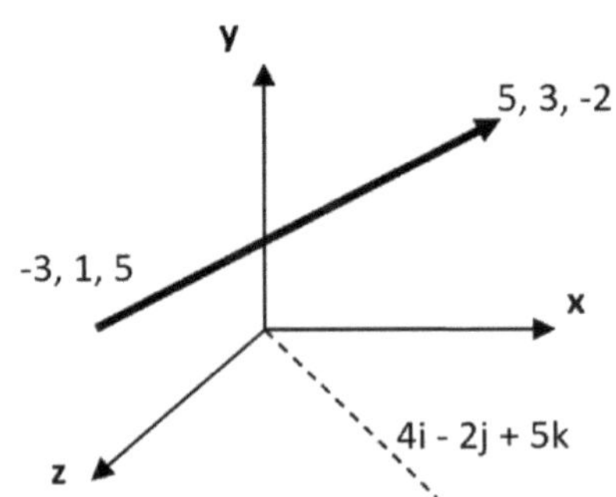

Example 1.14

The polar form of a vector is (10, 130°). The rectangular form is

A. $7.66 + 6.43j$

B. $6.43 - 7.66j$

C. $-6.43 + 7.66j$

D. $-7.66 - 6.43j$

Solution This can be worked out by using

$$x = r\cos\theta = 10\cos 130 = -6.43$$
$$y = r\sin\theta = 10\sin 130 = 7.66$$

Therefore, the vector is $-6.43 + 7.66j$

Answer is C.

NOTE However, if one notices that the angle of 130° puts this vector in the second quadrant and the only answer choice in the second quadrant is *C*, then no computations are necessary.

Example 1.15

What are the direction cosines of the vector $5i + 6j - 3k$?

A. $u_x = 0.60;\ u_y = 0.72;\ u_z = -0.36$

B. $u_x = 0.72;\ u_y = 0.36;\ u_z = -0.56$

C. $u_x = 0.60;\ u_y = 0.72;\ u_z = 0.36$

D. $u_x = 0.72;\ u_y = 0.36;\ u_z = 0.56$

Solution Magnitude of the vector: $\sqrt{5^2 + 6^2 + 3^2} = 8.367$

Unit vector: $\underline{u} = \dfrac{5\boldsymbol{i} + 6\boldsymbol{j} - 3\boldsymbol{k}}{8.367} = 0.598\boldsymbol{i} + 0.717\boldsymbol{j} - 0.359\boldsymbol{k}$

Answer is A.

Dot Product

The dot product, also known as the scalar product, between two vectors **A** and **B** is defined as

$$\mathbf{A} \cdot \mathbf{B} = |\mathbf{A}||\mathbf{B}|\cos\theta \tag{1.42}$$

where θ is the angle between the vectors **A** and **B**. In Cartesian coordinates, the dot product is calculated as

$$\mathbf{A} \cdot \mathbf{B} = A_x B_x + A_y B_y + A_z B_z \tag{1.43}$$

The result of the dot product operation is a scalar, which is why it is also called the scalar product. In the field of mechanics, the most common application for the dot product is the calculation of the component of a force in a particular direction. Note that the dot product only gives the magnitude of this component. For Fig. 1.15, in order to calculate the component of the force **F** along the dashed line 4**i** – 2**j** + 5**k**, one must calculate the unit vector (direction only) of the line, which is given by

$$\mathbf{u}_L = \frac{4\mathbf{i}-2\mathbf{j}+5\mathbf{k}}{6.708} = 0.596\mathbf{i}-0.298\mathbf{j}+0.716\mathbf{k}$$

The component of the force (calculated previously) is given by

$$F_L = \mathbf{F}\cdot\mathbf{u}_L = (185\mathbf{i}+46.3\mathbf{j}-161.8\mathbf{k})\cdot(0.596\mathbf{i}-0.298\mathbf{j}+0.716\mathbf{k}) = -19.39$$

The (vector) component of the force **F** along the line L is then given by

$$\mathbf{F}_L = F_L\mathbf{u}_L = -19.39(0.596\mathbf{i}-0.298\mathbf{j}+0.716\mathbf{k}) = -11.56\mathbf{i}+5.78\mathbf{j}-13.88\mathbf{k}$$

Example 1.16

What is the angle between the lines AB and CD shown in the figure?

A. 26°

B. 74°

C. 106°

D. 116°

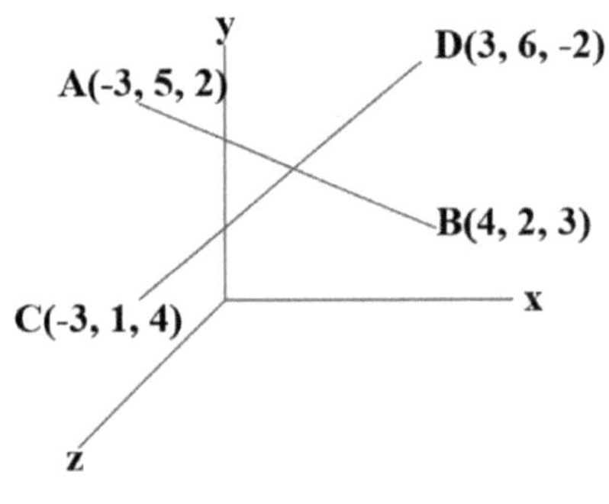

Solution The dot product is a convenient way to calculate the angle between two vectors, since the result of the operation is a scalar.

Vector **AB** = 7**i** – 3**j** + **k**
Vector **CD** = 6**i** + 5**j** – 6**k**
AB·**CD** = (7)(6) + (–3)(5) + (1) (–6) = 21

$$|AB| = \sqrt{7^2 + 3^2 + 1^2} = \sqrt{59}$$

$$|CD| = \sqrt{6^2 + 5^2 + 6^2} = \sqrt{97}$$

$$\cos\theta = \frac{21}{\sqrt{59}\sqrt{97}} = 73.9°$$

Answer is B.

Cross Product

The cross product, also known as the vector product, between two vectors **A** and **B** is calculated as

$$\mathbf{A} \times \mathbf{B} = [A_y B_z - A_z B_y]\mathbf{i} + [A_z B_x - A_x B_z]\mathbf{j} + [A_x B_y - A_y B_x]\mathbf{k} \tag{1.44}$$

This can also be expressed in the form of the determinant

$$\mathbf{A} \times \mathbf{B} = \begin{vmatrix} \mathbf{i} & \mathbf{j} & \mathbf{k} \\ A_x & A_y & A_z \\ B_x & B_y & B_z \end{vmatrix} \tag{1.45}$$

In the field of mechanics, the most common application for the cross product is the calculation of the moment of a force about a point, which is given by

$$\mathbf{M}_o = \mathbf{r} \times \mathbf{F} \tag{1.46}$$

where $\boldsymbol{r}$ is the (vector) distance from the pivot point O to any point on the line of action of the force.

In the example shown in Fig. 1.15, the moment of the force about the origin is therefore calculated as the cross product of the distance from O (0, 0, 0) to either (–3, 1, 5) or (5, 3, –2) and the force **F**.

$$\mathbf{r} = -3\mathbf{i} + \mathbf{j} + 5\mathbf{k}$$

$$\mathbf{F} = 185\mathbf{i} + 46.3\mathbf{j} - 161.8\mathbf{k}$$

$$\mathbf{r} \times \mathbf{F} = \begin{vmatrix} \mathbf{i} & \mathbf{j} & \mathbf{k} \\ -3 & 1 & 5 \\ 185 & 46.3 & -161.8 \end{vmatrix} = -393.3\mathbf{i} + 439.6\mathbf{j} - 323.9\mathbf{k}$$

Example 1.17

What is the area of the triangle ABC shown below?

A. 9.58

B. 12.35

C. 16.25

D. 32.50

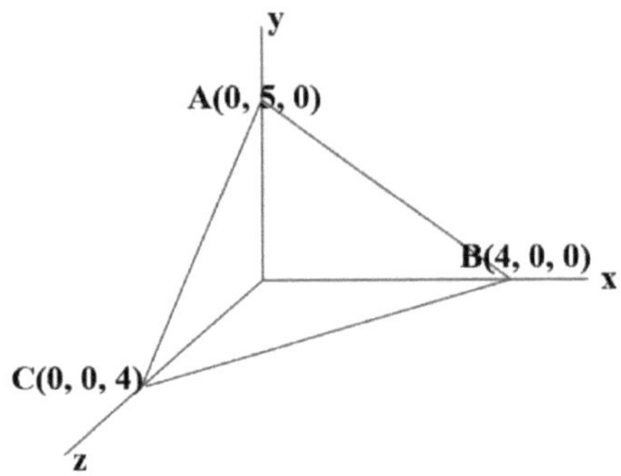

Solution

Vector **AB** = 4**i** – 5**j**

Vector **AC** = –5**j** + 4**k**

AB × **AC** = –20**k** – 16**j** – 20**i**

$|AB \times AC| = \sqrt{20^2 + 16^2 + 20^2} = \sqrt{1056} = 32.5$

Therefore, area of the triangle = ½(32.5) = 16.25.

Answer is C.

NOTE The cross product is a convenient way to calculate the area of a trapezoid whose sides can be expressed as vectors. The magnitude of the cross product is the area of the trapezoid and the unit vector of the cross product is the vector normal to the plane of the trapezoid.

Example 1.18

A parallelopiped is formed by drawing edges OA, OB, and OC radiating out from the origin O and constructing opposite edges parallel to these rays. Coordinates are in cm. What is the volume of the parallelopiped?

A. 60 cm^3

B. 80 cm^3

C. 96 cm^3

D. 120 cm^3

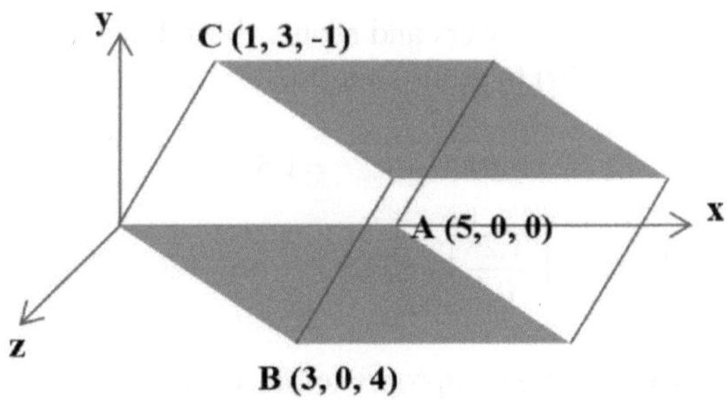

Solution

$OA = 5i$

$OB = 3i + 4k$

$OC = i + 3j - k$

The volume of the parallelopiped is calculated as the scalar triple product (absolute value). The scalar triple product is calculated as

$$\overrightarrow{OA} \cdot (\overrightarrow{OB} \times \overrightarrow{OC}) = \begin{vmatrix} 5 & 0 & 0 \\ 3 & 0 & 4 \\ 1 & 3 & -1 \end{vmatrix} = -60 \text{ cm}^3$$

Volume of the parallelopiped is 60 cm^3.

Answer is A.

Complex Numbers

A complex number $a + jb$ is one which has a real part (a) and an imaginary part (b). The unit along the real number line is the scalar number 1, whereas the unit along the imaginary number line is the square root of negative 1, commonly depicted as i (or j). In this book, we will use j.

Example 1.19

Calculate the following: $(-0.9+1.2j)^7$

A. $16.7 + 3.5j$

B. $-16.7 + 3.5j$

C. $16.7 - 3.5j$

D. $-16.7 - 3.5j$

Solution Anything to do with powers and roots is best done using polar coordinates. The first step is to convert (–0.9 + 1.2*j*) to polar coordinates:

$$\sqrt{0.9^2+1.2^2}=1.5$$

$$\theta=\tan^{-1}\left(\frac{1.2}{-0.9}\right)=-53.13^\circ \quad \text{or} \quad 126.87^\circ$$

The second quadrant angle (126.87°) is appropriate, since the vector is a second quadrant vector. However, the angle *must* be expressed in radians.

$$\theta=126.87^\circ=2.2143 \text{ rad}$$

$$-0.9+1.2j=1.5e^{2.2143j} \Rightarrow (-0.9+1.2j)^7=(1.5e^{2.2143j})^7=17.086e^{15.5j}$$

$$e^{15.5j}=-0.9785+0.2065j$$

$$17.086e^{15.5j}=17.086(-0.9785+0.2065j)=-16.72+3.53j$$

Answer is B.

NOTE All four answer choices have the same magnitude. So, just calcualting the angle of the resulting vector is enough.

The angle of $-0.9 + 1.2j$ is 126.87°;

$$126.87\times 7=888.1^\circ$$

If we remove two complete revolutions (720°), we have the angle 168.1°, which is a second quadrant angle. Only choice (B) is in the second quadrant.

Roots of Complex Equations

A complex number z can be expressed in rectangular form, polar form, or exponential form. The exponential form is best suited to calculate powers and roots of complex expressions. Given a complex number

$$z=Re^{i\theta} \tag{1.47}$$

The nth power of z can be written

$$z^n=R^n e^{in\theta} \tag{1.48}$$

and the nth root can be written

$$z^{1/n}=R^{1/n}e^{i\theta/n} \tag{1.49}$$

However, the root given by Eq. (1.49) is only one (the fundamental) root of the complex number z. The other $n-1$ roots are given by

$$z^{1/n}=R^{1/n}e^{i(\theta+2\pi k)/n} \qquad k=1,2,\ldots,n-1 \tag{1.50}$$

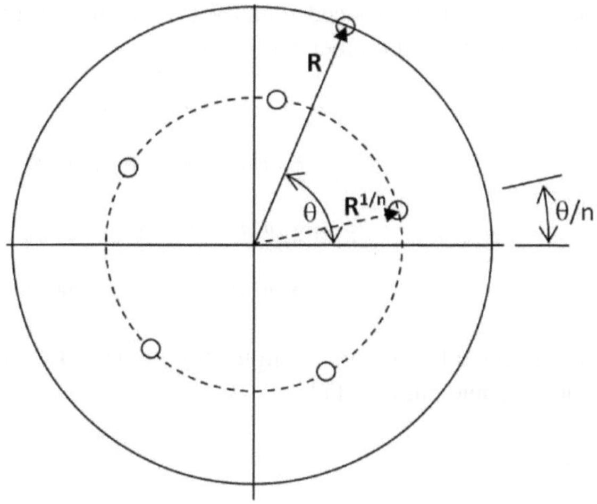

Figure 1.16 Roots of a complex number.

Visually, this means that the n roots are distributed equally around a circle. The case with $n = 5$ is shown in Fig. 1.16. The original complex number z has magnitude R and is represented by a vector to the solid circle. This vector is associated with angle θ. The five roots (represented by the open circles) are distributed equally ($360° \div 5 = 72°$ apart) around the dashed circle, which has radius = $R^{1/n}$. The first or fundamental root is shown as the dashed line vector at an angle of θ/n.

Example 1.20

Which of the following choices is a root of the following equation?

$$z^5 + 22.63 - 22.63j = 0$$

A. $z = -1.98 + 0.31j$

B. $z = 1.3 + 2.3j$

C. $z = 3 + 4j$

D. $z = 1.41 + 1.41j$

Solution

$$z^5 = -22.63 + 22.63j$$

The magnitude of the complex number on the right = 32 and the angle = $\tan^{-1}(22.63/-22.63) = 135 = 3\pi/4$.

Therefore,

$$z^5 = 32e^{j\frac{3\pi}{4}}$$

The first root is

$$z_1 = 32^{1/5}e^{j3\pi/20} = 2e^{j3\pi/20}$$

This can be represented by a vector at angle of 27°.

The other roots will be spaced equally (72° apart) around the 2π radians of the circle (spaced $2\pi/5$). Therefore, the other roots are

$$z_2 = 2e^{j(3\pi/20+2\pi/5)} = 2e^{j11\pi/20}$$

$$z_3 = 2e^{j(3\pi/20+4\pi/5)} = 2e^{j19\pi/20}$$

$$z_4 = 2e^{j(3\pi/20+6\pi/5)} = 2e^{j27\pi/20}$$

$$z_5 = 2e^{j(3\pi/20+8\pi/5)} = 2e^{j35\pi/20}$$

These are represented by vectors at angles 99°, 171°, 243°, and 315°, respectively. Option A has magnitude = 2 and angle = 171°.

Answer is A.

NOTE As soon as it is obvious that the magnitude of the complex number is 32 and the fifth root will have magnitude = fifth root of 32 = 2, you should scan the answer choices. In this case, this examination will eliminate choices B and C.

Note that your calculator will yield $\tan^{-1}(22.63/-22.63) = -45$. It is important to recognize that the angle should actually be taken as $-45 + 180 = 135$, which is a second quadrant angle, as appropriate for the complex number $-22.63 + 22.63j$.

Also note that for the exponential form of complex numbers, it is essential to express the angle in radians.

Matrix Algebra

Some of the fundamental concepts of matrix algebra are the calculation of (a) the cofactor matrix, (b) the classical adjoint, (c) the determinant, and (d) the matrix inverse.

Cofactor

Every term of a square matrix has a cofactor, which is the determinant of the matrix that remains if both the row and the column of the term are eliminated. The sign of the determinant is changed if the term in question is in an odd position (sum of row and column number is an odd number).

For example, for the 3 × 3 matrix P shown below

$$P = \begin{bmatrix} a & b & c \\ d & e & f \\ g & h & i \end{bmatrix}$$

the cofactor of the "f" term (which is located at the intersection of the second row and the third column and could therefore be referred to as P_{23}) is the determinant of the 2 × 2 matrix Q

$$Q = \begin{vmatrix} a & b \\ g & h \end{vmatrix} = ah - bg$$

However, since the sum 2 + 3 = 5, which is odd, the term P_{23} is considered an "odd" position, and the cofactor of P_{23} is changed to $-(ah-bg)=bg-ah$.

Example 1.21

What is the cofactor matrix of the matrix *A* shown below?

$$A=\begin{bmatrix} 2 & 3 & 1 \\ 7 & -2 & 4 \\ 5 & -4 & 8 \end{bmatrix}$$

A. $\begin{bmatrix} 0 & 36 & -18 \\ 28 & 11 & -23 \\ 14 & 1 & -25 \end{bmatrix}$

B. $\begin{bmatrix} 0 & 28 & 14 \\ 36 & 11 & 1 \\ -18 & -23 & -25 \end{bmatrix}$

C. $\begin{bmatrix} 0 & -28 & 14 \\ -36 & 11 & -1 \\ -18 & 23 & -25 \end{bmatrix}$

D. $\begin{bmatrix} 0 & -36 & -18 \\ -28 & 11 & 23 \\ 14 & -1 & -25 \end{bmatrix}$

Solution Choose one of the positions on the original matrix for which the entries in the answer choices are unique. This will allow you to choose the correct answer with the minimum amount of work.

For example, for the number 3 in the (1, 2) position (row 1, column 2) in the matrix *A*, the four choices for cofactor are 36, 28, –28, and –36, which are all different.

The cofactor of 3 is [with a change in sign because (1, 2) is an odd position] the determinant $-\begin{vmatrix} 7 & 4 \\ 5 & 8 \end{vmatrix}=-36$.

Answer is D.

NOTE In any problem where the answer choices contain multiple entities (as in each matrix contains nine elements), look for an entity that has unique choices among the four answer choices. This will allow you to choose the correct answer without solving the entire problem.

Classical Adjoint

The classical adjoint is the transpose (result of interchanging rows with columns) of the matrix of cofactors.

Example 1.22

What is the classical adjoint of the matrix A shown below?

$$A = \begin{bmatrix} 2 & 3 & 1 \\ 7 & -2 & 4 \\ 5 & -4 & 8 \end{bmatrix}$$

A. $\begin{bmatrix} 0 & 36 & -18 \\ 28 & 11 & -23 \\ 14 & 1 & -25 \end{bmatrix}$

B. $\begin{bmatrix} 0 & 28 & 14 \\ 36 & 11 & 1 \\ -18 & -23 & -25 \end{bmatrix}$

C. $\begin{bmatrix} 0 & -28 & 14 \\ -36 & 11 & -1 \\ -18 & 23 & -25 \end{bmatrix}$

D. $\begin{bmatrix} 0 & -36 & -18 \\ -28 & 11 & 23 \\ 14 & -1 & -25 \end{bmatrix}$

Solution Choose one of the positions on the four given answer choices for which the entries are unique. This will allow you to choose the correct answer with the minimum amount of work.

For example, for the (3, 2) position (row 3, column 2), the four choices for the adjoint are 1, −23, 23, and −1, which are all different.

Now, remember that the adjoint is obtained by transposing the cofactor matrix. Therefore, entries in the (3, 2) position of the adjoint will originate from the (2, 3) position of the original matrix. The cofactor of $A_{23} = 4$ is [with a change in sign because (2, 3) is an odd position] the determinant $-\begin{vmatrix} 2 & 3 \\ 5 & -4 \end{vmatrix} = 23.$

This number will occupy the (2, 3) position in the cofactor matrix, but move to the (3, 2) position in the adjoint matrix.

Answer is C.

Determinant

The determinant of a matrix is calculated by expanding any row or column of the matrix in the form of a sum of the product of the terms and their cofactors. For example, for the matrix P

$$P = \begin{bmatrix} a & b & c \\ d & e & f \\ g & h & i \end{bmatrix}$$

the cofactors of the third column are

cofactor of $c = dh - eg$

cofactor of $f = bg - ah$ (demonstrated earlier)

cofactor of $i = ae - bd$

The determinant is given by $c(dh-eg)+f(bg-ah)+i(ae-bd)$.

The same could have been calculated by expanding any of the rows or columns.

Properties of Determinants

Property 1 If two rows (or columns) of the matrix are interchanged, it changes the sign of the determinant (keeping the magnitude unchanged).

Property 2 If one row (or column) is scaled (multiplied by a factor), the determinant gets scaled by the same factor.

Example 1.23

What is the determinant of the matrix shown below?

$$\begin{bmatrix} 2 & 1 & 3 \\ 7 & 4 & 2 \\ -2 & 2 & 5 \end{bmatrix}$$

A. 45

B. 59

C. 67

D. 97

Solution Instead of following the traditional approach of reducing a 3×3 matrix to a series of 2×2 matrices, the following approach is much more convenient for calculating the determinant. However, this method *will not* work for larger matrices.

Construct an augmented matrix by copying the first two columns of the matrix on the right of the original matrix as shown.

$$\left[\begin{array}{ccc|cc} 2 & 1 & 3 & \mathbf{2} & \mathbf{1} \\ 7 & 4 & 2 & \mathbf{7} & \mathbf{4} \\ -2 & 2 & 5 & \mathbf{-2} & \mathbf{2} \end{array}\right]$$

Diagonals are then drawn—one set starting at the top left corner and going down and to the right; another set starting at the top right corner and going down and to the left. The determinant is the difference between the products of the first set of diagonals and the products of the second set of diagonals.

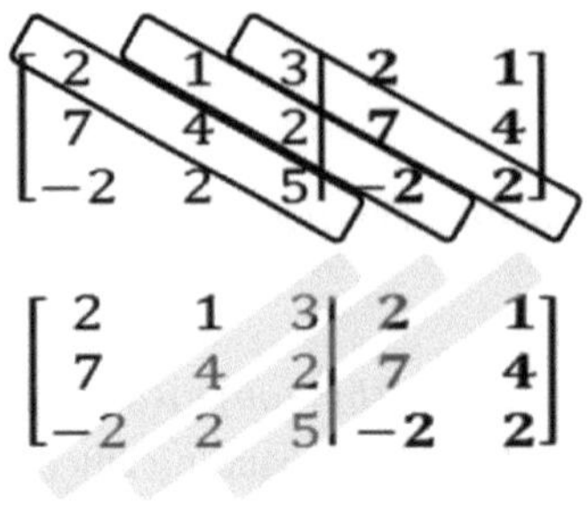

Determinant: $\Delta = (2\times4\times5+1\times2\times-2+3\times7\times2)-(3\times4\times-2+2\times2\times2+1\times7\times5)$
$= 78-19 = 59$

Answer is B.

Example 1.24

Given that the matrix A shown below has a determinant = 72, what is the determinant of the matrix B?

$$A = \begin{bmatrix} 1 & 2 & 3 & 4 \\ 5 & 6 & 7 & 8 \\ 1 & 3 & 2 & 1 \\ 3 & 4 & 2 & 6 \end{bmatrix}; \; B = \begin{bmatrix} 1 & 3 & 2 & 1 \\ 5 & 6 & 7 & 8 \\ 2 & 4 & 6 & 8 \\ 3 & 4 & 2 & 6 \end{bmatrix}$$

A. 72

B. −72

C. 144

D. −144

Solution The two matrices are closely related in the following way: If row 1 and row 3 from matrix A are interchanged, and then row 3 is doubled (every cell in the row is doubled), then we get matrix B.

A row interchange causes the determinant to change sign. Multiplying a row (or column) by a factor causes the determinant to get multiplied by that factor. The set {1, 2, 3, 4} that occurs in row 3 after the row interchange is doubled to the set {2, 4, 6, 8}.

Therefore, determinant of matrix B is $72\times-1\times2=-144$.

Answer is D.

Polynomials of Higher Order—Interrelationships of Roots

A polynomial of order n is shown in Eq. (1.51)

$$x^n + a_1x^{n-1} + a_2x^{n-2} + a_3x^{n-3} + \cdots + a_n = 0 \tag{1.51}$$

Note that if the roots of a polynomial are x_1, x_2, x_3 … then the coefficients of the polynomial are

- Alternating in sign
- Sum of roots, taken one at a time, two at a time, three at a time, etc.

$$x_1 + x_2 + \cdots + x_n = -a_1$$
$$x_1x_2 + x_1x_3 + \cdots + x_{n-1}x_n = +a_2$$
etc.

Second-Order Polynomial (Quadratic)

$$(x - x_1)(x - x_2) = x^2 - (x_1 + x_2)x + x_1x_2 = 0$$

- Roots are x_1 and x_2.
- Second coefficient is "negative $x_1 + x_2$" (sum of roots taken one at a time).
- Third coefficient is "positive x_1x_2" (sum of roots taken two at a time).

Third-Order Polynomial (Cubic)

$$(x - x_1)(x - x_2)(x - x_3) = x^3 - (x_1 + x_2 + x_3)x^2 + (x_1x_2 + x_2x_3 + x_1x_3)x - x_1x_2x_3 = 0$$

- Roots are x_1, x_2, and x_3.
- Second coefficient is "negative $x_1 + x_2 + x_3$" (sum of roots taken one at a time).
- Third coefficient is "positive $x_1x_2 + x_2x_3 + x_1x_3$" (sum of roots taken two at a time).
- Fourth coefficient is "negative $x_1x_2x_3$" (sum of roots taken three at a time).

Fourth-Order Polynomial (Quartic)

$$(x - x_1)(x - x_2)(x - x_3)(x - x_4) = x^4 - (x_1 + x_2 + x_3 + x_4)x^3 + (x_1x_2 + x_1x_3 + x_1x_4 + x_2x_3 + x_2x_4 + x_3x_4)x^2 - (x_1x_2x_3 + x_2x_3x_4 + x_1x_3x_4 + x_1x_2x_4)x + x_1x_2x_3x_4 = 0$$

- Roots are x_1, x_2, x_3, and x_4.
- Second coefficient is "negative $x_1 + x_2 + x_3 + x_4$" (sum of roots taken one at a time).
- Third coefficient is "positive $x_1x_2 + x_1x_3 + x_1x_4 + x_2x_3 + x_2x_4 + x_3x_4$" (sum of roots taken two at a time).
- Fourth coefficient is "negative $x_1x_2x_3 + x_2x_3x_4 + x_1x_3x_4 + x_1x_2x_4$" (sum of roots taken three at a time).
- Fifth coefficient is "positive $x_1x_2x_3x_4$" (sum of roots taken four at a time).

Example 1.25

What are the roots of $2x^3 - 8x - 3 = 0$?

A. –7.9, –3, –0.38

B. –3, –2, 2

C. –3, –0.38, 2

D. –1.78, –0.39, 2.17

Solution The second coefficient is zero (x^2 term). Therefore, the roots must be such that $(a + b + c = 0)$. The only one that works is (D).

Newton's Method of Root Extraction

A function $f(x)$ can have zeros at various values of $x = x_1, x_2, x_3, \ldots$. These values are called the roots of the equation $f(x) = 0$. If exact methods do not exist to solve for these roots, then the roots may be extracted using a variety of iterative numerical algorithms. One of these methods is the Newton-Raphson algorithm, which works as follows:

Starting with a trial value $x = a^j$, the next estimate a^{j+1} is calculated as

$$a^{j+1} = a^j - \left.\frac{f(x)}{\frac{df(x)}{dx}}\right|_{x=a^j} \tag{1.52}$$

The iterations continue until the solution converges to one of the actual roots of the equation (not guaranteed). This method works as long as the function $f(x)$ has a continuous first derivative at $x = a^j$. Also, in order for the algorithm to converge to a root, the initial trial value (a^0) must be close to the root.

Example 1.26

The quartic function $f(x)$ is given below. Find a root starting at (a) $x = 0$, (b) $x = 1$, and (c) $x = -2$.

$$f(x) = -6x^4 - 5x^3 + 16x^2 + 5x - 3$$

Solution The derivative of the function is

$$f'(x) = -24x^3 - 15x^2 + 32x + 5$$

(a) If we set the initial guess to $a^0 = 0$, it takes only four iterations to converge to the true root $x = 0.319$.

The iterations proceed as follows:

a^j	$f(a^j)$	$f'(a^j)$	$a^{j+1} = a^j - f(a^j)/f'(a^j)$
0	−3.000	5.000	0.600
0.600	3.902	13.616	0.313
0.313	−0.073	12.817	0.319
0.319	0.000	12.904	0.319

Converged to $x = 0.319$

(b) If we set the initial guess to $a^0 = 1$, it takes nine iterations to converge to the true root $x = 1.367$.

The iterations proceed as follows:

a^j	$f(a^j)$	$f'(a^j)$	$a^{j+1} = a^j - f(a^j)/f'(a^j)$
1.000	7.000	−2.000	4.500
4.500	−2572	−2342	3.401
3.401	−801	−1004	2.604
2.604	−246	−437	2.042
2.042	−72.98	−197	1.671
1.671	−20.06	−95.35	1.460
1.460	−4.440	−55.02	1.380
1.380	−0.518	−42.44	1.368
1.368	−0.011	−40.67	1.367
1.367	0.000	−40.63	1.367

Converged to $x = 1.367$

(c) If we set the initial guess to $a^0 = -2$, it takes only two iterations to converge to the true root $x = -1.924$.

The iterations proceed as follows:

a^j	$f(a^j)$	$f'(a^j)$	$a^{j+1} = a^j - f(a^j)/f'(a^j)$
−2.000	−5.000	73.000	−1.932
−1.932	0.446	60.173	−1.924
−1.924	−0.005	58.86	−1.924
−1.924	0.000	58.84	−1.924

Converged to $x = -1.924$

The true roots of this equation are at $x = -1.924$, -0.596, 0.319, and 1.367.

Differential Calculus

Limits

In calculus, a limit is a value that a function of a variable/variables approaches as the independent variable(s) approaches some specific value. The statement

$$\lim_{x \to a} f(x) = b$$

is read as "the limit of the function $f(x)$ as x approaches the value a is equal to b." This means that as the variable x approaches closer and closer to the value a, the value of the function $f(x)$ approaches closer and closer to the value b.

In general, limits can be evaluated by simply substituting the value of the variable into the function. For example

$$\lim_{x \to a} f(x) = f(a) \tag{1.53}$$

Also, the point $x = a$ can be approached from the left ($x \to a^-$) or from the right ($x \to a^+$). The function is continuous at $x = a$ if

$$\lim_{x \to a^-} f(x) = f(a^-) = \lim_{x \to a^+} f(x) = f(a^+) \tag{1.54}$$

The Derivative

The derivative of a function $y = f(x)$ at the point $x = a$ is defined as

$$\left.\frac{dy}{dx}\right|_{x=a} = \lim_{\Delta \to 0} \frac{f(a+\Delta) - f(a)}{\Delta} \tag{1.55}$$

The operator d/dx is also designated by the letter D or by using a prime (′) accent, as below

$$\frac{dy}{dx} = Dy = y' \tag{1.56}$$

The derivative y' is the slope of the curve $y = f(x)$. As may be seen in Fig. 1.17, the true slope at $x = a$ (shown marked by the tangent) is only approximated by the "rise over run" gradient between points $x = a$ and $x = a + \Delta$. However, as the limit of this gradient is taken, as $\Delta \to 0$, the calculated value approaches the true slope of the function at $x = a$.

Example 1.27

Given the function: $y(x) = (2x)^x$. What is the derivative $\frac{dy}{dx}$?

A. $\frac{dy}{dx} = x^{2x}[1 + \ln(2x)]$

B. $\frac{dy}{dx} = (2x)^x[1 + \ln(2x)]$

C. $\frac{dy}{dx} = x^x[1 + 2\ln(x)]$

D. $\frac{dy}{dx} = (2x)^x[1 + 2\ln(2x)]$

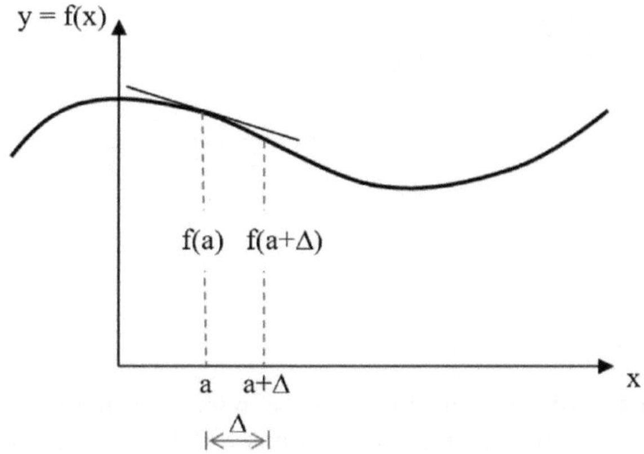

Figure 1.17
The derivative.

Solution

$$y=(2x)^x \Rightarrow \ln y = x\ln(2x)$$

Taking derivatives of both sides (and using the Chain Rule):

$$\frac{1}{y}\frac{dy}{dx}=\ln(2x)+x\frac{1}{2x}2=1+\ln(2x)$$

$$\frac{dy}{dx}=y[1+\ln(2x)]=(2x)^x[1+\ln(2x)]$$

Answer is B.

L'Hôpital's Rule

For evaluating the limit of a rational fraction of the form $f(x)/g(x)$ as x approaches a specific value a, the process of evaluating the limit by simply substituting $x = a$ will not work if

$$f(a)=g(a)=0 \text{ or } \infty$$

For this condition, the limit becomes

$$\lim_{x\to a}\frac{f(x)}{g(x)}=\frac{f(a)}{g(a)}=\frac{0}{0} \text{ or } \frac{\infty}{\infty} \tag{1.57}$$

Both of these forms are indeterminate.

In such cases, the limit may be found using the L'Hôpital rule, which requires replacing both the numerator and denominator functions $f(x)$ and $g(x)$—with their derivatives $f'(x)$ and $g'(x)$, respectively. If this still results in one of the indeterminate forms, then derivatives must be taken again—$f''(x)$ and $g''(x)$—until the form is no longer indeterminate.

Example 1.28

What is the limit shown below?

$$\lim_{x\to 1}\frac{4x^4-2x^3+x^2-3}{5x^3-4x^2-1}$$

A. 11/9

B. 12/7

C. 10/3

D. 14/5

Solution At $x = 1$, this limit has the indeterminate form 0/0. Therefore, one must replace the numerator and denominator functions with their derivatives.

$$\lim_{x\to 1}\frac{4x^4-2x^3+x^2-3}{5x^3-4x^2-1}=\lim_{x\to 1}\frac{16x^3-6x^2+2x}{15x^2-8x}=\frac{12}{7}$$

Answer is B.

Example 1.29

Evaluate the following limit:

$$\lim_{x\to 0^+} x\ln x$$

A. 0

B. 1

C. ½

D. ∞

Solution The L'Hôpital rule only applies to the quotient form, not the product form. However, we can convert the product to a quotient form as below:

$$\lim_{x\to 0^+} x\ln x=\lim_{x\to 0^+}\frac{\ln x}{1/x}$$

This form has the ∞/∞ form. Again, taking derivatives

$$\lim_{x\to 0^+}\frac{\ln x}{1/x}=\lim_{x\to 0^+}\frac{1/x}{-1/x^2}=\lim_{x\to 0^+}(-x)=0$$

Answer is A.

Tests for Maxima and Minima

The first derivative can be used to test for the maxima and minima of a function. The conditions that must exist at all true maxima and minima are

$$\text{Maximum:} \quad y' = 0 \ \text{ and } \ y'' < 0 \qquad (1.58a)$$

$$\text{Minimum:} \quad y' = 0 \ \text{ and } \ y'' > 0 \qquad (1.58b)$$

These tests, applied to a function defined over an infinite domain ($-\infty < x < \infty$), will result in the identification of *all* true maxima and minima. However, when a function is defined over a limited domain, there may not be any of the true maxima or minima within the given domain. In such a case, the function must be evaluated at the two boundaries to yield the extreme value.

Example 1.30

What is the maximum value of the function $y(x)$ in the interval $0 < x < 3$?

$$y = 2x^3 - 8x^2 + 8x - 12$$

A. –9.63

B. +9.63

C. –6.0

D. 12.4

Solution The first derivative dy/dx is

$$\frac{dy}{dx} = 6x^2 - 16x + 8$$

Equating dy/dx to zero, we get

$$3x^2 - 8x + 4 = 0 \Rightarrow x = 0.667, 2.0$$

Both of these extrema are within the given domain $0 < x < 3$.

The second derivative d^2y/dx^2 is

$$\frac{d^2y}{dx^2} = 12x - 16$$

At $x = 0.667$, $d^2y/dx^2 < 0$. This is a *maximum.* At $x = 2.0$, $d^2y/dx^2 > 0$. This is a *minimum.*

To find the maximum value of the function, evaluate $y(0.667) = -9.63$.

NOTE However, since a closed domain is given, one must also evaluate the function at the domain boundaries ($x = 0$ and $x = 3$).

At $x = 0$, $y = -12$.

At $x = 3$, $y = -6$. (Thus, if we are looking for the absolute maximum of the function within this domain, the value at $x = 3$ is the desired answer, even though it is not a *true* extremum.)

Answer is C.

Chain Rule for Derivatives

Given the following cascade of functions:

$$f(u) = v$$
$$g(v) = w$$
$$h(w) = y$$

and so on, the following derivatives can be defined:

$$f'(u) = \frac{dv}{du}$$
$$g'(v) = \frac{dw}{dv}$$
$$h'(w) = \frac{dy}{dw}$$

Then, the chain rule suggests that

$$\frac{dy}{du} = \frac{dy}{dw}\frac{dw}{dv}\frac{dv}{du} = h'(w)g'(v)f'(u) \tag{1.59}$$

Example 1.31

Given the function $y(x)$ below, find the derivative dy/dx

$$y = \ln(\sin(x^2))$$

Solution

The cascade of functions is	x	$\to x^2$	$\to \sin(x^2)$	$\to \ln(\sin(x^2))$
These can be represented as	x	$\to u = x^2$	$\to v = \sin(u)$	$\to y = \ln(v)$

Thus,

$$\frac{dy}{dx} = \frac{dy}{dv}\frac{dv}{du}\frac{du}{dx} = \frac{1}{v}\cos u\, 2x = \frac{2x\cos(x^2)}{\sin(x^2)} = \frac{2x}{\tan(x^2)}$$

Derivative of an Exponent

The standard form of the derivative of the exponent x^n is nx^{n-1}

$$\frac{d}{dx}x^n = nx^{n-1} \tag{1.60}$$

By the chain rule, the derivative of the exponent of a function $u(x)$ is $nu^{n-1}u'$

$$\frac{d}{dx}u^n = nu^{n-1}\frac{du}{dx} \tag{1.61}$$

Derivative of a Product of Several Functions

The derivative of the product of n functions is calculated as the sum of n terms, with each term evaluated by taking the derivative of one function while holding the other functions constant in the product. Thus,

$$\frac{d}{dx}[u(x)v(x)w(x)] = \frac{du}{dx}v(x)w(x) + \frac{dv}{dx}u(x)w(x) + \frac{dw}{dx}u(x)v(x) \tag{1.62}$$

Derivative of a Quotient of Two Functions

The derivative of a quotient of two functions $u(x)$ and $v(x)$ is given by

$$\frac{d}{dx}\left[\frac{u(x)}{v(x)}\right] = \frac{\frac{du}{dx}v(x) - \frac{dv}{dx}u(x)}{[v(x)]^2} \tag{1.63}$$

Curvature

For a curve describing the function $y = f(x)$, the curvature changes continuously, as does the location of the center of curvature. At points of inflection, the curvature is zero. As we pass through a point of inflection, the center of curvature "switches" from one side of the curve to the other. We commonly describe this by differentiating between "convex" and "concave" portions of the curve.

For a function described as $x = f(y)$, the curvature is calculated as

$$K = -\frac{x''}{[1+(x')^2]^{3/2}} \tag{1.64}$$

where $x' = \frac{dx}{dy}$ and $x'' = \frac{d^2x}{dy^2}$

For a function described as $y = f(x)$, the curvature is calculated as

$$K = \frac{y''}{[1+(y')^2]^{3/2}} \tag{1.65}$$

where $y' = \frac{dy}{dx}$ and $y'' = \frac{d^2y}{dx^2}$.

In either case, the radius of curvature is the absolute value of the inverse of the curvature

$$\rho = \left|\frac{1}{K}\right| \tag{1.66}$$

Example 1.32

What are the coordinates of the inflection point of the curve defined by the equation below?

$$y = 3x^3 - 4x^2 - 9x + 5$$

A. $x = 0.44$, $y = 0.48$

B. $x = 0.65$, $y = 8.34$

C. $x = 0.44$, $y = -7.39$

D. $x = 0.65$, $y = -5.00$

Solution At the point of inflection, the curvature = 0. Therefore, the second derivative is zero.

$$y = 3x^3 - 4x^2 - 9x + 5$$

$$\frac{dy}{dx} = 9x^2 - 8x - 9$$

$$\frac{d^2y}{dx^2} = 18x - 8$$

Setting the second derivative to zero, $x = 0.444$.

Therefore, $y = 0.478$.

Answer is A.

Taylor's Series

The Taylor's series for a function expanded about a point a is given by

$$f(x) = f(a) + \frac{f'(a)}{1!}(x-a) + \frac{f''(a)}{2!}(x-a)^2 + \cdots + \frac{f^n(a)}{n!}(x-a)^n + \cdots \tag{1.67}$$

When $a = 0$, the Taylor series is called the Maclaurin series

$$f(x) = f(0) + \frac{f'(0)}{1!}x + \frac{f''(0)}{2!}x^2 + \cdots + \frac{f^n(0)}{n!}x^n + \cdots \tag{1.68}$$

The only condition required for these series to be defined is that the derivatives exist and are continuous.

Example 1.33

What is the Taylor series for the function $f(x)$?

$$f(x) = \sin^2(x)$$

A. $f(x) = x + x^2 - \frac{x^3}{6} + \cdots$

B. $f(x) = x - x^2 - \frac{x^3}{3} + \cdots$

C. $f(x) = x^2 - \frac{x^4}{6} + \frac{2x^6}{15} \cdots$

D. $f(x) = x^2 - \frac{x^4}{3} + \frac{2x^6}{45} \cdots$

Solution

$f(x) = \sin^2(x)$

$f'(x) = 2\sin(x)\cos(x) = \sin(2x)$

$f''(x) = 2\cos(2x)$

$f'''(x) = -4\sin(2x)$

$f''''(x) = -8\cos(2x)$

and so on …

At $x = 0$, $f(0) = 0$; $f'(0) = 0$; $f''(0) = 2$; $f'''(0) = 0$; $f^{IV}(0) = -8$, and so on …

Taylor's series (about $a = 0$) is

$$f(x) = 0 + \frac{0}{1!}(x-0) + \frac{2}{2!}(x-0)^2 + \frac{0}{3!}(x-0)^3 + \frac{-8}{4!}(x-0)^4 + \cdots$$

$$f(x) = x^2 - \frac{x^4}{3} \cdots$$

Answer is D.

Integral Calculus

Standard Integrals

Page 29 of the *Handbook* has a list of standard integrals.

Example 1.34

What is the value of the following definite integral?

$$\int_0^4 \frac{dx}{3x^2 - 5x + 3}$$

A. 2.38

B. 1.44

C. 82.33

D. 49.40

Solution This is one of the standard forms in the *Handbook*.
$a = 3$; $b = -5$; $c = 3 \rightarrow 4ac - b^2 = 11 > 0$. This is case 27*a*.

$$I = \frac{2}{\sqrt{4ac-b^2}} \tan^{-1} \frac{2ax+b}{\sqrt{4ac-b^2}} = \frac{2}{\sqrt{11}} \tan^{-1}\left(\frac{6x-5}{\sqrt{11}}\right)$$

Evaluated between the limits $x = 0$ and $x = 4$, we get

$$I = \frac{2}{\sqrt{11}} \left[\tan^{-1}\left(\frac{19}{\sqrt{11}}\right) - \tan^{-1}\left(\frac{-5}{\sqrt{11}}\right)\right] = 1.437$$

Answer is B.

NOTE Your calculator must be in RADIAN mode when evaluating this integral.

Integration by Parts

The differential of a product of two functions $u(x)$ and $v(x)$ is given by

$$d(uv) = udv + vdu \tag{1.69}$$

We can then write

$$udv = d(uv) - vdu \tag{1.70}$$

$$\int udv = uv - \int vdu \tag{1.71}$$

Thus, if we can cast an integral in the form $\int udv$, Eq. (1.71) above can be used to evaluate it.

In applying this principle, the most important question is "In a product of two functions, which one should be designated as $u(x)$ and which one as $v(x)$?" To help with this question, the following mnemonic may be used—ILATE. The letters stand for

- Inverse (as in inverse trigonometric such as $\sin^{-1}$, $\cos^{-1}$, etc.)
- Logarithmic
- Algebraic
- Trigonometric
- Exponential

Compare the two functions $u(x)$ and $v(x)$. The one that ranks higher on the list should be classified as $u(x)$.

Example 1.35

What is the following integral?

$$\int x^2 e^{5x} dx$$

A. $e^{5x}\left(\frac{x^2}{5}+\frac{2x}{25}+\frac{2}{125}\right)$

B. $e^{5x}\left(\frac{x^2}{5}+\frac{2x}{25}-\frac{2}{125}\right)$

C. $e^{5x}\left(\frac{x^2}{5}-\frac{2x}{25}+\frac{2}{125}\right)$

D. $e^{5x}\left(\frac{x^2}{5}-\frac{2x}{25}-\frac{2}{125}\right)$

Solution The two functions are *algebraic* and *exponential*. Therefore, we use $u = x^2$ and $dv = e^{5x}dx$.

Integrating dv: $v = \int dv = \int e^{5x}dx = \frac{1}{5}e^{5x}$

$\int udv = uv - \int vdu$

$$\int x^2 e^{5x}dx = (x^2)\left(\frac{1}{5}e^{5x}\right) - \int \frac{1}{5}e^{5x}2xdx = \frac{x^2}{5}e^{5x} - \frac{2}{5}\int xe^{5x}dx$$

Now we must perform the integral

$$\int xe^{5x}dx$$

Once again, using $u = x$ and $dv = e^{5x}dx \rightarrow v = 1/5\ e^{5x}$

$$\int xe^{5x}dx = x\frac{1}{5}e^{5x} - \int \frac{1}{5}e^{5x}dx = \frac{x}{5}e^{5x} - \frac{1}{5}\int e^{5x}dx = \frac{x}{5}e^{5x} - \frac{1}{25}e^{5x}$$

Therefore, the original integral is

$$\int x^2 e^{5x}dx = \frac{x^2}{5}e^{5x} - \frac{2}{5}\int xe^{5x}dx = \frac{x^2}{5}e^{5x} - \frac{2}{5}\left(\frac{x}{5}e^{5x} - \frac{1}{25}e^{5x}\right) = e^{5x}\left(\frac{x^2}{5} - \frac{2x}{25} + \frac{2}{125}\right)$$
$$= e^{5x}\left(\frac{25x^2 - 10x + 2}{125}\right)$$

Answer is C.

NOTE A shortcut for the case where the *u* function is a polynomial is illustrated in the table below.
Step 1: List the polynomial as the *u* function and the other function as *v*.
Step 2: Successively differentiate *u* and integrate *v* until the derivative of the polynomial (*u*) becomes zero.
Step 3: Starting at the top of the table, assign alternating signs (+ and –) to the expressions.
Step 4: Starting with the upper left corner ($+x^2$ in this case), form products diagonally by multiplying it with the function in the cell "below and to the right." Continue accumulating these products until the zero is reached in the *u* column.

	u (differentiate)	*v* (integrate)
+	x^2	e^{5x}
–	$2x$	$\frac{1}{5}e^{5x}$
+	2	$\frac{1}{25}e^{5x}$
–	0	$\frac{1}{125}e^{5x}$

$$+(x^2)\left(\frac{1}{5}e^{5x}\right)-(2x)\left(\frac{1}{25}e^{5x}\right)+(2)\left(\frac{1}{125}e^{5x}\right)$$

Integration by Substitution

A change of variable and subsequent substitution often simplifies the process of integration. A few examples are shown below.

Example 1.36

What is the indefinite integral?

$$\int(1+\cos^2 x)\sin x\,dx$$

Solution If we assume $u = \cos x$, then the differential $du = -\sin x\,dx$ and the integral reduces to

$$-\int(1+u^2)\,du=-\left(u+\frac{1}{3}u^3\right)=-\left(\cos x+\frac{1}{3}\cos^3 x\right)$$

If this had been a definite integral, as below

$$\int_0^{\pi}(1+\cos^2 x)\sin x\,dx$$

It seems to follow that it should be evaluated as

$$-\left(\cos x + \frac{1}{3}\cos^3 x\right)\Bigg|_0^{\pi} = \frac{8}{3}$$

Integration by Separable Variables

When a derivative has the form where the rational expression can be separated into x and y functions as shown in Eq. (1.72)

$$\frac{dy}{dx} = \frac{f(x)}{g(y)} \tag{1.72}$$

the equation can be written

$$f(x)\,dx = g(y)\,dy \tag{1.73}$$

Equation (1.73) can then be solved by integrating both sides.

Example 1.37

Solve the following differential equation:

$$\frac{dy}{dx} = \frac{\sin(5x)}{y^3}$$

Solution The equation can be rewritten as

$$\sin(5x)\,dx = y^3 dy$$

Integrating both sides, $-\frac{1}{5}\cos(5x) = \frac{y^4}{4} + c$, which can also be written $5y^4 + 4\cos(5x) = \text{constant}$

Numerical Integration: Euler's Rule, Trapezoidal Rule, and Simpson's Rule

The three most common numerical integration schemes—the zeroth-order forward rectangular rule (Euler), the first-order trapezoidal rule, and the second-order Simpson's rule—are summarized on page 36 of the *Handbook*. Simpson's rule is restricted to an even number of subintervals.

Example 1.38

What is the estimate of the integral shown below if the trapezoidal rule is used, dividing the domain into six equal intervals?

$$\int_0^6 f(x)\,dx = \int_0^6 (2x^3 - 8x^2 + 8x - 12)\,dx$$

A. 154

B. 144

C. 120

D. 58

Solution The domain $0 < x < 6$ is to be divided into six equal intervals. Therefore, the function must be evaluated at $x = 0, 1, 2, 3, 4, 5, 6$.

$f(a) = f(0) = -12$

$f(b) = f(6) = 180$

$f(1) = -10$

$f(2) = -12$

$f(3) = -6$

$f(4) = 20$

$f(5) = 78$

$$\int_0^6 f(x)\,dx = \frac{1}{2}\left[-12 + 2(-10 - 12 - 6 + 20 + 78) + 180\right] = 154$$

Answer is A.

Example 1.39

What is the estimate of the integral shown below if the forward rectangular rule is used, dividing the domain into six equal intervals?

$$\int_0^6 f(x)\,dx = \int_0^6 (2x^3 - 8x^2 + 8x - 12)\,dx$$

A. 154

B. 144

C. 120

D. 58

Solution The domain $0 < x < 6$ is to be divided into six equal intervals. Therefore, the function must be evaluated at $x = 0, 1, 2, 3, 4, 5$.

$f(a) = f(0) = -12$

$f(1) = -10$

$f(2) = -12$

$f(3) = -6$

$f(4) = 20$

$f(5) = 78$

$$\int_0^6 f(x)\,dx = 1 \times [-12 - 10 - 12 - 6 + 20 + 78] = 58$$

Answer is D.

Example 1.40

What is the estimate of the integral shown below if Simpson's rule is used, dividing the domain into six equal intervals?

$$\int_0^6 f(x)\,dx = \int_0^6 (2x^3 - 8x^2 + 8x - 12)\,dx$$

A. 154

B. 144

C. 120

D. 58

Solution The domain $0 < x < 6$ is to be divided into six equal intervals. Therefore, the function must be evaluated at $x = 0, 1, 2, 3, 4, 5, 6$.

$f(a) = f(0) = -12$
$f(b) = f(6) = 180$
$f(1) = -10$
$f(2) = -12$
$f(3) = -6$
$f(4) = 20$
$f(5) = 78$

$$\int_0^6 f(x)\,dx = \frac{1}{3}\times[-12+180+4\times(-10-6+78)+2\times(-12+20)] = 144$$

Answer is B.

Partial Fractions

Consider a rational fraction of the form $P(x)/Q(x)$, where $P(x)$ and $Q(x)$ are polynomials. Such rational fractions do not easily lend themselves to standard derivatives or integrals. However, if the rational fraction can be decomposed into partial fractions, each of which are of a standard form, then any operations such as integrations, are made easier.

The rational expressions are classified based on the factorization of the denominator polynomial $Q(x)$.

Type 1: The n-order polynomial $Q(x)$ can factored into n distinct factors *and* $P(x)$ is of order lower than $Q(x)$:

$$Q(x) = C(x - x_1)(x - x_2)\cdots(x - x_n)$$

Example 1.41

What is the proper decomposition of the following function into partial fractions?

$$f(x) = \frac{3x - 7}{(x-4)(2x+1)}$$

Denominator polynomial $Q(x)$ is quadratic with two distinct linear factors. $P(x)$ is linear and therefore satisfies the condition that the order of $P(x)$ is less than that of $Q(x)$. In this case, the numerator in each case is chosen to be one order lower than the denominator.

$$\frac{3x-7}{(x-4)(2x+1)}=\frac{A}{(x-4)}+\frac{B}{(2x+1)}$$

which can be converted to

$$\frac{A}{(x-4)}+\frac{B}{(2x+1)}=\frac{A(2x+1)+B(x-4)}{(x-4)(2x+1)}$$

Comparing like terms

$$2A+B=3$$

$$A-4B=-7$$

Solving these equations, we have $A = 5/9$ and $B = 17/9$. Therefore,

$$\frac{3x-7}{(x-4)(2x+1)}=\frac{5/9}{(x-4)}+\frac{17/9}{(2x+1)}$$

Example 1.42

What is the proper decomposition of the following function into partial fractions?

$$f(x)=\frac{3x-7}{(x-4)(2x^3+1)}$$

Denominator polynomial $Q(x)$ is quartic with two distinct factors—one linear and one cubic. $P(x)$ is linear and therefore satisfies the condition that the order of $P(x)$ is less than that of $Q(x)$. In this case, the numerator in each case is chosen to be one order lower than the denominator.

$$\frac{3x-7}{(x-4)(2x^3+1)}=\frac{A}{(x-4)}+\frac{Bx^2+Cx+D}{(2x^3+1)}$$

which can be converted to

$$\frac{A}{(x-4)}+\frac{Bx^2+Cx+D}{(2x^3+1)}=\frac{A(2x^3+1)+(Bx^2+Cx+D)(x-4)}{(x-4)(2x^3+1)}$$

Comparing like terms

x^3 terms: $2A+B=0$

x^2 terms: $-4B+C=0$

x terms: $-4C+D=3$

constant: $A-4D=-7$

Solving these equations, we have $A = 0.039$, $B = -0.078$, $C = -0.310$, and $D = 1.760$. Therefore,

$$\frac{3x-7}{(x-4)(2x^3+1)} = \frac{0.039}{(x-4)} + \frac{-0.078x^2 - 0.310x + 1.760}{(2x^3+1)}$$

Type 2: The polynomial $Q(x)$ can be factored into a number of nonunique factors *and* $P(x)$ is of order lower than $Q(x)$. One or more factors may be repeated more than once.

Example 1.43

What is the proper decomposition of the following function into partial fractions?

$$f(x) = \frac{3x-7}{(x-4)^2(2x^3+1)}$$

A. $f(x) = \dfrac{A}{x-4} + \dfrac{Bx+C}{(x-4)^2} + \dfrac{Dx^2+Ex+F}{(2x^3+1)}$

B. $f(x) = \dfrac{A}{x-4} + \dfrac{B}{(x-4)^2} + \dfrac{Cx^2+Dx+E}{(2x^3+1)}$

C. $f(x) = \dfrac{Ax+B}{(x-4)^2} + \dfrac{Cx^2+Dx+E}{(2x^3+1)}$

D. $f(x) = \dfrac{A}{x-4} + \dfrac{B}{(x-4)^2} + \dfrac{C}{(2x^3+1)}$

Solution The denominator can be decomposed into two factors: $(x - 4)$ repeated twice and $(2x^3 + 1)$ repeated once.

Partial fractions originating from the two repeat instances of $(x - 4)$ will be $\dfrac{A}{x-4}$ and $\dfrac{B}{(x-4)^2}$.

Partial fractions originating from the one instance of $(2x^3 + 1)$ will be $\dfrac{Cx^2+Dx+E}{(2x^3+1)}$ with numerator one order lower than the denominator.

Answer is B.

Ordinary Differential Equation with Constant Coefficients

The general equation for an ordinary differential equation with constant coefficients is

$$b_n \frac{d^n y(x)}{dx^n} + \cdots + b_1 \frac{dy(x)}{dx} + b_o y(x) = f(x) \tag{1.74}$$

where $b_0, b_1, \ldots, b_n$ are constant coefficients.

Homogeneous Differential Equations

When the function $f(x) = 0$, the equation is called a homogeneous differential equation.

A homogeneous ordinary differential equation with constant coefficients (of any order n) admits the exponential solution of the form

$$y_h(x) = e^{rx} \tag{1.75}$$

Inserting this trial solution into the differential equation yields the characteristic equation

$$b_n r^n + b_{n-1} r^{n-1} + \cdots + b_1 r + b_o = 0 \tag{1.76}$$

Solving the characteristic equation yields the roots $r_1, r_2, \ldots, r_n$. If the roots are distinct, the homogeneous solution $y_h(x)$ is of the form

$$y_h(x) = C_1 e^{r_1 x} + C_2 e^{r_2 x} + \cdots + C_n e^{r_n x} \tag{1.77}$$

where the coefficients $C_1, \ldots, C_n$ are determined from boundary conditions.

First-Order Linear Homogeneous Differential Equation with Constant Coefficients

The first-order, linear, homogeneous differential equation shown below

$$b_1 \frac{dy(x)}{dx} + b_o y(x) = 0 \tag{1.78}$$

has the solution

$$y = Ce^{-\frac{b_0}{b_1}x} \tag{1.79}$$

where the constant of integration C, obtained from the boundary condition, is

$$C = y\big|_{x=0} \tag{1.80}$$

First-Order Linear Nonhomogeneous Differential Equation with Constant Coefficients

The first-order, linear, nonhomogeneous differential equation shown below

$$\tau \frac{dy(t)}{dt} + y(t) = Kx(t) \qquad x(t) = \left\{ \begin{array}{l} A \text{ for } t < 0 \\ B \text{ for } t > 0 \end{array} \right\} \tag{1.81}$$

The equation represents a system subjected to a "step" input $Kx(t)$, where $x(t)$ is as shown below.

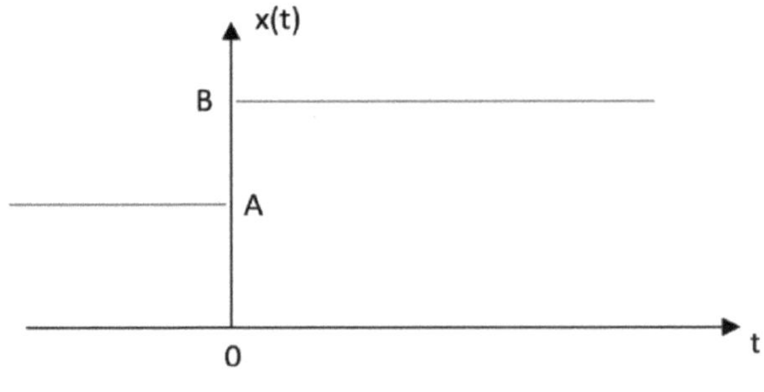

The response of the system to such a step input is an asymptotic exponential that starts at $y = KA$ and asymptotes to $y = KB$ as shown below. The rate of growth to the saturation level depends on the time factor τ (a smaller time factor τ causes a steeper climb to $y = KB$).

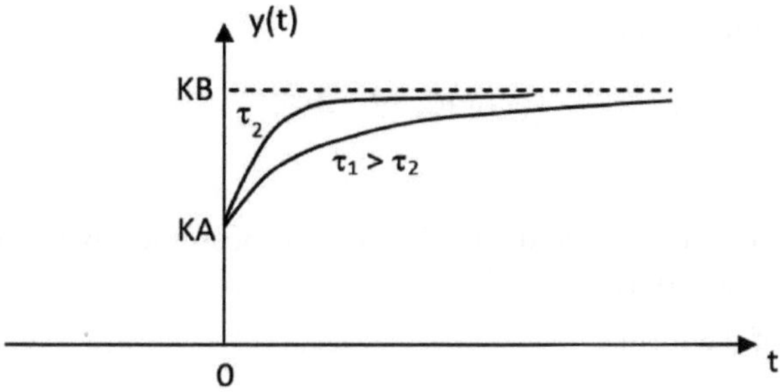

$$y(t) = KA + (KB - KA)\left(1 - e^{-\frac{t}{\tau}}\right) \tag{1.82}$$

where K is called the gain factor, and τ is called the time constant.

Second-Order Linear Homogeneous Differential Equation with Constant Coefficients

The general equation for the second-order homogeneous D.E. can be written

$$b_2 \frac{d^2y(x)}{dx^2} + b_1 \frac{dy(x)}{dx} + b_0 y(x) = 0 \tag{1.83}$$

which can be simplified to

$$\frac{d^2y(x)}{dx^2} + a\frac{dy(x)}{dx} + by(x) = 0 \tag{1.84}$$

where the constants $a = b_1/b_2$ and $b = b_0/b_2$.

The characteristic equation for this (Eq. 1.84) differential equation (by trying the exponential solution $y_h = e^{rx}$) is

$$r^2 + ar + b = 0 \tag{1.85}$$

which yields the roots

$$r_{1,2} = \frac{-a \pm \sqrt{a^2 - 4b}}{2} \tag{1.86}$$

Case 1: $a^2 > 4b$.

Such a system is called an overdamped system. It possesses two distinct real roots r_1 and r_2. The solution is of the form

$$y(x) = C_1 e^{r_1 x} + C_2 e^{r_2 x} \tag{1.87}$$

Case 2: $a^2 = 4b$.

Such a system is called a critically damped system. It possesses two identical real roots $r_1 = r_2$.

The solution is of the form

$$y(x) = (C_1 + C_2 x)e^{r_1 x} \quad (1.88)$$

Case 3: $a^2 < 4b$.

Such a system is called an underdamped system. It possesses two complex roots, which are conjugates of each other ($\alpha \pm i\beta$).

The solution is of the form

$$y(x) = e^{\alpha x}(C_1 \cos \beta x + C_2 \sin \beta x) \quad (1.89)$$

where $\alpha = -\frac{a}{2}$.

$$\beta = \frac{\sqrt{4b - a^2}}{2}$$

Example 1.44

Which is a solution to the differential equation below?

$$4\frac{d^2y}{dx^2} - 7\frac{dy}{dx} + 3y = 0$$

A. $y = 7e^{3x} - 3e^{-4x}$

B. $y = 7e^{0.75x} - 3e^{x}$

C. $y = 7e^{3x} - 3e^{4x}$

D. $y = 7e^{-0.75x} - 3e^{-4x}$

Solution This is a homogeneous, ordinary differential equation with constant coefficients. Trying the exponential solution $y = e^{rx}$, the characteristic equation is

$$4r^2 - 7r + 3 = 0$$

Roots are $r = \frac{7 \pm \sqrt{7^2 - 4 \times 4 \times 3}}{2 \times 4} \Rightarrow r = \frac{3}{4},\ 1$

$$y = Ae^{0.75x} + Be^{x}$$

With the given information (i.e., without any boundary conditions), there is no way to determine coefficients—only choice B has the correct pattern.

Answer is B.

Nonhomogeneous Linear Ordinary Differential Equation: Particular Solution

When the function $f(x)$ in Eq. (1.74) is nonzero, it generates a particular solution $y_p(x)$, thus creating the complete solution

$$y(x) = y_h(x) + y_p(x) \tag{1.90}$$

The form of the particular solution $y_p(x)$ is consistent with the form of the function $f(x)$ as given in the table below.

$f(x)$	$y_p(x)$
A	B
$Ae^{\alpha x}$, $\alpha \neq r_n$	$Be^{\alpha x}$
$Ae^{\alpha x}$, $\alpha = r_n$	$Bxe^{\alpha x}$
$A_1 \sin wx + A_2 \cos wx$	$B_1 \sin wx + B_2 \cos wx$

Example 1.45

What is a valid solution to the differential equation below?

$$\frac{d^2y}{dx^2} + 6\frac{dy}{dx} + 25y = 4\sin x$$

A. $y = e^{-3x}(A\cos 4x + B\sin 4x) + Cx\sin x$

B. $y = e^{-3x}(A\cos 4x + B\sin 4x) + C\sin x$

C. $y = e^{3x}(A\cos 4x + B\sin 4x) + C\sin x$

D. $y = e^{3x}(A\cos 4x + B\sin 4x) + Cx\sin x$

Solution This is a nonhomogeneous ordinary differential equation with constant coefficients. For the homogeneous solution, trying the exponential solution $y = e^{rx}$, the characteristic equation is

$$r^2 + 6r + 25 = 0$$

Roots are $r = \dfrac{-6 \pm \sqrt{6^2 - 4 \times 1 \times 25}}{2 \times 1} \Rightarrow r = -3 \pm 4i$.

This can be expressed in four different ways (the *Handbook* only shows the third, but it is important to recognize that all of the following forms are equivalent).

$$y_H = Ae^{(-3+4i)x} + Be^{(-3-4i)x}$$
$$y_H = e^{-3x}\left(Ae^{4ix} + Be^{-4ix}\right)$$
$$y_H = e^{-3x}\left(A_1 \cos 4x + B_1 \sin 4x\right)$$
$$y_H = Ce^{-3x}\cos(4x + \varphi)$$

The particular solution will have the same form as the right-hand side of the differential equation

$$y_P = C\sin x + D\cos x$$

The complete solution will be of the form

$$y = y_H + y_P = e^{-3x}(A\cos 4x + B\sin 4x) + C\sin x + D\cos x$$

With the given information (i.e., without any boundary conditions), there is no way to determine coefficients—only choice B has the correct pattern.

Answer is B.

Example 1.46

What is a valid solution to the differential equation below?

$$\frac{d^2y}{dx^2} + 16y = 4\sin 4x$$

A. $y = (A + Bx + Cx^2)\cos 4x$

B. $y = (A + Bx + Cx^2)\sin 4x$

C. $y = (A + Bx)\cos 4x + (C + Dx)\sin 4x$

D. $y = (A + Bx)\cos x + (C + Dx)\sin x$

Solution This is a nonhomogeneous ordinary differential equation with constant coefficients. For the homogeneous solution, trying the exponential solution $y = e^{rx}$, the characteristic equation is:

$$r^2 + 16 = 0$$

Roots are $r = \pm 4i$.

This can be expressed in three different ways:

$$y_H = Ae^{4ix} + Be^{-4ix}$$
$$y_H = A_1\cos 4x + B_1\sin 4x$$
$$y_H = C\cos(4x + \varphi)$$
$$y_H = A\cos 4x + B\sin 4x$$

The particular solution will have the same form as the right-hand side of the differential equation $y_P = C\sin 4x + D\cos 4x$. However, this function already exists in the homogeneous solution. To make it unique, we multiply it with the simplest polynomial in the series x, x^2, x^3, etc.

The complete solution will be of the form:

$$y = y_H + y_P = A\sin 4x + B\cos 4x + Cx\sin 4x + Dx\cos 4x$$

With the given information (i.e., without any boundary conditions), there is no way to determine coefficients—only choice C has the correct pattern.

Answer is C.

CHAPTER 2

Probability and Statistics

Permutations and Combinations

In calculating probabilities, it is often necessary to calculate the number of ways in which certain events can take place. Two underlying concepts are (a) permutations and (b) combinations.

A permutation is a particular sequence of a set of objects. Each sequence is unique. On the other hand, a combination is a particular set of objects created without regard for the sequence of objects. As an example, choosing a list of three speakers from a list of 10 possible candidates is a combination, whereas creating a program with three speakers in a particular order is a permutation.

The number of different permutations of r objects from a population of n distinct objects is given by

$$P(n,r) = \frac{n!}{(n-r)!} \tag{2.1}$$

The number of different combinations of r objects from a population of n distinct objects is given by

$$C(n,r) = \frac{n!}{(n-r)!r!} \tag{2.2}$$

Some calculators identify these functions as nPr and nCr, respectively.

Example 2.1

A community organization is putting together an event where four speakers will be on the program. The committee can choose from a pool of seven candidates. (a) How many groups of speakers can be created? (b) How many different unique programs can be created?

Solution

(a) Number of groups of four speakers (order does not matter) = $C(7, 4) = 35$

(b) Number of unique four-speaker lists (order matters) = $P(7, 4) = 840$

Example 2.2

A bag contains 10 green, 6 yellow, and 4 red balls. How many unique sequences of 20 balls can be made by drawing from the bag?

A. 21.9 million

B. 27.9 million

C. 38.8 million

D. 42.1 million

Solution Arrangements (order matters) of n objects when they appear in k groups are calculated as

$$P(n;n_1,n_2,n_3,\ldots,n_k)=\frac{n!}{n_1!n_2!\cdots n_k!}$$

Therefore, with 20 objects (n = 20 balls) in three groups (k = 3 colors), the number of unique arrangements (permutations) is

$$\frac{20!}{10!6!4!}=38,798,760$$

Answer is C.

Example 2.3

A bridge has 12 lighting pedestals as part of its superstructure design. Each pedestal is capable of supporting one light post. If six identical light posts are available at a certain stage of the project, how many unique lighting arrangements can be created?

A. 72

B. 256

C. 924

D. 665,280

Solution Each lighting pedestal is in a unique position. Imagine them to be numbered 1 to 12. So, the question boils down to "How many different ways are there to choose 6 locations out of 12?". Since the light posts are identical, the order of the choice is immaterial. Thus, the answer is ${}^{12}C_6 = 924$.

Answer is C.

Laws of Probability

The laws of probability govern events that are uncertain to some degree. The probability of an event A is generally written $P(A)$ and is a real number in the range 0 to 1, with a probability of zero assigned to an impossible event and a probability of one assigned to a certain one.

The law of total probability (of two related events) is stated as

$$P(A \text{ and/or } B) = P(A) + P(B) - P(A \text{ and } B)$$

This is usually represented as

$$P(A \cup B) = P(A) + P(B) - P(A \cap B) \tag{2.3}$$

This states that the probability that either A or B (or both) occurring is equal to the sum of the individual probabilities of A and B less the probability of them occurring simultaneously.

Independent Events

When two events are independent (i.e., they do not influence each other), the probability of their occurring simultaneously is simplified to

$$P(A \cap B) = P(A) \cdot P(B) \tag{2.4}$$

Example 2.4

A six-sided die is being thrown onto a normal chessboard (32 white squares and 32 black squares). What is the probability that either a six will be thrown or it will land on a white square?

Solution Let A = a six gets thrown on the die. $P(A) = 0.167$
Let B = die lands on a white square. $P(B) = 0.5$
Since the two events are independent, the probability of throwing a six onto a white square is

$$P(A \cap B) = P(A) \cdot P(B) = 0.083$$

Therefore, $P(A \cup B) = P(A) + P(B) - P(A \cap B) = 0.167 + 0.5 - 0.083 = 0.583$

Conditional Probability—Related Events

For events that are not independent, the conditional probabilities are defined as

$P(A \mid B)$ = the probability that A occurs given that B has already occurred

$P(B \mid A)$ = the probability that B occurs given that A has already occurred

Example 2.5

A bag contains 11 balls which are identical except for their color: 5 are red, 4 are yellow, and 2 are black. If balls are drawn from this bag, with replacement, what is the probability of drawing a yellow ball followed by a red one?

Solution Probability of drawing a yellow ball first = $P(Y) = 4/11$.
After the yellow ball has been drawn and not replaced, there are 10 balls, 5 of which are red. $P(R \mid Y) = 5/10 = 0.5$.

Therefore, the probability of a yellow ball followed by a red one = $P(Y) \cdot P(R \mid Y) = 2/11$.

Bayes' Theorem

The Bayes theorem describes the probability of an event, based on prior knowledge of conditions that may be related to the event. The theorem in its simplest form, relating a single event A to a single event B, may be stated as

$$P(A|B) = \frac{P(B|A)P(A)}{P(B)} \tag{2.5}$$

For a population of events B_j ($j = 1, \ldots, n$), the above may be extended to

$$P(B_j|A) = \frac{P(A|B_j)P(B_j)}{\sum_{i=1}^{n} P(A|B_i)P(B_i)} \tag{2.6}$$

where $P(B_i)$ is the probability of event B_i within the population of B. $P(A|B_i)$ is the probability of event A given event B_i has occurred.

Example 2.6

ABC Bolts gets its bolts manufactured at three factories: A, B, and C. The three factories have the following fraction defective—A: 3%, B: 5%, C: 5%. Annual production at the factories (in thousands) are—A: 250, B: 400, C: 500. If a bolt is discovered to be defective, what is the probability that it came from factory B?

Solution Based on the production numbers, the probabilities of any bolt being from factories A, B, and C are 250/1150, 400/1150, and 500/1150, respectively.

Probability of being defective given bolt from factory A = $P(D|A) = 0.03$

Probability of being defective given bolt from factory B = $P(D|B) = 0.05$

Probability of being defective given bolt from factory C = $P(D|C) = 0.05$

Probability that a bolt came from factory B, given it is defective:

$$\begin{aligned} P(B|D) &= \frac{P(D|B)P(B)}{P(D|A)P(A) + P(D|B)P(B) + P(D|C)P(C)} \\ &= \frac{0.05 \times 400/1150}{0.03 \times 250/1150 + 0.05 \times 400/1150 + 0.05 \times 500/1150} \\ &= 0.38 \end{aligned}$$

Measures of Central Tendency

The arithmetic mean, the median, and the mode are all measures of central tendency of a set of data.

Mean

The arithmetic mean (average) of a set of discrete data $\{x_1, x_2, \ldots, x_n\}$ is given by

$$\bar{x} = \frac{x_1 + x_2 + \cdots + x_n}{n} = \frac{\sum_1^n x_i}{n} \tag{2.7}$$

If the data is grouped, with each discrete value x_i having frequency f_i, then the mean is the weighted arithmetic average

$$\bar{x} = \frac{f_1x_1 + f_2x_2 + \cdots + f_nx_n}{n} = \frac{\sum_1^n f_ix_i}{\sum_1^n f_i} \tag{2.8}$$

Example 2.7

The table below shows summary of speed data for 100 cars observed on the highway. What is the mean speed?

Speed Range (km/h)	Frequency
40–46	12
46–52	19
52–58	41
58–64	21
64–70	7

A. 53.7 km/h

B. 54.5 km/h

C. 57.1 km/h

D. 58.7 km/h

Solution Representing each range by its midpoint:

Speed (km/h)	Frequency
43	12
49	19
55	41
61	21
67	7

The grouped mean is given by

$$\bar{X} = \frac{\sum f_iX_i}{\sum f_i} = \frac{12\times 43 + 19\times 49 + 41\times 55 + 21\times 61 + 7\times 67}{100} = 54.52 \text{ km/h}$$

Answer is B.

Variance

The variance is a measure of dispersion of the data. It is defined as the arithmetic mean of the square of the deviations from the mean. If the arithmetic mean of a population (size n) is known to be μ, then the variance (which is the square of the standard deviation σ) is calculated as

$$\sigma^2 = \frac{\sum_i^n (x_i - \mu)^2}{n} \tag{2.9}$$

The variance is unaffected by translation. In other words, if an entire set of numbers is shifted to the right (by adding a constant to each value) or to the left (by subtracting a constant), the variance will remain unchanged.

However, if each number in a data set is scaled (multiplied by a scale factor α), then the variance will be scaled by the factor α^2.

Example 2.8

The variance of the set of numbers {2, −3, 0, 2, 5, 12} is 22. What is the variance of the set {102, 97, 100, 102, 105, 112}?

Solution We see that the new set is the result of translating (adding 100 to) the original set of numbers. Therefore, the variance will remain the same (22).

Example 2.9

The variance of the set of numbers {2, −3, 0, 2, 5, 12} is 22. What is the variance of the set {1.6, −2.4, 0, 1.6, 4, 9.6}?

Solution We see that the new set is the result of scaling (multiplying by 0.8) the original set of numbers. Therefore, the variance of the new set will be

$$\sigma^2 = 0.8^2 \times 22 = 14.08$$

Example 2.10

The table below shows summary of speed data for 100 cars observed on the highway. What is the standard deviation of the speed data?

Speed Range (km/h)	Frequency
40–46	12
46–52	19
52–58	41
58–64	21
64–70	7

A. 3.7 km/h

B. 4.5 km/h

C. 5.9 km/h

D. 6.4 km/h

Solution Representing each range by its midpoint:

Speed Range (km/h)	Frequency
43	12
49	19
55	41
61	21
67	7

The grouped mean is given by

$$\overline{X} = \frac{\sum f_i X_i}{\sum f_i} = \frac{12\times 43 + 19\times 49 + 41\times 55 + 21\times 61 + 7\times 67}{100} = 54.52 \text{ km/h}$$

The deviations from the mean are listed in the table below:

Speed (km/h)	Frequency	Deviation
43	12	−11.52
49	19	−5.52
55	41	0.48
61	21	6.48
67	7	12.48

The variance is given by

$$\sigma^2 = \frac{\sum f_i (X_i - \overline{X})^2}{\sum f_i}$$

$$= \frac{12\times 11.52^2 + 19\times 5.52^2 + 41\times 0.48^2 + 21\times 6.48^2 + 7\times 12.48^2}{100}$$

$$= 41.53 \text{ (km/h)}^2$$

Standard deviation $\sigma = 6.44$ km/h

Answer is D.

Combinations of Random Variables

If a random variable y is defined as the sum of n random variables $x_1, x_2, \ldots, x_n$, the variance of y is given by

$$\sigma_y^2 = \sigma_1^2 + \sigma_2^2 + \cdots + \sigma_n^2 \qquad (2.10)$$

When a random variable y is defined as a linear combination of n random variables $x_1, x_2, \ldots, x_n$

$$y = a_1x_1 + a_2x_2 + \cdots + a_nx_n \tag{2.11}$$

the expected value of y is given by

$$\mu_y = a_1\mu_1 + a_2\mu_2 + \cdots + a_n\mu_n \tag{2.12}$$

the variance of y is given by

$$\sigma_y^2 = a_1^2\sigma_1^2 + a_2^2\sigma_2^2 + \cdots + a_n^2\sigma_n^2 \tag{2.13}$$

Example 2.11

The cost function (thousands of dollars) for a construction project has been established empirically as

$$C = 50X_1 + 30X_2 - 20X_3$$

where X_1, X_2, and X_3 are relevant random variables (e.g., number of labor hours, number of material units consumed, number of embedded safety features, etc.). These three random variables have the following parameters:

X_1: Normally distributed—mean = 250,000, standard deviation = 12,000
X_2: Normally distributed—mean = 200,000, standard deviation = 10,000
X_3: Normally distributed—mean = 60,000, standard deviation = 5,000

What is the probability that the overall project cost exceeds $18.5 million?

Solution (see also section on normal distribution) Since the cost C is a linear combination of three normally distributed random variables, it will be normally distributed as well. The parameters of the distribution of random variable C are

$$\mu_C = a_1\mu_{X_1} + a_2\mu_{X_2} + a_3\mu_{X_3} = 50\times 250 + 30\times 200 - 20\times 60 = 17{,}300 \text{ (thousands of dollars)}$$

Mean cost, μ_C = $17.3 million

$$\sigma_C^2 = a_1^2\sigma_{X_1}^2 + a_2^2\sigma_{X_2}^2 + a_3^2\sigma_{X_3}^2 = 50^2\times 12^2 + 30^2\times 10^2 + 20^2\times 5^2 = 460{,}000$$

Standard deviation of cost, $\sigma_C = 678$

The probability that $C > 18{,}500$ is given by calculating the standard normal variable

$$Z = \frac{18{,}500 - 17{,}300}{678} = 1.77$$

The cumulative area under the standard normal curve to Z = 1.77 is 0.9616 (96.16%). Therefore, the probability of cost exceeding $18.5 million is 3.84%.

Sample Mean and Sample Variance

When a sample of size n is taken from a population, the sample mean is the arithmetic average

$$\bar{x} = \frac{\sum_1^n x_i}{n} \tag{2.14}$$

and the sample variance (which is the square of the sample standard deviation s) is

$$s^2 = \frac{\sum_1^n (x_i - \bar{x})^2}{n-1} \tag{2.15}$$

Example 2.12

A bolt manufacturing process is being set up for production of 25-mm-diameter bolts. A sample of seven bolts is taken and the diameters are measured as 24.6, 24.8, 24.9, 25.1, 25.2, 25.0, and 25.0 mm. What is the sample variance?

A. 0.0395 mm^2

B. 0.0583 mm^2

C. 0.0750 mm^2

D. 0.0917 mm^2

Solution The sample variance is defined as

$$s^2 = \frac{\sum (X_i - \bar{X})^2}{n-1}$$

The sample mean is

$$\bar{X} = \frac{\sum X_i}{n} = 24.943$$

The (absolute) deviations from the mean are

0.343, 0.143, 0.043, 0.157, 0.257, 0.057, 0.057

Sum of the squares of the deviations = 0.237
Sample variance = 0.237 mm^2/(7 − 1) = 0.0395 mm^2

Answer is A.

NOTE All of the permitted calculators do statistical computations such as sample mean, standard deviation, etc. You should know how to use your calculator to quickly do what is otherwise a tedious calculation.

Median

Unlike the mean, which reflects an average (and therefore, a central tendency), the median (which is also a measure of central tendency) represents the *center* of the data set, when arranged in ascending or descending order. Thus, the median reflects a center based on rank whereas the mean represents a center based on value.

Example 2.13

What are (a) the arithmetic mean and (b) median of the following set of numbers: {2, −3, 0, 2, 5, 12}?

Solution The arithmetic mean = 18 ÷ 6 = 3

Arranging the numbers in ascending order, we have {−3, 0, 2, 2, 5, 12}. Since this is a set of an even number of values, there is no "central value." The median is the average of the *two* values in the center. In this case, both of those numbers are 2. So the median of this set is 2.

If the set had been changed to {2, −3, 0, 2, 5, 72}, the mean (which is strongly influenced by the value of the last number changing from 12 to 72) would change to 13, but the median (which is only influenced by rank) would remain at 2.

Example 2.14

The table below shows summary of speed data for 100 cars observed on the highway. What is the median speed?

Speed Range (km/h)	Frequency
40–46	12
46–52	19
52–58	41
58–64	21
64–70	7

A. 53 km/h

B. 55 km/h

C. 59 km/h

D. 61 km/h

Solution Constructing a cumulative frequency column:

Speed Range (km/h)	Frequency	Cumulative Frequency
40–46	12	12
46–52	19	31
52–58	41	72
58–64	21	93
64–70	7	100

With a 100 observations, the median should be at a cumulative frequency of 50 (strictly speaking, it should be the average of the 50th and 51st observations), which falls in the range 52 to 58 km/h. The cumulative frequency corresponding to speed of 52 km/h is 31 and that corresponding to speed of 58 km/h is 72. Since the cumulative frequency of 50 is almost exactly halfway between these two values, the corresponding speed is midway between 52 and 58 km/h.

So, the median speed is 55 km/h.

Answer is B.

Median for Continuously Distributed Data

For data where observations can be continuously distributed, the median is that value for which the cumulative probability is 0.5.

Example 2.15

A random variable x has the probability density function shown below. What is the median?

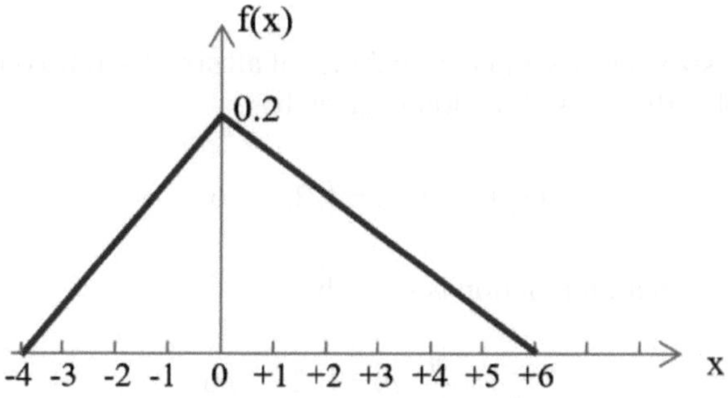

Solution To find the median, we must find that value which splits the area under the probability density function into two halves of area 0.5. The triangle on the left ($-4 < x < 0$) is 0.4 and the one on the right is 0.6. In order to obtain a triangle of area 0.5 cut from the triangle on the right, the base needs to be 5. Thus, the median is $x = 1$.

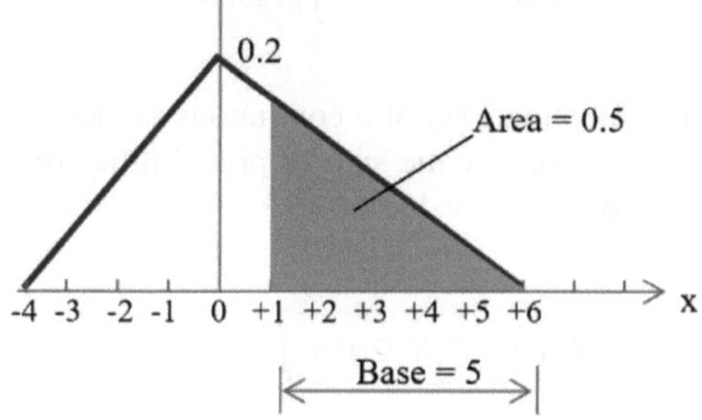

Probability Functions

A random variable X can be either discrete (can take on only discrete values $x_1, x_2, \ldots, x_n$) or continuous (can take on any value within a range $x_{min} < x < x_{max}$).

Discrete Random Variable

For a discrete random variable, the probability of a single outcome $X = x_k$ has a distribution called the probability mass function $f(x_k)$

$$f(x_k) = P(X = x_k),\ k = 1, 2, \ldots, n \tag{2.16}$$

The cumulative distribution function $F(x)$ of a discrete random variable X with the probability mass function $f(x_k)$ is given by the sum of probabilities of the variable X taking on values less than or equal to a particular value

$$F(x_m) = \sum_{k=1}^{m} P(x_k) = P(X \le x_m),\ m = 1, 2, \ldots, n \tag{2.17}$$

An unbiased die with six sides has equal probability of all six (discrete) outcomes (1, 2, 3, 4, 5, and 6). Thus, the probability mass function is given by

$$f(x_k) = 1/6,\ k = 1, 2, \ldots, 6$$

and the cumulative distribution function is given by

$$F(x_m) = \frac{m}{6},\ m = 1, 2, \ldots, 6$$

Continuous Random Variable

For a continuous random variable, the probability density function $f(x)$ is such that

$$P(a \le X \le b) = \int_a^b f(x)\,dx \tag{2.18}$$

The cumulative distribution function $F(x)$ of a continuous random variable X with the probability density function $f(x)$ is given by the sum of probabilities of the variable X taking on values less than or equal to a particular value

$$F(a) = P(X \le a) = \int_{-\infty}^{a} f(t)\,dt \tag{2.19}$$

Normal (Gaussian) Distribution

The outcome of many independent experiments (typically greater than 30) on a random variable usually follows the normal or Gaussian probability distribution. The Gaussian probability density function (bell-shaped curve) is given by

$$f(x) = \frac{1}{\sqrt{2\pi}\sigma} \exp\left[-\frac{1}{2}\left(\frac{x-\mu}{\sigma}\right)^2\right] = \frac{1}{\sqrt{2\pi}\sigma} \exp\left[-\frac{1}{2}Z^2\right] \tag{2.20}$$

where Z is the standard normal variable, which results from the shifting and scaling of the random variable X, as given below

$$Z = \frac{X-\mu}{\sigma} \tag{2.21}$$

In Eq. (2.21), μ is the mean and σ is the standard deviation of the normal variable X. The standard normal variable Z has mean 0 and standard deviation 1.

For any random variable which is expected to follow the normal distribution, the first step is to transform the relevant values of the random variable to a corresponding value of Z (see Example 2.11).

The areas associated with different Z values (positive values only) are listed on page 46 of the *FE Reference Handbook,* version 9.4. The following convention is used:

$f(x)$	Ordinate of the normal probability density function
$F(x)$	Cumulative distribution function (shaded in Fig. 2.1)
$R(x)$	Area complementary to $F(x)$. This means $R(x) = 1 - F(x)$
$W(x)$	Area complementary to $2R(x)$. This means $W(x) = 1 - 2R(x)$

Note that, due to the symmetry of the Gaussian curve, the area corresponding to a negative Z value (i.e., $X < \mu$) is given by

$$F(-Z) = P(z < -Z) = 1 - P(z < Z) = 1 - F(Z) = R(Z) \tag{2.22}$$

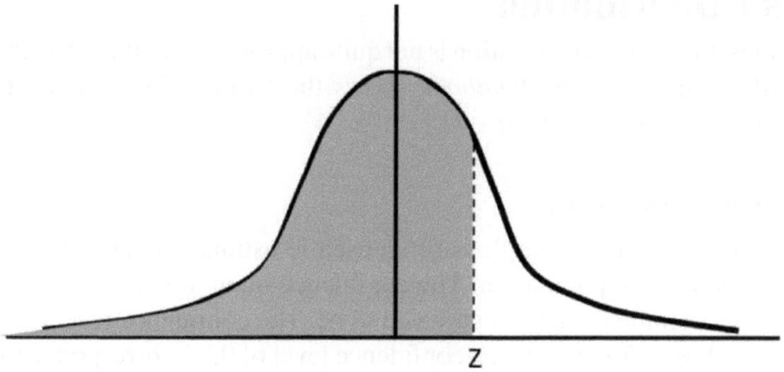

Figure 2.1 Cumulative probability for a normal distribution $F(x)$.

Example 2.16

A long record ($N > 30$) of damage data to a small jurisdiction from coastal flooding due to hurricanes is assumed to follow the Gaussian (normal) distribution. The mean cost (thousands of dollars) is 680 and the standard deviation is 45.

(a) What is the probability of a storm causing more than \$750,000 of damage?

(b) What is the probability that the damage is between \$600,000 and \$700,000?

Solution

(a) The specific value of X needs to be converted to their corresponding Z values.

$$X_1 = 750 \Rightarrow Z_1 = \frac{750-680}{45} = +1.556$$

From the table $P(Z \leq 1.556) = 0.9401$ (interpolating between 0.9394 and 0.9406)

Therefore, $P(Z > 1.556) = 1.0 - 0.9401 = 0.0599$ (5.99%). This is also $R(1.556)$.

(b) Again, transform the specific values of X to their corresponding Z values.

The probability that X is between 600,000 and 700,000 is given by $P(-1.778 \leq Z \leq +0.444)$. However, since the table in the *Handbook* does not contain negative values of Z, one needs to use the symmetry properties of the normal distribution.

$$P(-1.778 \leq Z \leq +0.444) = F(0.444) - F(-1.778) = F(0.444) - [1 - F(+1.778)]$$
$$= F(+0.444) + F(+1.778) - 1 = 0.6714 + 0.9623 - 1 = 0.6337$$

Therefore, $P(-1.778 \leq z \leq +0.444) = 0.6337$ (63.37%)

Student's *t*-Distribution

For small samples, the normal distribution is not quite appropriate. In that case, the *t*-distribution should be used. Page 47 of the *Handbook* shows the *t*-values for various α (significance) values and degrees of freedom ($\nu = n - 1$).

Confidence Intervals

For a random variable, a finite sample is often used to estimate a particular parameter (such as the mean) of the infinite population. The confidence interval is a range of values between which a population parameter can be estimated to be. The confidence level is the complement of the significance level. For example, a confidence level of 95% corresponds to a significance level of 5%. In the extreme, to predict a range at the 100% confidence level, the confidence interval would extend from $-\infty$ to $+\infty$, but such a result would have no significance. As the range gets narrower, the associated confidence level also gets smaller.

Confidence Interval for the Mean (μ) of a Normal Distribution

Case 1—Standard Deviation σ Known

The confidence interval for the population mean μ is

$$\bar{X} - Z_{\alpha/2}\frac{\sigma}{\sqrt{n}} \le \mu \le \bar{X} + Z_{\alpha/2}\frac{\sigma}{\sqrt{n}}$$

Example 2.17

A random variable X is normally distributed with known standard deviation = 6.8. A sample of 21 observations of X results in a sample mean = 28.5. What is the 90% confidence interval for the population mean?

A. $25.59 \le \mu \le 31.41$

B. $26.60 \le \mu \le 30.40$

C. $25.94 \le \mu \le 31.06$

D. $26.06 \le \mu \le 30.94$

Solution Since the population standard deviation σ is known and we seek the 90% confidence interval, $\alpha/2 = (1 - 0.9)/2 = 0.05$. The corresponding Z value is 1.6449. The limits of the confidence interval are

$$\bar{X} \pm Z_{\alpha/2}\frac{\sigma}{\sqrt{n}} = 28.5 \pm 1.6449 \times \frac{6.8}{\sqrt{21}} = 28.5 \pm 2.44$$

Therefore, confidence interval limits are

$$26.06 \le \mu \le 30.94$$

Answer is D.

Case 2—Standard Deviation σ Not Known

The confidence interval for the population mean (μ) is

$$\bar{X} - t_{\alpha/2}\frac{s}{\sqrt{n}} \le \mu \le \bar{X} + t_{\alpha/2}\frac{s}{\sqrt{n}}$$

where X is sample mean, and s is sample standard deviation.

Example 2.18

A group of 10 students from elementary school have their height measured. The sample mean is 102 cm and the sample standard deviation is 1.56 cm. What is the 90% confidence interval about the mean?

A. 100 to 104 cm

B. 100.5 to 103.5 cm

C. 101.1 to 102.9 cm

D. 101.6 to 102.4 cm

Solution With a small sample ($N < 30$), we can assume that the probability distribution is the student's t-distribution. Since we seek the 90% confidence interval, $\alpha = (1 - 0.9)/2 = 0.05$. For degrees of freedom $= N - 1 = 9$ and $\alpha = 0.05$, the t-value is 1.833:

$$\frac{ts}{\sqrt{N}} = \frac{1.833 \times 1.56}{\sqrt{10}} = 0.9$$

90% confidence interval for the mean is

$$\bar{X} - \frac{ts}{\sqrt{n}} \le \mu \le \bar{X} + \frac{ts}{\sqrt{n}}$$

$$102 - 0.9 \le \mu \le 102 + 0.9$$

$$101.1 \le \mu \le 102.9$$

Answer is C.

Binomial Distribution

The binomial distribution may be applied whenever a probabilistic event has only two possible mutually exclusive outcomes, each with a well-defined probability. For example, if the occurrence of a specific design flood has an annual probability P, then the annual probability of *nonoccurrence* is $(1 - P)$. The probability of x occurrences in N trials is given by

$$P(x; N) = C(N, x)P^x(1 - P)^{N-x} \tag{2.23}$$

where $C(N, x)$ is the combination function given by

$$C(N, x) = \frac{N!}{x!(N - x)!} \tag{2.24}$$

The binomial distribution has two parameters: mean nP and variance $nP(1 - P)$.

Example 2.19

What is the probability of three 20-year floods occurring during the next 10 years?

A. 0.5%

B. 0.8%

C. 1.0%

D. 2.2%

Solution Since the 20-year flood has an annual probability of 0.05 (by definition), we can designate $P = 0.05$ and $(1 - P) = 0.95$.

The probability of three floods over the next 10 years is therefore

$$P(3;10) = C(10,3)(0.05^3)(0.95^7) = 120 \times (0.05^3)(0.95^7) = 0.01$$

Thus, there is a 1% probability of exactly three (not necessarily consecutive) 20-year floods occurring during the next 10 years.

Answer is C.

Example 2.20

A slightly biased coin has a 53% probability of showing heads. If the coin is tossed 90 times, what is the probability that heads will turn up exactly 40 times?

A. 7.5%

B. 4.6%

C. 2.3%

D. 1.9%

Solution Defining p = probability of heads = 0.53 and q = probability of tails = 0.47. This is an example of binomial distribution.

Probability of 40 "heads" and 50 "tails" is calculated as

$$C(90,40) \times 0.53^{40} \times 0.47^{50} = 0.02255$$

Answer is C.

Example 2.21

In a parlor game, a competitor throws a ball at a target. On each throw, the average person has a 15% probability of hitting the target. What is the probability that a person will be able to make at least three hits on seven throws?

A. 7%

B. 5%

C. 3%

D. 1%

Solution This is an example of a binomial distribution. Each throw can be considered an independent experiment. Probability of success on each throw = 0.15. Probability of failure = 0.85. Probability of at least three hits = $P(3) + P(4) + P(5) + P(6) + P(7)$.

However, it is easier to calculate as $1 - P(0) - P(1) - P(2)$

$$P(r \geq 3) = 1 - P(0) - P(1) - P(2)$$

$$= 1 - C(7, 0)\, 0.15^0 0.85^7 - C(7, 1)\, 0.15^1 0.85^6 - C(7, 2)\, 0.15^2 0.85^5$$

$$= 1 - 0.32 - 0.396 - 0.21 = 0.07$$

Answer is A.

Example 2.22

A biased coin has a 53% probability of showing heads. If the random variable X represents the number of head when the coin is tossed 120 times, what is the standard deviation of the random variable X?

A. 4.28

B. 5.00

C. 6.00

D. 5.47

Solution As defined, the random variable X has a binomial distribution with parameters $n = 120$ and $p = 0.53$.

The variance of a binomial distribution is

$$\sigma^2 = np(1 - p) = 120 \times 0.53 \times 0.47 = 29.9$$

Standard deviation: $\sigma = 5.47$

Answer is D.

Example 2.23

A power plant is to be designed with a design life of 30 years. If the acceptable cumulative risk of failure due to seismic overload is 2%, what should be the return period of the design earthquake?

A. 30 years

B. 50 years

C. 500 years

D. 1500 years

Solution Let the annual probability (of occurrence) of the design earthquake (or greater) be p.

Then, the probability of nonoccurrence (in any given year) of the design earthquake = $1 - p$.
The probability of nonoccurrence of the design earthquake for the entire 30-year life of the plant = $(1 - p)^{30}$.
The probability of occurrence of the design earthquake during the entire 30-year life of the plant = $1 - (1 - p)^{30}$.
Therefore, $1-(1-p)^{30} = 0.02 \Rightarrow p = 0.000673$
The return period is defined as the inverse of the annual probability of occurrence:

$$N = \frac{1}{p} = \frac{1}{0.000673} = 1485 \text{ years}$$

Answer is D.

Expected Value

Discrete Variable

For a discrete random variable X having a probability mass function $f(x_k)$, the expected value of the random variable is defined as

$$\mu = E[X] = \sum_{k=1}^{n} x_k f(x_k) \tag{2.25}$$

The variance is given by

$$\sigma^2 = V[X] = \sum_{k=1}^{n} (x_k - \mu)^2 f(x_k) \tag{2.26}$$

Example 2.24

A loaded dice has six possible outcomes, with the following probability mass function.

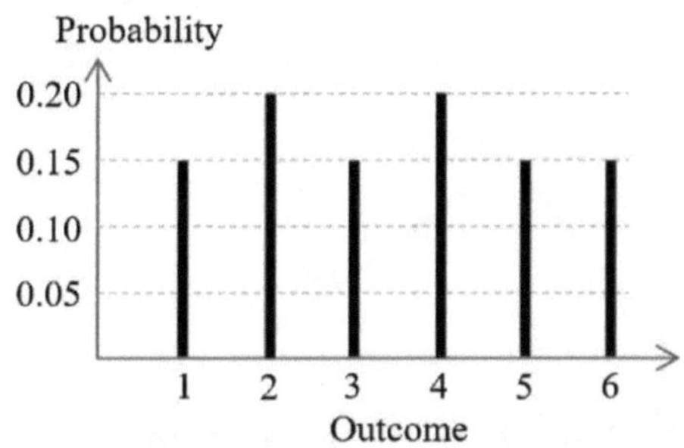

What is the expected value on a single throw of the dice?

A. 3.15

B. 3.45

C. 3.65

D. 3.75

Solution The expected value of a random variable with finite number of discrete outcomes is $\mu = E[X] = \sum x_k f(x_k) = 1\times 0.15 + 2\times 0.20 + 3\times 0.15 + 4\times 0.20 + 5\times 0.15 + 6\times 0.15 = 3.45$

Answer is B.

Continuous Variable

For a continuous random variable X having a probability density function $f(x)$, the expected value of the random variable is defined as

$$\mu = E[X] = \int_{-\infty}^{\infty} x f(x) dx \tag{2.27}$$

and the variance is defined as

$$\sigma^2 = V[X] = \int_{-\infty}^{\infty} (x-\mu)^2 f(x) dx \tag{2.28}$$

The expected value of a function $g(x)$ is given by

$$E[g(x)] = \int_{-\infty}^{\infty} f(x) g(x) dx \tag{2.29}$$

The median of a continuously distributed data is that value for which the cumulative frequency is 0.5.

Example 2.25

A random variable x has the probability density function shown below. What is the expected value of the quadratic function $y(x) = 2x^2$?

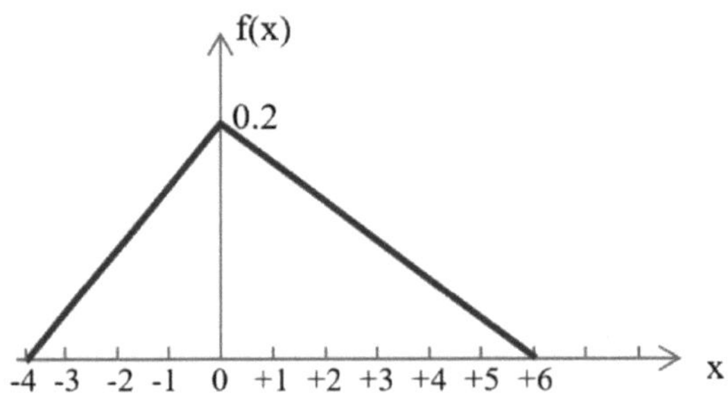

A. 9.33

B. 7.20

C. 5.08

D. 2.13

Solution The expected value of a function $g(x)$ is given by

$$E[g(x)] = \int_{-\infty}^{\infty} f(x) g(x) dx$$

The probability density function can be expressed as

$$f(x) = 0.05x + 0.2 \quad \text{for } -4 < x < 0$$
$$f(x) = -0.033x + 0.2 \quad \text{for } 0 < x < 6$$

Therefore, the expected value of $y(x) = 2x^2$ is calculated as

$$E[y(x)] = \int_{-4}^{0} 2x^2(0.05x + 0.2)\,dx + \int_{0}^{6} 2x^2(-0.033x + 0.2)\,dx = 2.13 + 7.20 = 9.33$$

Answer is A.

Use of Expected Value in Decision Making

When faced with a choice between two (or more) alternatives based on risk, the expected value can be computed for each choice. The alternative with the highest expected value is then chosen.

Hypothesis Testing

A hypothesis test is a statistical test to infer if the data sample contains enough evidence that a certain condition (the null hypothesis, H_0) is true for the entire population. For example, the null hypothesis could be that the true mean (μ) is equal to a specific value (μ_o).

Null hypothesis $H_0 : \mu = \mu_o$

Alternative hypothesis $H_1 : \mu \neq \mu_o$

Test on Means of Normal Distribution—Variance Known

The test statistic used for this test is

$$Z_o = \frac{\bar{X} - \mu_o}{\sigma/\sqrt{n}} \tag{2.30}$$

(i) If the null hypothesis (H_0) and alternate hypothesis (H_1) are as below:

Null hypothesis $H_0 : \mu = \mu_o$

Alternative hypothesis $H_1 : \mu \neq \mu_o$

then the criterion for rejection of the null hypothesis is

$$|Z_o| > Z_{\alpha/2}$$

(ii) If the null hypothesis (H_0) and alternate hypothesis (H_1) are as below:

Null hypothesis $H_0 : \mu = \mu_o$

Alternative hypothesis $H_1 : \mu < \mu_o$

then the criterion for rejection of the null hypothesis is

$$Z_o < -Z_\alpha$$

Example 2.26

The county school board sets the math proficiency score for 8th graders as 80. A sample of 45 randomly selected students yields a sample mean score = 75. The population standard deviation is 3.5. At the 1% significance level, does the data suggest that 8th graders are proficient in math?

Solution Minimum score for proficiency = 80. Therefore, the test is a left tailed test.

Null hypothesis $H_0 : \mu = 80$

Alternative hypothesis $H_1 : \mu < 80$

The criterion for rejection of the null hypothesis is

$$Z_o < -Z_{0.01} = -2.5758$$

$$\text{Test statistic: } Z_o = \frac{\bar{X} - \mu_o}{\sigma/\sqrt{n}} = \frac{75 - 80}{3.5/\sqrt{45}} = -9.58$$

Since $Z_o < -Z_{0.01}$, the null hypothesis is rejected. The population is *not* proficient in math.

If the null hypothesis (H_0) and alternate hypothesis (H_1) are as below:

Null hypothesis $H_0 : \mu = \mu_o$

Alternative hypothesis $H_1 : \mu > \mu_o$

then the criterion for rejection of the null hypothesis is

$$Z_o > Z_\alpha$$

Linear Regression

Given a data set containing *x*, *y* paired data, linear regression analysis uses the concept of minimum least square error to estimate the parameters of the "line of best fit."

The equation of this line of best fit is designated

$$y = \hat{a} + \hat{b}x \tag{2.31}$$

The two parameters slope and *y*-intercept ($\hat{a}$) of this line of best fit are given by

$$\hat{b} = \frac{S_{xy}}{S_{xx}} = \frac{\sum x_i y_i - 1/n \sum x_i \sum y_i}{\sum x_i^2 - 1/n\left(\sum x_i\right)^2} \tag{2.32}$$

$$\hat{a} = \bar{y} - \hat{b}\bar{x} \tag{2.33}$$

where the computed arithmetic average of the series *x* and *y* are

$$\bar{x} = \frac{\sum x_i}{n} \qquad \bar{y} = \frac{\sum y_i}{n}$$

The standard error (S_e) of this estimate is given by

$$S_e^2 = \frac{S_{xx}S_{yy} - S_{xy}^2}{S_{xx}(n-2)} = MSE \tag{2.34}$$

The confidence interval for the intercept ($\hat{a}$) is given by

$$\hat{a} \pm t_{\alpha/2,n-2}\sqrt{\left(\frac{1}{n} + \frac{\overline{x}^2}{S_{xx}}\right)MSE} \tag{2.35}$$

The confidence interval for the slope ($\hat{b}$) is given by

$$\hat{b} \pm t_{\alpha/2,n-2}\sqrt{\frac{MSE}{S_{xx}}} \tag{2.36}$$

The correlation coefficient (Pearson) is given by

$$R = \frac{S_{xy}}{\sqrt{S_{xx}S_{yy}}} \tag{2.37}$$

and the coefficient of determination is given by R^2

$$R^2 = \frac{S_{xy}^2}{S_{xx}S_{yy}} \tag{2.38}$$

Example 2.27

Given the following data on two random variables X and Y, determine (a) the equation of the line of best fit, (b) the correlation coefficient, (c) the standard error, and (d) the 90% confidence intervals of the slope and the intercept estimates.

Solution

X	Y	XY	X²	Y²
5	10	50	25	100
12	32	384	144	1024
23	45	1035	529	2025
30	56	1680	900	3136
32	65	2080	1024	4225
ΣX = 102	ΣY = 208	ΣXY = 5229	ΣX² = 2622	ΣY² = 10510

Mean values: $\overline{x} = \frac{102}{5} = 20.4$ and $\overline{y} = \frac{208}{5} = 41.6$

$$S_{xx} = \sum x_i^2 - \frac{1}{n}\left(\sum x_i\right)^2 = 2622 - \frac{1}{5} \times 102^2 = 541.2$$

$$S_{xy} = \sum x_i y_i - \frac{1}{n}\left(\sum x_i\right)\left(\sum y_i\right) = 5229 - \frac{1}{5} \times 102 \times 208 = 985.8$$

$$S_{yy} = \sum y_i^2 - \frac{1}{n}\left(\sum y_i\right)^2 = 10{,}510 - \frac{1}{5} \times 208^2 = 1857.2$$

(a) Slope and intercept estimates

$$\hat{b} = \frac{S_{xy}}{S_{xx}} = \frac{985.8}{541.2} = 1.8215$$

$$\hat{a} = \bar{y} - \hat{b}\bar{x} = 41.6 - 1.8215 \times 20.4 = 4.441$$

Therefore, the line of best fit is given by

$$y = 4.441 + 1.8215x$$

(b) The correlation coefficient (Pearson) is given by

$$R = \frac{S_{xy}}{\sqrt{S_{xx} S_{yy}}} = \frac{985.8}{\sqrt{541.2 \times 1857.2}} = 0.983$$

(c) The standard error (MSE or S_e) of this estimate is calculated from

$$S_e^2 = \frac{S_{xx} S_{yy} - S_{xy}^2}{S_{xx}(n-2)} = \frac{541.2 \times 1857.2 - 985.8^2}{541.2 \times (5-2)} = 20.52$$

Therefore, the standard error is $S_e = 4.53$

(d) For 90% confidence, $\alpha/2 = 0.05$ and $t_{\alpha/2,\, n-2} = t_{0.05,\, 3} = 2.353$

The confidence interval for the intercept ($\hat{a}$) is given by

$$\hat{a} \pm t_{\alpha/2, n-2} \sqrt{\left(\frac{1}{n} + \frac{\bar{x}^2}{S_{xx}}\right) MSE} = 4.44 \pm 2.353 \times \sqrt{\left(\frac{1}{5} + \frac{20.4^2}{541.2}\right) \times 20.52}$$

$$= 4.44 \pm 10.49 = -6.05, 14.93$$

The confidence interval for the slope ($\hat{b}$) is given by

$$\hat{b} \pm t_{\alpha/2, n-2} \sqrt{\frac{MSE}{S_{xx}}} = 1.8215 \pm 2.353 \times \sqrt{\frac{20.52}{541.2}} = 1.8215 \pm 0.4582 = 1.3633,\ 2.2797$$

Example 2.28

Given the following x, y data, what is the equation of the least squares line of best fit?

x	Y
0	12
1	23
2	32
3	40

A. $y = 8.6x + 12.0$

B. $y = 9.3x + 11.8$

C. $y = 0.11x - 1.36$

D. $y = 9.3x + 12.8$

Solution

$n = 4$; $\sum x = 6$; $\sum y = 107$; mean $x = 1.5$; mean $y = 26.75$
$\sum x^2 = 14$
$\sum xy = 207$

$$S_{xy} = \sum x_i y_i - \frac{1}{n}\sum x_i \sum y_i = 207 - \frac{6\times 107}{4} = 46.5$$

$$S_{xx} = \sum x_i^2 - \frac{1}{n}\left(\sum x_i\right)^2 = 14 - \frac{6^2}{4} = 5$$

Slope: $\hat{b} = S_{xy}/S_{xx} = 46.5/5 = 9.3$

Intercept: $\hat{a} = \overline{y} - \hat{b}\overline{x} = 26.75 - 9.3\times 1.5 = 12.8$

Answer is D.

NOTE It is worthwhile to learn to do these statistical calculations on your calculator.

CHAPTER 3

Computational Tools

Number Systems

Binary (Base 2)

The binary number system uses the symbols 0 and 1. Each position has a place value which is a power of the base 2. Thus, starting from the right, the first (right-most) position has place value $2^0 = 1$, the second position has place value $2^1 = 2$, and so on. See Table 3.1.

Table 3.1 Place Values and Symbols Utilized for Different Number Systems

			Position and Place Value					
Number System and Base		Symbols Used	5	4	3	2	1	0
Binary	2	0, 1	2^5	2^4	2^3	2^2	2^1	2^0
Octal	8	0–7	8^5	8^4	8^3	8^2	8^1	8^0
Decimal	10	0–9	10^5	10^4	10^3	10^2	10^1	10^0
Hexadecimal	16	0–9, A–F	16^5	16^4	16^3	16^2	16^1	16^0

Octal (Base 8)

The octal system utilizes the symbols 0 to 7. Each position has a place value which is a power of the base 8. A number represented as $(275)_8$ has the decimal equivalent $= 2 \times 8^2 + 7 \times 8^1 + 5 \times 8^0 = (189)_{10}$.

Decimal (Base 10)

The familiar decimal system utilizes the symbols 0 to 9. Each position has a place value which is a power of the base 10. Thus a number represented as $(748)_{10}$ has the breakdown $= 7 \times 10^2 + 4 \times 10^1 + 8 \times 10^0 = 748$.

Hexadecimal (Base 16)

The hexadecimal system utilizes the symbols 0 to 9 and A to F, with the alpha symbols A to F having value 10 to 15, respectively. Each position has a place value which is a power of the base 16. Thus a number represented as $(2A5)_{16}$ has the decimal equivalent $= 2 \times 16^2 + 10 \times 16^1 + 5 \times 16^0 = (677)_{10}$.

Table 3.1 shows place values and symbols utilized for the four-number systems.

In EXCEL, functions such as BIN2OCT, DEC2HEX, OCT2HEX, etc., can be used to convert between the number systems.

Example 3.1

Convert the decimal number 1438 to (a) binary, (b) octal, and (c) hexadecimal.

Solution

Binary

The powers of 2 run as—1, 2, 4, 8, 16, 32, 64, 128, 256, 512, 1024, … .

Removing 1024 from 1438, we have the remainder 414.

Removing 256 from 414, we have the remainder 158.

Removing 128 from 158, we have the remainder 30.

Removing 16 from 30, we have the remainder 14.

Removing 8 from 14, we have the remainder 6.

Removing 4 from 6, we have the remainder 2.

Therefore, $1438 = 1024 + 256 + 128 + 16 + 8 + 4 + 2 = 2^{10} + 2^8 + 2^7 + 2^4 + 2^3 + 2^2 + 2^1$

Therefore, the binary representation of the decimal number is 10110011110.

Octal

The powers of 8 run as—1, 8, 64, 512, 4096, … .

Removing 2 times 512 from 1438, we have the remainder 414.

Removing 6 times 64 from 414, we have the remainder 30.

Removing 3 times 8 from 30, we have the remainder 6.

Therefore, $1438 = 2 \times 512 + 6 \times 64 + 3 \times 8 + 6 \times 1 = 2 \times 8^3 + 6 \times 8^2 + 3 \times 8^1 + 6 \times 8^0$

Therefore, the octal representation of the decimal number is 2636.

Hexadecimal

The powers of 16 run as—1, 16, 256, 4096, … .

Removing 5 times 256 from 1438, we have the remainder 158.

Removing 9 times 16 from 158, we have the remainder 14.

Therefore, $1438 = 5 \times 256 + 9 \times 16 + 14 \times 1 = 5 \times 16^2 + 9 \times 16^1 + 14 \times 16^0$

Therefore, the hexadecimal representation of the decimal number is 59E.

Spreadsheets as an Analysis and Design Tool

Spreadsheets can be used very effectively as tools for analysis and design. Analysis is typically one-directional—for example, we could have a given polynomial function $f(x)$ in the variable x. We can use a spreadsheet to create a column of x values and use the given function to populate a column of y values, where $y = f(x)$. A plot of this function can then answer various questions about the polynomial, such as: Where are the zeros located? Where are maxima, minima, and inflection points located? Etc.

Spreadsheets are also very effective in design. The unique feature of a design problem is that design parameters must be *tweaked* until some specific design objective is achieved. For example, the design objective could be to choose the coefficients of the polynomial such that the value of the function is always within certain given limits. A spreadsheet makes such an exercise very easy since any change made to these design parameters are automatically reflected in all dependent calculations.

In a spreadsheet, data is stored in cells. Each cell is at the intersection of a row (numbers 1, 2, 3, …) and column (letters A, B, … , Z, AA, AB, …). Thus cell B12 lies at the intersection of row 12 and column B. A formula can be entered into a cell by typing into the entry bar. An equal sign must precede the formula, otherwise it will be interpreted as plain text.

Copying Formulas

If a formula is copied from one cell to another, the formula is modified by the offset between the source and the destination cells. For example, if a formula is copied from cell B12 to cell F9, then the row offset is 9 – 12 = –3 and the column offset is F – B = +4. Every single term in the formula copied will then be modified the same way (row offset = –3 and column offset = +4. Thus, if the formula in cell B12 is C15^2 – D11*E8 and it is copied into cell F9, then the formula in cell F9 will be G12^2 – H8*I5.

The exception to this is using a $ symbol in front of either/both row and column labels. Any row or column identifier preceded by the $ symbol does not change when the formula is copied to another location. Thus, if the formula in cell B12 is C$15^2 – $D11*$E$8 and it is copied into cell F9, then the formula in cell F9 will be G$15^2 – $D8*$E$8.

Some Useful Functions in EXCEL

Function	Utility	Syntax
ABS	This function calculates the absolute value of a number |X|.	ABS(X)
EXP	This function evaluates the function e^x.	EXP(X)
LOG	This function calculates the logarithm of any number with respect to any base. If the second argument is omitted, the base is assumed to be 10.	LOG(X,[base])
LOG10	This function calculates the base-10 logarithm of any number. It is the same as LOG(X).	LOG10(X)
LN	This function calculates the natural (base-e) logarithm of any number.	LN(X)
DEGREES	Converts an angle from radians to degrees.	DEGREES(X)

Function	Utility	Syntax
RADIANS	Converts an angle from degrees to radians.	RADIANS(X)
SIN COS TAN	These trigonometric functions can be used to calculate the trigonometric functions. Other functions are SEC, CSC, COT. The angle must be in radians.	SIN(X)
ASIN ACOS ATAN	These trigonometric functions can be used to calculate the inverse trigonometric functions. Other functions are ASEC, ACSC, ACOT. The resulting angle is in radians.	ASIN(X)
ATAN2	Unlike the ATAN function, which takes a single argument and produces an acute angle in the range $-\pi/2$ to $+\pi/2$, the ATAN2 function takes two arguments (Δ_x, Δ_y) and produces an angle in the range 0 to 2π.	ATAN2(X,Y)
AVERAGE	This calculates the average of the numeric values in the specified range (ignoring empty cells and nonnumeric values).	AVERAGE(A1:A12)
SUM	This calculates the sum of all the numbers in the given range. Nonnumeric elements are ignored.	SUM(A1:A12)
MAX, MIN	These pick out the maximum or minimum value out of the entries in the specified range.	MAX(A1:A12) MIN(A1:A12)
SUMIF	Over a range of cells, this function calculates the sum of those cell whose contents meet a certain criterion. For example, SUMIF(D1:D20, ">0") calculates the sum of those cells in the range D1:D20 that contain positive values.	SUMIF(D1:D20,">0")
COUNT	This counts the number of cells in the range that have numbers in them. This will only count cells with numeric entries.	=COUNT(A1:A12)
COUNTA	This counts the number of nonempty cells in the range. This will look for both numeric and other characters.	=COUNTA(A1:A12)
COUNTIF	This counts the number of occurrences that a given criterion is met within the specified range.	=COUNTIF(range, criteria)
IF	The IF function uses a logical condition to choose between one of two actions. If the condition is true, the first action is taken. If the condition is false, the second action is taken.	=IF(condition, action if true, action if false)
LEFT	LEFT returns the specified number of characters from the left of the text string in the target cell.	=LEFT(cell, number of characters)
MID	MID gives you the specified number of characters from the middle of the word.	=MID(cell, start number, number of characters)
RIGHT	RIGHT returns the specified number of characters from the right of the text string in the target cell.	=RIGHT(cell, number of characters)
VLOOKUP	This looks for a value in the leftmost column of a table, and then returns a value in the same row from a column you specify. If it finds a match of the "lookup_value" in the left column of the "table_array" it will return the value in the column you specify using the "index_num." The "index_num" is relative to the left most column (which is counted as 1). "range_lookup" =TRUE will give you the closest match. "range_lookup" = FALSE will only give you an exact match.	=VLOOKUP(lookup_value, table_array, index_num, range_lookup)

Function	Utility	Syntax
FLOOR	This function rounds number down, toward zero, to the nearest multiple of significance. For example, FLOOR(2.75,0.1) rounds 2.75 down to the nearest multiple of 0.1, i.e., 2.7.	FLOOR(X,significance)
CEILING	This function rounds number up, away from zero, to the nearest multiple of significance. For example, FLOOR(2.75,0.05) rounds 2.75 up to the nearest multiple of 0.05, i.e. 2.8.	CEILING(X,significance)
RANK	This function gives the descending rank (high to low) of a number within an array (range of cells).	RANK(range,X)
RAND	This function does not have any arguments. RAND() generates a random number between 0 and 1.	RAND()
RANDBE-TWEEN	This function generates a random integer between two integers.	RANDBETWEEN(M,N)

Example 3.2

In a spreadsheet, the following formula is entered into cell A2: =A1+2 and copied through the range A3:A10. The formula "= A$1^2+A1" is then typed into cell B1 and copied through the range B2:B10.

What is the value in cell B10?

	A	B	C	D
1	5			
2	=A1+2			
3				
4				
5				

A. 23

B. 48

C. 462

D. 552

Solution The numbers in the range A1:A10 form an arithmetic series with a = 5 and d = 2. Cell A10 will contain the 10th term in the series = $5 + 9 \times 2 = 23$.

The formula A$1^2 + A1 will get copied as A$1^2 +A10 from cell B1 into cell B10.

Therefore, the entry in cell B10 will be $5^2 + 23 = 48$.

Answer is B.

Example 3.3

In a spreadsheet, cells A1 and B1 are initialized with values 2 and 6, respectively. The following formula is entered into cell A2: = A1+B$1 and copied through the range A3:A10. The formula "= A$1^2+A1" is then typed into cell B2 and copied through the range B3:B10.

What is the value in cell B10?

	A	B	C	D
1	2	6		
2	=A1+B$1	=A$1^2+A1		
3				
4				
5				

A. 54

B. 48

C. 462

D. 552

Solution Since the A column uses the formula A2=A1+B$1, it populates each cell in column A with the previous value incremented by the content of cell B1 (fixed cell addressing), which is 6. Thus, the A column develops as 2, 8, 14, 20, … , 50, 56.

Cell B10 would have the formula: B10 = A$1^2+A9.

A1 contains the value 2 and A9 has 50. Therefore, B10 will have the value $2^2 + 50 = 54$.

Answer is A.

Example 3.4

In a spreadsheet, cell A1 contains the value 5. The following formula is entered into cell A2: = 2*A1+7 and copied through the range A3:A10. The formula "= A2^2+$A1" is then typed into cell B2 and copied through the range B3:B10.

What is the value in cell B10?

	A	B	C	D
1	5			
2	=2*A1+7	=A$1^2+A1		
3				
4				
5				

A. 37,668,906

B. 37,667,128

C. 37,666,452

D. 37,665,834

Solution Since the A column uses the formula A2=2*A1+7, it populates each cell in column A with *double the previous value plus 7*. Thus, the A column develops as 5, 17, 41, 89, 185, 377, 761, 1529, 3065, 6137.

Cell B2 has the formula: B2 = A2^2+$A1.

After copying, cell B10 would have the formula: A10^2+$A9. Thus, B10 will have the value $6137^2 + 3065 = 37{,}665{,}834$.

Answer is D.

Example 3.5

In a spreadsheet, the values 5 and 8 are entered in cells C1 and D1, respectively. The formula =2*C$1+8*$D1 is then entered into cell A1. The contents of cell A1 are then copied to cell B7. What is the value in cell B7?

	A	B	C	D
1			5	8
2				
3				
4				
5				

A. 12

B. 16

C. 56

D. 74

Solution In cell A1, the formula is "=2*C$1 + 8*$D1"

Upon copying it to cell B7, the formula becomes "=2*D$1 + 8*$D7"

Since cell D1 contains the value 8 and cell D7 has no entry, cell B7 will have the value $2 \times 8 = 16$.

Answer is B.

Example 3.6

Student exam scores are stored and grades are calculated in a spreadsheet. Columns B, C, and D contain test scores from tests 1, 2, and 3. Column E contains a factored score (out of 100). In column F, the following grade rule is to be applied:

Total > 90	Grade = A
90 > Total > 80	Grade = B
80 > Total > 70	Grade = C
70 > Total > 60	Grade = D
60 > Total	Grade = F

What formula should be entered in cell F2 for the purpose of copying down through the range F3:F6?

	A	B	C	D	E	F
1		Test 1	Test 2	Test 3	Total	Grade
2	Adams					
3	Bray					
4	Carlos					
5	David					
6	Eugene					

A. F2:=IF(E2>90,A,IF(E2>80,B,IF(E2>70,C,IF(E2>60,D,F))))

B. F2: =IF(E2>90,"A",(IF(E2>80,"B",(IF(E2>70,"C",(IF(E2>60,"D","F")))))))

C. F2: =IF(E2>90,A,(IF(E2>80,B,(IF(E2>70,C,(IF(E2>60,D,F)))))))

D. F2: =IF(E2>90,'A',IF(E2>80,'B',IF(E2>70,'C',IF(E2>60,'D,'F'))))

Solution The proper structure of the IF-ELSE structure is that in choice B. Choice D also has the same structure, but alpha character values like letter grades should be enclosed in double quotes, not single quotes.

Answer is B.

Structured Programming

The questions in this category will involve pseudocode. In other words, a fragment of code will be given—not in any specific programming language—and the question will typically ask for the value of a particular variable at the end of the code fragment. The following features are typically encountered in these questions.

Declaration

Usually the very first thing in a computer program, a set of declarations specifies properties of entities such as variables, constants, functions, and classes. In addition to defining the name of the entity, declaration statements define data type and size when required (such as for arrays). In some programming languages, an entity must be declared before its use. In other languages, an implicit declaration is provided the first time the variable is encountered by the complier.

Initialization

Usually at the very beginning of the program, certain variables and data objects are assigned initial values. A common instance is a counter that is initially set to zero but is then incremented in the body of the program.

Counters

A very typical example of using a counter is to keep track of the number of times a particular condition is met. For example, given an array of integers, we might be interested in counting the number of times a *negative number* is encountered. The counter can be initialized to zero and a loop can be then used to test whether the next element of the array meets the test or not. If a *negative number* is found, the counter can be incremented by 1; otherwise the next value can be tested. If this continues to the end of the loop, the value stored in the counter variable will be the number of negative numbers that were found in the array.

Loops

A loop is used whenever the same operation needs to be carried out over and over. Loops are of two broad types. The first type repeats a fixed number of times, as a loop counter increments from 1 to *N*. For example, given an array of *N* integers, we want the sum of all the numbers in the array. In this case, a variable called SUM can be initialized to zero and the loop reads in all *N* values from the array one at a time, adding each value into SUM for each loop iteration. At the end of the loop, the variable SUM holds the sum of all the integers $A(i)$, $i = 1 \ldots N$.

```
SUM = 0.0
FOR I = 1 to N
READ A(I)
SUM = SUM + A(I)
END
```

The second type of a loop is one that may terminate prematurely, depending on whether a specific condition is met or not. For example, given an array of *N* integers, let us say that we are interested in finding out when the first negative number is encountered.

```
COUNTER = 0
READ A
DO WHILE (A(I) > = 0)
COUNTER = COUNTER + 1
END DO
IF COUNTER = N
PRINT ('NO NEGATIVE NUMBERS FOUND')
ELSE
COUNTER = COUNTER + 1
END
```

In the code fragment above, the DO WHILE loop is executed as long as a nonnegative or negative number is found in the array A. Each time, the counter is incremented. So, when the first negative number is encountered, the program exits the loop. However, the last number stored in COUNTER is the count of the last nonnegative value, indicating that the next value turned out to be negative. Therefore, COUNTER is increased by 1.

Note that the example above also has a "safety" in case the array did not contain any negative numbers.

IF ELSE Expressions

In the previous example, an "IF ELSE" expression was used to determine whether any negative numbers were found in the array. An "IF ELSE" uses a logical test to control the flow of the program.

"IF ELSE" structures can be nested. Nested if-else statements are used in situations where there are more than one condition to evaluate. A SWITCH statement, on the other hand, evaluates only one variable and starts to work there. You can think of a SWITCH statement as a simplified version of an IF ELSE block.

Following is an example of a nested IF ELSE structure

```
if(x == 0) {
// execute this code option 1
} else if(y == 1) {
// execute this code option 2
} else if(x == 1 && y == 0) {
// execute this code option 3
}
```

Sum

In order to compute the sum of an array of numbers, the following steps need to be followed:

1. A variable, which is to serve as a bin, needs to be created and initialized to zero. A convenient name for this variable can be "SUM."
2. Using a loop, the values to be added can be either computed or read, one by one.
3. The variable in step 1 () is updated by adding in the value in step 3. A couple of specific means of doing this are shown below

   SUM = SUM + 1

   SUM++
4. When the loop terminates, the variable SUM will hold the result of the cumulative addition.

Logical Tests

A logical test in an algorithm does not compute or assign values to variables, but is used to determine if a particular condition is met or not. While the exact syntax is specific to a particular program, conditions such as "greater than," "less than," etc. are likely to be represented as the generic ">" and "<" operators in a pseudocode fragment.

Example 3.7

A fragment of a computer program is given below. What is the expression whose value is stored in the variable *A* at the end of the program?

```
INPUT X
A = 1
S = 1
FOR I = 1 TO 20
S = S*I
A = A + X^I/S
END
```

A. $A = 1 + \frac{x}{1} + \frac{x^2}{2} + \cdots + \frac{x^{20}}{20}$

B. $A = 1 + \frac{x}{1!} + \frac{x^2}{2!} + \cdots + \frac{x^{20}}{20!}$

C. $A = 1 - \frac{x}{1} + \frac{x^2}{2} - \cdots + \frac{x^{20}}{20}$

D. $A = 1 - \frac{x}{1!} + \frac{x^2}{2!} - \cdots + \frac{x^{20}}{20!}$

Solution The line "S = S*I" builds the factorial function FACT(I). There is no sign switching from one term to the next.

Answer is B.

Example 3.8

A fragment of a computer program is given below. What is the expression whose value is stored in the variable S at the end of the program?

```
S = 0
A = 2
FOR I = 1 TO 10
B = A^2
S = S+B
A = A+1
END
```

A. 201

B. 257

C. 309

D. 505

Solution This is a classical algorithm for calculating the sum of a series. S is the series sum which is initialized to zero. The line S = S + B keeps adding "B" terms, which in this case is A^2, starting with A = 2 and A being incremented by 1 every time. The length of the series is 10 terms.

Therefore, S will hold the sum: $S = 2^2 + 3^2 + 4^2 + \cdots + 11^2 = 505$.

Example 3.9

A fragment of a computer program is given below. What is the value of variable *B* at the end of the program?

```
B = 0
N = 6
INPUT ARRAY X = {3, 1, 0, -2, 5, 7, 2, 0, 3, 5}
FOR I = 1 TO N
B = B + X(I)
END
B = B/N
```

A. 2.40

B. 2.33

C. 14.0

D. 24.0

Solution The code fragment above computes the sum of the first *N*(6) elements of the array X. In the last line, this sum is divided by *N*. Thus, at the end of the program, the variable *B* stores the average of the first 6 elements of the array X.

The sum of the first 6 terms is 14. The average is 2.33.

Answer is B.

Example 3.10

A fragment of a computer program is given below. What is the value of X(5) at the end of the program?

```
N = 10
INPUT ARRAY X = {3, 1, 0, -2, 5, 7, 2, 0, 3, 5}
FOR I = 1 TO N
TEMP = X(I)
FOR J = I+1 TO N
IF X(J)< X(I)
X(I)= X(J)
X(J) = TEMP
TEMP = X(I)
ELSE
CONTINUE
END
```

A. 0

B. 2

C. 3

D. 5

Solution This is a classic example of an algorithm that sorts a set of numbers into ascending order. Starting with each number in the series, the program scans all the following numbers and if a smaller number is found, it is put near the beginning of the series. As a result, smaller numbers are moved to the front of series and larger numbers are pushed back. At the end of the program, the array X = {−2, 0, 0, 1, 2, 3, 3, 5, 5, 7}. The fifth element is 2.

Answer is B.

Example 3.11

A fragment of a computer program is given below. What is the value of variable S at the end of the program?

```
S = 0
A = 4
B = 1.5
FOR I = 1 TO 10
T = A*B^(I-1)
S = S + T
END
```

A. 102.52

B. 153.77

C. 299.55

D. 453.32

Solution This is a classic algorithm for calculating the sum of a geometric series. The first term of the series is $A = 4$ and the ratio between the successive terms is $B = 1.5$. The variable T stores the current value of the term of the series. The variable S stores the sum of first 10 terms in the series, by adding each value of T.

The sum of n terms in a geometric series is given by $S_n = \dfrac{a(r^n - 1)}{r - 1} = \dfrac{4 \times (1.5^{10} - 1)}{1.5 - 1} = 453.32.$

Answer is D.

Example 3.12

A fragment of a computer program is shown below. What is the value stored in variable N at the end of the code fragment?

```
L = 0
M = 0
N = 0
A = {4,5,6,2,-3,2,9,0,5,-2,7,5}
B = {1,7,-1,5,-3,6,2,3,1,8,3,2}
FOR I = 1 TO 7
IF A(I)>B(I)
L = L + 1
ELSE IF A(I)== B(I)
M = M + 1
ELSE
N = N + 1
END
```

A. 3

B. 5

C. 7

D. 9

Solution In this program, L, M, and N are counters, which are initialized to zero at the beginning of the program. Every time an element of array A is found to be greater than the corresponding element of array B, counter L is incremented. Otherwise, for every instance where an element of array A is found to be equal to the corresponding element of array B, counter M is incremented. Failing these two tests means that an element of array A is found to be less than the corresponding element of array B, which is counted by counter N.

Thus, at the end of this code fragment, variable N will store the number of instances where an element of array A is found to be less than the corresponding element of array B. Because the loop counter I is equal to 7, only the first 7 pairs of A and B are compared. Out of the first 7, the number of instances where $A(I) < B(I)$ is 3. If the entire data is scanned, N would be 5.

Answer is A.

CHAPTER 4

Ethics and Professional Practice

Code of Ethics

There are several documents that outline rules of professional conduct expected of engineers. The most comprehensive of these are the Code of Ethics for Engineers (National Society of Professional Engineers), the National Council of Examiners for Engineering and Surveying (NCEES) Rules of Professional Conduct (found in Section 240 of the Model Rules), and the American Society of Civil Engineers (ASCE) Code of Ethics.

National Society of Professional Engineers Code of Ethics

The Fundamental Canons of the National Society of Professional Engineers (NSPE) Code of Ethics state:

Engineers, in the fulfillment of their professional duties, shall:

1. Hold paramount the safety, health, and welfare of the public. (A1, A2, A3)
2. Perform services only in areas of their competence. (B1, B2, B3)
3. Issue public statements only in an objective and truthful manner. (A4, A5)
4. Act for each employer or client as faithful agents or trustees. (B4)
5. Avoid deceptive acts. (A6, A7, A9, B6, B9, C1, C2)
6. Conduct themselves honorably, responsibly, ethically, and lawfully so as to enhance the honor, reputation, and usefulness of the profession. (A8, A10, B5, B7, B8, C3, C4)

National Council of Examiners for Engineering and Surveying Rules

The clauses referenced in parentheses are the articles of the NCEES Rules of Professional Conduct, which are grouped into the following three headers:

A. Licensee's obligation to the public

B. Licensee's obligation to employer and clients

C. Licensee's obligation to other licensees

These are found in expanded language in the *FE Reference Handbook*, version 9.4.

NSPE Rules of Practice

The following is a summary of the NSPE Rules of Practice.

1. *Engineers shall hold paramount the safety, health, and welfare of the public.* Engineers shall approve only those engineering documents that are in compliance with applicable standards. If an engineer's judgment is overruled under circumstances that endanger life or property, they shall notify their employer and/or client and such other authority as may be appropriate.

   Engineers shall not reveal facts, data, or information without the prior consent of the client or employer, nor permit the use of their name in fraudulent or dishonest business ventures. Engineers having knowledge of any alleged violation of this code shall report it to and cooperate with appropriate professional bodies and also to public authorities.

2. *Engineers shall perform services only in the areas of their competence.* Engineers shall undertake assignments only when qualified by education and/or experience in the specific technical fields involved. They shall not affix their signatures to any plans or documents dealing with subject matter in which they lack competence, nor to any plan or document not prepared under their direction and control. The only "exception" to this rule is that engineers may accept assignments and assume responsibility for coordination of an entire project and sign and seal the engineering documents for the entire project, provided that each technical segment is signed and sealed only by the qualified engineers who prepared the segment.

3. *Engineers shall issue public statements only in an objective and truthful manner.* Engineers shall be objective and truthful in professional reports, statements, or testimony. They shall include all relevant and pertinent information in such reports, statements, or testimony.

   Engineers may publicly issue technical opinions that are founded upon first-hand knowledge of the facts and competence in the subject matter. Engineers shall issue no statements, criticisms, or arguments on technical matters that are inspired or paid for by interested parties, unless they have prefaced their comments by explicitly identifying the interested parties on whose behalf they are speaking.

4. *Engineers shall act for each employer or client as faithful agents or trustees.* This standard applies to avoiding conflicts of interest. An engineer should practice full

disclosure to their employer and client when two different roles could come into conflict and influence (or appear to influence) their judgment. It is unethical for the engineer to accept compensation from two different parties for the same effort, unless these relations are fully disclosed to all concerned parties.

Engineers in public service as members, advisors, or employees of government agencies shall not participate in decisions on services provided by them or their organizations in private or public engineering practice. Engineers shall not solicit or accept a contract from a governmental body on which a principal or officer of their organization serves as a member.

5. *Engineers shall avoid deceptive acts.* This covers the issues of "false advertising" and "lobbying." Engineers shall not falsify or misrepresent the qualifications and past accomplishments of individuals or firms. Engineers shall neither receive nor give, either directly or indirectly, any contribution to influence the award of a contract by public authority. They shall not offer any gift or pay a commission, percentage, or brokerage fee in order to secure work, except to established commercial or marketing agencies retained by them.

Professional Obligations (as Listed in the NSPE Code of Ethics)

1. *Engineers shall be guided in all their relations by the highest standards of honesty and integrity.* Engineers shall acknowledge their errors and shall not distort or alter the facts. Engineers shall advise their clients or employers when they believe a project will not be successful. Engineers shall not accept outside employment to the detriment of their regular work or interest. Before accepting any outside engineering employment, they will notify their employers. Engineers shall not attempt to attract an engineer from another employer by false or misleading pretenses. Engineers shall not promote their own interest at the expense of the dignity and integrity of the profession.
2. *Engineers shall at all times strive to serve the public interest.* Engineers are encouraged to participate in civic affairs; career guidance for youths; and work for the advancement of the safety, health, and well-being of their community. Engineers shall not complete, sign, or seal plans and/or specifications that are not in conformity with applicable engineering standards. If the client or employer insists on such unprofessional conduct, they shall notify the proper authorities and withdraw from further service on the project. Engineers are encouraged to extend public knowledge and appreciation of engineering and its achievements. Engineers are encouraged to adhere to the principles of sustainable development in order to protect the environment for future generations.
3. *Engineers shall avoid all conduct or practice that deceives the public.* Engineers shall avoid the use of statements containing a material misrepresentation of fact or omitting a material fact. Consistent with the foregoing, engineers may advertise for recruitment

of personnel. Consistent with the foregoing, engineers may prepare articles for the lay or technical press, but such articles shall not imply credit to the author for work performed by others.

4. *Engineers shall not disclose, without consent, confidential information concerning the business affairs or technical processes of any present or former client or employer, or public body on which they serve.* Engineers shall not, without the consent of all interested parties, promote or arrange for new employment or practice in connection with a specific project for which the engineer has gained particular and specialized knowledge. Engineers shall not, without the consent of all interested parties, participate in or represent an adversary interest in connection with a specific project or proceeding in which the engineer has gained particular specialized knowledge on behalf of a former client or employer.

5. *Engineers shall not be influenced in their professional duties by conflicting interests.* Engineers shall not accept financial or other considerations, including free engineering designs, from material or equipment suppliers for specifying their product. Engineers shall not accept commissions or allowances, directly or indirectly, from contractors or other parties dealing with clients or employers of the engineer in connection with work for which the engineer is responsible.

6. *Engineers shall not attempt to obtain employment or advancement or professional engagements by untruthfully criticizing other engineers, or by other improper or questionable methods.* Engineers shall not request, propose, or accept a commission on a contingent basis under circumstances in which their judgment may be compromised. Engineers in salaried positions shall accept part-time engineering work only to the extent consistent with policies of the employer and in accordance with ethical considerations. Engineers shall not, without consent, use equipment, supplies, laboratory, or office facilities of an employer to carry on outside private practice.

7. *Engineers shall not attempt to injure, maliciously or falsely, directly or indirectly, the professional reputation, prospects, practice, or employment of other engineers. Engineers who believe others are guilty of unethical or illegal practice shall present such information to the proper authority for action.* Engineers in private practice shall not review the work of another engineer for the same client, except with the knowledge of such engineer, or unless the connection of such engineer with the work has been terminated. Engineers in governmental, industrial, or educational employ are entitled to review and evaluate the work of other engineers when so required by their employment duties. Engineers in sales or industrial employ are entitled to make engineering comparisons of represented products with products of other suppliers.

8. *Engineers shall accept personal responsibility for their professional activities, provided, however, that engineers may seek indemnification for services arising out of their practice for other than gross negligence, where the engineer's interests cannot otherwise be protected.* Engineers shall conform with state registration laws in the practice of engineering. Engineers shall not use association with a non-engineer, a corporation, or partnership as a "cloak" for unethical acts.

9. Engineers shall give credit for engineering work to those to whom credit is due, and will recognize the intellectual property rights of others. See the following section.

Intellectual Property

Engineers shall, whenever possible, name the person or persons who may be individually responsible for designs, inventions, writings, or other accomplishments. Designs supplied by a client remain the property of the client and may not be duplicated by the engineer for others without express permission. When working on improvements or modifications to work protected by copyrights or patents, the engineer should enter into a positive agreement regarding ownership. Engineers' designs, data, records, and notes referring exclusively to an employer's work are the employer's property. The employer should indemnify the engineer for use of the information for any purpose other than the original purpose.

Intellectual property rights are typically protected by four different types of products:

1. Trademark: a word, symbol, or name that identifies the source of an intellectual asset, gives exclusive right of use to the holder and provides protection against unauthorized use.
2. Patent: a detailed set of claims that must be new, useful and inventive that gives right of use to the owner of the patent.
3. Copyright: for original literary, dramatic, musical or artistic works, a copyright provides protection against use without permission.
4. Industrial design: provides protection for a product including all distinguishing functional as well as visual features such as shape, color, etc.

Professional Liability

Once certified as a professional engineer, subject to the limitations of their license, an engineer has the authority to sign and seal plans and other documents. The following language from the NCEES Model Law should be noted:

> Licensees shall sign and seal only those plans, surveys, and other documents that conform to accepted engineering and surveying standards and that safeguard the health, safety, and welfare of the public.
>
> Licensees shall, to the best of their knowledge, include all relevant and pertinent information in an objective and truthful manner within all professional documents, statements, and testimony.
>
> Licensees shall comply with the licensing laws and rules governing their professional practice in each of the jurisdictions in which they practice.
>
> Licensees shall undertake assignments only when qualified by education or experience in the specific technical fields of engineering or surveying involved.
>
> Licensees shall not affix their signatures or seals to any plans or documents dealing with subject matter in which they lack competence, nor to any such plan or document not prepared under their responsible charge.
>
> Licensees may accept assignments and assume responsibility for coordination of an entire project, provided that each technical segment is signed and sealed by the licensee responsible for preparation of that technical segment.

Licensure

Engineers, especially those whose work affects public safety and welfare, must be certified by a central body as evidence of their professional competence. Because a civil engineer's role is primarily to design, create, and maintain public infrastructure elements, certification is especially important. Once certified, they are officially recognized as a "professional engineer." It is an official recognition of the engineer's ability to competently serve an employer, a client and more broadly, the general public.

In the United States, engineering licensure is granted by "boards of engineering" in each state. The requirements to get certified can vary from state to state. The basic requirement, common to all 50 states, is passing the 8-hour Principles and Practice (PE) examination in the applicable discipline. Some states impose additional requirements over and above the "8-h test." For example, in California, civil engineers must pass two additional examinations (seismic and surveying) in addition to the 8-h examination.

Engineering Intern

Before getting certified as a professional engineer, most states require that engineering graduates achieve the status of an "engineering intern" by passing the Fundamentals of Engineering (FE) examination. As of 2014, this examination has changed from a paper-based test to a computer-based test that can be scheduled throughout the year (instead of the prior April and October test cycle). This examination can be taken immediately after completing an undergraduate degree in engineering or related discipline. Some states have a "grandfather clause" that waives the FE examination for candidates with enough experience.

Portfolio Review

Most state boards of engineering licensure will require that an aspiring licensee submit a portfolio of their experience, duly signed by supervisors, that demonstrates that they have engaged in meaningful, responsible, and relevant engineering work. Usually, the supervisors who "sign off" on such experience must be licensed engineers themselves. Once the state board determines that a candidate has the necessary education and experience, the candidate will be approved to sit for one or more examinations.

Examination(s)

In the United States, the basic examination that a candidate must pass to become certified as a PE is the 8-hour Principles & Practice examination. Currently, this examination is offered twice a year (April and October). The format of the examination is a breadth-depth format, wherein the first half of the examination is a breadth examination, whereas the second half is more focused on a particular subspecialty.

Certification of Professional Competence

Engineers shall continue their professional development throughout their careers and should keep current in their specialty fields by engaging in professional practice, participating in continuing education courses, reading in the technical literature, and attending professional meetings and seminars. In most states, registered professional engineers must show evidence of earning such continuing education at the time of license renewal.

General versus Specific Types of Licensure

Passing the 8-h civil examination certifies an engineer as a "civil engineer" with no restrictions on the area of civil engineering in which they practice, except as dictated by the professional ethics clause which states that an engineer "shall undertake assignments only when qualified by education or experience in the specific technical field of engineering or surveying involved." However, depending on the specific area of practice and the requirements of the project (as dictated by the client), other certifications may be necessary for one or more engineers on the team.

Some specialized licenses (for civil engineers) are as follows:

Structural engineer (SE). Some states require that a structural engineer sign and seal structural drawings. Such engineers must possess the SE license, which is earned by passing the 16-h SE examination.

Professional traffic operations engineer (PTOE). Certification as a Professional Traffic Operations Engineer® (PTOE) is a powerful demonstration of requisite knowledge, skill, and ability in the specialized application of traffic operations engineering.

Geotechnical engineer license. This is a certification desired of a specialized geotechnical engineer and is typically required above and beyond the basic civil engineer's license. For example, the California Board of Professional Engineers has the following language for the required experience to appear for the 8-hour Geotechnical Engineer license examination:

> Four (4) years of responsible charge experience in geotechnical engineering after licensure as a Civil Engineer, less a maximum of one year credit for post-graduate education (if not previously used). Applicants must have a valid, unexpired license as a Civil Engineer in California.

Environmental engineer license. The 8-h examination for the environmental engineer license is offered concurrently with the 8-h civil engineering examination. It tests examinees solely in the field of environmental engineering.

Sustainability and Sustainable Design

The process of design is always a constrained process. The designer is subject to the limits of cost (the budget), time (the schedule), and technology (constructability). In current times, with increasing pressures of increasing population, diminishing resources and a greater awareness of environmental impacts, the designer must pay attention to not only immediate costs but also to life-cycle costs, such as the potential harm to the local and global environment and secondary impacts on future development.

Sustainable development is about meeting the current needs without adversely impacting the needs of the future. The three pillars of sustainability are social, environmental, and economic. A project designed solely based on profit (economic) motives will very likely have adverse impacts, which increase the true cost of the proposed solution manyfold. Similarly, a design that is based purely on minimizing environmental impact may be so expensive as to have an irreversible and devastating impact on the economics of the region. The social pillar supports initiatives like promoting peace and social justice and reducing poverty.

Economic sustainability is the ability to support a defined level of *economic* production indefinitely. Social *sustainability* is the ability of a social system, such as a country, to function at a defined level of social well-being indefinitely. Sustainable development involves the simultaneous pursuit of economic prosperity, environmental quality, and social equity. Companies aiming for sustainability need to be guided not by a single, financial bottom line but by this triple bottom line. Over time, social values change. What may have been seen as "just and equitable" in the past may have changed. Project planners need to be in touch with current as well as evolving social values when conceiving projects.

Indicators of Sustainability

An indicator is something that helps us understand the current status of a system. If there is a well-defined goal of where we want to be, this gives us qualitative and quantitative measures of how far we are from the desired target. A good indicator alerts us to a problem before it gets too bad and helps us recognize what needs to be done to fix the problem. Indicators of a sustainable community point to areas where the links between the economy, environment, and society are weak. They allow us to see where the problem areas are and help show the way to fix those problems.

Indicators of sustainability are different from traditional indicators of economic, social, and environmental progress. Traditional indicators, such as stockholder profits, asthma rates, and water quality, measure changes in one part of a community as if they were entirely independent of the other parts. Sustainability indicators reflect the reality that the three different factors are highly interconnected.

Raw materials necessary for production deplete the environment. In addition, if resources such as water and air need to meet certain purity standards before being used in production, it adds extra costs to the cost of production. Production processes produce pollution and impact health. Geographic location of economic activity creates inequities in employment production and may lead to pockets of poverty and crime.

Sustainability requires this type of integrated view of the world—it requires multidimensional indicators that show the links among a community's economy, environment, and society. For example, the Gross Domestic Product (GDP), a traditional economic indicator, measures the amount of money being spent in a country. It is generally reported as a measure of the country's economic well-being: the more money being spent, the better the overall economic well-being is assumed to be. However, because GDP reflects only the amount of economic activity, regardless of the effect of that activity on the community's social and environmental health, GDP can go up when overall community health goes down.

Index of Sustainable Economic Welfare

An alternative is a sustainability indicator such as the Index of Sustainable Economic Welfare (ISEW). The key difference between the GDP and ISEW is that the latter takes into account the links between environment, economy, and society. In order to get a more complete picture of economic progress, the ISEW subtracts from the GDP corrections for harmful consequences of economic activity and adds to the GDP corrections for significant activities such as projected creation of domestic labor. For instance, the ISEW accounts for air pollution by estimating the cost of damage per ton of five key air pollutants. It accounts for depletion of resources by estimating the cost to replace a barrel of oil equivalent with the same amount of energy from a renewable source. It estimates the cost of climate change due to greenhouse gas emissions per ton of emissions. The cost of ozone depletion is also calculated per ton of ozone depleting substance produced. Additionally, adjustments are made to reflect concern about unequal income distribution. Some health expenses are considered as not contributing to welfare, as well as some education expenses.

Indicators of sustainable community are useful to different communities for different reasons. For a healthy, vibrant community, indicators help monitor that health so that negative trends are caught and dealt with before they become a problem. For communities with economic, social, or environmental problems, indicators can point the way to a better future. For all communities, indicators can generate discussion among people with different backgrounds and viewpoints, and, in the process, help create a shared vision of what the community should be.

Example 4.1

You and your design group are competing for a multidisciplinary concept project. Your firm is the lead group in the design professional consortium formed to compete for the project. Your consortium has been selected to be the first to enter fee negotiations with the project owner. During negotiations, the amount you have to cut from your fee to be awarded the contract will require dropping one of the consortium members whose staff has special capabilities not found in the staff of the remaining consortium members.

Is your consortium's response in the negotiations ethical?

A. No, not if the owner is left with the impression that the consortium is still fully qualified to perform all the required tasks.

B. Yes, if your remaining consortium members hire a few new, lower cost employees to do the special work originally intended to be provided by the consortium member dropped.

C. No, because an engineer may not accept a contract to coordinate a project with other professional firms providing capabilities and services not under the engineer's direct control.

D. Yes, if in accepting an assignment to coordinate a project, a single person will sign and seal all the documents in the entire consortium work.

Solution It is never ethical to accept a contract if you or the organization you are negotiating for cannot complete the work. Answer (A) does not tell us if the scope of the contract will be reduced so the consortium can complete the work without the eliminated partner, or if the consortium has some other way of completing the work without the special skills of the eliminated partner, so we cannot really say that answer (A) is correct, but it is the best option provided.

Therefore, the answer is (A).

Example 4.2

Under what circumstances can a registered engineer sign and seal plans or documents he/she did not prepare?

A. Under no circumstances.

B. Registered engineers can coordinate projects that include segments that they are not competent in if a qualified registered engineer signs and seals plans or documents for those segments of the project.

C. If the plans or documents were prepared by someone under the registered engineer's direct supervision and the registered engineer is an expert in the subject matter.

D. When practicing in a state different than the one in which the engineer is registered.

Solution Answer (B) is a true statement, but it has nothing to do with the problem statement. Plans or documents prepared under the direct supervision of a registered engineer where the engineer is an expert, can be signed and sealed by the engineer assuming he or she has reviewed the plans or documents.

Answer is C.

Example 4.3

A registered engineer is retained as an expert witness by one of the parties in a civil case where the public safety is not involved. In investigating the technical data in the case, the engineer makes findings that are not favorable to the party who retained her.

The engineer should:

A. Inform the party who retained her of the findings.

B. Inform the judge of the findings.

C. Inform the opposing party of the findings.

D. Say nothing about the findings until called to testify.

Solution In this case, the engineer has only an obligation to her client until she is called to the stand, so she should report her findings to the party who retained her and that party will decide whether or not to call the engineer to testify.

Therefore, (A) is correct.

Example 4.4

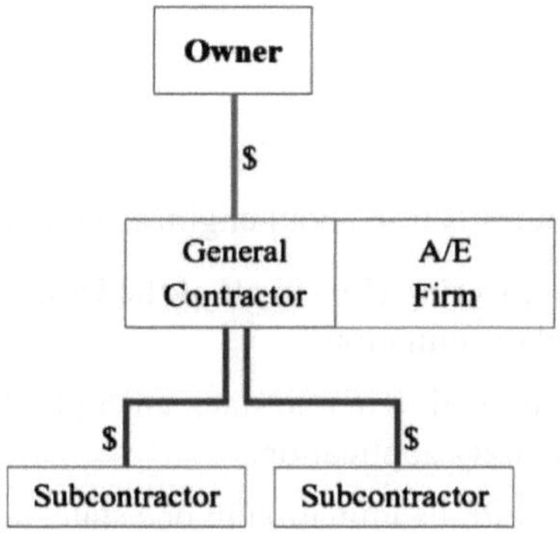

What kind of contractual relationship is shown in the diagram above?

A. Traditional

B. Design-build

C. Direct force

D. Owner CM

Solution The relationship between a team of general contractor and A&E firm, which work in tandem, rather than in a hierarchy, indicates a design build contract.

Answer is B.

Example 4.5

You work for a multinational company that is venturing into a country where it has never done business. On one of the preliminary visits to one of the production facilities, you observe poor working conditions in the factory, compared to working conditions back in the home country. Upon inquiring, you are told that these conditions are the norm at all similar facilities. What is the course of action you should pursue?

A. You should recommend to your supervisors that the contract must have language that ensures the same work conditions as prevalent in the home country.

B. You should ignore the working conditions and do nothing.

C. You should recommend that all negotiations for the contract be discontinued.

D. You should recommend that the contract be negotiated as it stands.

Solution Working conditions vary significantly from country to country. It is hasty to make a rash judgment of a culture that one does not understand. For this reason, option C is incorrect. On the other hand, to do nothing (option B) is not ethical either. Option D is basically the same as option B. The only viable alternative is A—while achieving "the same" working conditions is an unrealistic objective, it is the ethical alternative.

Answer is A.

Example 4.6

Which of the following statements is true about engineering licensure in the United States?

A. A license to practice engineering (PE) in all of the United States is awarded upon passing a single national examination.

B. After passing the PE examination in one state, an engineer can obtain a PE license in other states by filing a comity application.

C. After passing the national PE examination in one state, additional examination may be required in order to receive a PE license in other states.

D. Each state has additional examinations/requirements in addition to the national examination.

Solution A is incorrect because of the word "all." Some states have additional requirements.

D is incorrect because of the word "each." Some states have additional requirements, not all.

B is incorrect because licensure by comity is not automatic. If the comity state has extra requirements, they must be satisfied, as stated in C.

Answer is C.

Example 4.7

Which of the following could an engineer accept with the least likelihood of violating ethical standards?

A. A pen set embossed with the name of an office supplies company.

B. A fruit basket sent by a previous client in appreciation of a previous job that was successfully completed.

C. A trip to a resort where visiting engineers will attend technical seminars about products built by the sponsoring company.

D. A $100 gift card from a company that manufactures bolts used for structural connections.

Solution The dollar-value associated with a gift is immaterial. Therefore, A, C, and D all fall in the same category. They are all gifts from a company with a "vested interest." Since the fruit basket gift is a gift based on a "past" relationship and not to secure a future relationship, it seems the one least likely to violate ethical standards.

Answer is B.

Example 4.8

A professional engineer with long experience in hydrogeology reads a news report about a proposal seeking public funding for a dam. Based on her experience, she can identify several aspects of the proposed design that will pose severe environmental hazard. What is the most appropriate course of action that she should pursue?

A. The engineer should write a letter to the editor urging lawmakers to cancel funding for the project.

B. The engineer should do nothing because other qualified experts who know more specific information have taken, what must be, a good decision.

C. The engineer should contact the project manager in charge of the proposal to seek more detailed information.

D. The engineer should start a citizens' petition collecting signatures with the sole purpose of stalling funding for the project.

Solution B is incorrect because the engineer DOES have relevant experience and doing nothing is "passing the buck." Option D could be an ordinary citizen's response, but the engineer, based on education and experience, should do more. Option A can be an initial response, but C is more specific.

Answer is C.

Example 4.9

Which of the following statements is true?

I. Ethical standards vary with time.

II. Ethical standards are the same as legal standards.

III. Proper ethical code is conditional on local culture.

A. I and II

B. I and III

C. II and III

D. I, II, and III

Solution Ethical standards are not absolute in space and time. They are a function of social norms, which are in turn a function of the time and location. Also, ethical standards are based on moral standards and are often higher than legal standards. Therefore, I and III are true.

Answer is B.

Example 4.10

An engineer has a full-time position as a structural engineer at an A&E firm. During the weekends, the engineer works on privately obtained consulting jobs. Occasionally, for the consulting job, the engineer uses a Structural Analysis and Design software, which is licensed to his employer, to which he has access using a remote VPN connection from home. This action is most probably unethical because:

A. The engineer's employment contract prohibits misuse and misappropriation of the software license.

B. The company has not given the engineer permission to use the software for outside contracts.

C. The access gives the engineer an unfair economic advantage.

D. The use of the software on private computers can make the software vulnerable to computer viruses and render it unoperational.

Solution Choices A, B, and D could all be true, except that we have to make assumptions of facts that are not stated in the question.

Choice C is definitely true because the engineer has not paid for the license and therefore has lower overhead costs compared to other individuals/companies competing for the same contracts.

Answer is C.

Example 4.11

An engineer serves as the sole point of contact with a state agency accepting bids for an upcoming contract. The bid submitted by one of the contractors is accompanied by two Superbowl tickets. What is the most appropriate course of action for this engineer?

A. Return the tickets and reject the bid.

B. Accept the tickets and accept the bid.

C. Report the company, return the tickets, and accept the bid.

D. Report the company, return the tickets, and reject the bid.

Solution The engineer must not accept the tickets. This rules out option B. Since no specific rules governing the bid process are stated, this cannot be the basis for rejecting the bid. This rules out options A and D. Option C mentions "report the company, return the tickets, and accept the bid." This seems to be the clearest course of action. Making a formal report of the tickets inclusion with the bid seems to be governed by prudence (rather than ethics).

Answer is C.

Example 4.12

While pursuing your professional duties as a professional engineer, you become aware of some instances of willful disregard of rules of professional conduct (promulgated by the state society of professional engineers) by a colleague, who is also a professional engineer. The colleague in question is working on a project where you have no involvement. Which of the following is the most appropriate course of action on your part?

A. Do nothing, because you have no responsibility in said project.

B. Report the behavior to the project manager overseeing your colleague.

C. Report the behavior to your direct supervisor and expect him to follow up.

D. Report the behavior to the society of professional engineers.

Solution If the behavior violates rules of professional conduct set up by the state society of professional engineers, then you should report directly to that professional society.

Answer is D.

Example 4.13

A company designs, builds, and sells prefabricated commercial buildings. During a recent heavy snowstorm, the roof of one of the buildings (designed and constructed by this company) failed, leading to significant damage to the buildings and its contents. Which of the following reasons is the most likely reason the building owner may not recover damages from the company?

A. Insurance policy has an "act of God" clause.

B. Statute of limitations.

C. The contents of the building are not covered by insurance policies.

D. None of the above. The safety of the building and its contents is the legal responsibility of the company that designed the roof.

Solution Standard contracts provide protection against design and construction defects and claims arising out of them for only a limited time (statute of limitations). All other choices require us to make assumptions about the coverage provided by the policy.

Answer is B.

Example 4.14

Following an airplane crash, the preliminary statement from the NTSB implies structural failure of the fuselage. You are a licensed structural engineer, specializing in the design of steel bridges and buildings. A local radio station wants to interview you to seek your opinion about the possible causes and ramifications of the failure. What should be your course of action?

A. Agree to do the interview, since your expertise in structural engineering qualifies you to comment on the event.

B. Agree to do the interview, but make it clear that your expertise is only with buildings and bridges and not aircraft.

C. Decline the interview because your expertise is only with buildings and bridges and not aircraft.

D. Decline the interview because the NTSB investigation prohibits you from expressing an opinion that might conflict with the official version.

Solution You should decline to do the interview specifically because your area of expertise is not in aircraft design and even though you have general expertise in structural design, any opinion you express may be misinterpreted by the general public as the final word of an "expert."

If your training had been specifically in structural design of aircraft systems, you would have been perfectly within your rights to express your opinion, whether they agreed with the official statement or not.

Answer is C.

CHAPTER 5

Engineering Economics

Types of Cash Flow

The primary objective of engineering economic analyses is to evaluate various economic alternatives based on a comparison of costs and benefits. While the nominal value of an amount of money may remain the same at various times (e.g., a \$100 currency note today versus the same \$100 note 10 years in the future), the actual value is different because of the devaluation of money with time. Economic analysis must therefore adjust different types of money transactions based on the timeframe in which they occur.

The four types of idealized cash flow discussed in this chapter are

1. *Present value (P)*: A one-time cash flow that occurs now (i.e., at $t = 0$). Any economic transactions (cost or revenue) that occur as a single lump sum at the very beginning of the timeline is considered as present value. Initial capital costs are the most common example of this type of cash flow.
2. *Future value (F)*: A one-time cash flow that occurs after a finite duration ($t = n$). A very common example of this type of cash flow is salvage value or replacement cost.
3. *Annuity (A)*: A constant value cash flow that starts at $t = 1$ and recurs for n time periods. The number of periods n is usually considered finite. For example, a 5-year car loan, paid in monthly instalments, has $n = 60$ periods. A 30-year home loan, paid in monthly instalments, has $n = 360$. In each of these cases, the interest rate must correspond to the period. For a loan paid in monthly instalments, the interest rate must be the monthly interest rate.

   An annuity can also be of infinite length. For example, a community management association may be planning for landscaping costs over a long period (in other words, "forever"). The limit of the P/A factor, as $n \to \infty$, is called the capitalization factor. For example, if the annual interest rate is 7%, then the capitalization cost of an infinite annuity (e.g., annual maintenance costs of \$100 forever) is calculated as 100/0.07 = \$1428.57. In other words, an initial ($t = 0$) investment (capital) of \$1428.57 will, at an annual interest rate of 7%, yield an annual interest of \$100, which will pay for the annuity without drawing down on the initial investment. It then becomes a case of simple interest on a year to year basis, with the annual interest generated exactly paying for the annuity, keeping the principal amount unchanged.

4. *Gradient series (G)*: A uniformly increasing (arithmetic) finite cash flow series that has value zero at $t = 1$, value G at $t = 2$, $2G$ at $t = 3$, etc. ending at a value $(n - 1)G$ at $t = n$. A common example of this type of series is a pattern of constant increases to maintenance costs. Two things to remember about the gradient series are
   - This is not the same as a geometric series, which is characterized by a pattern of increases that occur at a constant *rate*, such as 3% traffic growth every year. The pattern in that case is exponential, whereas for a gradient series, the pattern is linear.
   - For a gradient series of length n, the actual number of nonzero increases are $n - 1$, since the zero increase in period 1 is also considered a part of the series (Fig. 5.1).

These cash flows are shown graphically in Fig. 5.1. The analytical relationships to transform one type of cash flow to another are summarized in Eqs. (5.1) through (5.9). In these equations, the following variables are used:

P = present worth (single lump sum)

A = an annuity (constant installment at the end of every period)

F = future worth (single lump sum)

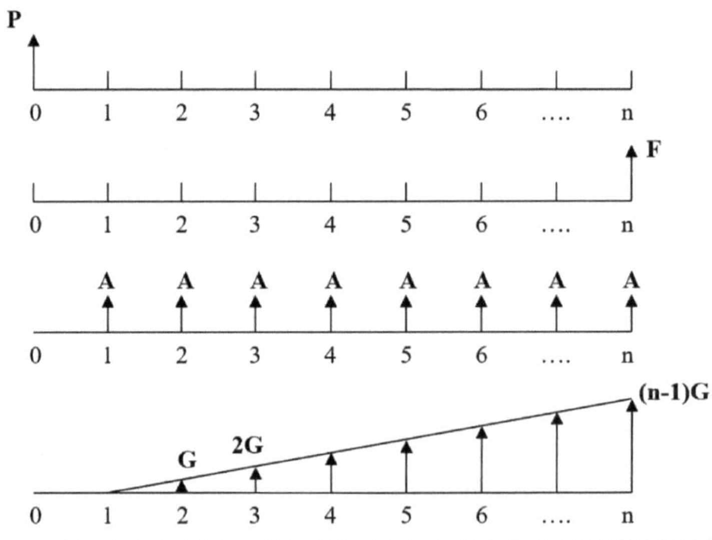

Figure 5.1 Types of cash flow.

G = gradient series starting at zero for the first period and increasing by constant increment G every period

i = nominal interest rate per period (MARR). This is often expressed as a percentage, but must be used as a decimal value in the equations.

n = the number of compounding intervals (periods)

Single Payment Compounded—Symbol (*F/P, i%, n*)—Converts to *F* given *P*

The F/P factor is used to convert a present sum of money (at $t = 0$) to its equivalent value at some future time ($t = n$). For example, if the MARR = 6% (annual), the F/P factor for 10 years is written (F/P, 6%, 10 years). From page 35 of the *FE Reference Handbook*, version 9.4, we get this factor to be 1.7908. This means that \$100 at $t = 0$ and \$179.08 at $t = 10$ years have the same value, if the expected rate of investment growth were to be 6% over that time.

$$(F/P,i,n) = (1+i)^n \tag{5.1}$$

Note that these engineering economics factors follow the rules of algebra, that is, the value of the F/P factor is the arithmetic inverse of the P/F factor (for the same values of i and n). By the same argument, the chain rule of algebra is also valid. For example, $P/G = P/A \times A/F \times F/G$

Single Payment Present Worth—Symbol (*P/F, i%, n*)—Converts to *P* given *F*

$$(P/F,i,n) = (1+i)^{-n} \tag{5.2}$$

Uniform Series Sinking Fund—Symbol (*A/F, i%, n*)—Converts to *A* given *F*

This factor is used to relate an annuity, which is defined as a finite series of equal payments, to a single lump-sum payment occurring at the end of the annuity period.

$$(A/F,i,n) = \frac{i}{(1+i)^n - 1} \tag{5.3}$$

An application may be that a certain amount is desired in a retirement fund 30 years from now, and what's desired is the constant amount that must be saved every year to reach that goal.

Capital Recovery—Symbol (*A/P, i%, n*)—Converts to *A* given *P*

This factor is used to relate an annuity, which is defined as a finite series of equal payments, to a single lump-sum payment occurring at the beginning of the annuity period. In other words, it yields the required annuity (installment) to pay off a certain amount of present value. For this reason, it is the factor used to calculate the instalment on a loan whose present value is known.

$$(A/P,i,n) = \frac{i(1+i)^n}{(1+i)^n - 1} \tag{5.4}$$

For example, an auto loan of \$30,000 with an annual interest rate of 6% to be repaid over a 5-year period in equal monthly instalments would have the following data:

$P = \$30{,}000$

$n = 60$ months

$i = 6\%$ per year $= 0.5\%$ per month $= 0.005$

$$(A/P, 0.005, 60) = \frac{0.005 \times 1.005^{60}}{1.005^{60} - 1} = 0.0193$$

Annuity: $A = \frac{A}{P} \times P = 0.0193 \times 30{,}000 = 579.96$

The \$30,000 auto loan will be repaid after 60 equal monthly payments of \$579.96.

Uniform Series Compounded—Symbol (*F/A, i%, n*)—Converts to *F* given *A*

$$(F/A, i, n) = \frac{(1+i)^n - 1}{i} \tag{5.5}$$

Uniform Series Present Worth—Symbol (*P/A, i%, n*)—Converts to *P* given *A*

$$(P/A, i, n) = \frac{(1+i)^n - 1}{i(1+i)^n} \tag{5.6}$$

Gradient Series

A gradient series is a linearly varying pattern of finite length. It can be an increasing or decreasing pattern. If the length of the series is n, the value at $t = 1$ is zero (see Fig. 5.1). Therefore, a series that starts at a nonzero value and varies linearly each period should be represented as the sum of an annuity and a gradient series, both of the same length. In Fig. 5.2, the linearly varying series of cash flows (length $= n$) is separated into an annuity A and a gradient series G, both of the same length $= n$.

Uniform Gradient Present Worth—Symbol (*P/G, i%, n*)—Converts to *P* given *G*

$$(P/G, i, n) = \frac{(1+i)^n - 1}{i^2(1+i)^n} - \frac{n}{i(1+i)^n} \tag{5.7}$$

Uniform Gradient Future Worth—Symbol (*F/G, i%, n*)—Converts to *F* given *G*

$$(F/G, i, n) = \frac{(1+i)^n - 1}{i^2} - \frac{n}{i} \tag{5.8}$$

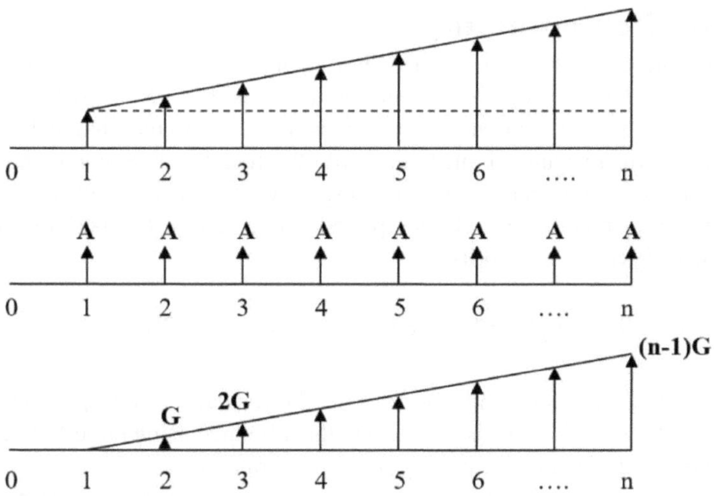

Figure 5.2 Gradient series.

Uniform Gradient Uniform Series—Symbol (*A/G, i%, n*)—Converts to *A* given *G*

$$(A/G,i,n)=\frac{1}{i}-\frac{n}{(1+i)^n-1} \tag{5.9}$$

Year-End Accounting Convention

Equations and tables in this chapter are based on the year-end convention of accounting, in which all expenses incurred and revenues earned in a particular year are represented as cash flows at the end of that year.

Minimum Attractive Rate of Return

Equations (5.1) through (5.9) are based on the concept of *time value of money*, which is the anticipated or expected growth of an investment, due to being compounded at the prevalent or expected rate of return. This rate of return is called the minimum attractive rate of return (MARR). In the formulations in this chapter, this rate of return is represented by the variable *i*. MARR is one of the most important concepts in economic analysis. It is the lowest rate of return that investors will accept by balancing the investment risk or the opportunity to invest elsewhere for possibly greater returns.

In economic analysis, this interest rate is then used to convert one type of cash flow to another. The interest rate is used to calculate the time value of money, which is based on the fundamental idea that money that one has now is worth more than the same amount of money in the future. This is because money has an earning potential through investment.

Nonannual Compounding

Note that these interest rates are per period. For example, if the duration is $n = 3$ years, and the effective annual interest rate is 6%, then the table yields (P/F, $i = 6\%$, $n = 3$) = 0.8396. However, if the time period is measured in months (e.g., $n = 36$ months) and the interest rate is still given as an effective annual interest rate of 6%, then the tables cannot be used directly. This becomes a case of nonannual compounding. The equivalent monthly interest rate i_M is that rate of increase (monthly) which, compounded over a year (12 periods) will yield 6%. Thus, the value of i_M can be solved from

$$(1+i_M)^{12} = 1.06 \Rightarrow i_M = 0.00487 \quad (0.487\%)$$

Noting the absence of tables for this interest rate, the formula for (P/F, $i = 0.00487$, $n = 36$) yields the value 0.8396. This value is identical to that obtained using three 1-year periods. This is because the 6% is designated as the *effective* annual rate.

However, if the annual interest rate of 6% were quoted as a *nominal* rate (instead of an effective rate), then the equivalent monthly interest rate would be 6%/12 = 0.5% = 0.005. This would have yielded a different P/F factor. (P/F, $i = 0.005$, $n = 36$) = 0.8356.

Example 5.1

A bank offers a 5% nominal interest rate per annum, compounded daily. What is the effective annual rate?

Solution Since the interest is compounded daily, we must find the nominal daily rate i_d

$$i_d = 5\% \div 365 = 0.0164\% = 1.6438 \times 10^{-4}$$

The F/P factor (compounded daily for 1 year = 365 days) is given by

$$F/P = (1+i)^n = (1+0.00016438)^{365} = 1.0618$$

Present Worth

The present worth of a project or a proposal is the cumulative value of all financial transactions—costs and revenue—involved in the project, after converting all of them to their equivalent at $t = 0$. In these calculations, one must be careful to assign appropriate signs to all cash flows. Positive cash flows such as revenue and salvage value must be considered positive while negative cash flows such as costs and measurable losses must be considered as negative.

Capital expenditures that occur at $t = 0$ do not need to be converted. However, some projects involve deferred capital expenditures (with some expenditures occurring at $t = 0$ and other stages incurring lump-sum expenditures at various times), which must be converted to present worth using the P/F factor.

Example 5.2

A wastewater treatment plant can be installed using one of two options:

Option A: An initial investment of \$12 million
Salvage value (after 20 year design life) = \$2.5 million

Option B: Initial investment = \$8 million
Capital improvements (after 6 years) = \$8 million
Salvage value (after 20 year design life) = \$5 million

If the MARR = 8% and benefits for both options are equal, which option is less expensive?

A. Option A is cheaper by \$1.2 million

B. Option B is cheaper by \$1.2 million

C. Option A is cheaper by \$0.5 million

D. Option B is cheaper by \$0.5 million

Solution Present worth of option A (million \$):

$$PW = -12 + 2.5 \times \left(\frac{P}{F}, 8\%, 20 \text{ years}\right) = -12 + 2.5 \times 0.2145 = -11.464$$

Present worth of option B (million \$):

$$PW = -8 - 8 \times \left(\frac{P}{F}, 8\%, 6 \text{ years}\right) + 5 \times \left(\frac{P}{F}, 8\%, 20 \text{ years}\right)$$

$$= -8 - 8 \times 0.6302 + 5 \times 0.2145 = -11.969$$

Option A is cheaper by \$505,000.

Answer is C.

Transactions that occur at a constant rate per period (like a uniform instalment) are called annuities and must be converted to an equivalent present worth by multiplying with the *P*/*A* factor. This factor is normally read as "*P* given *A*." An annuity must start in year 1 and continue throughout the entire duration of the timeline (i.e., up to $i = n$). The origin of such an annuity is at $i = 0$. Converting an annuity to an equivalent present worth converts it to a lump-sum value at $i = 0$.

An annuity that starts at any other time (than $i = 1$) may be called a deferred annuity. For an example of a deferred annuity, see Example 5.16.

Gradient series that represent a pattern of uniform increases should be converted to present worth using the *P*/*G* factor.

Single lump-sum transactions that occur at a future date can be converted to their present worth (P) by multiplying by the (P/F) factor.

Example 5.3 illustrates the conversion of a series of cash flows to an equivalent "present worth." It can be used to determine "today's dollars" equivalent of a combination of capital costs, annuities, future sums, etc.

Example 5.3

Annual maintenance costs for a water treatment plant for a small community are \$30,000 in year 2001 and expected to increase by \$500 each year during 2002–2020. If the MARR is 6%, what is the present worth of all maintenance costs during the period 2001–2020?

A. \$695,000

B. \$623,450

C. \$511,050

D. \$387,700

Solution This is an example of a gradient series (arithmetic progression). The payment series is 30k, 30.5k, 31k, ...

This can be decomposed into an annuity of 30k and a gradient series 0, 0.5k, 1k, ... The present worth of all payments is

$$P = 30{,}000 \times \left(\frac{P}{A}, 6\%, 20 \text{ years}\right) + 500 \times \left(\frac{P}{G}, 6\%, 20 \text{ years}\right)$$

$$= 30{,}000 \times 11.4699 + 500 \times 87.2304 = \$387{,}712$$

Answer is D.

NOTE A common mistake is to miscount the number of periods. In Example 5.3, the first year is given as 2001. This is to be represented by a cash flow *at the end* of the year 2001. Therefore, the origin ($t = 0$) for this series is the end of the year 2000. This makes the length of the series = 20 years and not 19 years (as it might seem to be from the difference 2020 − 2001 = 19).

Example 5.4

What amount must be invested in a fund with a 10% effective annual interest rate to yield \$2000.00 at the end of 3 years, assuming that interest is compounded half yearly?

Solution Let the half yearly interest rate be i.

Then the effective annual rate, after two half-year periods, would be $(1+i)^2$. However, since the effective annual rate is known to be 10%

$$(1+i)^2 = 1 + 0.1 = 1.1 \Rightarrow i = 0.04881$$

3 years = 6 half-year periods

Since there are no tables for $i = 4.88\%$, we have to use the formulas to determine the P/F factor

$$P/F(i\%, n \text{ periods}) = (1 + i)^{-n} = 1.04881^{-6} = 0.7513$$

$$P = (P/F) \times F = 0.7513 \times \$2000 = \$1502.60$$

Note that since the interest rate is 10% effective annual rate, and the total period is an exact multiple of a year, we can also do the following (*this would not have worked if the total duration was, say, 30 months*):

From the tables: (P/F, 3 years, 10%) = 0.7513

Therefore, $P = \$2000 \times 0.7513 = \$1502.60$

Example 5.5

Investment A costs \$45,000 upfront investment and pays back \$58,000 after 3 years. Investment B costs \$25,000 upfront and pays back \$12,000 each year for 3 years. If the MARR is 6%, which investment is superior?

Solution Converting all cash flows to present worth—costs as negative cash flow and revenues as positive.

The present worth of investment A:

$$PW_A = -45{,}000 + 58{,}000 \times (P/F, 6\%, 3 \text{ years}) = -45{,}000 + 58{,}000 \times 0.8396 = 3696.80$$

The present worth of investment B:

$$PW_B = -25{,}000 + 12{,}000 \times (P/A, 6\%, 3 \text{ years}) = -25{,}000 + 12{,}000 \times 2.6730 = 7076.00$$

Therefore, investment B is superior by virtue of having a greater present worth.

Principal in a Sinking Fund

Example 5.6

To create a capital fund for future expenditures, a company invests \$125,000 annually into a fund accumulating interest rate of 5% (compounded annually). What is the value of the fund immediately after the 5th payment?

Solution The timeline for this problem is 5 years. The annual payments form an annuity. The value of the fund at the end of year 5, that is, at $t = 5$, is calculated as the future worth (F)

$$125{,}000 \times (F/A, 5\%, 5 \text{ years}) = 125{,}000 \times 5.5256 = \$690{,}700$$

Capitalized Cost

Paying out of a fund depletes it. However, in some applications, such as a maintenance fund or a scholarship fund, it is desirable to not draw down the original investment. In such cases, the interest gained during the year can be used to pay out the annuity. If each year's accumulated interest is used to pay the annuity, the initial investment returns to its original value and the cycle repeats itself every year. This pattern can therefore (theoretically) continue for an infinite length of time ($n \to \infty$).

The annuity (A) is therefore equal to the simple interest accumulated over 1 year.

$$A = Pi \Rightarrow P = \frac{A}{i}$$

This is called the capitalized cost of an annuity.

As $n \to \infty$, the factor P/A approaches $1/i$.

Example 5.7

What is the capitalized cost of a project that has an initial cost of \$50 million and requires annual maintenance cost of \$2.5 million? Assume that the effective interest rate is 10%.

Solution The capitalization cost of an annuity is the principal that generates the annuity to perpetuity. Therefore, the capitalization cost factor is given by (P/A) as $n \to \infty$ and is given by

$$(P/A, n \to \infty) = \frac{1}{i} \Rightarrow P_{\text{cap}} = \frac{A}{i}$$

$$\begin{aligned}\text{Capitalization cost} &= \text{initial cost} + \text{annual cost} \times (P/A, 10\%, \infty)\\ &= \$50\text{M} + \$2.5\text{M} \times (1/0.1) = \$75\text{M}\end{aligned}$$

Therefore, the capitalization cost for the project is \$75 million.

Equivalent Uniform Annual Cost

When two alternatives are being compared and they have unequal time lines, that is, unequal useful life, they cannot be compared based on present worth or future worth. They must be compared in terms of annual cost. All costs must then be converted to annuities using the appropriate length of time. This cost is called the equivalent uniform annual cost (EUAC).

Example 5.8

A sidewalk construction project has two design alternatives—brick and concrete. Brick will last 12 years, have an initial cost of \$600,000 and annual maintenance cost of \$2000. Concrete will last 20 years, have an initial cost of \$1,000,000 and annual maintenance cost of \$1000. Which option is superior from an economy point of view? The interest rate is 6%.

Solution We calculate the equivalent uniform annual cost (EUAC) for both alternatives:

$$\text{EUAC}_{\text{brick}} = 600{,}000(A/P, 6\%, 12 \text{ years}) + 2000 = 600{,}000 \times 0.1193 + 2000 = 73{,}580$$

$$\text{EUAC}_{\text{concrete}} = 1{,}000{,}000(A/P, 6\%, 20 \text{ years}) + 1000 = 1{,}000{,}000 \times 0.0872 + 1000 = 88{,}200$$

Therefore, the annual cost of constructing in brick is 17% cheaper than concrete.

Depreciation

Depreciation is an accounting device that is used to reduce the book value of capital assets. The reduction in book value represents the wear and tear on the equipment. The primary reason for using the device of depreciation is to lower the value of capital assets on the balance sheet in order to lower tax liabilities.

Of the various models for depreciation, the two that are outlined in the *Handbook* are (1) straight-line method and (2) the modified accelerated cost recovery system (MACRS).

Straight-Line Method

The straight-line method of depreciation distributes the "write-off" amount evenly over the expected life of the asset. The write off in value is taken as the reduction in value from the initial cost C to the salvage value S. Thus, according to the straight-line method, the annual depreciation claimed on a property is given by

$$D = \frac{C - S}{n} \tag{5.10}$$

where C = initial value of the property
S = salvage value at the end of design life
n = design life (number of accounting periods)
D = depreciation per period

The book value at the end of j years (periods) BV_j, is given by

$$BV_j = C - \sum_{i=1}^{j} D_i \tag{5.11}$$

Example 5.9

An asset has a purchase price of $25,000. It has a 10-year useful life and a salvage value of $8000. Find the book value at the end of year 3 using straight-line depreciation.

Solution Annual depreciation D = (25,000 – 8000)/10 = $1700 per year
Book value = 25,000 – 3 × 1700 = $19,900

Modified Accelerated Cost Recovery System

Under the modified accelerated cost recovery system (MACRS), the owner of the property is allowed to depreciate the property by published percentages that are given in a table on page 132 of the *Handbook*. The table lists assets with recovery periods of 3, 5, 7, and 10 years. For each recovery period, the schedule publishes an additional number which is the residual value after the value written off during the recovery period. This number is, therefore, the equivalent of the salvage value. For example, for the 3-year recovery period, the schedule dictates the following depreciation percentages—33.33% during year 1, 44.45% during year 2, and 14.81% during year 3. This adds up to 92.59%, which leaves 7.41% as the residual or salvage value.

Example 5.10

An asset has a purchase price of $25,000. It has a 10-year useful life. Find the book value at the end of year 3, using the MACRS method of depreciation.

Solution For a 10-year asset, depreciation percentages in years 1, 2, and 3 = 10%, 18%, and 14.4%, respectively

$$\text{Book value} = 25{,}000 \times [1 - 0.1 - 0.18 - 0.144] = 25{,}000 \times 0.576 = \$14{,}400$$

Tax Issues

Example 5.11

A corporation pays 48% in income tax on profits. It purchases an asset that will produce a revenue stream of $5000 for 10 years. The cost of the asset is $30,000. Annual expenses are $1500. Salvage value (at the end of year 10) of the equipment is $1200. Ignore depreciation. Use a MARR of 8%. What is the after-tax present worth of the asset?

Solution

$$\text{Annual revenue} = \$5000$$

$$\text{Annual expenses} = \$1500$$

$$\text{Annual profit} = \$3500$$

$$\text{Annual taxes} = 0.48 \times \$3500 = \$1680$$

$$\text{NET annual profit} = \$1820$$

$$\text{Present worth} = -30000 + 1820(P/A, 8\%, 10\text{ years}) + 1200(P/F, 8\%, 10\text{ years})$$

$$= -30000 + 1820 \times 6.7101 + 1200 \times 0.4632 = -17231.78$$

Bonds

Example 5.12

What is the maximum an investor should pay for a 25-year bond with a $20,000 face value and an 8% coupon rate (with interest paid semiannually)? The bond will be held to maturity. The effective annual rate for comparison is 10%.

Solution If the interest is paid semiannually, the payment is 4% of 20,000 = \$800.

The effective MARR = 10%. This must be converted to an equivalent 6-month rate, using

$$(1+i)^2 - 1 = 0.10 \Rightarrow i = 0.0488$$

Bond is held to maturity, which is 25 years, or fifty 6-month periods.

The present worth of fifty \$800 payments at an interest rate 0.0488

$$\$800 \times \left(\frac{P}{A}, 4.88\%, 50\right) = \$800 \times \frac{1.0488^{50} - 1}{0.0488(1.0488^{50})} = \$800 \times 18.6 = 14{,}879.76$$

The present worth of a future sum of \$20,000 is

$$\$20{,}000 \times \left(\frac{P}{F}, 4.88\%, 50\right) = \$20{,}000 \times \frac{1}{1.0488^{50}} = \$20{,}000 \times 0.0923 = 1846.70$$

The total present worth of all sources of income from the bond, if held to maturity, is \$16,726.46. If the purchase price is equal to this amount, the investor breaks even. If purchased for a price less than this present worth, the bond is profitable to the investor.

Break-Even Analysis

Large capital expenditures for a system are only worthwhile if the benefits accrued from the investment are equal to or greater than the incurred cost. For example, for a manufacturing setting, if an initial investment is made into a factory with the objective of producing an item for sale, then the break-even analysis involves the calculation of the number of units that must be produced for the revenue from the resulting profits to equal the initial investment in value.

In other words, for an investment to "break even," the present worth of all cash flows must be zero.

Example 5.13

A company has a choice between two processes to manufacture a product. The data for both options (A and B) are given below. How many units must be produced annually to justify process A? Assume the MARR = 7%.

	Process A	Process B
Initial cost (\$)	50,000	28,000
Fixed annual cost (\$)	2500	1300
Annual production cost (\$/unit)	20	30
Life (years)	10	8
Salvage value of equipment (\$)	4000	0

Solution Let annual production be x units.

Since length of timeline is different for options A and B, use EUAC (equivalent uniform annual cost).

EUAC of all costs (option A) in thousands of dollars

$$50(A/P, 7\%, 10 \text{ years}) + 2.5 + 0.020x - 4(A/F, 7\%, 10 \text{ years})$$
$$= 50(0.1424) + 2.5 + 0.020x - 4(0.0724) = 9.3304 + 0.02x$$

EUAC of all costs (option B) in thousands of dollars

$$28(A/P, 7\%, 8 \text{ years}) + 1.3 + 0.030x = 28(0.1675) + 1.3 + 0.030x = 5.99 + 0.03x$$

In order to justify option A, we must have

$$9.3304 + 0.02x \le 5.99 + 0.03x \Rightarrow x \ge 334.04$$

At least 335 units must be produced annually in order to justify option A.

Return on Investment

The return on investment is that specific rate of interest which makes the present worth = 0. In calculating the present worth, all costs are considered negative and all benefits positive. Thus, the rate of return is that discount rate at which the investment breaks even. This is distinct from the MARR and could be higher (investment "beats the market") or lower than the MARR.

Example 5.14

A company invests $200,000 in a piece of equipment that manufactures a widget called X3P2. The useful life of the equipment is 10 years, at the end of which the salvage values is zero. If the annual overhead expenses are $12,000 and the company generates annual net revenue of $42,000, what is the rate of return on investment? Use MARR = 6%.

A. 6.1%

B. 7.4%

C. 8.1%

D. 9.2%

Solution

Initial investment, $P = 200{,}000$

Net annual return, $A = 42{,}000 - 12{,}000 = 30{,}000$

$$A/P = 30{,}000/200{,}000 = 0.150$$

For what rate of interest is $A/P = 0.150$ for $n = 10$ years?

By trial and error:

For $i = 6\%$, $A/P = 0.1359$

For $i = 8\%$, $A/P = 0.1490$

Therefore, the interest rate for which present worth = 0 is slightly greater than 8%.

Answer is C.

NOTE Exact solution (unnecessary for multiple choice exam) is 8.15%.

Example 5.15

A company invests $200,000 in a piece of equipment that manufactures a widget called X3P2. The useful life of the equipment is 10 years and the salvage value is $50,000. The cost for raw materials per piece of X3P2 is $3 and the sale price is $4.50. If annual overhead expenses are $18,000 and the company manufactures 28,000 units annually, what is the rate of return on investment? Use MARR = 6%.

A. 5.4%

B. 6.4%

C. 8.1%

D. 9.2%

Solution Let us convert all quantities to annual basis.

Annual profit: $A = (4.5 - 3.0) \times 28{,}000 = 1.5 \times 28{,}000 = 42{,}000$

For what rate of interest are expenses = revenue (break-even condition)?

$$200{,}000\left(\frac{A}{P}, i, 10 \text{ years}\right) - 50{,}000\left(\frac{A}{F}, i, 10 \text{ years}\right) + 18{,}000 - 42{,}000 = 0$$

By trial and error:

For $i = 5\%$, LHS = $200{,}000 \times 0.1296 - 50{,}000 \times 0.0796 - 24{,}000 = -2060$

For $i = 6\%$, LHS = $200{,}000 \times 0.1359 - 50{,}000 \times 0.0759 - 24{,}000 = -615$

For $i = 8\%$, LHS = $200{,}000 \times 0.149 - 50{,}000 \times 0.069 - 24{,}000 = 2350$

Therefore, the interest rate for which present worth = 0 is close to and slightly greater than 6%. Exact solution (unnecessary for multiple choice exam) is 6.42%.

Answer is B.

Benefit-Cost Analysis

A project is considered beneficial if benefits exceed costs. This may be expressed mathematically as

$$\frac{B}{C} > 1 \quad \text{or} \quad B - C > 0 \tag{5.12}$$

If the first approach is taken, all benefits and costs should be converted to equivalent quantities (i.e., all expressed as present worth, or all converted to annuities, etc.). For the second approach, it is typical to express the differential as the net present worth of all benefits and costs.

Example 5.16

A local municipality is contemplating the feasibility of repairing a bridge. The estimated useful life of the bridge is 20 years. The MARR is 6%. The relevant cost estimates are shown below. All costs are shown in thousands of dollars.

	Existing	Repair
Initial cost	0	50
Annual cost		
Years 1–10	8	3
Years 11–20	11	4
Salvage value	5	20

The benefit:cost ratio of making repairs is most nearly:

A. 1.15

B. 1.21

C. 1.32

D. 1.45

Solution The repair option has the following:
Initial cost = $50,000

Years 1 to 10: Savings in annual maintenance = $5000

Years 11 to 20: Savings in annual maintenance = $7000

The first is an annuity (A). The second is a shifted annuity (A). We can visualize the shifted annuity (1) as the difference between two annuities as shown below: (2) and (3). Annuities 2 and 3 are easier to deal with because their origin is at $t = 0$, and therefore the tables can be used.

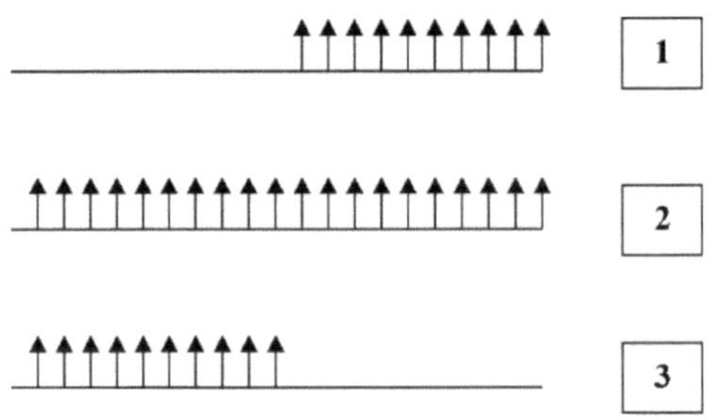

This can be expressed as a 20-year annuity of \$7k *minus* a 10-year annuity of \$2k. Benefit (converted to present value) is

$$7\text{k}\left(\frac{P}{A}, 20 \text{ years}, 6\%\right) - 2\text{k}\left(\frac{P}{A}, 10 \text{ years}, 6\%\right)$$

$$= 7000 \times 11.4699 - 2000 \times 7.3601 = 65{,}569.10$$

Extra salvage (\$15,000) converted to present worth:

$$15\text{k}\left(\frac{P}{F}, 20 \text{ years}, 6\%\right) = 15{,}000 \times 0.3118 = 4677$$

Benefit cost ratio: $B/C = \dfrac{65{,}569.10}{50{,}000 - 4677} = 1.45$

Answer is D.

NOTE An alternative calculation that may appeal to some is to count the salvage value as a benefit, which results in a benefit:cost ratio given by: $\dfrac{B}{C} = \dfrac{65{,}569 + 4877}{50{,}000} = 1.41.$ However, this approach is not recommended. The salvage value should be considered as a negative cost (cost offset), as in the first calculation.

Uncertainty (Expected Value and Risk)

Uncertainty deals with outcomes that are possible but unknown. On the other hand, risk is quantifiable. Risk is calculated as the expected loss resulting from a finite number of outcomes that have the potential to cause a quantifiable loss. Thus, if adopting a particular design has three specific outcomes, each with a defined probability P_i and a consequential loss L_i, then the risk can be calculated as the expected value

$$\text{Risk} = \sum_{1}^{3} P_i L_i \tag{5.13}$$

Decision making under conditions of risk should seek to identify, quantify, and mitigate risk whenever possible.

Example 5.17

The design of a dam has the following possible outcomes and associated probabilities and costs. What is the risk?

Outcome	Probability	Consequence (Cost)
Overtopping	0.05	Flood damage $650,000
Structural failure of vanes	0.01	Replacement cost $200,000
Electrical system failure	0.02	User costs $720,000

Solution The quantifiable risk is calculated as the expected value of the outcomes.

$$\text{Risk} = 0.05 \times 650 + 0.01 \times 200 + 0.02 \times 720 = 48.9$$

Risk = $48,900

CHAPTER 6

Statics

For mechanical systems in static equilibrium, the following (vector) equations must be satisfied. These conditions of static equilibrium imply that the combined effect of all forces acting on the system is zero and the combined effect of all moments about any arbitrary point A is also zero.

$$\sum \mathbf{F} = \underline{0} \tag{6.1}$$

$$\sum \mathbf{M}_A = \underline{0} \tag{6.2}$$

Vectors

A force (or a moment) is a vector; that is, it needs both magnitude and direction to be specified in order to be complete. In addition, if the effect of the force on a structure is to be computed, the location of the force also needs to be known precisely. For more on vectors, see also Chap. 1: Mathematics.

In Cartesian coordinates, a vector force is described as

$$\mathbf{F} = F_x\mathbf{i} + F_y\mathbf{j} + F_z\mathbf{k} \tag{6.3}$$

where F_x, F_y, and F_z are the scalar components of the force and **i**, **j**, and **k** are the unit vectors in the three Cartesian directions x, y, and z, respectively.

The magnitude of the force is calculated as

$$F = \sqrt{F_x^2 + F_y^2 + F_z^2} \tag{6.4}$$

and the force can then be decomposed into the magnitude F and the unit vector $\mathbf{u}_F$

$$\mathbf{F} = F\mathbf{u}_F = F[\cos\theta_x\mathbf{i} + \cos\theta_y\mathbf{j} + \cos\theta_z\mathbf{k}] \tag{6.5}$$

where the angles θ_x, θ_y, and θ_z are the angles between the force and the x, y, and z axes, respectively. Thus, the direction cosines are calculated as

$$\cos\theta_x = \frac{F_x}{F}; \quad \cos\theta_y = \frac{F_y}{F}; \quad \cos\theta_z = \frac{F_z}{F} \tag{6.6}$$

Definitions

Concurrent forces: A system of forces is called concurrent if all the forces pass through a single common point.

Coplanar forces: A system of forces is called coplanar if all the forces lie in the same plane.

Collinear forces: Two forces are called collinear if they share the same line of action.

Resultant of a System of Forces (2D)

When a system of n concurrent forces acts on a particle and each force F_i can be decomposed into Cartesian components $F_{x,i}$ and $F_{y,i}$, then the magnitude of the resultant is calculated as

$$R = \sqrt{\left(\sum_{i=1}^{n} F_{x,i}\right)^2 + \left(\sum_{i=1}^{n} F_{y,i}\right)^2} \tag{6.7}$$

and the angle made by the resultant (with the reference x-axis, where $\theta = 0$) is

$$\theta = \tan^{-1}\left(\frac{\sum_{i=1}^{n} F_{y,i}}{\sum_{i=1}^{n} F_{x,i}}\right) \tag{6.8}$$

Care must be taken to use the correct angle for each force. The angle must be measured (a) at the tail of the vector, (b) with respect to the same reference axis (commonly the x axis), and (c) in the same direction (counterclockwise for the right-handed system). See Example 6.1.

Example 6.1

A system of concurrent forces is shown below. Find (a) the resultant force magnitude and (b) the angle made by the resultant with the x axis.

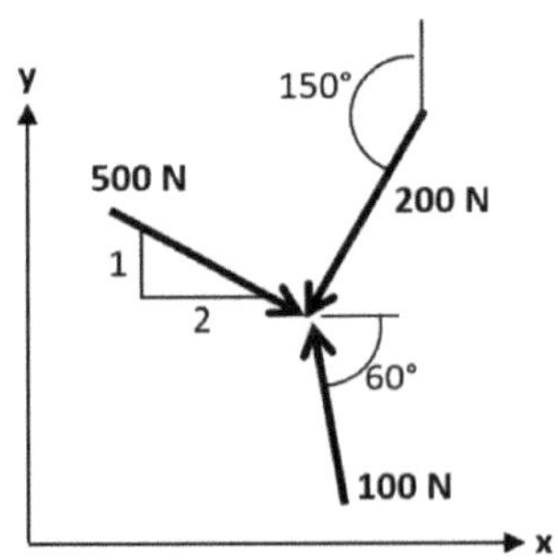

Solution The first step is to write each force as a vector with its Cartesian components.

$$F_1(200N, 240°) = 200[\cos 240\mathbf{i} + \sin 240\mathbf{j}] = -100\mathbf{i} - 173.2\mathbf{j}$$

$$F_2 = 500\left[\frac{2}{\sqrt{5}}\mathbf{i} - \frac{1}{\sqrt{5}}\mathbf{j}\right] = 447.2\mathbf{i} - 223.6\mathbf{j}$$

$$F_3(100N, 120°) = 100[\cos 120\mathbf{i} + \sin 120\mathbf{j}] = -50\mathbf{i} + 86.6\mathbf{j}$$

$$R = \sum F_i = 297.2i - 310.2j$$

Resultant: $|R| = \sqrt{297.2^2 + 310.2^2} = 430N$

Angle between resultant and x axis: $\theta = \tan^{-1}\left(\frac{-310.2}{297.2}\right) = -46.2°$ or $313.8°$

NOTE The angle chosen is the fourth quadrant angle because of the signs of the components of R (x positive, y negative).

Moment of a Force

Moment about a Point

The moment of a force about a point O is calculated the cross product

$$\mathbf{M}_o = \mathbf{r} \times \mathbf{F} \tag{6.9}$$

where **r** is the (vector) distance from point O to any point on the line of action of the force.

The cross product of two vectors $\underline{\mathbf{r}}$ and $\underline{\mathbf{F}}$ can be written as the determinant

$$M = \mathbf{r} \times \mathbf{F} = \begin{vmatrix} \mathbf{i} & \mathbf{j} & \mathbf{k} \\ r_x & r_y & r_z \\ F_x & F_y & F_z \end{vmatrix} \tag{6.10}$$

In terms of the scalar components, the moment can be written as

$$M = r \times F = (x\mathbf{i} + y\mathbf{j} + z\mathbf{k}) \times (F_x\mathbf{i} + F_y\mathbf{j} + F_z\mathbf{k}) = (yF_z - zF_y)\mathbf{i} + (zF_x - xF_z)\mathbf{j} + (xF_y - yF_x)\mathbf{k}$$

Example 6.2

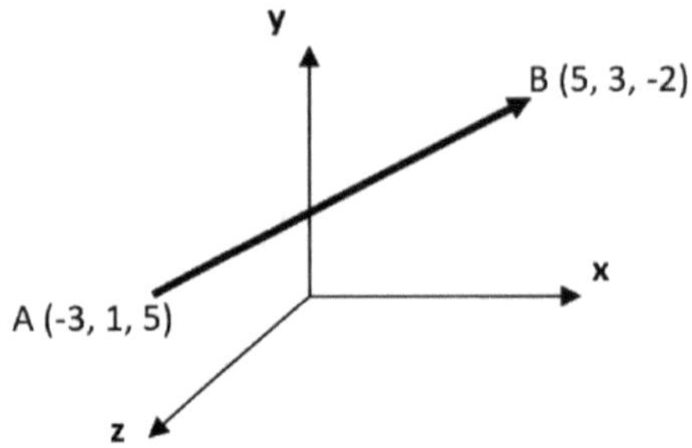

In the figure above, a force of magnitude 250 lb acts along the line directed from (−3, 1, 5) to (5, 3, −2). All coordinates are in feet. What is the moment of the force *F* about the origin (0, 0, 0)?

Solution The moment of the force about the origin is calculated as the cross product of the distance from *O* (0, 0, 0) to a point on the line of action of force *F*, that is, either *A*(−3, 1, 5) or *B*(5, 3, −2).

$$\mathbf{r} = \underline{OA} = -3\mathbf{i} + \mathbf{j} + 5\mathbf{k}$$

The length of the line *AB* is given by $|AB| = \sqrt{8^2 + 2^2 + 7^2} = 10.817$.

The unit vector (dimensionless) along the line *AB* is given by

$$\mathbf{u}_{AB} = \frac{8\mathbf{i} + 2\mathbf{j} - 7\mathbf{k}}{10.817} = 0.740\mathbf{i} + 0.185\mathbf{j} - 0.647\mathbf{k}$$

If the magnitude of the force is 250 lb, then the force *F* can then be expressed as

$$F = 250(0.740\mathbf{i} + 0.185\mathbf{j} - 0.647\mathbf{k}) = 185\mathbf{i} + 46.3\mathbf{j} - 161.8\mathbf{k}$$

The moment of the force about the origin *O* is then calculated as

$$M = \mathbf{r} \times \mathbf{F} = \begin{vmatrix} \mathbf{i} & \mathbf{j} & \mathbf{k} \\ -3 & 1 & 5 \\ 185 & 46.3 & -161.8 \end{vmatrix} = -393.3\mathbf{i} + 439.6\mathbf{j} - 323.9\mathbf{k}$$

Moment about a Line (Axis)

The moment of a force about a line (*AB*) is calculated the scalar triple product

$$M_{AB} = \mathbf{u}_{AB} \cdot (\mathbf{r} \times \mathbf{F}) \tag{6.11}$$

where **r** is the (vector) distance from *any point* on line *AB* to any point on the line of action of the force.

$$\mathbf{u}_{AB} = \text{unit vector of line } AB$$

The result of the triple product in Eq. (6.11) is scalar; that is, it only yields the magnitude of the moment M_{AB}. If the vector moment about line AB ($\mathbf{M}_{AB}$) is required, it needs to be multiplied by the unit vector $\mathbf{u}_{AB}$

$$\mathbf{M}_{AB} = \mathbf{M}_{AB}\mathbf{u}_{AB} \tag{6.12}$$

Example 6.3

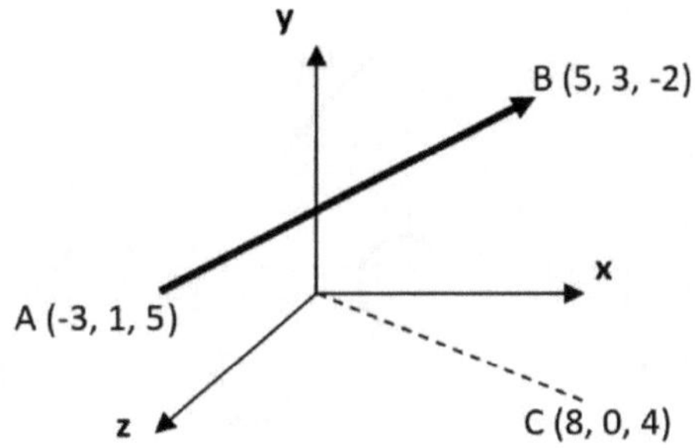

In the figure above, a force of magnitude 250 lb acts along the line directed from (−3, 1, 5) to (5, 3, −2). All coordinates are in feet. What is the moment of the force F about the line OC?

Solution The following can be calculated (for more details, see Example 6.2)

$$\mathbf{r} = \underline{OA} = -3\mathbf{i} + \mathbf{j} + 5\mathbf{k}$$

$$\mathbf{F} = 185\mathbf{i} + 46.3\mathbf{j} - 161.8\mathbf{k}$$

The moment of the force about the origin O: $M_o = \mathbf{r} \times \mathbf{F} = -393.3\mathbf{i} + 439.6\mathbf{j} - 323.9\mathbf{k}$

$$\underline{OC} = 8\mathbf{i} + 4\mathbf{k}$$

$$\mathbf{u}_{OC} = 0.8944\mathbf{i} + 0.4472\mathbf{k}$$

$$M_{OC} = \mathbf{u}_{OC} \cdot M_o = (0.8944\mathbf{i} + 0.4472\mathbf{k}) \cdot (185\mathbf{i} + 46.3\mathbf{j} - 161.8\mathbf{k}) = 93.1$$

Note that this result is a scalar; that is, it only yields the magnitude of the moment. If the vector component along line OC is required, it needs to be multiplied by the unit vector $\mathbf{u}_{OC}$.

$$M_{OC} = |M_{OC}|\mathbf{u}_{OC} = 93.1(0.8944\mathbf{i} + 0.4472\mathbf{k}) = 83.3\mathbf{i} + 41.6\mathbf{k}$$

Equivalent Force Systems

The analysis of force systems is greatly simplified by reducing a system of forces to an equivalent system of a single force and single couple. The fundamental principle of replacing one group (system) of forces and/or moments with another group is to equate the sum of forces in group 1 to the sum of forces in group 2. The same must be done for moments. An illustrative example is shown below.

Example 6.4

Replace the force system shown below with a force-couple pair at the support *A*.

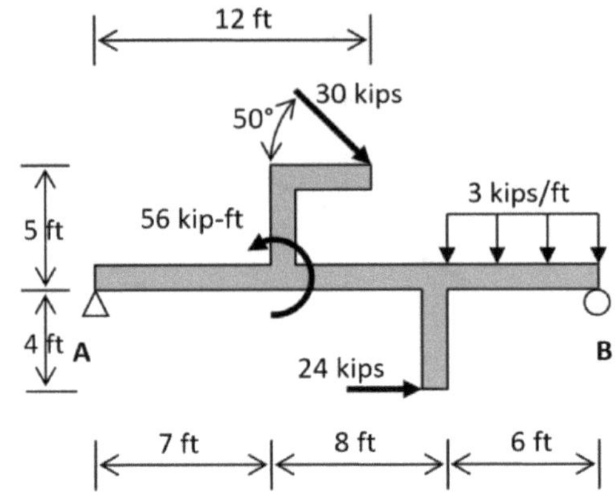

Solution If the force system is to be replaced by a force-couple pair at the left support, the two systems must be equivalent in terms of the resultant force $\Sigma\mathbf{F}$ and the resultant moment about the left support $\Sigma\mathbf{M}_A$.

For the original system, the summation of (vector) forces is

$$\sum\mathbf{F} = 30\cos 50\mathbf{i} - 30\sin 50\mathbf{j} + 24\mathbf{i} - 18\mathbf{j} = (19.28\mathbf{i} - 22.98\mathbf{j}) + (24\mathbf{i} - 18\mathbf{j}) = 43.28\mathbf{i} - 40.98\mathbf{j}$$

The summation of (vector) moments about support *A* is

$$\sum\mathbf{M}_A = (12\mathbf{i} + 5\mathbf{j}) \times (19.28\mathbf{i} - 22.98\mathbf{j}) + 56\mathbf{k} + (15\mathbf{i} - 4\mathbf{j}) \times 24\mathbf{i} + 18\mathbf{i} \times -18\mathbf{j} = -544.16\mathbf{k}$$

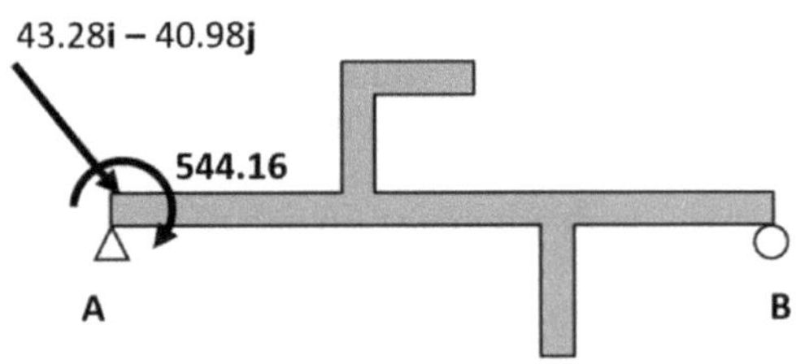

Note that according to the right-hand rule, which is being used here, the negative **k** component implies that the resultant moment is clockwise.

Example 6.5

A frame ABCDE is supported by a hinge at *A* and roller support at *D* as shown. If the 30 kN force at *C* and the distributed load over *DE* is to be replaced by a vertical (acting upward) force of 12 kN at *C*, while maintaining the same reaction at support *D*, what is the additional moment that must be applied at *C*?

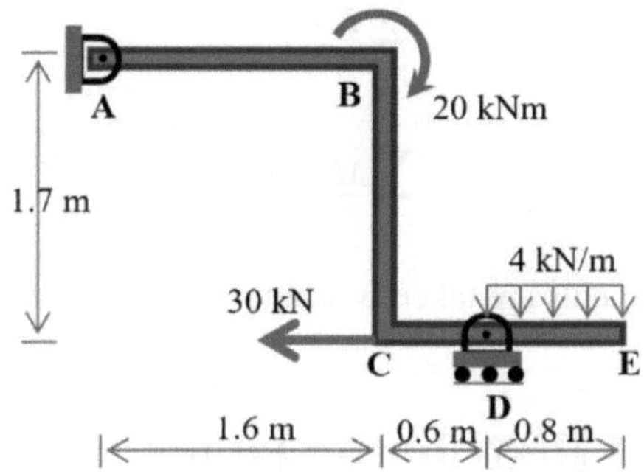

A. $M = 40.12$ kNm clockwise

B. $M = 40.12$ kNm anticlockwise

C. $M = 78.52$ kNm clockwise

D. $M = 78.52$ kNm anticlockwise

Solution The question states "the 30 kN force at *C* and the distributed load over *DE* is to be replaced by a vertical (acting upward) force of 12 kN at *C*." This means that the couple at *B* is unaffected. Also, the vertical reaction at *D* is unchanged. These two items can therefore be excluded from the equivalence of the two systems. Since the reactions at *A* will certainly change, we take moments about *A* so that they are excluded from the calculations.

For the existing system, sum of moments about *A*:

$$M_{A,1} = 30 \times 1.7 + 0.8 \times 4 \times 2.6 = 59.32 \text{ kNm (clockwise)}$$

For the new system, sum of moments about *A*:

$$M_{A,2} = 12 \times 1.6 = 19.2 \text{ kNm (anticlockwise)}$$

Therefore, an additional couple $M = 78.52$ kNm clockwise must be added.

Answer is C.

Equilibrium of Rigid Bodies

A rigid body in static equilibrium, must satisfy both force and moment equilibrium. These are expressed as the **vector** equations shown below [Eqs. (6.13) and (6.14)]. Each of these equations can be decomposed into scalar equations, each of those representing a different orthogonal component.

$$\sum \underline{F} = 0 \tag{6.13}$$

$$\sum \underline{M} = 0 \tag{6.14}$$

The point about which moments are taken is arbitrary.

Representation of Distributed Loads

For the purpose of calculating external effects (such as external reactions), a distributed load can be replaced by its resultant. The effective location of the resultant must be at the centroid of the shape of the distributed load. For example, a uniformly distributed load is represented by a rectangle, whose centroid is at mid-width (*L*/2), whereas a linearly varying distributed load may be represented by a triangle whose centroid is as shown below (2*L*/3). A linearly varying distributed load represented by a trapezoid can be broken down into a rectangle and a triangle. These are shown in Fig. 6.1.

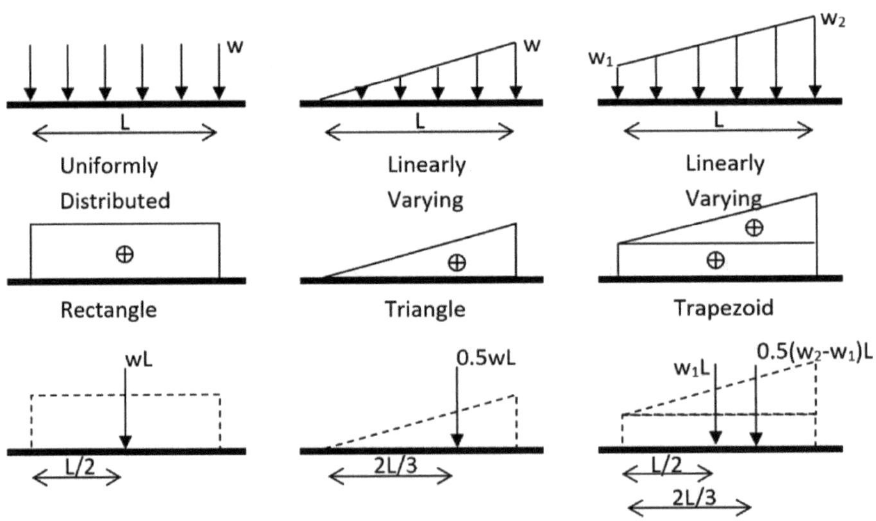

Figure 6.1 Distributed loads and their equivalents.

Rigid Frames: Degrees of Freedom, Determinacy, and Stability

A frame is an assemblage of rigid members connected by a combination of rigid and hinged connections. At a hinged joint (much like each node of an ideal truss), two pairs of equal and opposite internal forces exist. These forces are orthogonal. For an ideal truss, each member is subject to a single internal load—an axial force normal to the plane of the cross section. On the other hand, when a rigid member is sectioned, three internal loads—comprising of two orthogonal forces (as before) plus a moment must be incorporated into the free body diagram. These are shown in Fig. 6.2. Note the difference between local and global coordinates. In global coordinates, the internal forces are always aligned with the global axes, irrespective of the orientation of the member. In local coordinates, the internal forces are the normal force *N*, which is perpendicular to the cross-section plane and the shear force *V*, which is parallel to the cross-section plane.

Each node of a 2D rigid frame can have three degrees of freedom: (1) *x* translation, (2) *y* translation, and (3) rotation. Supports constrain some of these degrees of freedom. For example, a hinge support constrains both translational degrees of freedom, leaving only one rotational degree of freedom unconstrained. When enough compatibility equations exist to solve for all unknown degrees of freedom of the structure, it is said to be *kinematically* determinate. On the other hand, if actions are of interest (internal actions plus external reactions), then static determinacy must be considered. If the equilibrium equations are sufficient to solve for all actions, then the structure is called statically determinate.

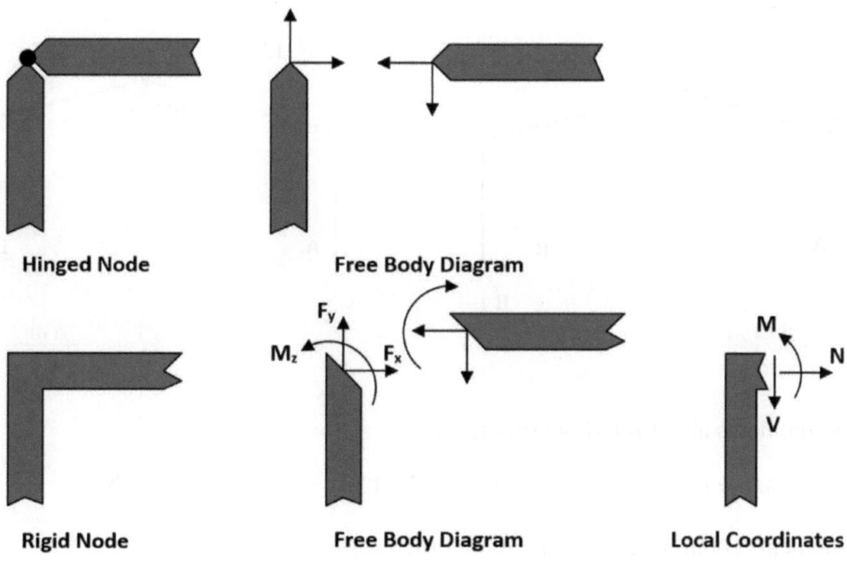

Figure 6.2 Free body diagrams of hinged and rigid nodes.

Example 6.6

For the three-hinged frame shown below, what is the reaction force at the support *A*? The crown *C* is hinged.

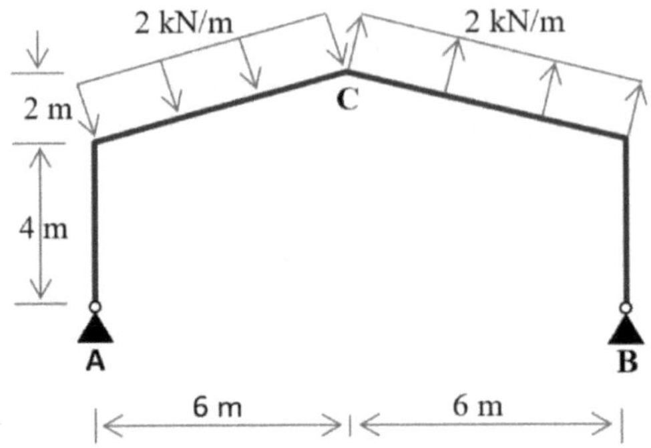

A. 2.67 kN

B. 3.33 kN

C. 4.0 kN

D. 4.8 kN

Solution The hypotenuse length of the gable roof is $\sqrt{(2^2 + 6^2)} = 6.325$ m. This results in the resultant on each half of the gable roof = 12.65 kN, which can in turn be resolved into components: 12 kN (vertical) and 4 kN (horizontal).

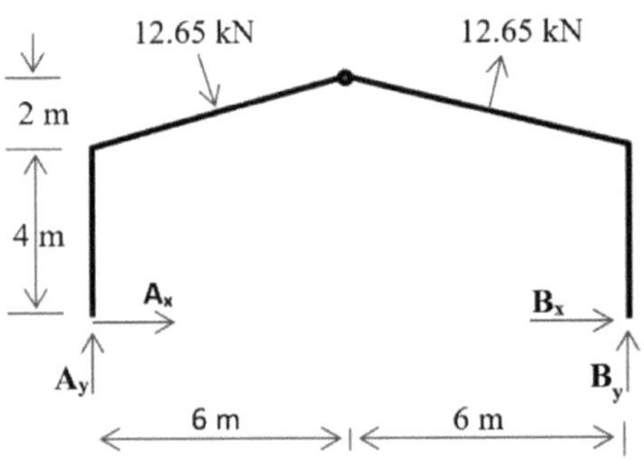

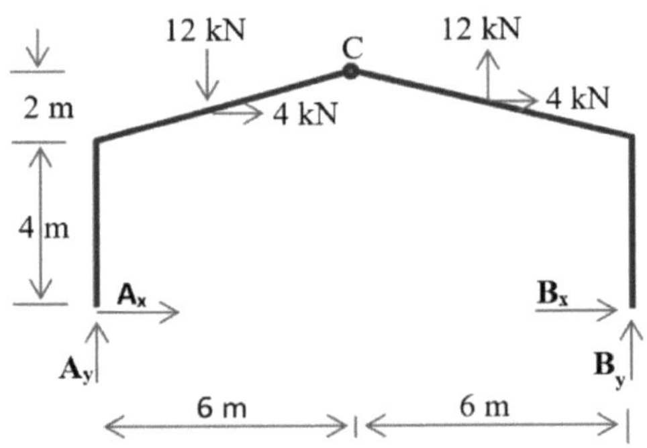

Taking moments about *B* (whole structure):

$$M_B = 12\times 3 + 4\times 5 - 12\times 9 + 4\times 5 + 12A_y = 0 \Rightarrow A_y = 2.67 \text{ kN}$$

Taking moments about hinge *C* (left half of structure):

$$M_{C,\text{left}} = -12\times 3 - 4\times 1 - 6A_x + 6A_y = 0 \Rightarrow -A_x + A_y = 6.67 \Rightarrow A_x = -4 \text{ kN}$$

Total reaction at A: $R_A = \sqrt{A_x^2 + A_y^2} = \sqrt{4^2 + 2.67^2} = 4.81$ kN

Answer is D.

Inclined Support

A fixed (fully rigid) support restrains all degrees of freedom of the structure. As a result, the orientation of the support does not matter. In a 2D structure, a fixed support should be represented by two orthogonal reaction forces (typically represented as R_x and R_y) representing the translational constraint, whose directions do not matter as long as they are perpendicular, and a moment reaction (typically represented as M_z) representing the rotational constraint. For a 3D structure, a fixed support must be represented by three orthogonal reaction forces (R_x, R_y, and R_z) and the accompanying reaction moments (M_x, M_y, and M_z).

For a pinned (hinged) support, translation is fully restrained while rotation is not at all. Thus, in 2D, a pinned support should be represented by two orthogonal reaction forces (typically represented as R_x and R_y) representing the translational constraint, whose directions do not matter as long as they are perpendicular. For a 3D structure, a pinned support must be represented by three orthogonal reaction forces (R_x, R_y, and R_z).

For a "roller" support, the only constraint offered by the support is the restraint against translation normal to the supporting surface (the "ground"). Therefore, a roller support should be represented by a single reaction force acting perpendicular to the supporting surface.

The 2D renditions of these three kinds of support are shown in Fig. 6.3.

The following example illustrates the use of an inclined roller support for a beam.

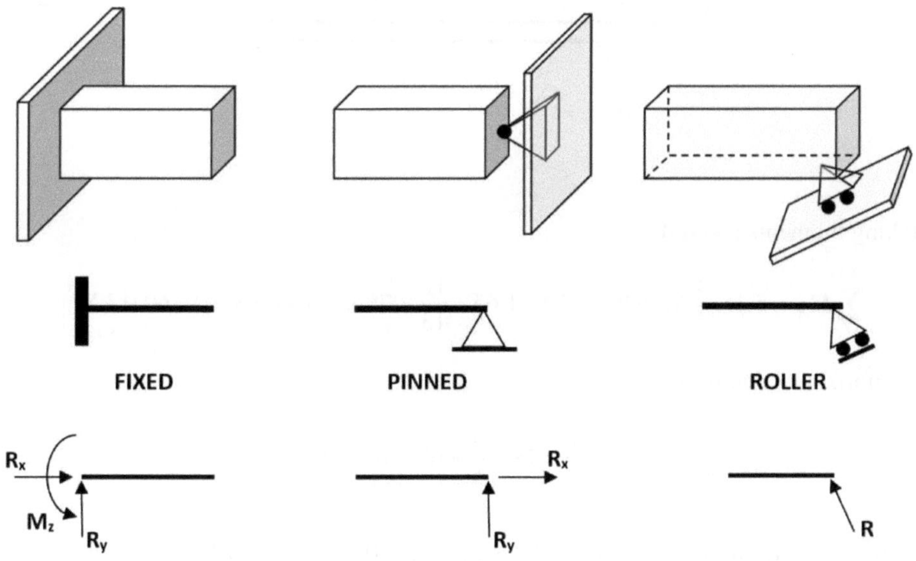

Figure 6.3 Various types of supports.

Example 6.7

A beam is supported by a hinged support at the left and by an inclined roller on the right as shown below. What is the reaction at the hinge support on the left?

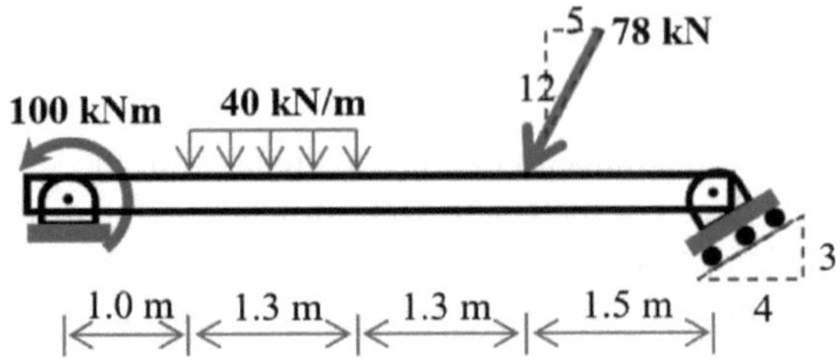

A. $R_B = 60$ kN

B. $R_B = 72$ kN

C. $R_B = 84$ kN

D. $R_B = 100$ kN

Solution The reaction at the roller is a single reaction perpendicular to the ground. Since the ground is sloped 4H:3V, the reaction force will be angled 3H:4V. The free body diagram is as shown below

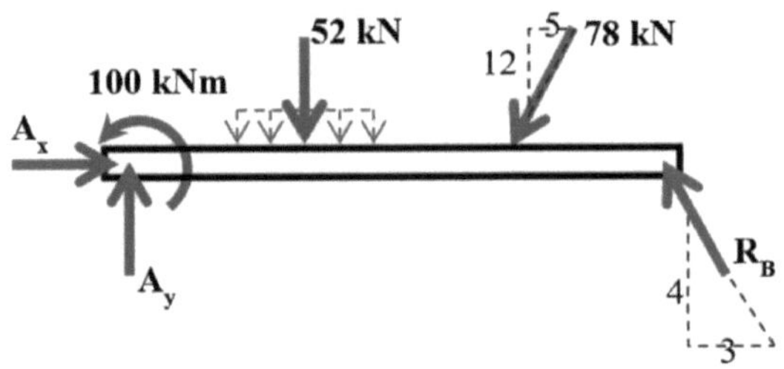

Taking moments about A:

$$\sum M_A = 5.1 \times \frac{4}{5} R_B + 100 - 52 \times 1.65 - \frac{12}{13} \times 78 \times 3.6 = 0 \Rightarrow R_B = 60.0 \text{ kN}$$

For horizontal equilibrium:

$$\sum F_x = A_x - \frac{5}{13} \times 78 - \frac{3}{5} \times 60 = 0 \Rightarrow A_x = 66 \text{ kN}$$

For vertical equilibrium, $A_y + 0.8R_B - 52 - \frac{12}{13} \times 78 = 0 \Rightarrow A_y = 76$ kN

Reaction at the hinge: $R_A = \sqrt{A_x^2 + A_y^2} = \sqrt{66^2 + 76^2} = 100.7\ \text{kN}$

Answer is D.

Example 6.8

A simply supported beam is shown below. What is the reaction at the roller support at *B*?

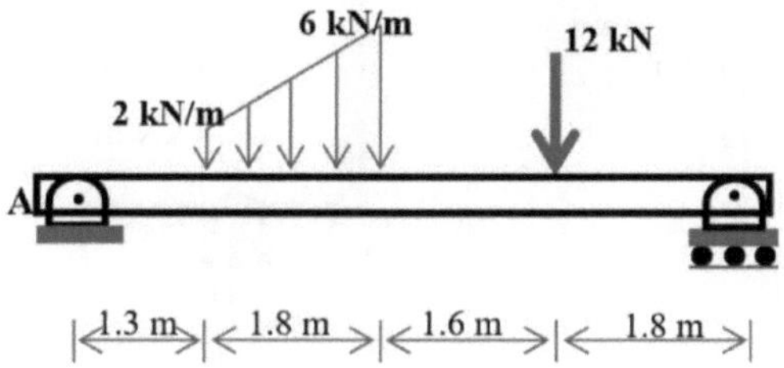

A. $B_y = 11.28\ \text{kN}$
B. $B_y = 11.97\ \text{kN}$
C. $B_y = 7.92\ \text{kN}$
D. $B_y = 9.03\ \text{kN}$

Solution Replacing each distributed load as a resultant acting at the centroid, we have the following FBD.

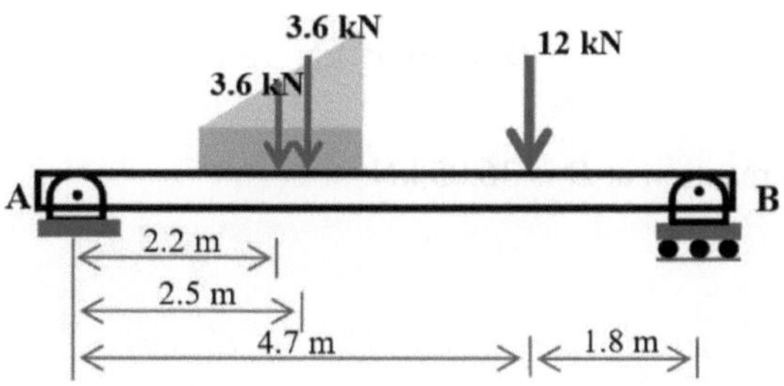

Now, taking moments about *A*:

$$\sum M_A = 3.6 \times 2.2 + 3.6 \times 2.5 + 12 \times 4.7 - 6.5 B_y = 0 \Rightarrow B_y = 11.28\ \text{kN}$$

Answer is A.

Example 6.9

A frame ABCDE is supported by a hinge at *A* and roller support at *D* as shown. The frame is subject to the following loads: a clockwise couple (20 kNm) on joint *B*, a 30-kN force at *C*, and a uniformly distributed load (4 kN/m) over *DE*. What is the vertical reaction at *D*?

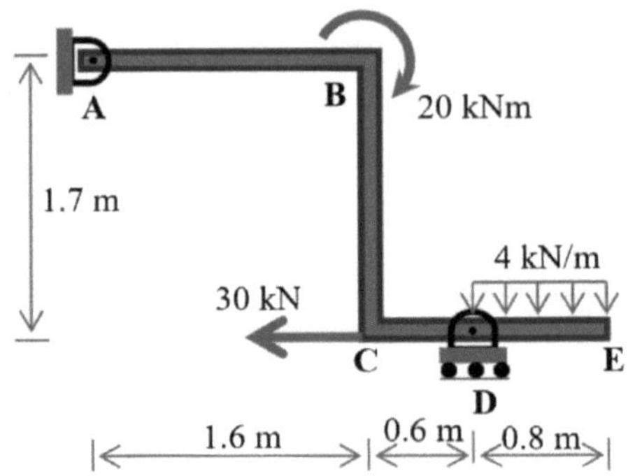

A. 22.8 kN

B. 27 kN

C. 36 kN

D. 41.3 kN

Solution Writing the sum of moments about *A*:

$$M_A = 20 + 30 \times 1.7 + 0.8 \times 4 \times 2.6 - 2.2D_y = 0$$

$$D_y = 36.05 \text{ kN}$$

Therefore, the vertical reaction at *D* is 36.05 kN.

Answer is C.

Example 6.10

A rectangular plate ABCD is acted upon by the forces indicated below. What is the resultant moment about the center of the plate?

A. 37 Nm

B. 41 Nm

C. 57 Nm

D. 62 Nm

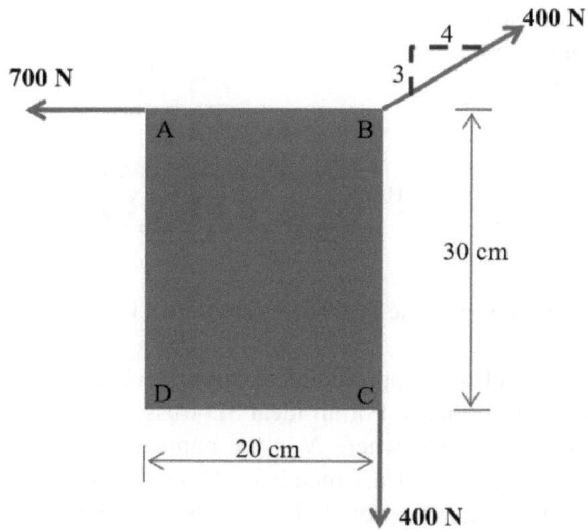

Solution The solution of this problem is greatly simplified by noticing that the three forces intersect at *B*. Therefore, the moment can be calculated by using a single **r** vector—from the center of the plate to point *B*. This vector (cm) is

$$\mathbf{r} = 10\mathbf{i} + 15\mathbf{j}$$

The total resultant force (*N*) acting at *B* is

$$F = -700\mathbf{i} + 400\left(\frac{4}{5}\mathbf{i} + \frac{3}{5}\mathbf{j}\right) - 400\mathbf{j} = -380\mathbf{i} - 160\mathbf{j}$$

Moment (N·cm) about center of plate:

$$\mathbf{M} = \mathbf{r} \times \mathbf{F} = (10\mathbf{i} + 15\mathbf{j}) \times (-380\mathbf{i} - 160\mathbf{j}) = -1600\mathbf{k} + 5700\mathbf{k} = 4100\mathbf{k}$$

Therefore, moment = 41 Nm

Answer is B.

Ideal Truss—Stability and Determinacy

An ideal truss is based on the following assumptions:

1. All members are connected at nodes via frictionless pinned connections.
2. At each node, lines of action through the centroid of the connection for each member are concurrent (see Fig. 6.4).
3. Loads are located (lumped) at nodes only—no member forces.

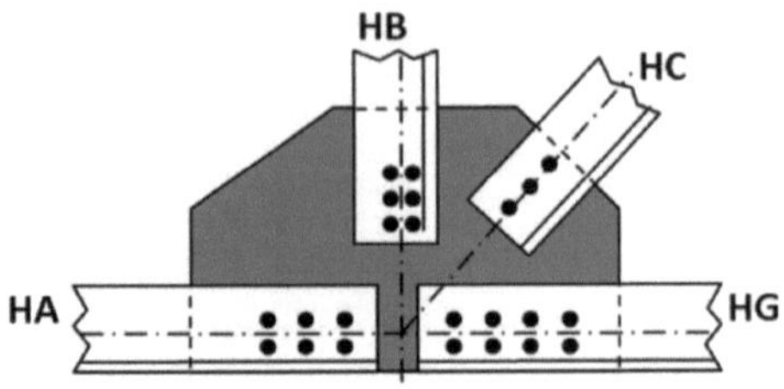

Figure 6.4 Ideal node configuration for a truss.

In Fig. 6.4, a magnified view of node *H* from Example 6.11 is shown such that condition 2 stated above is satisfied.

If these assumptions are closely approximated, truss members experience only axial loads (no shear force or bending moment). For an ideal 3D truss, the total number of equilibrium equations available is equal to 3*N*, where *N* is the number of nodes. The total number of unknowns is equal to the number of truss members, *M* (one axial force per member) plus the number of support reactions, *R*. For a truss to be classified as statically determinate, the number of equations is equal to the number of unknowns.

Statically determinate (3D) truss:	$3N = M + R$
Statically indeterminate (3D) truss:	$3N < M + R$
Unstable (3D) truss:	$3N > M + R$

Truss Member Forces: Method of Joints

The method of joints involves the listing of all equations of equilibrium, proceeding joint by joint. Each joint of a two-dimensional (plane) truss yields two equilibrium equations: $\Sigma F_x = 0$ and $\Sigma F_y = 0$, where *x* and *y* are two mutually perpendicular axes in the plane. On the other hand, the number of unknown quantities for the plane truss is the number of external reactions plus the number of member forces.

Example 6.11

For the plane truss shown below, calculate the force in member *EF*.

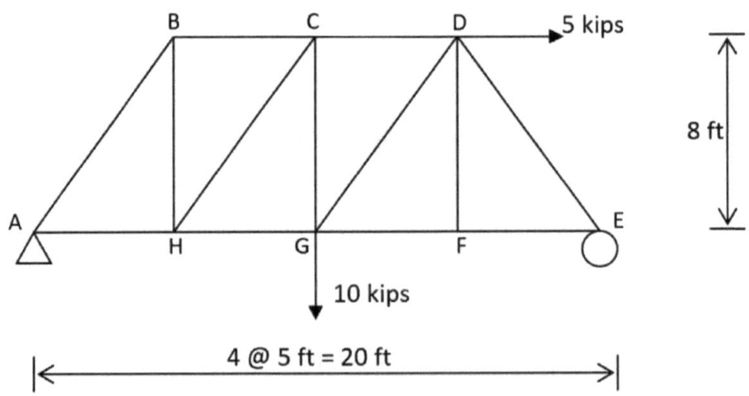

Solution For the truss shown above, $R = 3$, $M = 13$, and $J = 8$.
Thus, we have $R + M = 16 = 2J$. The truss is statically determinate.
Prior to solving for member forces, we need to solve for the external reactions. As long as there are no more than three external reactions, the truss is externally statically determinate.

$$\sum F_x = A_x + 5 = 0 \Rightarrow A_x = -5$$

$$\sum F_y = A_y - 10 + E_y = 0 \Rightarrow A_y + E_y = 10$$

$$\sum M_A = -(10 \times 10) - (8 \times 5) + (20 \times E_y) = 0 \Rightarrow E_y = 7$$

$$\therefore A_y = 10 - E_y = 3$$

The statement of equilibrium of a joint involves two equations of equilibrium. Thus, in order for the equations to be solvable, progressing from node to node, we would have to start at a node where there are no more than two unknowns. A and E—the two support nodes—are such supports.

Let us start solving from node E.

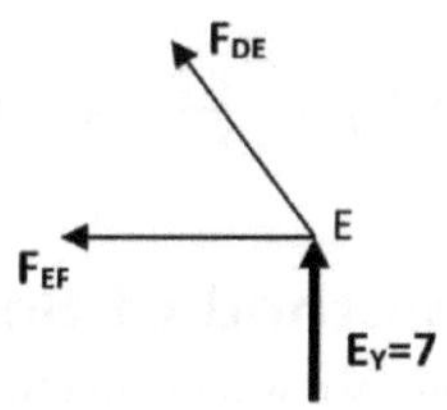

The free body diagram of node E is shown above. The support reaction E_y has been solved, leaving the two unknowns F_{DE} and F_{EF}. One would therefore solve joints in turn, with no more than two unknown member forces for each free body diagram

$$\sum F_x = -F_{EF} - F_{DE}\cos 58^\circ = 0 \Rightarrow F_{EF} = -F_{DE}\cos 58$$

$$\sum F_y = F_{DE}\sin 58^\circ + 7 = 0 \Rightarrow F_{DE} = -\frac{7}{\sin 58} = -8.255$$

$$F_{EF} = -F_{DE}\cos 58 = -(-8.255)\cos 58 = 4.374$$

If faced with a situation such as shown above, where there are only three forces in equilibrium, the solution of the equilibrium equations may also be done geometrically, rather than algebraically. The force polygon formed by three forces in equilibrium is a triangle, and the force triangle must be similar to the geometric triangle. The two triangles are shown in Fig. 6.5.

Note that in the force polygon on the left, the only force which is known a priori is the 7 kips vertical reaction at node E. This force is upward. If the forces are drawn "tip to tail" so

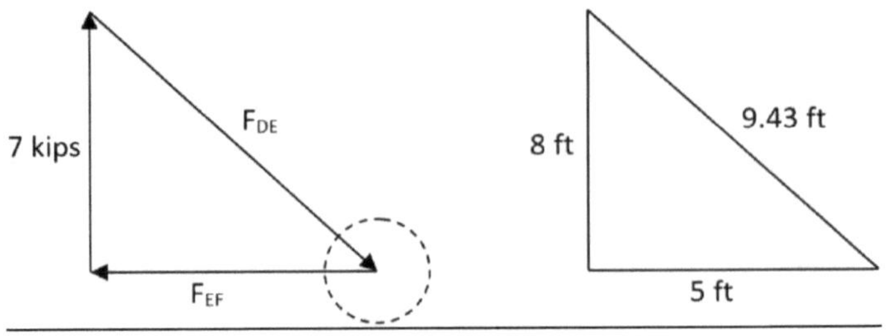

Figure 6.5 Similarity of force and geometric polygons.

as to complete the force polygon, we have the required directions of the unknown forces F_{DE} and F_{EF}. Comparing these forces with the original free body diagram, we see that F_{DE} must be compression (acting into the node, shown outlined) and F_{EF} must be tension (acting outward from the node).

The values of these forces may be obtained by corresponding ratios from the similar triangles. These yield simple equations rather than coupled simultaneous equations.

$$\frac{F_{DE}}{7} = \frac{9.43}{8} \Rightarrow F_{DE} = \frac{7 \times 9.43}{8} = 8.255$$

$$\frac{F_{EF}}{7} = \frac{5}{8} \Rightarrow F_{EF} = \frac{7 \times 5}{8} = 4.375$$

Truss Member Forces: Method of Sections

In order to apply the method of sections, we want to cut the structure with a section, thus dividing it into two parts. At the location of the cut, we replace each cut member with the member force (which is still unknown). As in the method of joints, we will insert these unknown forces as tension forces to begin with. The algebraic sign of the force, once solved, will tell us whether the actual internal force is tensile or compressive.

There are some rules for choosing a valid section or cut. For a "cut" to be valid—such that the substructures are solvable—one must follow certain rules in choosing the orientation of the cut, that is, which members it passes through.

1. The "cut" or section must separate the structure into two parts.
2. The "cut" must not pass through more than three unknown members (i.e., whose internal forces are unknown). The substructure is governed by the three equations of static equilibrium. Thus, any more than three unknowns will lead to an unsolvable system.
   Exception: In some situations, there is no single section that meets criterion 1 and 2. In those cases, it may be necessary to make two successive sections, producing a system of six simultaneous equations in six unknown member forces.
3. The unknown forces must not be concurrent. If these forces all pass through a common point, the problem will reduce to equilibrium of that point—producing two equations of equilibrium—$\Sigma F_x = 0$ and $\Sigma F_y = 0$. The equation $\Sigma M = 0$ will be satisfied identically ($0 = 0$) and will not yield any useful information.

Example 6.12

For the determinate plane truss shown below, calculate the force in member *CG*.

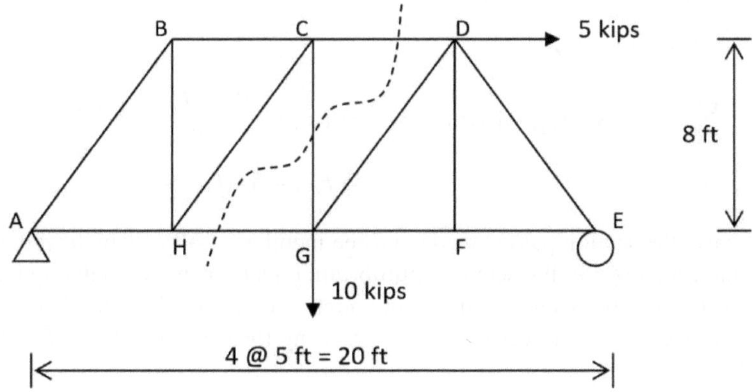

Solution In order to solve for F_{CG}, consistent with the rule given earlier, we choose a section that passes through the three members *CD*, *CG*, and *GH*. This section is shown as a dashed line in the diagram above. Verifying that this is consistent with the "rules," we note that

1. The section does separate the structure into two parts.
2. The section passes through (no more than) three members: *CD*, *CG*, and *GH*.
3. The three members mentioned above are not concurrent.

Analyzing the entire truss leads to the following values for the support reactions:

$A_x = -5$ (left)
$A_y = +3$ (up)
$E_y = +7$ (up)

Let us look at the left substructure.

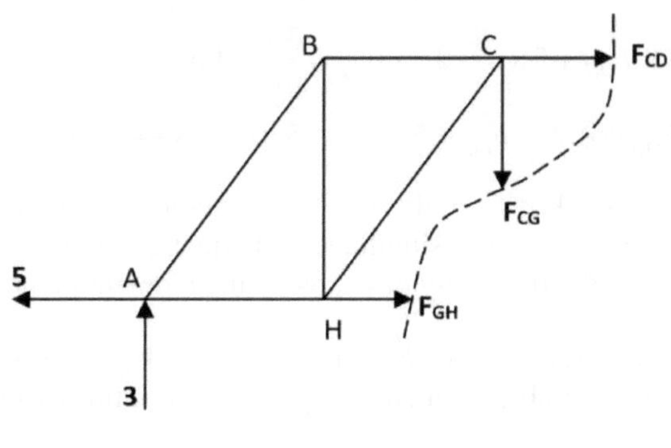

The equations of equilibrium are

$$\sum F_x = -5 + F_{GH} + F_{CD} = 0$$

$$\sum F_y = 3 - F_{CG} = 0 \qquad \Rightarrow F_{CG} = 3$$

$$\sum M_A = -8 \times F_{CD} - 10 \times F_{CG} = 0 \Rightarrow F_{CD} = -\frac{10 \times F_{CG}}{8} = -3.75$$

$$\Rightarrow F_{GH} = 5 - F_{CD} = 8.75$$

In the above case, the section passed through three members—all of which were either vertical or horizontal. As a result, the section equilibrium equations were relatively simple. None of the forces needed to be resolved into components. Let us consider another section of the same truss. Let us consider the vertical section passing through members *CD*, *DG*, and *GF*, as shown below.

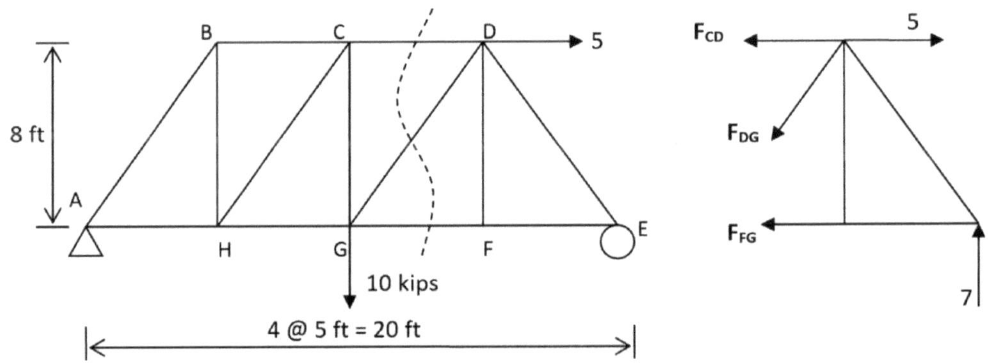

The free body diagram of the right substructure is shown. The equations of equilibrium for the substructure are

$$\sum F_x = -F_{CD} - F_{DG}\cos 58 - F_{FG} + 5 = 0$$

$$\sum F_y = -F_{DG}\sin 58 + 7 = 0 \Rightarrow F_{DG} = \frac{7}{\sin 58} = 8.255$$

$$\sum M_D = -8 \times F_{DG} + 5 \times 7 = 0 \Rightarrow F_{FG} = \frac{35}{8} = 4.375$$

$$\Rightarrow F_{CD} = -F_{DG}\cos 58 - F_{FG} + 5 = -8.255 \times \cos 58 - 4.375 + 5 = -3.75$$

In many situations, one is asked for only one of the member forces rather than all three. A complete solution of the three (often simultaneous) equations is not necessary. Noting that moment equations give us the flexibility of choosing the moment pivot, we see that

- If we are interested in solving for F_{CD}, the equation $\Sigma M_G = 0$ is algebraically simple, containing only the variable of interest F_{CD}. We achieve this by noting that we can eliminate the other two (F_{DG} and F_{FG}) unknown forces from our equation by taking the moments about point G, which is the intersection of these two forces.

- Similarly, in order to solve for F_{FG}, one would take moments about point D.
- If one wanted a simple equation in F_{DG}, one would attempt to eliminate F_{CD} and F_{FG}. This, however, cannot be done with a moment equation, as these two forces are parallel (horizontal). However, note that the $\Sigma F_y = 0$ equation would eliminate these two forces and yield an equation in only one variable F_{DG}.

Identification of Zero-Force Members

For an ideal truss (nodes are pinned, loads act only at nodes), each node is a system of concurrent forces. Some of these forces are collinear; that is, they share the same line of action. If there are nodes where all forces can be grouped into *only two lines of action*, we may say the following:

1. If one line of action consists of a single force, that force must be zero.
2. If one line of action consists of a pair of forces, they must be equal and opposite.

NOTE These rules apply if and only if the forces can be grouped into only two lines of action.

Example 6.13

Identify the zero-force members for the plane truss shown below.

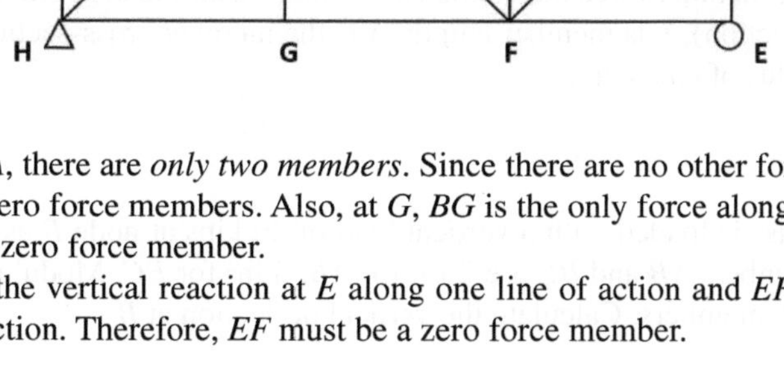

Solution At A, there are *only two members*. Since there are no other forces at A, both AB and AH must be zero force members. Also, at G, BG is the only force along its line of action. So, BG must be a zero force member.

At E, DE and the vertical reaction at E along one line of action and EF is the only force along its line of action. Therefore, EF must be a zero force member.

Once these zero force members are removed, we are left with the structure shown below.

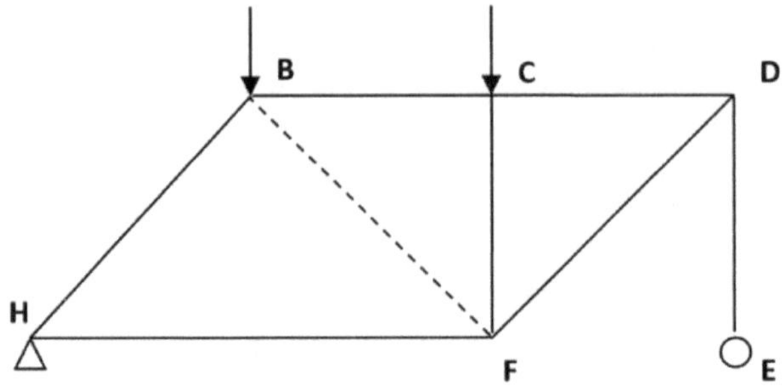

By external symmetry of structure and loading, both reactions (at H and E) must be 24k vertical. Therefore, the panel shear in the middle panel is zero. Only member in the middle panel that can contribute to panel shear is BF. Therefore, BF is a zero force member.

Truss Deflection: Method of Virtual Work

The principle of virtual work states that the external virtual work of virtual forces moving through real displacements is equal to the internal virtual work done by the stresses arising from the virtual forces moving through the strains corresponding to the real displacements. The words "real" and "virtual" in this statement may be interchanged.

The virtual load is an artificial load which is chosen to be consistent with the desired displacement. For example, if the vertical displacement is desired at point P on the structure, then the virtual load is a unit load applied vertically at point P.

For a truss, this yields the following expression for the deflection at any node (joint) of an ideal truss (axial loads only)

$$1 \cdot \Delta = \sum \mathbf{F}\left(\frac{\mathbf{f}L}{AE}\right) \Rightarrow \Delta = \sum\left(\frac{F\,\mathbf{f}L}{AE}\right) \tag{6.15}$$

where the summation occurs over all members in the truss, **F** is the member force due to the *actual load*, **f** is the member force due to the *virtual load* (a unit load applied in the same sense as the desired deflection), L is member length, A is the member cross-sectional area, and E is the member modulus of elasticity.

Example 6.14

A two-member truss is loaded with a vertical load of 20 kips at node B as shown. The cross sections of the members AB and BC are 2 in^2 for AB, 3 in^2 for BC. Modulus of elasticity E = 29,000 ksi for both members. Calculate the vertical deflection at B.

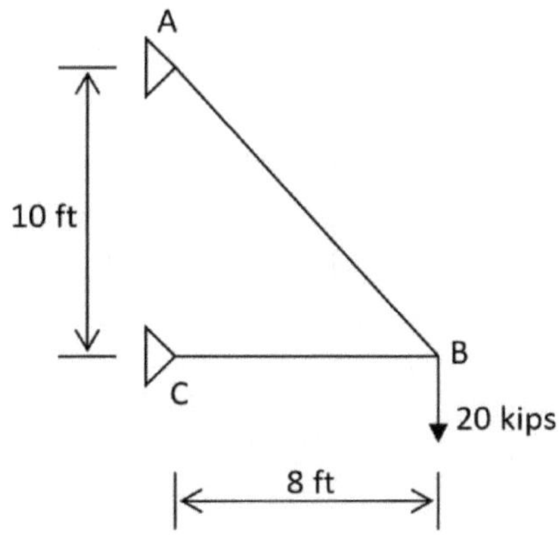

Solution Solve for truss member forces using equilibrium of joint *B*

$$F_{AB} = +25.6\text{k}\ (T)$$

$$F_{BC} = -16.0\text{k}\ (C)$$

Create the virtual load. Since the virtual load is consistent with the desired deflection, it will be a unit load applied vertically at node *B*.

Calculate the member forces due to virtual load. In this case, the real load and the virtual load look similar, so we can use scaling (factor of 20k)

$$f_{AB} = 25.6\text{k} \div 20\text{k} = +1.28\ (T)$$

$$f_{BC} = -16.0\text{k} \div 20\text{k} = -0.80\ (C)$$

$$\Delta = \sum\left(\frac{FfL}{AE}\right) = \frac{25.6 \times 1.28 \times (12.81 \times 12)}{2 \times 29{,}000} + \frac{-16 \times -0.8 \times (8 \times 12)}{3 \times 29{,}000} = 0.10\ \text{in}$$

NOTE The positive sign of the answer indicates that this deflection is in the same sense as the assumed unit load (downward).

Influence Line for Truss Member Force

The influence line for a particular effect (internal force, deflection, etc.) is a graphical representation of how that effect varies due to a traveling load. For a truss such as the one shown below, where *AE* is the truss deck, it is of interest as to how a particular effect (such as axial force in a particular member) varies as a unit load traverses the deck from *A* to *E*.

Note that whenever the unit load is at a support, the load is directly carried by the support and the structure (and all its members) essentially does nothing. Therefore, the ordinate of the influence diagram at the supports is zero.

Example 6.15

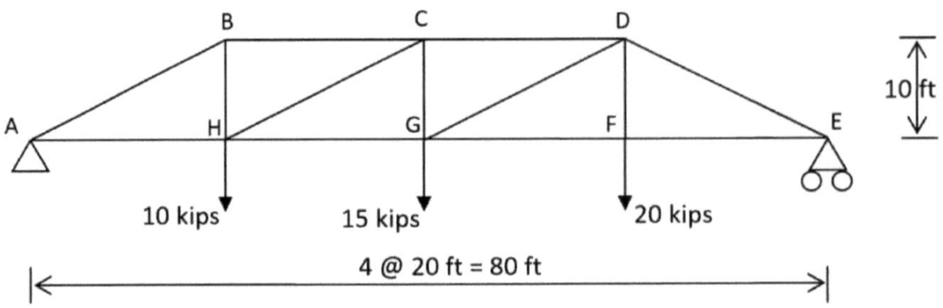

What is the influence line diagram for the web member *CH* in the truss shown above?

Solution Since the truss is statically determinate, the influence diagram will be composed of straight line segments. When a unit load traverses the truss deck from *A* to *E*, one must therefore consider the two load positions, at either end of the member *H* and *G*, which is directly below *C*. When the load is at the supports *A* and *E*, the member force F_{CH} is zero.

Unit load at *H*

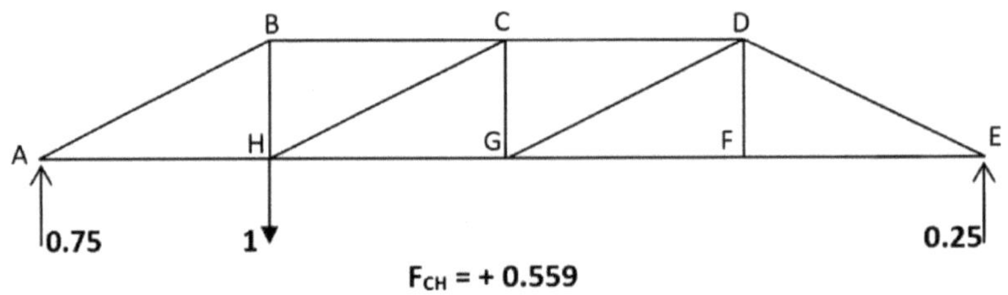

Using the section shown below:

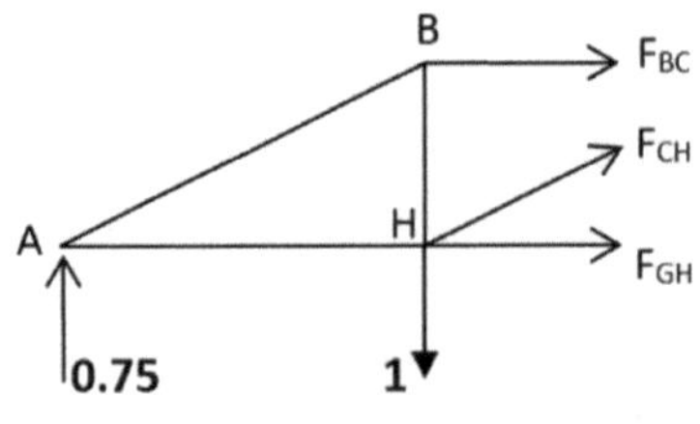

Summing vertical forces, $0.75 - 1.0 + F_{CH}\dfrac{1}{\sqrt{5}} = 0 \Rightarrow F_{CH} = 0.25\sqrt{5} = +0.559$

Unit load at *G*

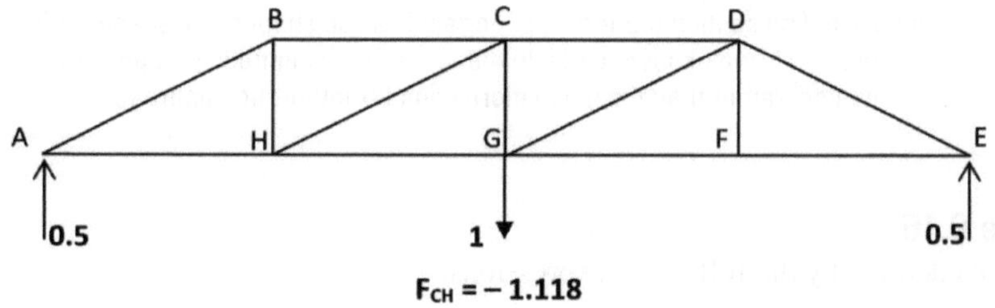

Using the section shown below:

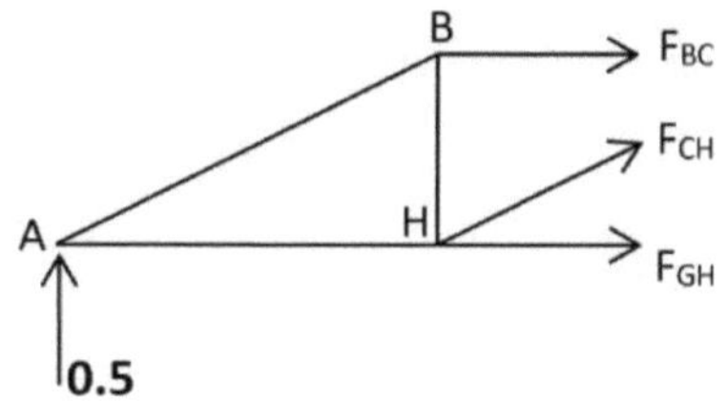

Summing vertical forces, $0.5 + F_{CH}\dfrac{1}{\sqrt{5}} = 0 \Rightarrow F_{CH} = -0.5\sqrt{5} = -1.118$

The influence line for F_{CH} is drawn by joining the following ordinates: at *A*, 0.0; at *H*, +0.559; at *G*, −1.118; at *E*, 0.0.

Influence line for F_{CH}

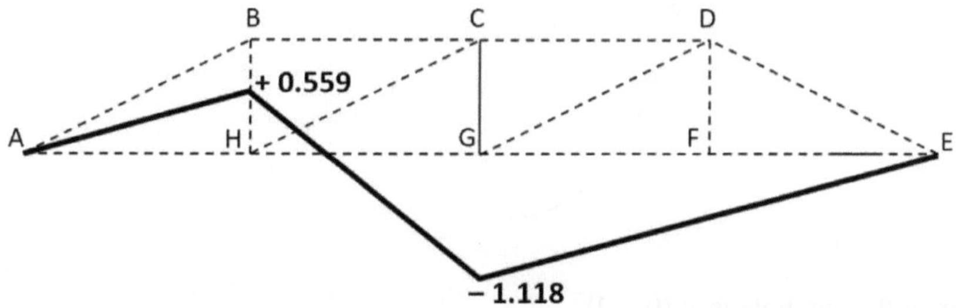

Centroid of an Area by Integration

The center of gravity of any object is calculated as the weighted average of the centers of gravity of subobjects that constitute the object. When the area is subdivided into infinitesimal elements, this weighted average is calculated by integration. For an area, the location of the centroid (x_c, y_c) is given by

$$x_c = \frac{\int x\,dA}{\int dA} \qquad y_c = \frac{\int y\,dA}{\int dA} \tag{6.16}$$

where *x* and *y* represent the coordinates of the center of the element *dA* being integrated.

NOTE Practical applications where the centroid location needs to be calculated, for example, to find section properties of certain built-up shapes, the shape is often composed of geometrically well-known parts. In that case, the weighted average of these parts can be computed, rather than the much more complex integration approach.

Example 6.16

Find the area defined by the following expressions:

(i) $3x^2 - 2 < y < 5$

(ii) $x > 0$

A. 7.1

B. 5.6

C. 4.8

D. 3.6

Solution The region is shown below:

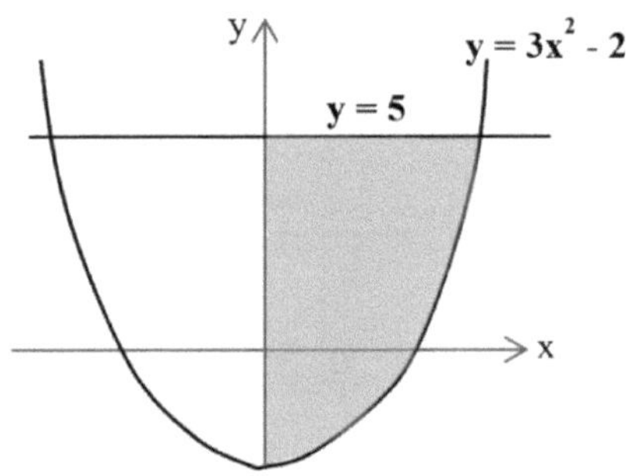

The vertex of the parabola is at (0, −2).
Point of intersection of the parabola and the line $y = 5$ is at (1.528, 5).

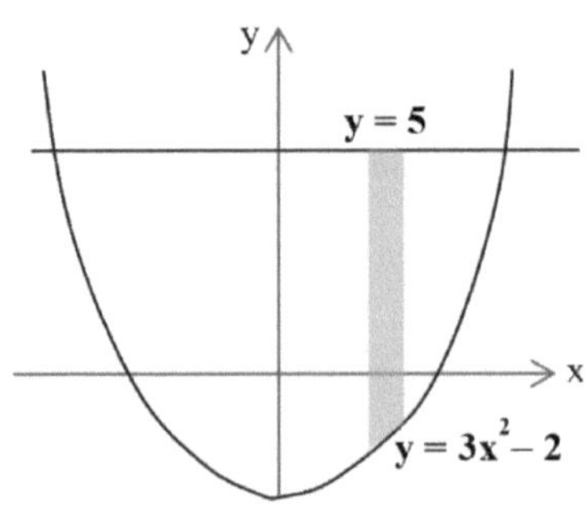

Height of vertical strips $y = 5 - (3x^2 - 2) = 7 - 3x^2$

$$A = \int dA = \int_0^{1.528} y\,dx = \int_0^{1.528} (7 - 3x^2)\,dx = (7x - x^3)_0^{1.528} = 7.128$$

Answer is A.

Centroid of a Compound Area—Weighted Average

When one is dealing with a compound shape (which is a combination of a finite number of known shapes), the coordinates of the centroid are given by the weighted average of the coordinates of individual parts

$$\bar{x} = \frac{x_1A_1 + x_2A_2 + \cdots + x_nA_n}{A_1 + A_2 + \cdots + A_n} = \frac{\sum_{i=1}^{n} x_iA_i}{\sum_{i=1}^{n} A_i} \quad \text{and} \quad \bar{y} = \frac{\sum_{i=1}^{n} y_iA_i}{\sum_{i=1}^{n} A_i} \tag{6.17}$$

Example 6.17

Consider the composite shape shown below. All dimensions in mm. The composite object is broken up into four component parts, each being a well-known shape (rectangle, triangle, and circle) with known expressions for area and centroid location. Note that the circle represents a hole in the object and is therefore treated as a negative area,

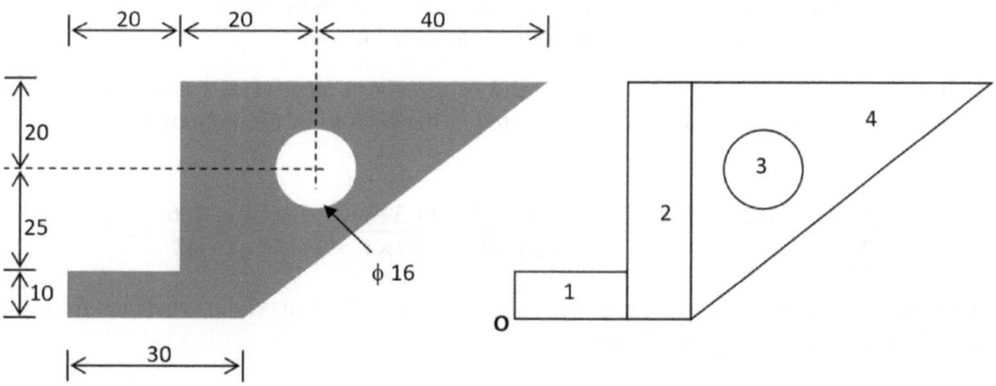

Shape	Area	x_i	y_i	x_iA	y_iA
1	(20)(10) = 200	10	5	2000	1000
2	(10)(55) = 550	25	27.5	13,750	15,125
3	$\pi(8)^2 = -201.1$	40	35	−8042.4	−7037.1
4	½(50)(55) = 1375	30+1/3(50) = 46.67	2/3(55) = 6.67	64,171.3	50,421.3
	ΣA = 1923.9			**ΣxA = 71,879**	**ΣyA = 59,509**

Thus, the centroid location for the composite area is

$$\bar{x} = \frac{\sum x_i A_i}{\sum A_i} = \frac{71{,}878.9}{1923.9} = 37.36$$

$$\bar{y} = \frac{\sum y_i A_i}{\sum A_i} = \frac{59{,}509.2}{1923.9} = 30.93$$

NOTE It is vitally important to select an origin and use it consistently throughout as the point of reference for all distances.

The in-plane axes passing through the geometric centroid are also called the elastic neutral axes (ENA) of the shape. Under the influence of bending moments, if the stresses remain elastic and linear, the bending strain and stress are zero at the neutral axis.

Various Section Properties

The following section properties are useful for calculating bending and shear stress—elastic centroid, moment of inertia, elastic, and plastic section modulus. For more details, see Chap. 8.

Example 6.18

Consider the singly symmetric I-shaped section shown in Fig. 6.6. Calculate (a) the coordinate of the centroid, (b) the moment of inertia, (c) the elastic section modulus, (d) the plastic neutral axis, (e) the plastic section modulus, and (f) the shape factor.

Solution The y coordinate of the centroid or the elastic neutral axis is calculated with respect to (an arbitrarily chosen) datum located at the bottom of the bottom flange. This is designated as $y = 0$

$$\bar{y} = \frac{\sum A_i y_i}{\sum A_i} = \frac{(12\times 1.5)\times 0.75 + (20\times 0.75)\times 11.5 + (10\times 1)\times 22}{12\times 1.5 + 20\times 0.75 + 10\times 1} = \frac{406}{43} = 9.44$$

Since the total depth of the section is 22.5 in, the top fiber is at maximum distance from the neutral axis, at distance $y_{max} = 13.06$ in.

Figure 6.6 I-Section used in Example 6.18.

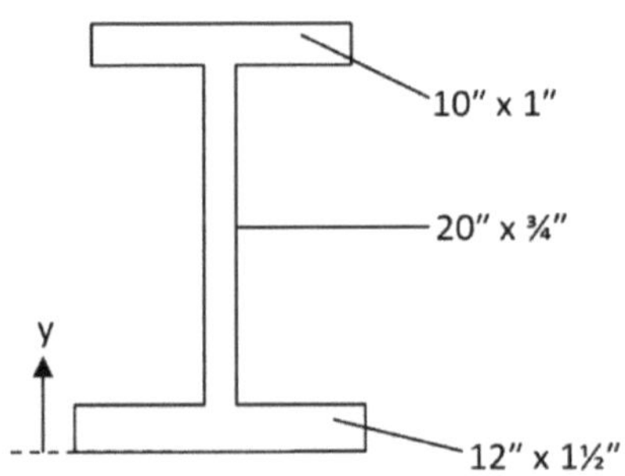

Moment of Inertia (*I*)

The moment of inertia with respect to any axis is calculated using the *parallel axis theorem* or the *moment of inertia transfer theorem*. This theorem states

$$I_x = I_{xc} + Ad^2$$

where I_{xc} is the moment of inertia with respect to the centroidal axis, A is the area of the cross section and d is the parallel shift between the centroidal axis and the axis with respect to which the moment of inertia is desired. Thus, for the section shown in Fig. 6.6, the moment of inertia about the neutral axis is calculated as the sum of the moments of inertia of the three rectangles: top flange, web, and bottom flange. For each of these rectangles, the local centroidal moment of inertia is given by $bh^3/12$.

The centroidal moments of inertia of shapes with straight sides (rectangles, triangles, trapezoids, and parallelograms) are given on page 69 of the *FE Reference Handbook*, version 9.4. Shapes with circular boundaries (circles, semicircles, circular sector, circular segment) are listed on page 70.

$$I_{NA} = \frac{1}{12} \times 12 \times 1.5^3 + 18 \times (0.75 - 9.44)^2$$

$$+ \frac{1}{12} \times \frac{3}{4} \times 20^3 + 15 \times (11.5 - 9.44)^2$$

$$+ \frac{1}{12} \times 10 \times 1^3 + 10 \times (22 - 9.44)^2 = 3504.7 \text{ in}^4$$

Note that for any shape, the moment of inertia about its centroidal axis has the minimum value. In other words, any other axis parallel to the centroidal axis would produce a greater magnitude of moment of inertia.

Polar Moment of Inertia (*J*)

At the centroid of an area, three orthogonal axes can be drawn. See Fig. 6.7. Two of these axes are in the plane of the area. These are traditionally labeled the x and y axes. The moments of the area about these axes are traditionally referred to as the moments of inertia of the area and are given by Eqs. (6.18) and (6.19)

$$I_x = \int y^2 dA \tag{6.18}$$

$$I_y = \int x^2 dA \tag{6.19}$$

The third axis is perpendicular (normal) to the cross section and is typically labeled as the z axis (also referred to as the polar axis). The moment of the area about the polar axis is referred to as the polar moment of inertia of the area (I_z or J) and is given by Eq. (6.20).

$$I_z = \int r^2 dA = \int (x^2 + y^2) dA = I_y + I_x \tag{6.20}$$

Figure 6.7
Centroidal axes.

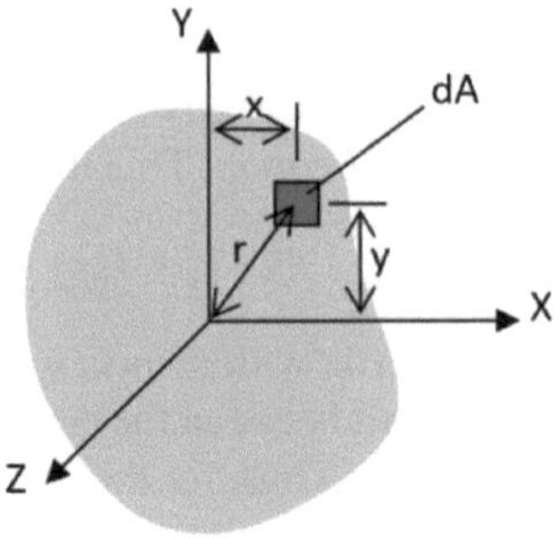

Product of Inertia

The product of inertia of an area located in the *x-y* plane (as shown in Fig. 6.8) is given by

$$I_{xy} = \int xydA \tag{6.21}$$

The parallel axis theorem also applies to the product of inertia, as follows:

$$I_{xy} = I_{x_c y_c} + Ad_x d_y \tag{6.22}$$

where d_x and d_y are the parallel shift between axes *x* and *y* and their corresponding centroidal axes.

For a shape with even one axis of symmetry, the centroidal product of inertia ($I_{x_c y_c}$) is zero.

Pages 69 to 71 of the *Handbook* list moments and products of inertia of various shapes. Page 69 lists shapes with straight edges, such as rectangles, triangles, and trapezoids. Page 70 lists shapes defined by a circular arc and page 71 lists shapes defined by a parabola.

Elastic Section Modulus (*S*)

The elastic section modulus of a section is defined as the ratio I/y_{max}, where y_{max} represents the distance from the elastic neutral axis to the fiber which is furthest from it. Thus, for the I-section shown in Fig. 6.6, the section modulus is given as

$$S_x = \frac{I_{xc}}{c} = \frac{3504.7}{13.06} = 268.4 \text{ in}^3$$

Figure 6.8
Parallel axis theorem applied to product of inertia.

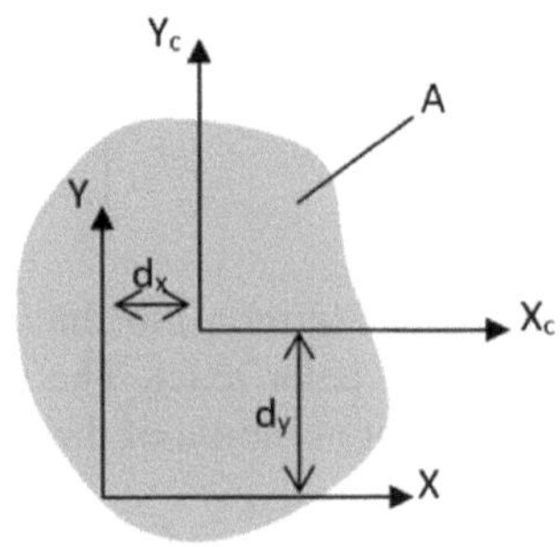

Plastic Neutral Axis

The plastic neutral axis (PNA) is the line that divides the section into two equal halves. For the section shown, the total area is

$$A = 18 + 15 + 10 = 43 \text{ in}^2$$

Therefore, half the area is 21.5 in^2.

Working upward from the bottom flange, since the area of the bottom flange is 18 in^2, the needed area from the web is $21.5 - 18.0 = 3.5$ in^2. Therefore, the depth of the web needed is $3.5/0.75 = 4.67$ in.

Measured from the bottom, the coordinate of the PNA is $y_{\text{PNA}} = 1.5 + 4.67 = 6.17$ in.

Plastic Section Modulus

The plastic section modulus (Z) is calculated as the sum of the first moments of the two "halves" of the section. Figure 6.9 shows the parts of the I-section with respect to the PNA ($y = 6.17$). It is convenient to split each T-shaped "half" into its two component rectangles.

Treating the section as a combination of four rectangles (see Fig. 6.9), the plastic section modulus is given by the first moment of these areas about the PNA

$$Z_x = (10 \times 1) \times 15.83 + \left(\frac{3}{4} \times 15.33\right) \times \frac{15.33}{2} + \left(\frac{3}{4} \times 4.67\right) \times \frac{4.67}{2} + (12 \times 1.5) \times (4.67 + 0.75)$$

$$= 352.2 \text{ in}^3$$

Shape Factor

The shape factor for a section is given as the ratio of the plastic section modulus and the elastic section modulus (Z/S). Thus, for the section shown, the shape factor is

$$\frac{Z}{S} = \frac{352.2}{268.4} = 1.31$$

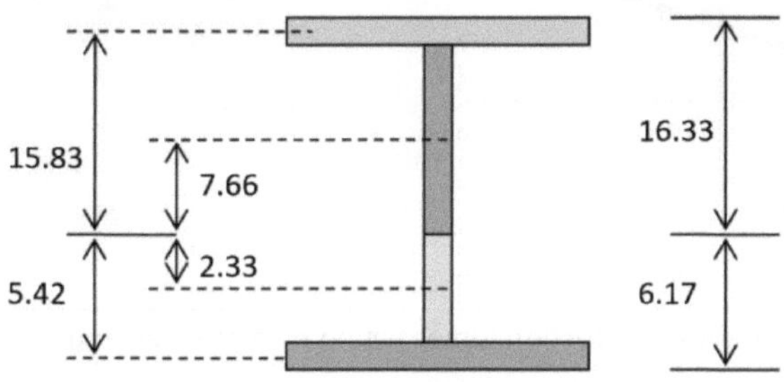

Figure 6.9 Computation of plastic section modulus.

The shape factor can give us some idea of the relative ductility of the member, since it is also the ratio of the plastic (or ultimate) moment capacity of the section to the moment that causes first yield.

Friction

Two rough surfaces in contact and subject to opposing forces experience frictional resistance forces at the interface (plane of contact). The magnitude of this force is proportional to the contact force N (acting normal to the interface). This proportionality is expressed as

$$F_f = \mu N \tag{6.23}$$

If the system is in static equilibrium, the coefficient used is called the coefficient of static friction μ_s whereas if the concerned objects are moving relative to each other (while remaining in contact), the coefficient is called the coefficient of kinematic friction μ_k. Typically, $\mu_s > \mu_k$, which means that it takes a greater force to initiate motion than it does to maintain it.

Frictional resistance will always act opposite to the direction of motion (or impending motion), even when the system is static. Figure 6.10 shows a block resting on a plane inclined at angle θ to the horizontal. With no other forces acting on the block, its downslope weight component will try to make it slide downhill. Thus, the impending motion is downslope. For this case, the friction force μN will tend to act upslope. The free body diagram of the block is shown on the right.

The system will stay static as long as the slope angle θ is less than $\tan^{-1}(\mu)$. This is why the friction between two surfaces can be also expressed in terms of a friction angle δ, where

$$\mu = \tan \delta \tag{6.24}$$

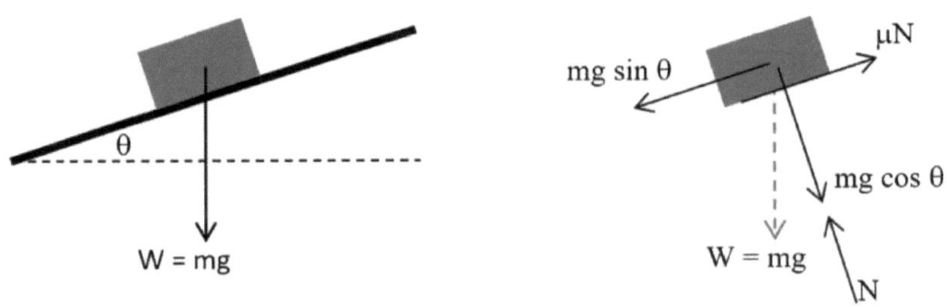

Figure 6.10 Gravity and friction forces acting on a block sliding on an inclined plane.

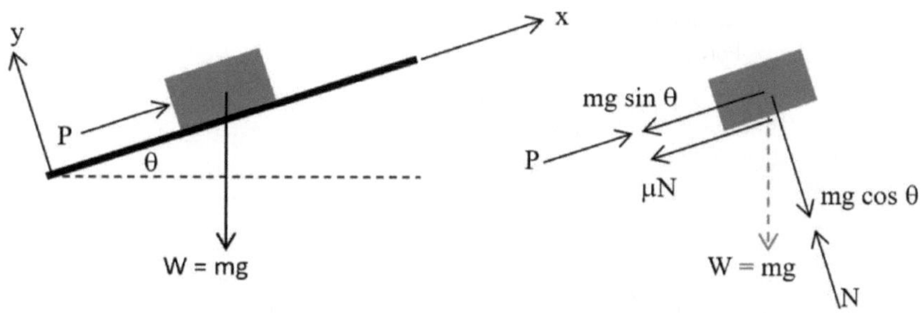

Figure 6.11 Gravity, friction, and external forces acting on a block sliding up inclined plane.

On the other hand, in Fig. 6.11, we see the same block, but with an additional force *P* acting on it. This force is trying to push the block upslope against gravity. The friction force will therefore act downslope in this case.

Once it has been determined what the impending motion is and the free body diagram is complete, the equations of motion can be written as follows. For the system shown in Fig. 6.11 (block subject to the upslope force *P*) and adopting the coordinate system as indicated below in Fig. 6.12, we have the following equations of static equilibrium

$$\sum F_y = N - mg\cos\theta = 0 \Rightarrow N = mg\cos\theta$$

$$\sum F_x = P - \mu N - mg\sin\theta = 0 \Rightarrow P = \mu N + mg\sin\theta = mg(\sin\theta + \mu\cos\theta)$$

This value of the force *P* is the *maximum* force for which the system remains static. When the force *P* exceeds this threshold, the block will accelerate upslope, according to

$$\sum F_x = P - \mu N - mg\sin\theta = ma$$

Figure 6.12 Gravity, friction, and external forces acting on a block sliding up inclined plane.

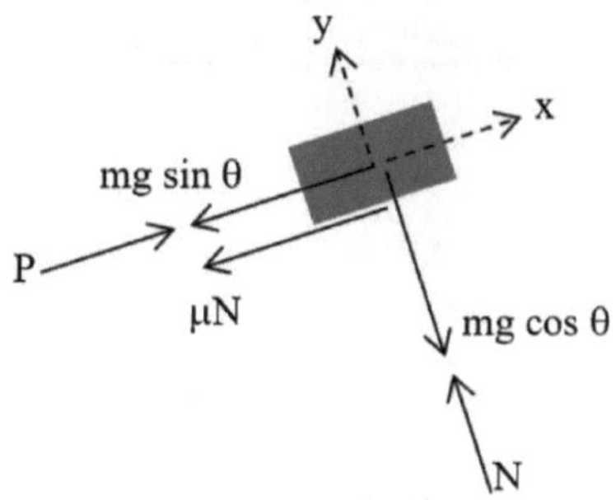

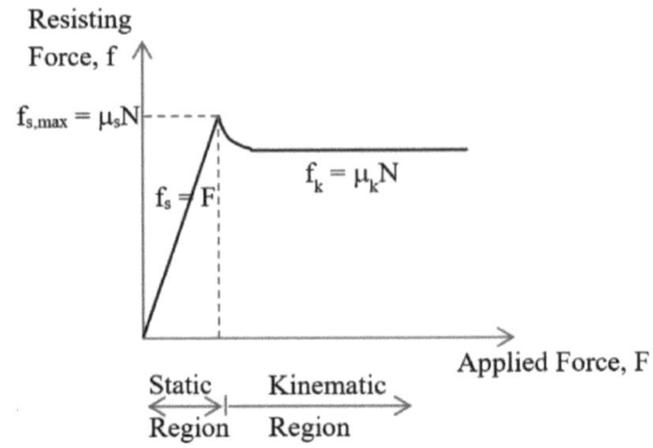

Figure 6.13 Resistive friction force as a function of externally applied force.

As the applied force on a body increases, the frictional resistance grows to resist it and the body is in static equilibrium. During this phase, the coefficient of friction is the coefficient of static friction. When the resisting force is equal to the applied force, the net unbalanced force is zero and the body has no acceleration. When the applied force is greater than the resisting force, it puts the system in motion and the coefficient of friction then drops down to the lower value m_k, the coefficient of kinematic friction. This is shown in Fig. 6.13.

Example 6.19

A 5-kg block on a 35° incline is acted upon by a horizontal force F as shown. What is the minimum force required to keep the block in static equilibrium?

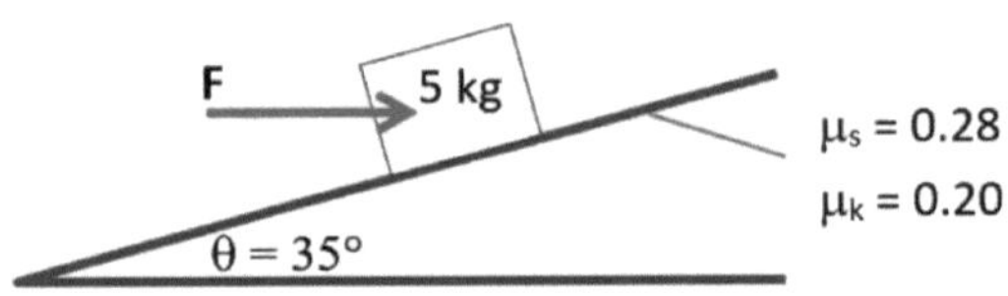

A. 12.2 N

B. 17.2 N

C. 46.2 N

D. 81.2 N

Solution The free body diagram of the block, resolving forces along, and perpendicular to the incline, is shown. The minimum force will prevent it from sliding *downslope*.

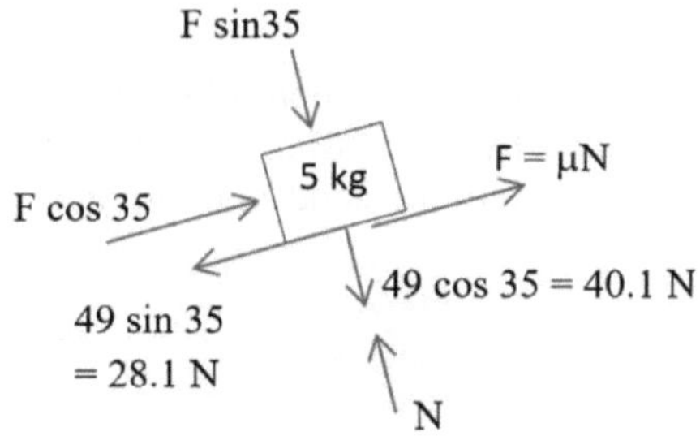

$$F_y = N - F\sin 35 - 40.1 = 0 \Rightarrow N = F\sin 35 + 40.1$$

$$F_x = F\cos 35 - 28.1 + \mu N = 0 \Rightarrow F\cos 35 - 28.1 + 0.28(F\sin 35 + 40.1) = 0 \Rightarrow F = 17.22\text{ N}$$

Answer is B.

NOTE If the question had asked for the maximum force *F* (for static equilibrium), we would solve for the condition just before the block starts sliding *upslope*. This decision will affect the choice of the friction force, which always opposes incipient motion.

Belt Friction

Pulleys are often analyzed as "frictionless" implying that the coefficient of friction between the pulley wheel and the rope (or belt) is negligible. The direct result of this assumption is that the tension in a continuous length of the rope is constant. However, in those cases where this friction is significant, the belt tension on one side of the pulley is greater than that on the other side. The difference in the tension depends on the angle of contact θ (radians) as follows:

$$\frac{T_1}{T_2} = e^{\mu\theta} \tag{6.25}$$

where T_1 = (greater) tension on the "driving" (or effort) side of the pulley. See Fig. 6.14. This is the direction of impending motion

T_2 = (lesser) tension on the "driven" (or load) side of the pulley

μ = coefficient of static friction between the belt and the pulley

θ = total angle of contact between belt and pulley (radians)

Figure 6.14 Belt friction.

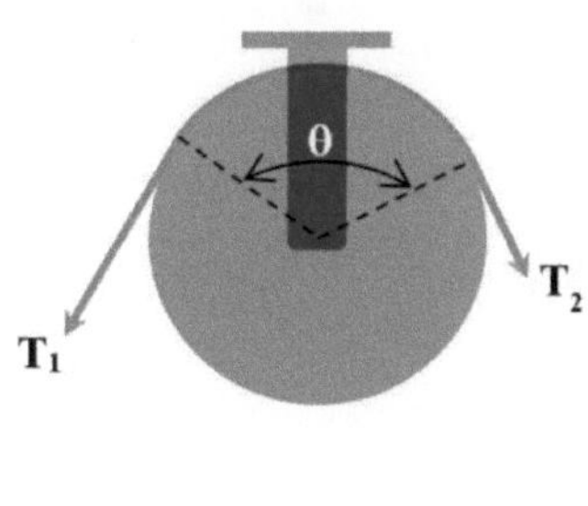

Example 6.20

A fixed and a movable pulley are used to support a load 120 kg as shown. If the pulleys are assumed to be massless and the coefficient of friction between the rope and the pulleys is 0.10, what is the *minimum* force *F* necessary for static equilibrium?

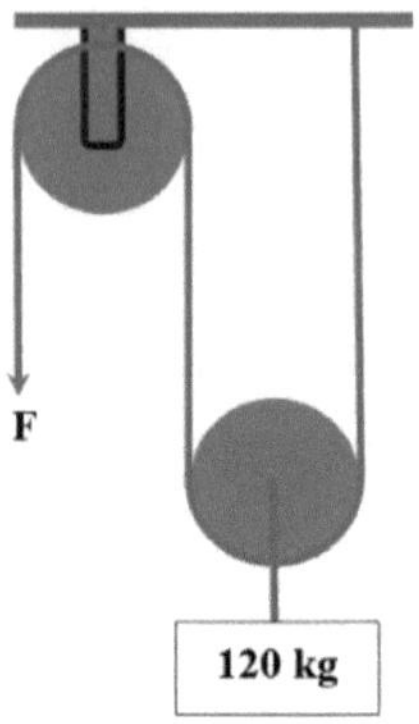

A. 360 N

B. 430 N

C. 590 N

D. 806 N

Solution The friction between each pulley and the rope (angle of contact = 180° = 3.142 rad). Since the *minimum* force *F* is asked for, we must solve for the case $F_3 > F_2 > F_1$

$$\frac{F_3}{F_2} = \frac{F_2}{F_1} = e^{\mu\theta} = e^{0.1\times3.142} = 1.37$$

From the free body diagram of the movable pulley,

$$F_3 + F_2 = 2.37F_2 = 1177.2 \rightarrow F_2 = 496.7 \text{ N}$$

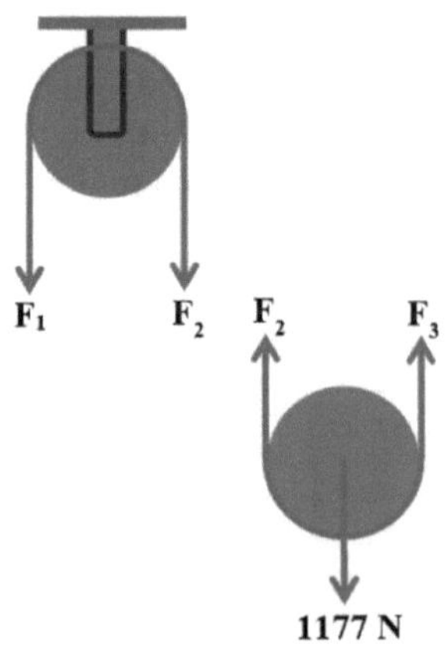

From the free body diagram of the fixed pulley,

$$F_2 = 1.37F_1 \rightarrow F_1 = 362.6 \text{ N}$$

Answer is A.

NOTE If friction had been neglected, the force F would have been $1178 \div 2 = 589$ N.

Screw Thread

A screw thread is shown in Fig. 6.15. The pitch of the thread is the distance advanced by the screw with a single complete turn of the thread. The pitch angle a is calculated from

$$\tan\alpha = \frac{\text{pitch}}{\text{pitch diameter}} \tag{6.26}$$

The coefficient of friction between the screw thread and the surrounding material is given by

$$\mu = \tan\phi \tag{6.27}$$

The external moment applied to the axis of the screw is given by

$$M = Pr\tan(\alpha \pm \phi) \tag{6.28}$$

where P = axial load applied
M = external moment applied to axis of the screw. The plus sign is for screw tightening and the minus sign is for loosening of the screw
$\phi = \tan^{-1}\mu$
μ = coefficient of friction
r = pitch radius

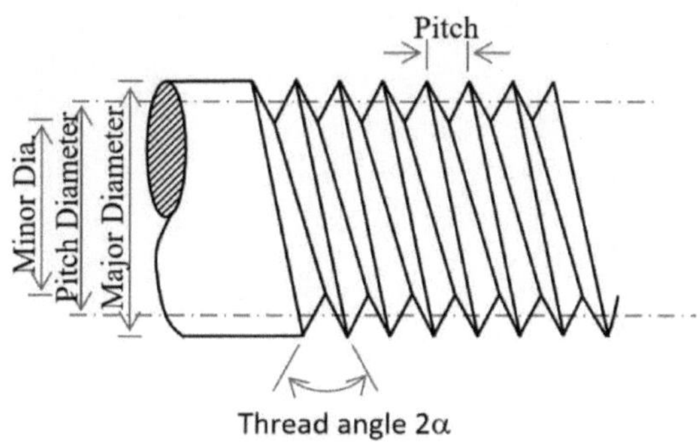

Figure 6.15 Screw thread.

Example 6.21

A square thread screw-jack with mean thread diameter = 34 mm and pitch = 8 mm is used to lift a load $F = 40$ kN. If the coefficient of friction is 0.16, what is the torque necessary to lift the load?

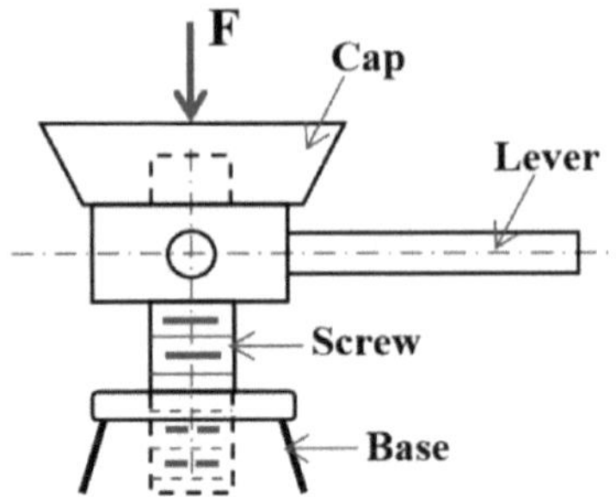

A. 120 Nm

B. 160 Nm

C. 200 Nm

D. 480 Nm

Solution The axial load $F = 40{,}000$ N.
With a pitch of 8 mm and a diameter of 34 mm, it means that the thread advances 8 mm over a circumference of $\pi D = 34\pi$

Pitch angle: $\tan\alpha = \dfrac{p}{\pi D} = \dfrac{8}{34\pi} = 0.075 \Rightarrow \alpha = 4.28°$

Coefficient of friction $\mu = 0.16$

Friction angle: $\phi = \tan^{-1} 0.16 = 9.09°$

To lift the load, the positive angle should be used.

Torque: $M = Pr\tan(\alpha + \phi) = 40{,}000 \times \dfrac{0.034}{2} \times \tan(4.28 + 9.09) = 161.6$ Nm

Answer is B.

CHAPTER 7

Dynamics

Kinematics

In dynamics, kinematics is the study of motion without involving any discussion of the forces that cause the motion. In other words, kinematics involves the description of time-dependent position, velocity, and acceleration of the system.

There are two broad types of systems that are discussed in this chapter—a particle and a rigid body. For a particle, tracking simply translation of a point is adequate. This translation may occur in one-dimensional (motion along a straight line), two-dimensional (motion in a plane), or three-dimensional space. On the other hand, for a rigid body, with a finite size and defined shape, both translation and rotation must be tracked. Once again, this may occur in 2D or 3D space.

Particle Kinematics

Cartesian Coordinates

The position of a particle in 3D space may be described using its position vector

$$\mathbf{r}(t) = x(t)\mathbf{i} + y(t)\mathbf{j} + z(t)\mathbf{k} \tag{7.1}$$

The velocity vector is then given by

$$\mathbf{v}(t) = \dot{x}(t)\mathbf{i} + \dot{y}(t)\mathbf{j} + \dot{z}(t)\mathbf{k} \tag{7.2}$$

and the acceleration vector is

$$\mathbf{a}(t) = \ddot{\mathbf{x}}(t)\mathbf{i} + \ddot{\mathbf{y}}(t)\mathbf{j} + \ddot{\mathbf{z}}(t)\mathbf{k} \tag{7.3}$$

Thus, since the unit vectors **i**, **j**, and **k** are fixed, taking successive time derivatives of the position vector function $\mathbf{r}(t)$ can produce the velocity and acceleration vectors.

Example 7.1

The position of a particle is described by the position vector $(3t^3 - 2t^2)\mathbf{i} + (4t^2 + 2t)\mathbf{j} + (6t^3 - 2t^2 + 3t)\mathbf{k}$. What are the velocity and acceleration at $t = 5$ s?

Solution

Position vector: $\mathbf{r}(t) = (3t^3 - 2t^2)\mathbf{i} + (4t^2 + 2t)\mathbf{j} + (6t^3 - 2t^2 + 3t)\mathbf{k}$

Taking time derivatives, we have

Velocity vector: $\mathbf{v}(t) = (9t^2 - 4t)\mathbf{i} + (8t + 2)\mathbf{j} + (18t^2 - 4t + 3)\mathbf{k}$

Acceleration vector: $\mathbf{a}(t) = (18t - 4)\mathbf{i} + (8)\mathbf{j} + (36t - 4)\mathbf{k}$

At $t = 5$ s, $\mathbf{v}(t) = 205\mathbf{i} + 42\mathbf{j} + 433\mathbf{k}$

Magnitude of velocity vector: $|v| = \sqrt{205^2 + 42^2 + 433^2} = 480.9$

At $t = 5$ s, $\mathbf{a}(t) = 86\mathbf{i} + 8\mathbf{j} + 176\mathbf{k}$

Magnitude of acceleration vector: $|a| = \sqrt{86^2 + 8^2 + 176^2} = 196.1$

Radial and Transverse Components for Plane Motion

Consider a particle traversing a path (in the x-y) plane as shown in Fig. 7.1. Using the radial and transverse $(r - \theta)$ coordinates, the following can be written (using the time dependent functions $r(t)$ and $\theta(t)$

$$\mathbf{r} = r\mathbf{e_r} \tag{7.4}$$

$$\mathbf{v}(t) = \dot{r}\mathbf{e_r} + r\dot{\theta}\mathbf{e}_\theta \tag{7.5}$$

$$\mathbf{a}(t) = [\ddot{r} - r\dot{\theta}^2]\mathbf{e_r} + [r\ddot{\theta} + 2\dot{r}\dot{\theta}]\mathbf{e}_\theta \tag{7.6}$$

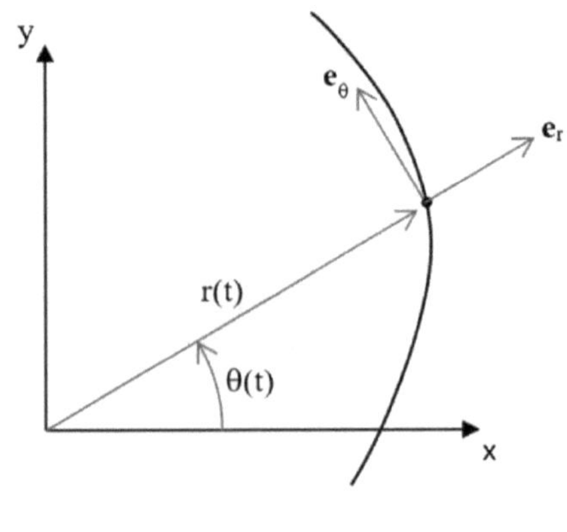

Figure 7.1 Radial and transverse components for plane motion.

Example 7.2

An airplane's path is tracked by radar using the distance $r(t)$ and angular elevation $\theta(t)$ functions given below (r measured in km, θ measured in radians, t in seconds). At $t = 5$ s, what is the speed of the aircraft?

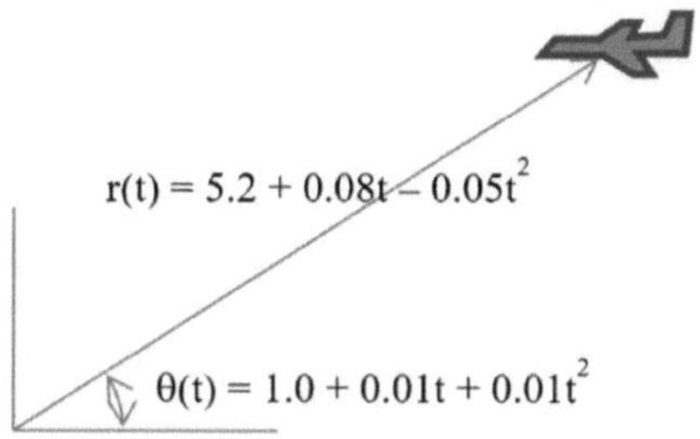

A. 1100 km/h

B. 1300 km/h

C. 1800 km/h

D. 2300 km/h

Solution Using $r - \theta$ coordinates, the velocity vector is

$$\mathbf{v}(t) = \dot{r}\mathbf{e_r} + r\dot{\theta}\mathbf{e_p} = (0.08 - 0.1t)\mathbf{e_r} + (5.2 + 0.08t - 0.05t^2)(0.01 + 0.02t)\mathbf{e_p}$$

At $t = 5$, this becomes

$$\mathbf{v}(t) = -0.42\mathbf{e_r} + 0.48\mathbf{e_p}$$

Magnitude of the velocity vector: $\mathbf{v} = \sqrt{0.42^2 + 0.48^2} = 0.638\dfrac{\text{km}}{\text{s}} = 2296\dfrac{\text{km}}{\text{h}}$

Answer is D.

Normal and Tangential Components

Since the velocity component is always directed along the tangential direction, the tangential and normal axes also make a natural choice for describing the motion of particle along a curvilinear path. The tangential unit vector $\mathbf{e_t}$ is directed in the direction of travel and the normal unit vector ($\mathbf{e_n}$) is directed along the inward normal (Fig. 7.2).

For a particle with tangential velocity and acceleration given by v_t and a_t, respectively, the velocity and acceleration vectors can be written

$$\mathbf{v}(t) = \frac{ds}{dt}\mathbf{e_t} = v_t\mathbf{e_t} \tag{7.7}$$

$$\mathbf{a}(t) = a_t\mathbf{e_t} + \frac{v_t^2}{\rho}\mathbf{e_n} \tag{7.8}$$

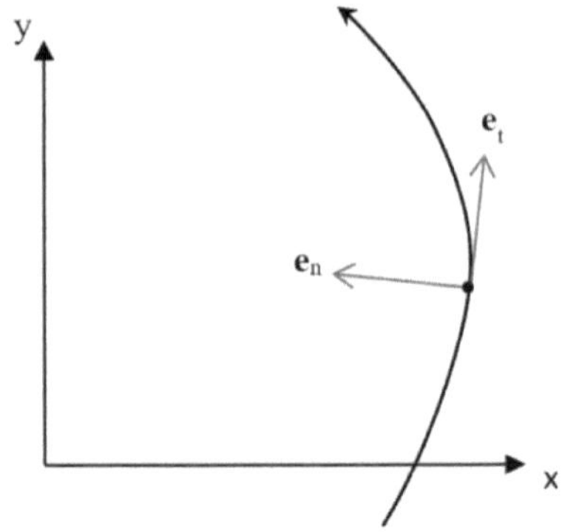

Figure 7.2 Normal and tangential components for plane motion.

Example 7.3

A car is traveling at the crest of a hill as shown below. At the crest, the hill has a radius of curvature = 240 m. If the tangential velocity = 120 km/h and acceleration = 5 km/h/s, what is the total acceleration?

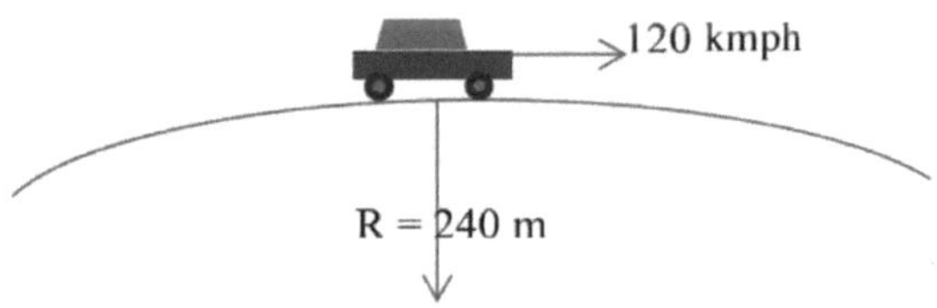

A. 1.39 m/s²

B. 4.63 m/s²

C. 4.83 m/s²

D. 5.98 m/s²

Solution

$v_t = 120$ km/h $= 33.3$ m/s
$a_t = 5$ km/h/s $= 1.39$ m/s²

Normal acceleration: $a_n = \dfrac{v_t^2}{\rho} = \dfrac{33.3^2}{240} = 4.63\ \dfrac{\text{m}}{\text{s}^2}$

The two components of acceleration are orthogonal. Therefore the magnitude of total acceleration is given by:

$$a = \sqrt{1.39^2 + 4.63^2} = 4.83\ \frac{\text{m}}{\text{s}^2}$$

Answer is C.

Circular Motion

The specific characteristic of circular motion is that the radial distance $r(t)$ from a fixed point is constant. This results in the time derivatives of $r(t)$ to be zero. The expressions for velocity and acceleration in the $r - \theta$ coordinate system [Eqs. (7.5) and (7.6)] become

$$\mathbf{v}(t) = r\dot{\theta}\mathbf{e}_\theta \tag{7.9}$$

$$\mathbf{a}(t) = [-r\dot{\theta}^2]\mathbf{e}_r + [r\ddot{\theta}]\mathbf{e}_\theta \tag{7.10}$$

The derivative $d\theta/dt$ is called the angular velocity, ω and the second derivative $d^2\theta/dt^2$ is called the angular acceleration, α. The expressions for velocity and acceleration then be written

$$\mathbf{v}(t) = r\omega\mathbf{e}_\theta \tag{7.11}$$

$$\mathbf{a}(t) = [-r\omega^2]\mathbf{e}_r + [r\alpha]\mathbf{e}_\theta = \left[-\frac{v_t^2}{r}\right]\mathbf{e}_r + [r\alpha]\mathbf{e}_\theta \tag{7.12}$$

Note that these expressions are essentially identical to the normal and tangential expressions for velocity and acceleration [Eqs. (7.7) and (7.8)], with the substitutions:

Tangential velocity: $v_t = r\omega$

Tangential acceleration: $a_t = r\alpha$

Example 7.4

A turntable starts spinning from rest (at $t = 0$) and accelerates uniformly at 1 rad/s² to an angular speed of 120 rev/min. How many revolutions does it take to reach maximum speed?

A. 11.5

B. 12.5

C. 13.5

D. 14.0

Solution Final angular velocity: $\omega = 120$ rev/min $= 2$ rev/s $= 4\pi$ rad/s

For constant acceleration:

$$\omega_f^2 = \omega_0^2 + 2\alpha\theta \Rightarrow \theta = \frac{\omega_f^2 - \omega_0^2}{2\alpha} = \frac{12.57^2 - 0}{2 \times 1} = 79 \text{ rad} = 12.6 \text{ revolutions}$$

Answer is B.

Motion at Constant Acceleration

The fundamental relationships of Newtonian kinematics are

$$v = \frac{ds}{dt} \tag{7.13}$$

$$a = \frac{dv}{dt} = v\frac{dv}{ds} \tag{7.14}$$

When acceleration a = constant, the Laws of Motion can be written as

$$v(t) = v_0 + a(t - t_0) \tag{7.15}$$

$$s(t) = s_0 + v_0(t - t_0) + \frac{1}{2}a(t - t_0)^2 \tag{7.16}$$

$$v^2 - v_0^2 = 2a(s - s_0) \tag{7.17}$$

where s_0 and v_0 are position and velocity at time t_0, respectively.

Projectile Motion

The motion of a launched projectile, subject to only the gravitational field during its flight, is a common application for the motion with constant acceleration.

The motion of the projectile occurs in a vertical plane shown as the x-y plane in Fig. 7.3. The initial position of the projectile is shown as (x_0, y_0). At $t = 0$, it is launched with a velocity v_0 at an angle α with the horizontal as shown. At $t = t_1$, it reaches the highest point in its trajectory. At this time, its vertical velocity component = 0. From $t = t_1$ to $t = t_2$, it accelerates downward, at constant acceleration = g (9.81 m/s² = 32.2 ft/s²). Given the y axis in the figure (upward positive), the vertical acceleration should actually be $a = -g$. At $t = t_2$, the projectile hits the ground ($y = 0$).

The quantities one might be interested in solving for are

- Time taken to reach the ground (t_2)
- Distance to the point of impact ($x_{max} - x_0$)
- Elevation of highest point reached by the projectile (y_{max})

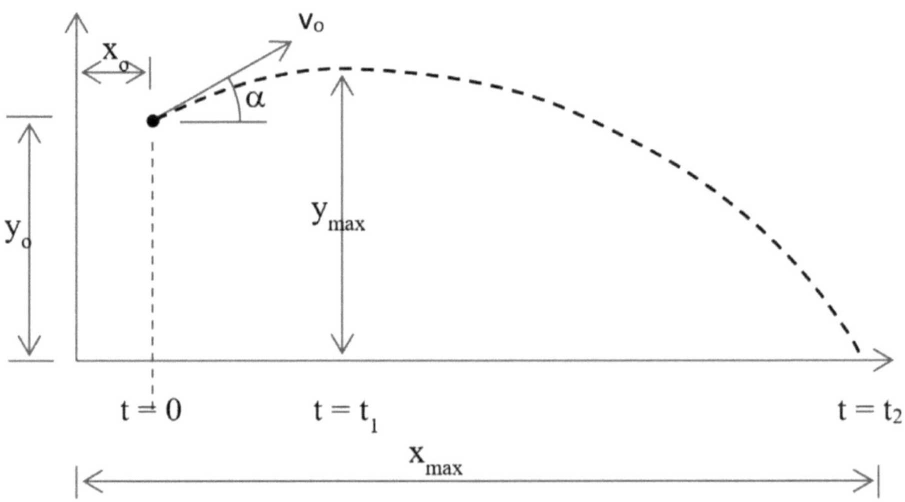

Figure 7.3 Projectile motion under gravity.

Motion in the *y* Direction

Initial velocity (vertical) = $v_0 \sin\alpha$

Initial position (at $t = 0$) = y_0

Final position (at $t = t_2$) = 0

Constant acceleration = $-g$

Using the second equation of motion from above [Eq. (7.16)]

$$0 = y_0 + v_0 \sin\alpha t_2 - \frac{1}{2} g t_2^2$$

In the equation above, once all parameters (y_0, v_0, α, g) are specified, the quadratic equation can be solved for the flight time t_2.

Once the flight time t_2 is known, the final velocity (vertical component only) can be found from

$$v_{2v} = v_0 \sin\alpha - g t_2$$

Motion in the *x* Direction Since the horizontal motion is at zero acceleration, the velocity remains constant at $v_0 \cos\alpha$.

$$v_{2h} = v_0 \cos\alpha$$

And the x coordinate of the point of impact is given by the sum of the initial position and the horizontal distance traveled

$$x_{max} = x_0 + (v_0 \cos\alpha)(t_2)$$

Example 7.5

A cannon fires a shot from the top of a cliff as shown. The muzzle velocity $v = 240$ m/s and the angle $\alpha = 40°$. What is the range (R) of the cannon?

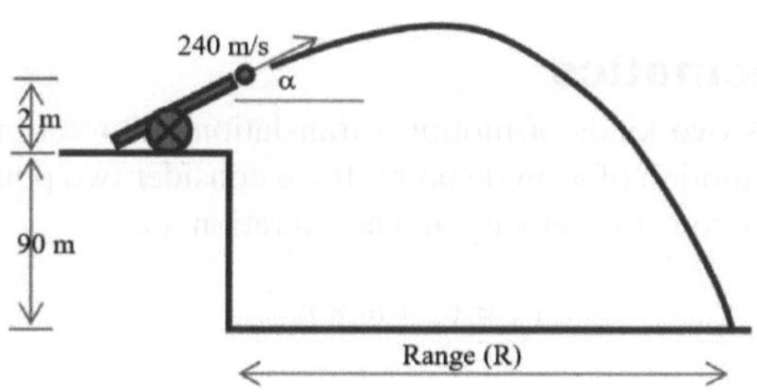

A. 4667 m

B. 5883 m

C. 6565 m

D. 7281 m

Solution This is a case of constant acceleration. Vertical acceleration = acceleration due to gravity.

Let the origin be located at the tip of the muzzle and the vertical (y) axis be measured positive upward, as shown below.

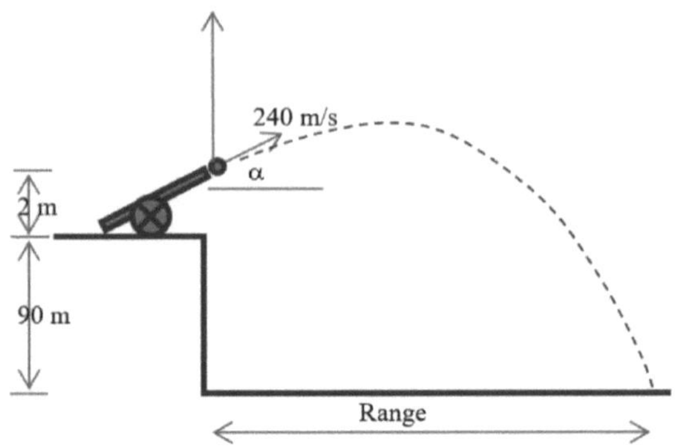

At $t = 0$, $y = 0$, $v_y = 240 \sin 40 = +154.3$ m/s
At the instant of hitting the ground, $t = T$, $y = -92$
With y axis pointing up, $a_y = -9.81$ m/s²

$$s = v_0 t + \frac{1}{2} a t^2$$

$$-92 = 154.3T - 4.9T^2$$

Solving this quadratic, $T = 32$ s

Since the horizontal velocity is constant (zero horizontal acceleration), $v_x = 240 \cos 40 = 183.9$ m/s,

$$\text{range} = 183.9 \times 32 = 5883 \text{ m}$$

Answer is B.

Rigid Body Kinematics

A rigid body undergoes two kinds of motion—translation and rotation. For this reason, it is insufficient to track the motion of a single point. If we consider two points (A and B) on a rigid body, the general expressions for velocity and acceleration are

$$v_A = v_B + \omega \times r_{A/B} \tag{7.18}$$

$$a_A = a_B + \alpha \times r_{A/B} + \omega \times (\omega \times r_{A/B}) \tag{7.19}$$

For 2D (planar motion in x-y plane), angular velocity ω and angular acceleration α are either clockwise or counterclockwise and therefore can be expressed as $\underline{\mathbf{k}}$ vectors and the $\mathbf{r}$ vector is in the x-y plane

$$\boldsymbol{\alpha} = \alpha \mathbf{k} \tag{7.20}$$

$$\boldsymbol{\omega} = \omega \mathbf{k} \tag{7.21}$$

$$\mathbf{r} = r_x\mathbf{i} + r_y\mathbf{j} \tag{7.22}$$

leading to

$$\mathbf{v}_A = \mathbf{v}_B - r_y\omega\mathbf{i} + r_x\omega\mathbf{j} \tag{7.23}$$

$$\mathbf{a}_A = \mathbf{a}_B + \alpha(-r_y\mathbf{i} + r_x\mathbf{j}) - \omega^2(r_x\mathbf{i} + r_y\mathbf{j}) = \mathbf{a}_B + \frac{\alpha}{\omega}\mathbf{v}_{A|B} - \omega^2\mathbf{r}_{A|B} \tag{7.24}$$

Mass Moment of Inertia

The mass moment of inertia (not to be confused with the second moment of an area) is defined as

$$I_x = \int (y^2 + z^2)dm \tag{7.25a}$$

$$I_y = \int (x^2 + z^2)dm \tag{7.25b}$$

$$I_z = \int (x^2 + y^2)dm \tag{7.25c}$$

A table listing moment of inertia formulas for some standard shapes is on page 79 of the *FE Reference Handbook*, version 9.4.

The mass radius of gyration is defined as

$$r_m = \sqrt{\frac{I}{M}} \tag{7.26}$$

Parallel Axis Theorem To compute the mass moment of inertia of a rigid body about any axis, we must first calculate the moment of inertia about a parallel axis passing through the centroid and then add a correction of the form Md^2, where M is the mass of the body and d is the perpendicular distance between the two parallel axes.

$$I = I_c + Md^2 \tag{7.27}$$

Example 7.6

A prismatic rod with mass 5 kg is supported by a frictionless hinge at a distance 65 cm from the left end as shown. The rod is held static in the horizontal position as shown and released from rest. What is the acceleration of the left end of the rod upon being released?

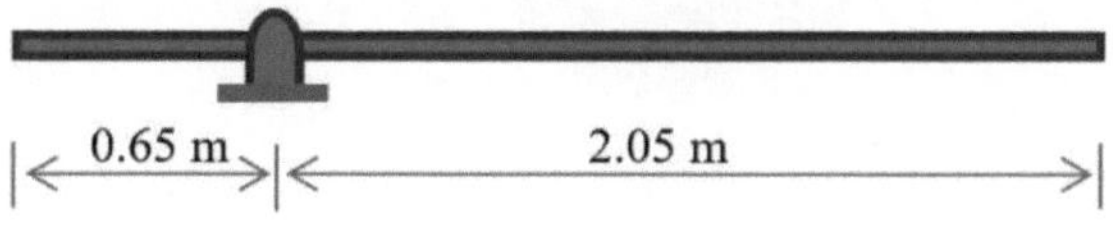

A. 0.64 m/s²

B. 1.25 m/s²

C. 4.06 m/s²

D. 6.25 m/s²

Solution The effective location of the weight (49 N) of the prismatic rod is at the midpoint, which is 1.35 m from the left end, or 0.7 m to the right of the hinge support. When the rod is released, this force has an unbalanced moment about the hinge. This moment is 49 × 0.7 = 34.3 Nm.

The equation of dynamic equilibrium is

$$M = I\alpha$$

where M is the resultant moment and I is the moment of inertia about the hinge.

The moment of inertia about the hinge is calculated using the parallel axis theorem.

$$I = \frac{mL^2}{12} + md^2 = \frac{5\times 2.7^2}{12} + 5\times 0.7^2 = 5.49\ \text{kg}\cdot\text{m}^2$$

Angular acceleration: $\alpha = \frac{M}{I} = \frac{34.3}{5.49} = 6.25\frac{\text{rad}}{\text{s}^2}$

At $t = 0$, the velocity is still zero. So the centrifugal acceleration (which would act horizontally toward the hinge) is still zero.

At the left end, the acceleration is given by

$$a = r\alpha = 0.65\times 6.25 = 4.06\frac{\text{m}}{\text{s}^2}$$

Answer is C.

Example 7.7

A metal disk of uniform thickness having mass = 3 kg is suspended from a frictionless pin as shown. If the force of 200 N acts as shown below, what is the instantaneous acceleration at the point of application of the force?

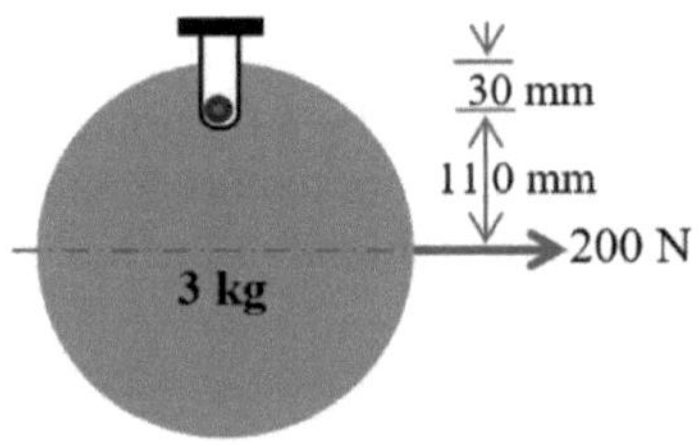

A. 40 m/s²

B. 50 m/s²

C. 60 m/s²

D. 70 m/s²

Solution The (mass) moment of inertia of the disk about the axis through the hinge is calculated using the parallel axis theorem. (The y and y_c axes come out of the plane.)

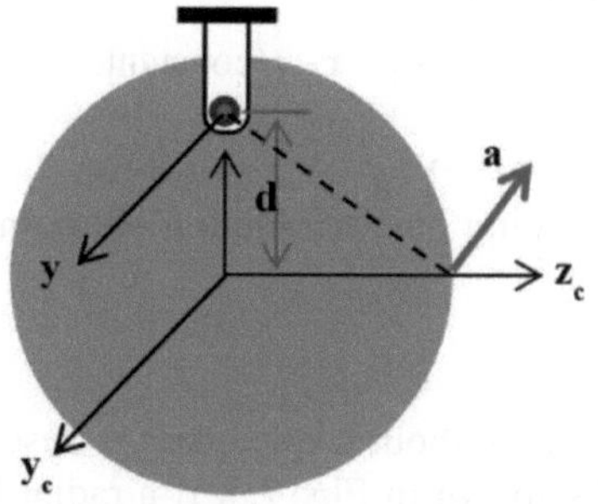

$$I_y = I_{y_c} + Md^2 = \frac{1}{2}MR^2 + Md^2 = \frac{1}{2}(3)(0.14^2) + (3)(0.11^2) = 0.0657 \text{ kg} \cdot \text{m}^2$$

Moment about the hinge: $M_y = 200 \times 0.11 = 22$ Nm

Angular acceleration: $\alpha = \frac{M_y}{I_y} = \frac{22}{0.0657} = 334.9 \text{ rad/s}^2$

Radial distance from hinge to point of application of force: $r = \sqrt{0.11^2 + 0.14^2} = 0.178$ m
Linear acceleration: $a = r\alpha = 59.6 \text{ m/s}^2$

Answer is C.

Pure Translation of a Rigid Body

In pure translation, a straight line connecting any two points on the rigid body remains unchanged in direction and magnitude. In Fig. 7.4, the rigid body undergoes translation.

Figure 7.4
Translation of a rigid body.

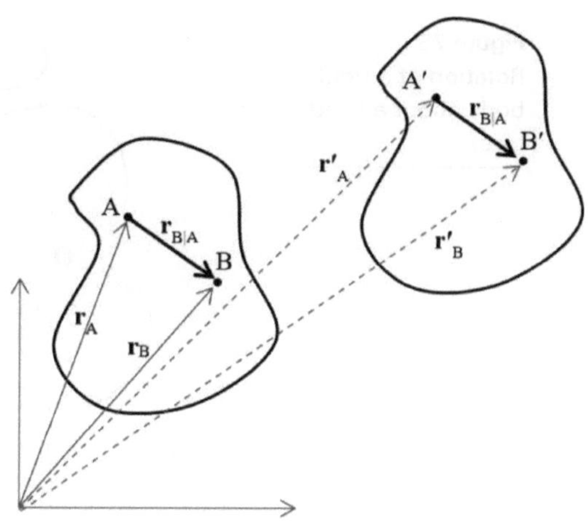

Points A and B change position in space, while the vector $\mathbf{r}_{B|A}$ (which represents the position of B relative to A) remains unchanged in magnitude (because the body is rigid) and direction (because the motion is pure translation).

$$\mathbf{r}_{B|A} = \mathbf{r}_B - \mathbf{r}_A = \text{constant}$$

Taking a time derivative: $\dot{\mathbf{r}}_B - \dot{\mathbf{r}}_A = 0 \rightarrow v_B = v_A \rightarrow a_B = a_A$

Thus, for a rigid body undergoing pure translation, all points have the same velocity and the same acceleration.

Rotation about a Fixed Axis

The motion of a rigid body rotating about a fixed axis is described in terms of the instantaneous angular position $\theta(t)$. As shown in Fig. 7.5, if a radial line is drawn from the axis of rotation to any point P, then the result of the rigid body rotation is that the radial line rotates to the new position OP'. Both vectors—OP and OP'—lie in the plane (shown shaded) which is normal to the axis of rotation.

Then, the angular velocity is given by $\omega = \dfrac{d\theta}{dt}$ and the angular acceleration by $\alpha = \dfrac{d^2\theta}{dt^2} = \omega \dfrac{d\omega}{d\theta}$.

Uniform Rotational Motion (Zero Angular Acceleration) If angular acceleration $\alpha = 0$, angular velocity $\omega =$ constant and angular displacement $\theta(t) = \theta_o + \omega t$.

Uniformly Accelerated Motion (Constant Angular Acceleration) If angular acceleration $\alpha =$ constant, then the angular velocity ω and angular displacement θ are given by

$$\omega(t) = \omega_0 + \alpha t \tag{7.28}$$

Figure 7.5
Rotation of a rigid body about a fixed axis.

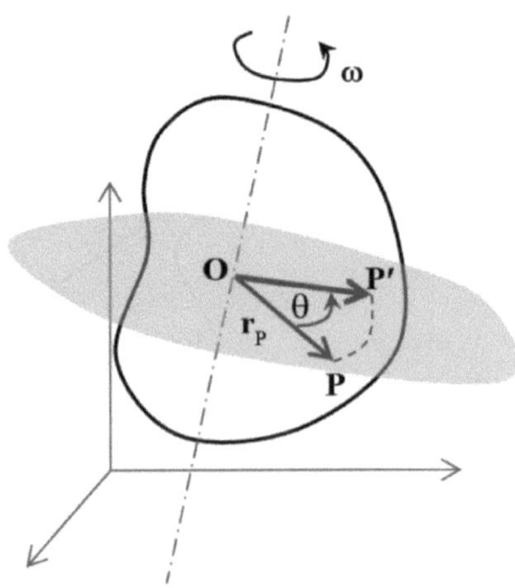

$$\theta(t) = \theta_0 + \omega_0 t + \frac{1}{2}\alpha t^2 \tag{7.29}$$

Kinetics

For particles or rigid bodies, kinetics equations relate the motion of an object (described by the kinematics expressions) to the forces and/or moments causing them.

Particle Kinetics

When it comes to the dynamics of a particle (i.e., an object whose mass can be assumed to be lumped at a single point), Newton's second law can be stated as "the force acting on a particle is equal to the rate of change of its momentum."

$$\underline{\mathbf{F}} = \frac{d}{dt}(m\underline{\mathbf{v}}) \tag{7.30}$$

where $\underline{\mathbf{F}}$ = resultant force acting on the particle
m = mass of the particle
$\underline{\mathbf{v}}$ = velocity of the particle

For a particle with constant mass, this becomes the familiar

$$\underline{\mathbf{F}} = m\frac{d\underline{\mathbf{v}}}{dt} = m\underline{\mathbf{a}} \tag{7.31}$$

where $\underline{\mathbf{a}}$ is acceleration of the particle.

Example 7.8

A 5-kg block on a 35° incline is acted upon by a horizontal force $F = 200$ N as shown. What is the acceleration (m/s²) of the block?

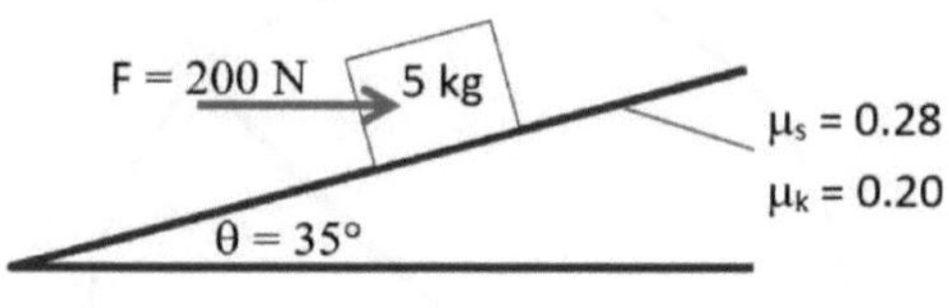

A. 11 m/s² up slope
B. 21 m/s² up slope
C. 11 m/s² down slope
D. 21 m/s² down slope

Solution The free body diagram of the block, with forces resolved along, and perpendicular to the incline, is shown.

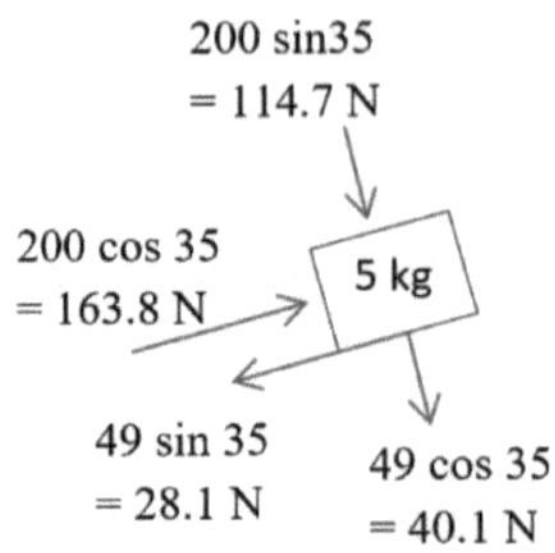

Since the up-plane force is more than the down-plane component of the weight, the block will have a tendency to move upslope. Therefore, the *friction force will act downslope.*

The total normal force perpendicular to the plane = 114.7 + 40.1 = 154.8 N

Friction force (with motion) = 0.2 × 154.8 = 30.96 N

Unbalanced up-plane force = 163.8 − 28.1 − 30.96 = 104.74 N

Acceleration = 104.74/5 = 20.95 m/s^2

Answer is B.

NOTE If the down-plane forces had been dominant, causing the block to slide *down* the incline, then everything would be unchanged except the friction force, which would act up-plane.

Example 7.9

A mass m = 1 kg has negligible size and is being swung in a vertical circular path (radius = 40 cm) using a string as shown. The angular velocity is 60 rev/min. At the position shown, what is the tension in the string?

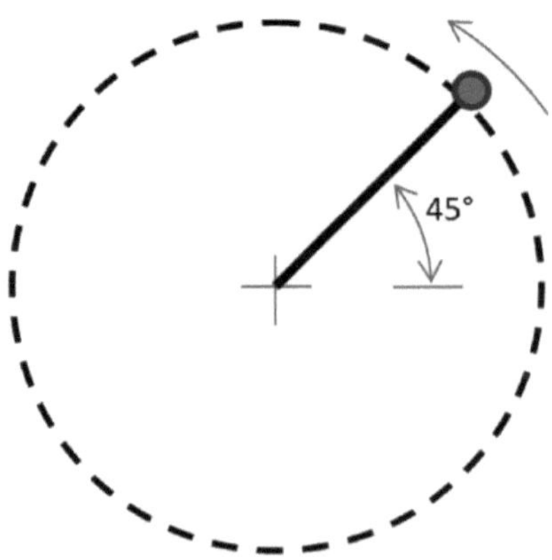

A. 5.98 N

B. 8.85 N

C. 18.86 N

D. 22.73 N

Solution The tension in the string acts inward (directed toward the center) on the mass. At constant velocity, the normal component of acceleration is directed inward and is equal to $v^2/r = \omega^2 r$

Angular velocity = 60 rev/min = $60(2\pi)/60 = 6.283$ rad/s

The free body diagram shows the forces acting on the mass—its weight and the string tension. In the second figure, the weight has been resolved into normal and tangential components.

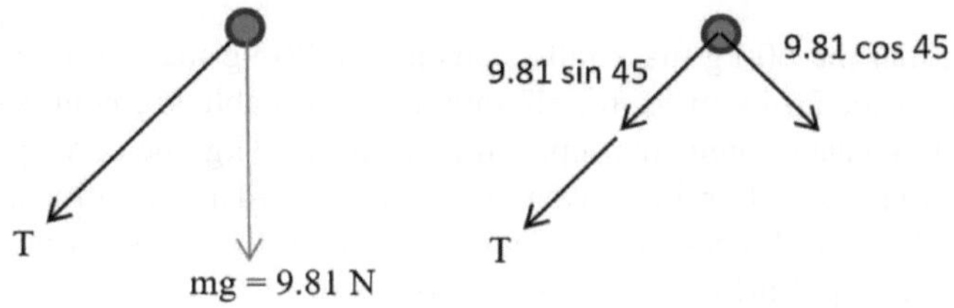

$$T + 9.81\cos 45 = m\omega^2 r = 1 \times 6.283^2 \times 0.40 \Rightarrow T = 8.85\ \text{N}$$

Answer is B.

Example 7.10

A pulley system consisting of a fixed and a movable pulley is loaded as shown. If the pulleys are assumed to be massless and frictionless, what is the vertical acceleration of the 60 kg mass?

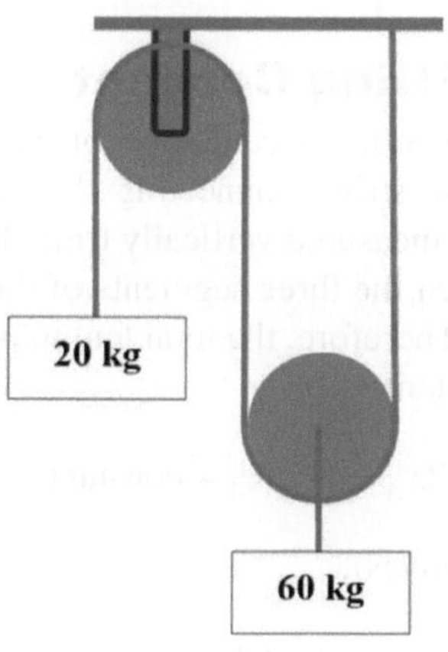

A. 2.2 m/s²

B. 2.4 m/s²

C. 2.6 m/s²

D. 2.8 m/s²

Solution Free body diagrams of the two blocks are shown below. Cable has constant tension T (due to "frictionless" assumption).

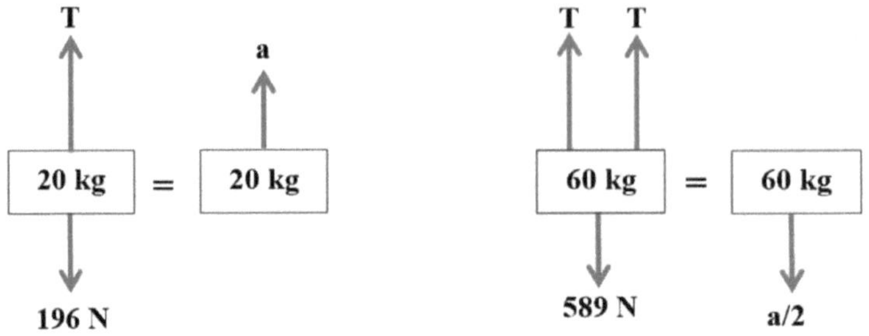

Let us assume that the 60-kg mass falls, causing the 20-kg mass to rise. Since for every unit displacement of the 20-kg mass, the shortening of the cable segment is compensated by the lengthening of *two* cable segments supporting mass of 60 kg, the corresponding displacement of the 60-kg mass is ½. For this reason, when the upward acceleration of the 20-kg mass is assumed to be "*a*," the downward acceleration of the 60-kg mass is written as $a/2$. The dynamic equations of equilibrium for the two masses are

$$T - 196 = 20a$$

$$589 - 2T = 60 \times \frac{a}{2} = 30a$$

Solving these two equations, $a = +2.81 \text{ m/s}^2$

Answer is D.

Constant Length of String Concept

The kinematic relationship between the acceleration of the masses A and B above can also be derived from the fact that the string connecting the two masses is of constant length (inextensible). If the coordinates (measured vertically from the fixed support above) of masses A and B are written x_A and x_B, then the three segments of the cable can be written as $x_A + c_1$, $x_A + c_2$, and $x_B + c_3$, respectively. Therefore, the total length of the cable is $2x_A + x_B + c_4$, where c_1, c_2, c_3, and c_4 are arbitrary constants.

$$2x_A + x_B + c_4 = \text{constant} \tag{7.32}$$

Leading to (by taking time derivatives),

$$2v_A + v_B = 0 \tag{7.33}$$

and

$$2a_A + a_B = 0 \tag{7.34}$$

This implies that the acceleration a_B is twice the magnitude and in opposite direction to a_A.

Example 7.11

A fixed pulley having radius = 100 mm and mass = 10 kg supports a smooth cable which supports two masses—of 20 kg and 60 kg—as shown in the figure below. What is the acceleration of the 60-kg mass?

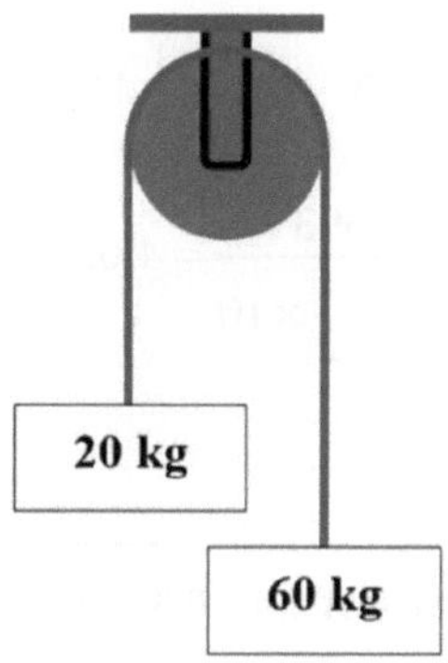

A. 4.32 m/s²

B. 4.46 m/s²

C. 4.62 m/s²

D. 4.85 m/s²

Solution When the pulley–cable friction can be neglected (smooth pulley) *and* the mass of the pulley is negligible, the tension in the cable is constant. However, in this case, the pulley mass is not negligible and the resulting inertia will lead to the tension in the two cable segments to be different. The free body diagrams of (1) the 20-kg mass, (2) the pulley, and (3) the 60-kg mass are shown

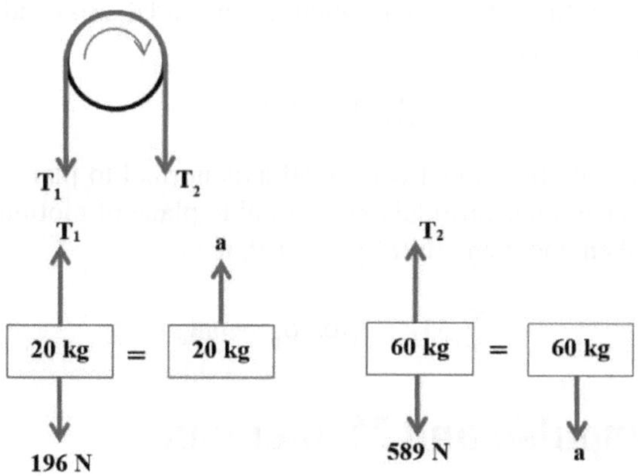

For mass 1: $T_1 - 196 = 20a$ (1)

For the pulley: $T_2 r - T_1 r = \frac{1}{2}Mr^2\alpha \Rightarrow T_2 - T_1 = \frac{1}{2}Mr\alpha = \frac{1}{2}Ma = 5a$ (2)

For mass 2: $589 - T_2 = 60a$ (3)

From Eqs. (1) and (3), we get

$$T_2 - T_1 = 589 - 60a - (196 + 20a) = 393 - 80a$$

Therefore

$$393 - 80a = \frac{1}{2}Ma$$

$$a = \frac{393}{80 + \frac{1}{2} \times 10} = 4.62 \text{ m/s}^2$$

Answer is C.

NOTE If the mass of the pulley were neglected, the acceleration of each mass would have been

$$a = \frac{(60 - 20) \times 9.81}{60 + 20} = 4.91 \text{ m/s}^2$$

Rigid Body Kinetics

For a rigid body, because of the distributed mass, the motion of the object can be decomposed into translation of the center of gravity and the rotation of the rigid body about a centroidal axis.

The relationship between the net force acting on the rigid body and the resulting acceleration is

$$\sum \underline{\mathbf{F}}_{\text{net}} = m\underline{\mathbf{a}}_{\text{CG}} \tag{7.35}$$

The relationship between the net moment (about a centroidal axis) and the resulting angular acceleration about the same axis is

$$\sum \underline{\mathbf{M}}_C = I_C \alpha \tag{7.36}$$

where α is angular acceleration about centroidal axis normal to plane of motion and I_C and mass moment of inertia about centroidal axis normal to plane of motion.

If the moment is taken about an arbitrary point P, then

$$\sum \underline{\mathbf{M}}_P = I_C \alpha + \boldsymbol{\rho}_{\text{pc}} \times m\mathbf{a}_{\text{c}} \tag{7.37}$$

Principle of Impulse and Momentum

The basis for kinetics is Newton's second law of motion, which says that the force acting on an object is equal to the rate of change of its linear momentum, $m\mathbf{v}$, the product of its mass m and velocity $\mathbf{v}$. Momentum, like velocity, is a vector quantity.

$$\mathbf{F} = \frac{d}{dt}(m\mathbf{v}) \tag{7.38}$$

This equation can also be written in integral form

$$\int \mathbf{F}dt = m\mathbf{v}\Big|_1^2 = (m\mathbf{v})_2 - (m\mathbf{v})_1 \tag{7.39}$$

The integral $\int \mathbf{F}dt$ is called the impulse of the force. Thus, Newton's second law can be also stated as "the impulse of the force is equal to the change in momentum." The impulse-momentum principle is useful for a system where the force is variable but known and we are interested only in solving for the velocity at a particular stage. By applying this principle between states 1 and 2, as shown in Eq. (7.36), we are only able to calculate the velocity at state 2, but not at all intermediate stages.

Applying the same principle to rotational motion, we can state that "the Impulse of the moment is equal to the change in angular momentum."

$$\int \mathbf{M}dt = \mathbf{H}\Big|_1^2 = \mathbf{H}_2 - \mathbf{H}_1 = (I\omega)_2 - (I\omega)_1 \tag{7.40}$$

Example 7.12

A force F of short duration acts on a stationary mass, m = 1.6 kg. The profile of force F is shown in the figure below. What is the velocity of the mass at t = 0.05 s?

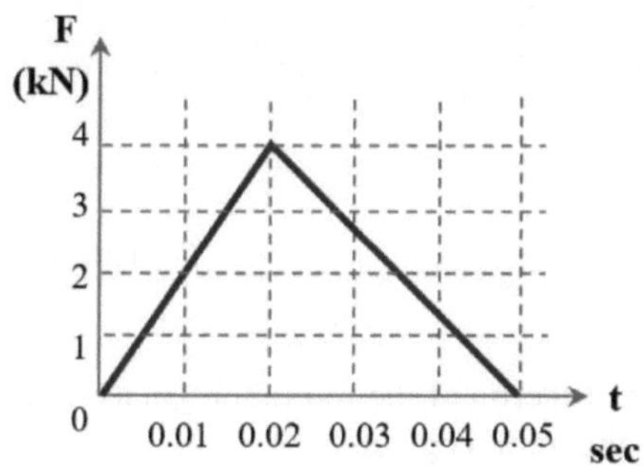

A. 125 m/s

B. 62.5 m/s

C. 31.3 m/s

D. 12.5 m/s

Solution The impulse of the force between $t = 0$ and $t = 0.05$ s is given by the integral

$$I = \int F\ dt$$

which is the area under the F-t curve.

$$I = \int F\ dt = \frac{1}{2} \times 4000 \times 0.05 = 100\ \text{Ns}$$

According to the impulse-momentum principle:

$$\Delta(mv) = \int F\,dt = 100\ \text{Ns} = mv_{0.05} - mv_0$$

Since the mass is stationary at $t = 0$, the velocity at $t = 0.05$ s is

$$v_{0.05} = \frac{100}{1.6} = 62.5\frac{\text{m}}{\text{s}}$$

Answer is B.

Example 7.13

A prismatic rod has mass m = 0.8 kg and length L = 80 cm is hinged at the upper end as shown. It is at rest at t = 0 and is then subjected to a moment that increases linearly from zero to 10 Nm at t = 5 s. What is the angular velocity at t = 10 s?

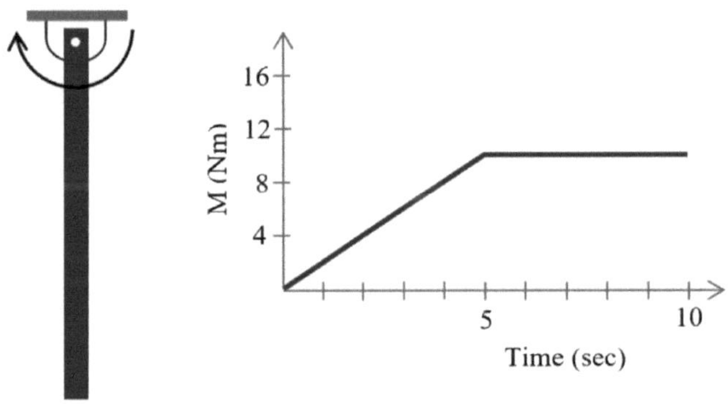

Solution Mass moment of inertia of the rod (about an axis through the hinged end of the rod), using the parallel axis theorem, is given by

$$I = \frac{ML^2}{3} = \frac{0.8 \times 0.8^2}{3} = 0.171\ \text{kg·m}^2$$

The integral $\int Mdt$ can be found as the area under the M-t curve (from t = 0 to t = 10 s) = 75 Nm·s.

Since the angular velocity at $t = 0$ is zero,

$$I(\omega_{10} - \omega_0) = \int_0^{10} Mdt = 75 \Rightarrow \omega_{10} = \frac{75}{0.171} = 439.5\frac{\text{rad}}{\text{s}} = 4197\ \text{rev/min}$$

Impact

When two elastic bodies collide, according to equilibrium principles, the force F_{12} exerted by body 1 on body 2 must be equal and opposite to the force (F_{21}) exerted by body 2 on body 1. Since the time of the collision is also the same, the impulse exerted on body 1 is equal and opposite to the impulse exerted on body 2. Considered as a system, the total impulse is therefore zero.

Since the impulse is equal to the change in momentum, the total (body 1 + body 2) change in momentum is zero. This is the statement of conservation of momentum before and after an impact. So, regardless of the nature of impact (ranging from plastic to elastic), linear momentum of the system is always conserved, even though energy is only conserved for elastic impact.

Coefficient of Restitution

The exact nature of impact can, however, vary. This is quantified in terms of the coefficient of restitution (e), which is defined as below.

$$e = \frac{(v_2')_n - (v_1')_n}{(v_1)_n - (v_2)_n} \tag{7.41}$$

The subscript "n" on all the quantities implies that these are velocity components normal to the plane of impact (see Fig. 7.6). The unprimed quantities in the denominator are the velocities before impact and the primed quantities in the numerator are the velocities after impact. Thus, the coefficient of restitution is the ratio of the relative velocity post-impact to the relative velocity pre-impact. Note the change in sign implicit in the equation.

Elastic versus Plastic Impact

The value of the coefficient of restitution (e) ranges from 0 (plastic impact) to 1 (elastic impact). When the colliding objects "stick" together after impact (such as a bullet fired into a block of wood and remaining embedded), the impact is termed plastic and $e = 0$. For such an impact, the relative velocity after impact = 0. On the other hand, when the impact is perfectly elastic ($e = 1$), the relative velocity remains unchanged in magnitude. An example would be a ball bouncing on a hard floor and bouncing back to exactly the height it was dropped from.

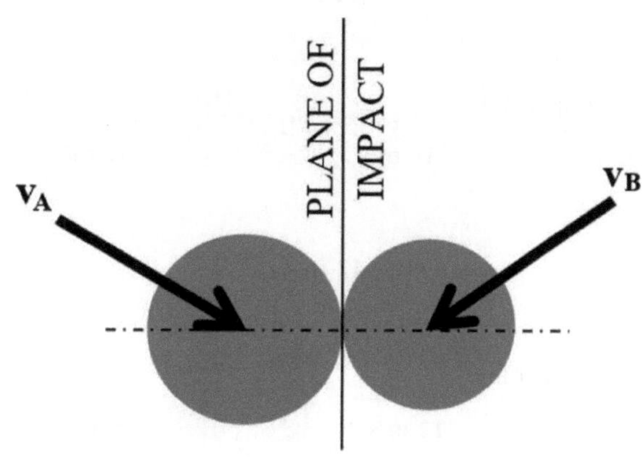

Figure 7.6 Plane of impact.

It is noteworthy that it is only the perfectly elastic impact for which energy is conserved, but momentum is always conserved (for all types of impact).

Simple case: When two objects of equal mass collide in a perfectly elastic impact ($e = 1$), they simply exchange velocities.

Example 7.14

Two identical billiard balls (mass = 170 g) collide on a horizontal table after traveling along the dashed lines as shown. Velocities just before impact are indicated on the diagram. If the impact is considered to be perfectly elastic, what are the velocities after impact?

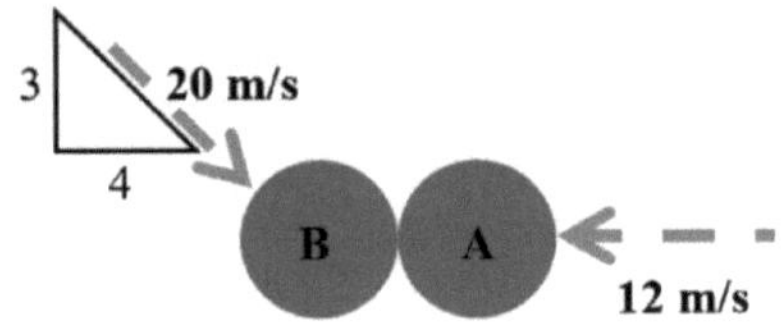

A. $v_A = -16\mathbf{i}$ and $v_B = -12\mathbf{i} + 12\mathbf{j}$
B. $v_A = 16\mathbf{i}$ and $v_B = -12\mathbf{i} - 12\mathbf{j}$
C. $v_A = 12\mathbf{i}$ and $v_B = -9\mathbf{i} - 12\mathbf{j}$
D. $v_A = -12\mathbf{i}$ and $v_B = -12\mathbf{i} - 12\mathbf{j}$

Solution When two identical objects (same mass) collide in a perfectly elastic impact ($e = 1$), the two masses "trade" velocities along the line of impact. For this simple solution, *both* conditions must exist.

The velocities along and perpendicular to the line of impact are shown. Since *both* conditions are met (identical mass *and* elastic impact), the horizontal velocity components of A and B are exchanged.

Therefore, after impact, $\mathbf{v}_{\mathbf{Ax}} = 16$; $\mathbf{v}_{\mathbf{Bx}} = -12$.

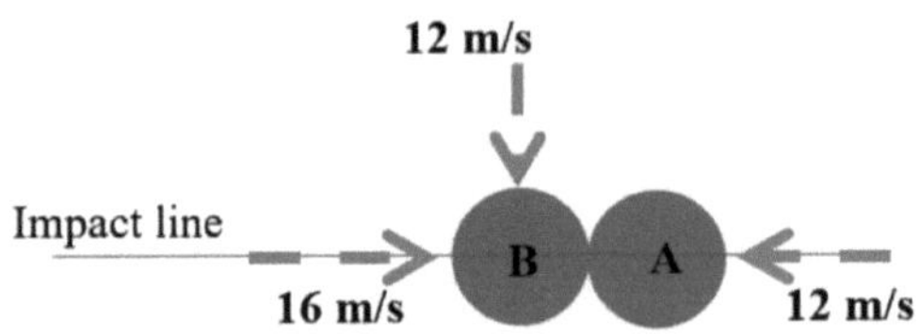

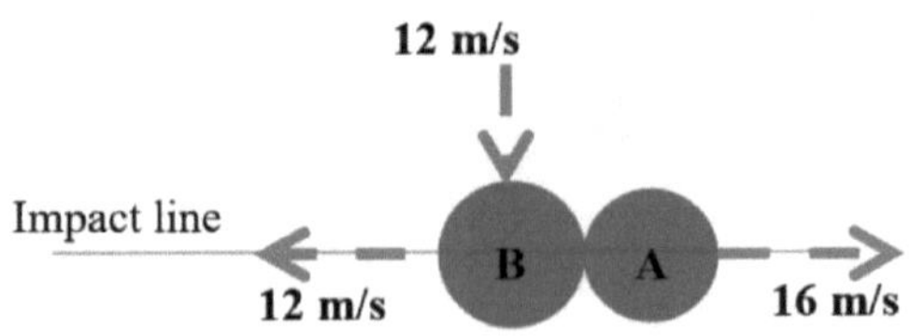

The y components of the velocity remain unchanged: $\mathbf{v}_{Ay} = 0$; $\mathbf{v}_{Bx} = -12$.

Answer is B.

Example 7.15

Two billiard balls ($m_A = 170$ g, $m_B = 200$ g) collide on a horizontal table after traveling along the dashed lines as shown. Velocities just before impact are indicated on the diagram. If the coefficient of restitution is 0.8, how much of the initial kinetic energy is lost due to the impact?

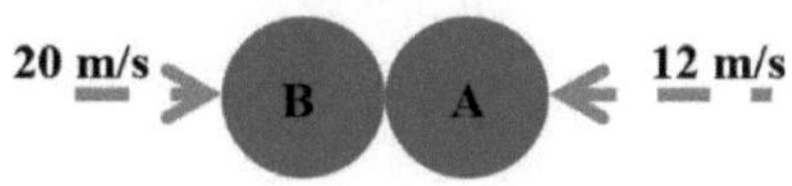

A. 20%

B. 32%

C. 0%

D. 26%

Solution The velocities after impact are given by

$$v'_A = \frac{m_B v_B (1+0.8) + (m_A - 0.8 m_B) v_A}{m_A + m_B} = \frac{200 \times 20 \times 1.8 + (170 - 0.8 \times 200) \times -12}{370} = 19.14 \text{ m/s}$$

$$v'_B = \frac{m_A v_A (1+0.8) - (0.8 m_A - m_B) v_B}{m_A + m_B} = \frac{170 \times -12 \times 1.8 - (0.8 \times 170 - 200) \times 20}{370} = -6.46 \text{ m/s}$$

Energy before impact: $E = \frac{1}{2} m_A v_A^2 + \frac{1}{2} m_B v_B^2 = \frac{1}{2} \times 0.17 \times 12^2 + \frac{1}{2} \times 0.2 \times 20^2 = 52.24$ J

Energy after impact: $E' = \frac{1}{2} m_A v_A'^2 + \frac{1}{2} m_B v_B'^2 = \frac{1}{2} \times 0.17 \times 19.14^2 + \frac{1}{2} \times 0.2 \times 6.46^2 = 35.3$ J

Energy loss = 16.94 J, which is 32.4% of the original 52.24 J.

Answer is B.

NOTE If the collision were perfectly elastic ($e = 1$), then no calculations would be necessary. Energy would have been conserved.

Example 7.16

A bullet with mass $m = 90$ g is fired into and gets embedded in a wooden block whose mass is $M = 3$ kg. The velocity of the bullet just before impact is 250 m/s. What is the distance traveled by the block before it comes to rest?

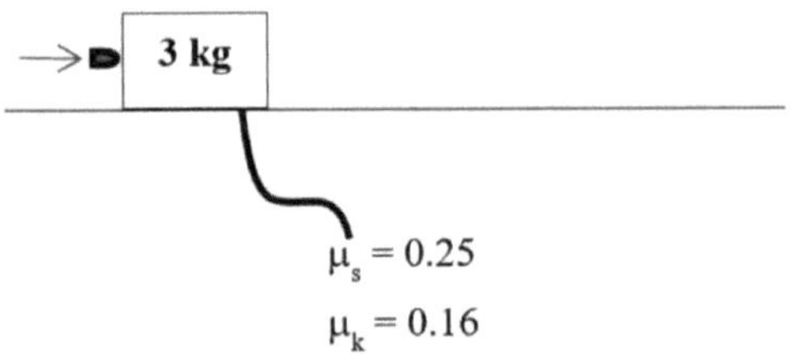

A. 17 m

B. 21 m

C. 8.5 m

D. 10.5 m

Solution Since the bullet gets lodged in the block (zero relative velocity after impact), this is a case of a plastic impact ($e = 0$). The two objects have a common velocity V after impact. Applying conservation of momentum, this is given by

$$m_1V_1 + m_2V_2 = (m_1 + m_2)V$$

$$V = \frac{m_1V_1 + m_2V_2}{m_1 + m_2} = \frac{0.09 \times 250 + 3 \times 0}{0.09 + 3} = 7.28\frac{\text{m}}{\text{s}}$$

The deceleration (due to friction) is $\mu_k g = 0.16 \times 9.81 = 1.57$ m/s^2

With initial velocity $V = 7.28$ m/s and acceleration $= -1.57$ m/s^2, the block-bullet combination will travel a distance d, calculated from kinematics equations for constant acceleration

$$d = \frac{V_f^2 - V_i^2}{2a} = \frac{0^2 - 7.28^2}{2 \times -1.57} = 16.9 \text{ m}$$

Answer is A.

Principle of Work and Energy

For a conservative system (one where total energy is conserved), the law of conservation of energy states that

$$T_1 + V_1 = T_2 + V_2 \tag{7.42}$$

where T_1 = kinetic energy of the system at state 1
V_1 = potential energy of the system at state 1
T_2 = kinetic energy of the system at state 2
V_2 = potential energy of the system at state 2

For a nonconservative system, there will be forces that do not satisfy Eq. (7.42). These forces do nonconservative work (W_{nc}) that must be accounted for as shown in Eq. (7.43). When dealing with dissipative forces such as friction, the term W_{nc} must be taken as negative, thereby *decreasing* the total energy from state 1 to state 2 in the equation below.

$$T_1 + V_1 + W_{nc} = T_2 + V_2 \tag{7.43}$$

Work

The work done by a force F translating through a distance dr, is given by the dot product

$$W = \int F \cdot dr \tag{7.44}$$

The work done by a moment M rotating through an angle $d\theta$, is given by the dot product

$$W = \int M \cdot d\theta \tag{7.45}$$

If the force remains constant, the work is given by

$$W = F \cos\theta \Delta s \tag{7.46}$$

where θ is the angle between the direction of the force F and the translation Δs.

Gravitational Potential Energy

The energy that a body possesses by virtue of being translated against the gravitational field is given by

$$V_g = Wh = mgh \tag{7.47}$$

where the elevation h can be defined with respect to an arbitrary datum. Since the gravity field direction is downward (directed toward the center of the earth), the gravitational potential energy is considered to be increasing with upward translation.

Elastic Potential Energy

For a linear, elastic spring whose spring constant (stiffness) is k, the force in the spring is directly proportional to the deformation (extension or shortening) of the spring s. The deformation is calculated as the deformed length minus the undeformed length

$$F_s = ks = k(l - l_0) \tag{7.48}$$

and the elastic potential energy in the spring is given by

$$V_e = \frac{1}{2}ks^2 \tag{7.49}$$

Kinetic Energy

Kinetic energy of an object that can be treated as a particle (i.e., one whose mass, m can be assumed to be concentrated at a point) is given by

$$V = \frac{1}{2}mV^2 \tag{7.50}$$

where V is velocity of the particle.

A rigid body of significant shape and size goes through a combination of translation and rotation. The kinetic energy of such an object is calculated as a sum of the energy due to translation of the centroid and that due to the rotation of the body about the centroidal axis.

$$V = \frac{1}{2}mV_c^2 + \frac{1}{2}I_c\omega_c^2 \tag{7.51}$$

where V_c = translational velocity of the center of mass
I_c = mass moment of inertia about the centroid
ω_c = angular velocity of the rigid body about the centroidal axis

Example 7.17

A uniform cylinder of radius R and mass M starts from rest at the top of an incline and rolls without slip down the incline through a vertical distance h, as shown below. The coefficients of static and kinetic friction between the cylinder and the plane are μ_s and μ_k. What is the velocity of the center of mass of the cylinder when it reached the bottom of the incline?

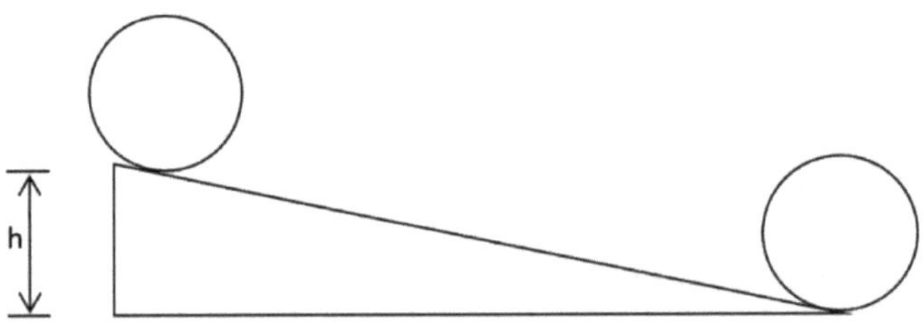

Solution Since the rolling occurs without slip, there is no work done (and lost) via friction. Applying principle of conservation of energy, we have (using the horizontal line through the downhill point as the datum)

Total energy at position A (at rest on top of incline), $E_A = Mgh$

If the angular velocity at the bottom of the incline = ω and the velocity (of the center of mass) parallel to the incline = V, then the total energy at the bottom of the incline is

$$E_B = PE + KE_{\text{translation}} + KE_{\text{rotation}} = 0 + \frac{1}{2}MV^2 + \frac{1}{2}I\omega^2$$

where I is mass moment of inertia of cylinder about the centroidal axis = ½ MR^2 and V is ωR. Therefore,

$$E_A = E_B \Rightarrow Mgh = \frac{1}{2}MV^2 + \frac{1}{2}I\omega^2 = \frac{1}{2}MV^2 + \frac{1}{2}\frac{1}{2}MR^2\omega^2 = \frac{3}{4}MV^2 \Rightarrow V = \sqrt{\frac{4}{3}gh}$$

Example 7.18

A 5-kg block traveling to the right has a velocity of 20 m/s when it is at a distance of 10 m from a mass-less bumper attached to a spring with spring constant $k = 100$ N/cm. The horizontal surface and the block have coefficient of static friction = 0.25 and kinetic friction = 0.20. What is the maximum compression of the spring?

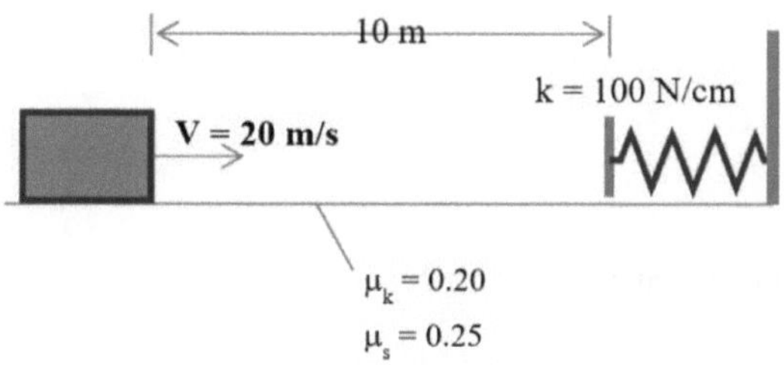

A. 44.7 cm

B. 42.5 cm

C. 90.0 cm

D. 32.6 cm

Solution Assume that the maximum compression of the spring is X. At the instant of maximum compression, the block will come to an instantaneous stop (velocity = 0) before elastic rebound of the spring will cause it to start moving to the left. The normal force between the block and the ground is constant $W = 49.0$ N. Therefore, the (kinetic) friction force is constant at $F_f = 0.2 \times 49 = 9.8$ N.

Using the principle of work and energy, initial kinetic energy = elastic energy stored in spring + energy lost (work done) due to friction

$$\frac{1}{2} \times 5 \times 20^2 = \frac{1}{2} \times 10{,}000 \times X^2 + 9.8 \times (10 + X)$$

Solving this quadratic, $X = 0.424$.

Answer is B.

Example 7.19

A metal bead (mass = 90 g) is released from rest at position *A* on a smooth circular guide rod (radius 35 cm) under the effect of a linear spring whose undeformed length is 25 cm. If there is no friction between the guide rod and the sliding mass, what is the velocity of the bead when it passes point *B*?

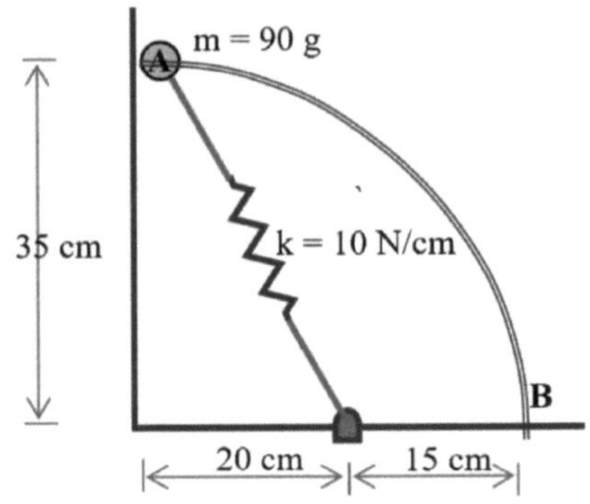

A. 9.5 m/s

B. 12.5 m/s

C. 16.8 m/s

D. Mass never reaches point *B*

Solution The stretched length of the spring (when the mass is at *A*) is given by $l = \sqrt{20^2 + 35^2} = 40.3$ cm

Therefore, at this position, the deformation of the spring is $\Delta = 40.3 - 25 = 15.3$ cm. Total energy in the system (using the horizontal line through *B* as the datum) when the mass is stationary at *A*:

$$E_A = 0.09 \times 9.81 \times 0.35 + \frac{1}{2} \times 10^3 \times 0.153^2 = 12 \text{ Nm}$$

When the mass is at *B*, the deformed length of the spring is 15 cm.
Therefore, at this position, the deformation of the spring is $\Delta = 15 - 25 = -10$ cm.
Total energy in the system (using the horizontal line through *B* as the datum) when the mass is at *B*:

$$E_B = \frac{1}{2} \times 10^3 \times 0.1^2 + \frac{1}{2} \times 0.09 \times V^2 = 5 + 0.045V^2$$

Equating $E_A = E_B$ yields $V = 12.47$ m/s

Answer is B.

CHAPTER 8

Mechanics of Materials

Shear Force and Bending Moment Diagrams

Figure 8.1 shows a determinate beam loaded with loads that are primarily transverse (to the axis of the beam).

In order to obtain the internal loads at any intermediate location of the beam, we make a section at this location. This section divides our structure into two substructures, as shown in Fig. 8.2. At the "cut" one must insert the set of internal loads as appropriate. There are three internal loads present at an arbitrary location of a two-dimensional beam. These are axial force N (longitudinal), shear force V (transverse), and bending moment M. Note how the internal loads occur in equal and opposite pairs on the two halves of the structure. In other words, the axial force exerted by the left half on the right is *equal and opposite* to that exerted by the right half on the left. The same holds for the shear force and bending moment.

The internal loads in Fig. 8.2 are drawn in keeping with a sign convention that is explained in more detail in Fig. 8.3. In that figure, the dashed line shows the original configuration of an element of the beam, whereas the solid lines show the type of deformation induced by these internal loads. Note that the internal forces on the element in Fig. 8.3 are consistent with the internal forces at a section shown in Fig. 8.2.

For any portion of the beam (except at discontinuities such as at location of point loads), the following relationships exist between the load function $w(x)$, the shear force function $V(x)$ and the bending moment function $M(x)$

$$\frac{dV}{dx} = -w(x) \tag{8.1}$$

$$\frac{dM}{dx} = V(x) \tag{8.2}$$

The equations shown above are in differential form. Geometrically, they have the following significance:

- The slope of the shear force function at any location is equal to the (negative) value of the load function at that location [Eq. (8.1)].

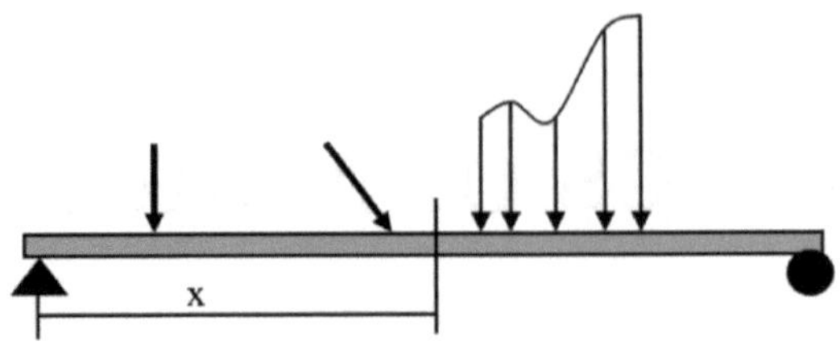

Figure 8.1 Transversely loaded beam.

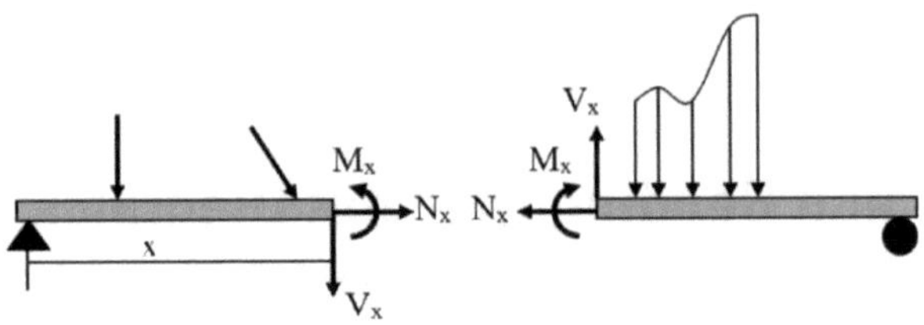

Figure 8.2 Internal loads for transversely loaded beam.

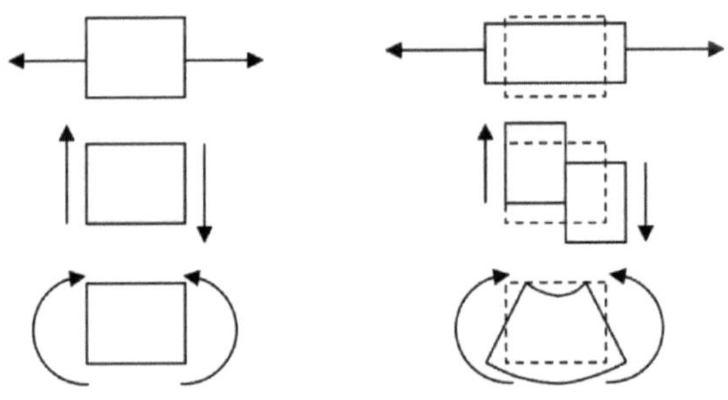

Figure 8.3 Adopted sign convention for internal loads.

- The slope of the bending moment function at any location is equal to the value of the shear force function at that location [Eq. (8.2)].

In integral form, these relationships may be expressed as

$$\Delta V = -\int w(x)\,dx \Rightarrow V_2 - V_1 = -\int_{x_1}^{x_2} w(x)\,dx \tag{8.3}$$

$$\Delta M = \int V(x)\,dx \Rightarrow M_2 - M_1 = \int_{x_1}^{x_2} V(x)\,dx \tag{8.4}$$

The geometric significance of the integral equations may be written as

- The change in shear force (between locations 1 and 2) is equal to the (negative) area under the load function between the same points [Eq. (8.3)].
- The change in bending moment (between locations 1 and 2) is equal to the area under the shear force function between the same points [Eq. (8.4)].

Shear force and bending moment diagrams can be constructed in two ways:

1. Using the geometric relationships [Eqs. (8.1) through (8.4)] between the load function $w(x)$, the shear force function $V(x)$, and the bending moment function $M(x)$, one can progress from the load diagram $w(x)$ to the shear force diagram $V(x)$ to the bending moment diagram $M(x)$.
2. To find the shear force function $V(x)$ or bending moment function $M(x)$ for a section of the beam, one makes a section or "cut" at that location, inserts the internal loads $V(x)$, $M(x)$ at the cut and solves for these functions.

The following geometric principles may be used to draw shear and bending moment diagrams, by first following the given load $w(x)$ to create the shear force $V(x)$ and then following the shear diagram to create the bending moment $M(x)$:

- Concentrated (point) loads push shear diagram up/down.
- Clockwise/counterclockwise couples (externally applied or reaction moment) push bending moment diagram up/down, respectively.
- Change in shear force between two locations = negative of the corresponding area under the load diagram.
- Change in bending moment between two locations = corresponding area under the shear diagram.
- Maximum/minimum bending moment occurs at location of zero shear.

Example 8.1

Draw the shear and bending moment diagrams for the beam shown below. The beam has an internal hinge at B (6 ft from the left support).

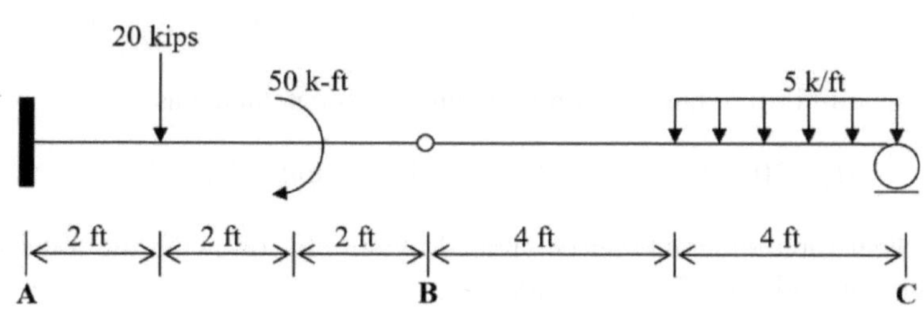

Solution The free body diagram for the entire beam is shown below.

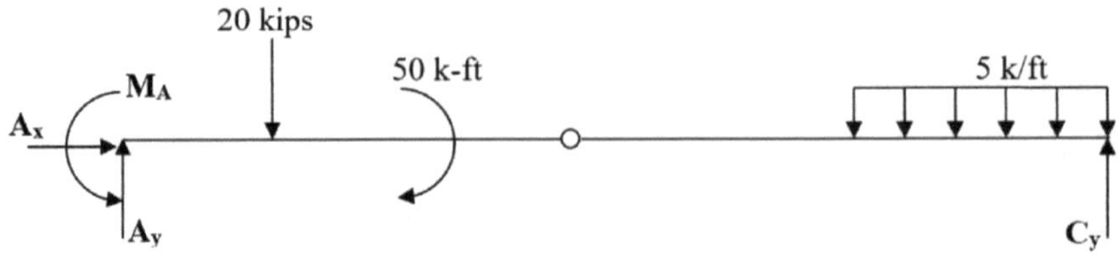

The first step would be to solve for the support reactions. The beam is statically determinate (three independent equilibrium equations + zero moment at the internal hinge). Even though a formal approach could be to write the four equations of equilibrium and solve for the four external reactions (A_x, A_y, M_A, and C_y), it is more convenient and time effective to separate the structure at the hinge B and start solving from the simple span BC (both ends moment-free).

Reactions: Since BC is a simple span, applying the lever rule

$$C_y = ¾ \times 20 = 15 \text{ kips and } B_y = ¼ \times 20 = 5 \text{ kips}$$

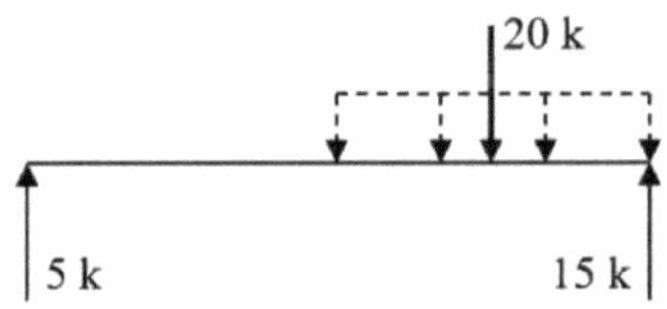

Since the total vertical load on the beam ABC is 40 kips, the vertical reaction at A is given by

$$A_y = 40 - 15 = 25 \text{ kips}$$

On segment AB, the vertical loads are 20 kips at a distance 5 ft from A and $B_y = 5$ kips at a distance 6 ft from A. These two forces produce a clockwise moment about A. In addition, there is an applied couple 50 kip · ft on segment AB.

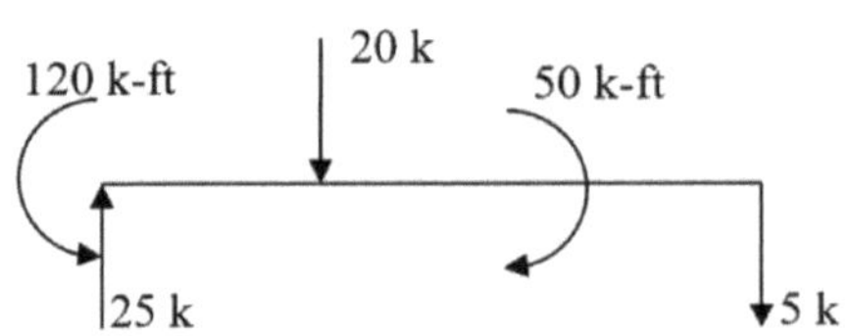

For AB to be in equilibrium, the reaction moment at A is calculated as

$$M_A = 20 \times 2 + 50 + 5 \times 6 = 120 \text{ kip·ft counterclockwise}$$

Next, the shear force diagram can be created by simply following the transverse (shear) loads such as point loads and distributed loads.

Shear Force Diagram

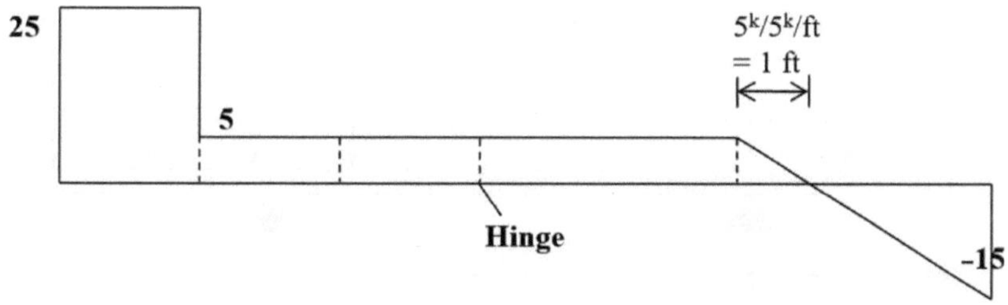

Note that at the internal hinge, there is an internal shear that exists as a pair of equal and opposite forces and therefore does not affect the diagram. However, the shear at this location (the hinge shear) is reflected in the diagram. Also, the uniformly distributed load has a resultant (area under the load) of 20 kips. The change in shear force between $x = 10$ ft and $x = 14$ ft is the negative of the area under the load. Thus, the shear force changes from $V = +5$ kips at $x = 10$ to $V = -15$ kips at $x = 14$ ft. This change occurs at a uniform rate (linear). The location at which the shear force goes through a zero is given by

$$\Delta x = \frac{\Delta V}{w} = \frac{5 \text{ kips}}{5 \text{ kips/ft}} = 1 \text{ ft}$$

Bending Moment Diagram

The bending moment diagram is constructed by "accumulating areas" from the shear force diagram. Since the bending moment function is obtained by integration of the shear function, it is one order more complex than the order of the corresponding shear diagram segment. In other words, if the shear function is linear, the corresponding moment function is quadratic.

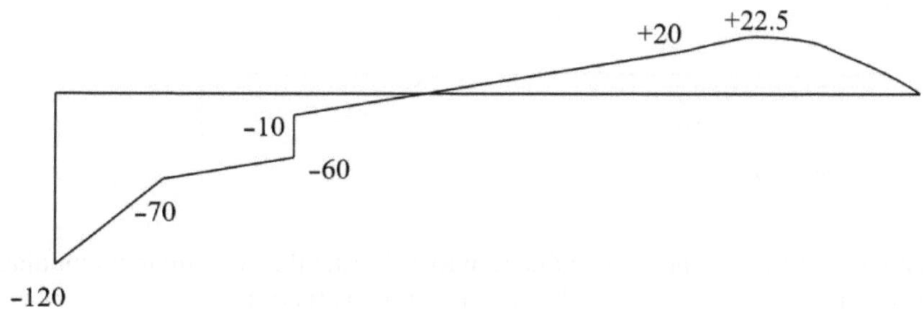

In the example above, there are two locations where a couple moment exists. At $x = 0$, there is a reaction moment $M = 120$ kip·ft (counterclockwise) and at $x = 4$ ft, there is an applied moment $M = 50$ kip·ft (clockwise). These introduce corresponding discontinuities in the moment diagram at these locations. The counterclockwise reaction moment pushes the BMD down and the clockwise applied moment pushes the BMD up.

Example 8.2

Draw the shear force diagram and the bending moment diagram for the simply supported beam with overhang loaded with a distributed load, as shown below.

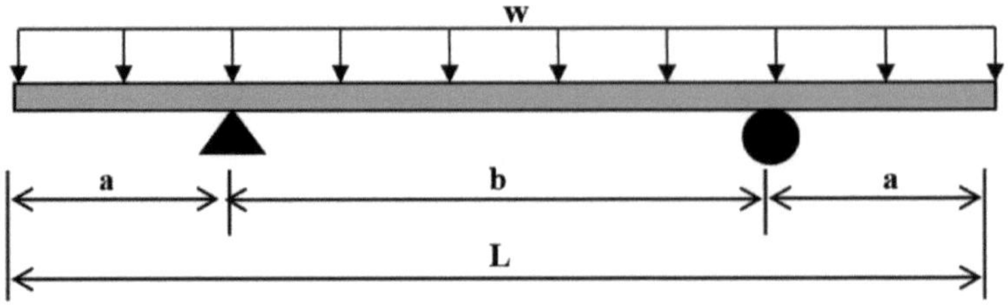

Method 1—Geometric Method

As before, we can write the equations of static equilibrium and solve for the support reactions. However, in this case, we can use symmetry arguments to say that due to symmetry of the structure and the loading, the load wL will be shared equally by the two supports. Next, we will follow the load to draw the shear force diagram.

The overhang to the left of the first support carries a total load of wa. The shear force diagram slopes down from a value of zero to $-wa$. At the left support, the reaction ($wL/2$) pushes the diagram up by $wL/2$ to a value $+wb/2$.

Between supports, the total load is wb. Thus, the shear force diagram goes from a value of $+wb/2$ to $-wb/2$, sloping down continuously at the same rate as the load, that is, w per unit length.

At the right support, the upward reaction $wL/2$ pushes the diagram up from a value of $-wb/2$ to $+wa$. Over the right overhang, the load wa pushes the diagram down to zero.

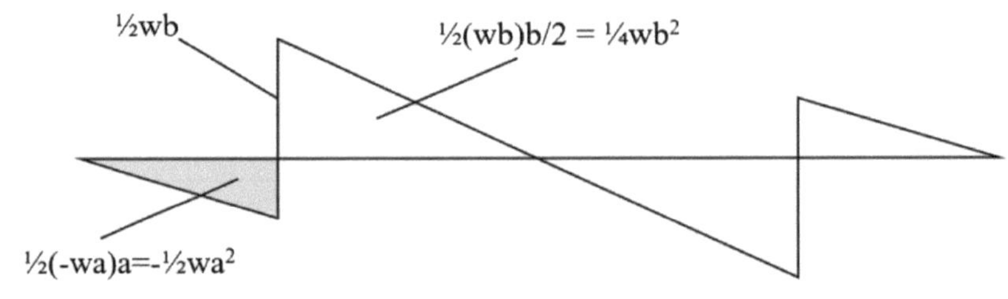

We can then follow the shear force diagram to construct the bending moment diagram. The following geometric facts can be used in our construction

- The value of the bending moment is zero at the free ends.
- The change in the bending moment from point A to point B is equal to the area under the shear force diagram between A and B.
- The order of the bending moment curve is one order higher than the shear force curve. (If the SFD is a straight line, i.e. x^1, then the BMD is a parabola, i.e. x^2.)

- The location of zero shear force, or more correctly, the location of change of sign of the shear force is the location of the maximum (or minimum) bending moment.

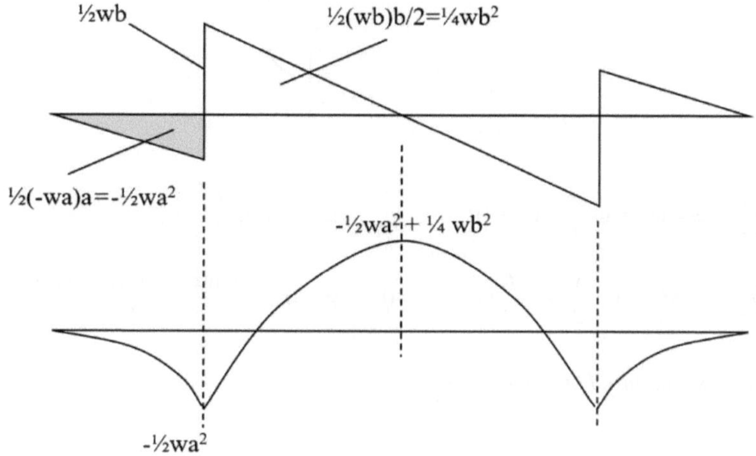

Method 2—Explicit Forms of *V*(*x*) and *M*(*x*) Functions

As we see from the figure, the loading on the beam changes at two points—the supports, where the reactions are introduced. These two points divide the beam into three segments: AB, BC, and CD.

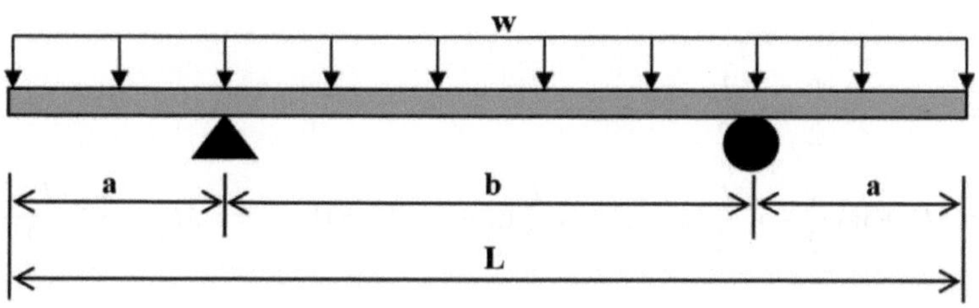

First section in segment AB ($0 < x < a$)

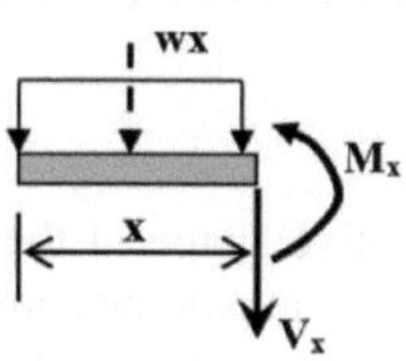

$$\sum F_y = -wx - V_x = 0 \qquad \Rightarrow V_x = -wx$$

$$\sum F_y = M_x + wx\frac{x}{2} = 0 \qquad \Rightarrow M_x = \frac{-wx^2}{2}$$

Second section in segment BC $(a < x < a + b)$

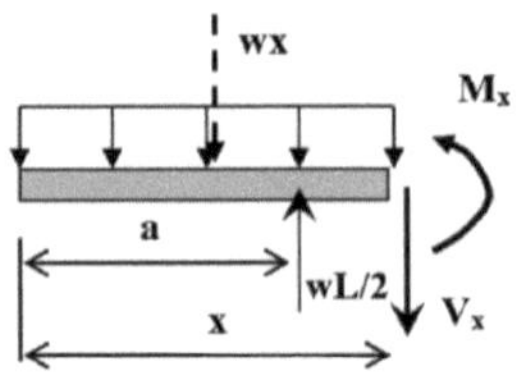

$$\sum F_y = -wx + w\frac{L}{2} - V_x = 0 \qquad \Rightarrow V_x = w\frac{L}{2} - wx$$

$$\sum F_y = M_x + wx\frac{x}{2} - w\frac{L}{2}(x - a) = 0 \qquad \Rightarrow M_x = \frac{-wx^2 + wLx - wLa}{2}$$

Third section in segment CD $(\ a + b < x < L\ )$

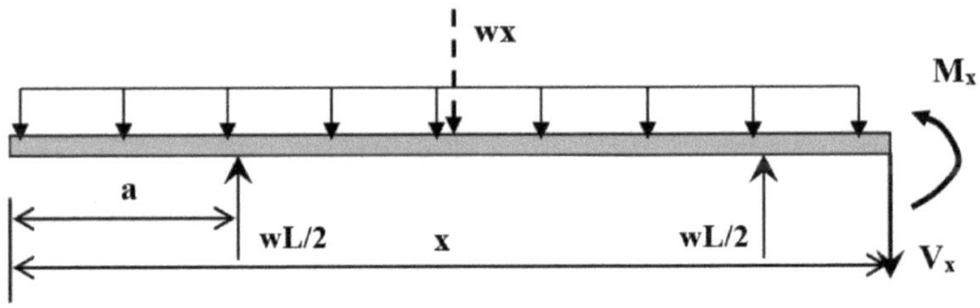

$$\sum F_y = -wx + w\frac{L}{2} + w\frac{L}{2} - V_x = 0 \qquad \Rightarrow V_x = wL - wx$$

$$\sum F_y = M_x + wx\frac{x}{2} - w\frac{L}{2}(x - a) - w\frac{L}{2}(x - a - b) = 0 \qquad \Rightarrow M_x = \frac{-w(L - x)^2}{2}$$

Example 8.3 illustrates the geometric principles behind drawing of shear force and bending moment diagrams. Note that specific values (of reactions, etc.) are not computed in this example, but attention is paid to consistency of the diagrams.

Example 8.3

Knowledge of the approximate qualitative shape of the bending moment diagram allows us to know the nature of the normal stress—tensile or compressive—at various locations of the beam. Consistent with the sign convention adopted here, a location with positive bending moment creates compression above the neutral axis and tension below. For negative bending moment, this pattern is reversed. For elements such as concrete beams in flexure, the inherent weakness of the concrete in tension makes it necessary to place steel reinforcement in the zones with maximum tensile stress.

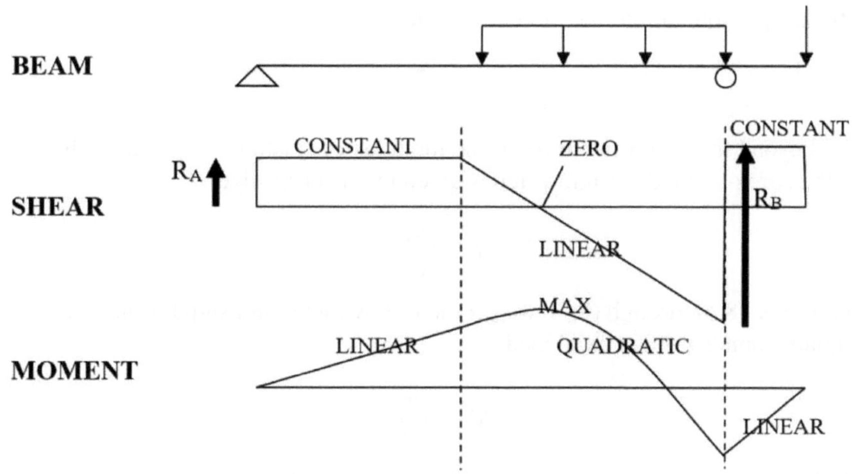

Normal Stress and Strain

When load (F) acts normal (perpendicular) to a surface and is assumed to be uniformly distributed on the cross-sectional area (A), the normal stress is defined as

$$\sigma = \frac{F}{A} \tag{8.5}$$

According to the Saint Venant principle, the uniform distribution assumption is valid at sections that are distant from points of application of loads. However, in the vicinity of the load application, the stress distribution is likely to be nonuniform.

The normal strain ε is defined as the proportional change of length in the direction of the load. In other words, if a prismatic rod is subjected to an axial load F, the rod will change length (extension for a tensile load and shortening for a compressive load). The normal strain is then calculated as the ratio of the change in length ΔL (with appropriate sign) and the original length L_0

$$\varepsilon = \frac{\Delta L}{L_0} \tag{8.6}$$

Hooke's Law

The Hooke law states that, for small values of stress, there is a linear relationship between the applied stress σ and the strain ε resulting from it. The constant of proportionality is named the Young's modulus or the modulus of elasticity E, which is the slope of the stress-strain diagram

$$\sigma = E\varepsilon \tag{8.7}$$

Within the proportional limit, the modulus of elasticity is given by

$$E = \frac{\sigma}{\varepsilon} \tag{8.8}$$

However, beyond the proportional limit, the modulus of elasticity is given by the (changing) slope of the stress-strain diagram, which is given by the derivative

$$E = \frac{d\sigma}{d\varepsilon} \tag{8.9}$$

Combining Eqs. (8.5) through (8.7), we get the following relationship for the change in length of a prismatic member under axial load

$$\Delta L = \frac{PL_o}{AE} \tag{8.10}$$

The deformed dimension in a particular direction can be written as

$$L' = L_o + \Delta L = L_o(1+\varepsilon) \tag{8.11}$$

Example 8.4

A composite cable (E = 180 GPa, diameter = 1 cm, original length = 150 cm) is subject to the following design requirements:

Maximum elongation = 2 mm

Allowable stress = 400 MPa

What is the maximum permissible axial tensile load on the cable?

Solution Based on the stress criterion (P/A <450 MPa), the maximum permissible load area is

$$P \le \sigma A = 400 \times 10^6 \times \frac{\pi}{4} \times 0.01^2 = 31{,}416 \text{ N} = 31.4 \text{ kN}$$

Based on the maximum elongation criterion (PL/AE <2 mm), the maximum permissible load area is

$$P \le \frac{AE\Delta}{L} = \frac{\frac{\pi}{4} \times 0.01^2 \times 180 \times 10^9 \times 0.002}{1.5} = 18{,}850 \text{ N} = 18.85 \text{ kN}$$

The second criterion governs. Maximum safe load = 18.85 kN.

Elastic versus Inelastic Behavior

Elasticity is the property of a material that allows it to return to its original (undeformed) state when loads are removed. In other words, an elastic material "recovers" *all* of its deformation

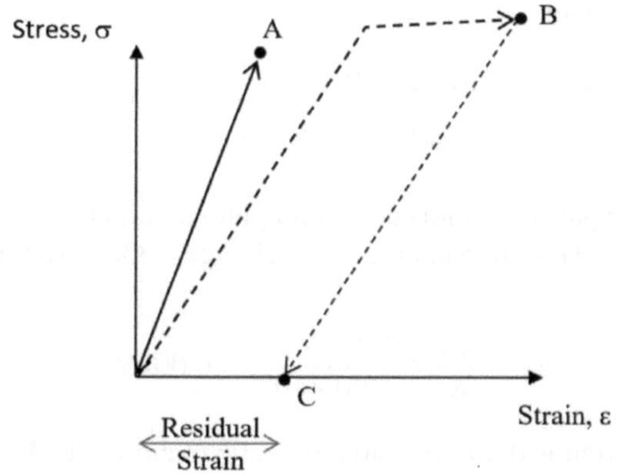

Figure 8.4 Elastic versus inelastic deformation.

(strain) when the applied stress is removed. On a plot of stress versus strain, this would mean that while loading takes the material from the origin (initial unloaded state) to the state indicated by point A in figure 8.4, removal of the load would cause the material to retrace the curve back to the origin. In this case, there is no residual strain. On the other hand, when a material is loaded past its elastic limit (as indicated by the dashed line) to point B, removal of the load results in the σ-ε curve following a slope parallel to the initial straight portion of the loading curve. This takes it back to point C when the load (and therefore, stress) has been completely removed. Therefore, upon unloading, the material does not return to its original state and has a certain amount of permanent deformation. This is also called residual strain. See Fig. 8.4.

NOTE Elastic behavior is *not* synonymous with linear behavior, even though most materials exhibit linear and elastic behavior at low levels of stress.

Example 8.5

A prismatic steel rod (F_y = 400 MPa, F_u = 550 MPa, E = 200 GPa, G = 78 GPa) has the following properties when unloaded:

Length = 40 cm

Diameter = 1 cm

The rod is subjected to axial tension until the rod elongates by 2 mm and then unloaded to a residual load of 5 kN. What is the final length of the rod?

Solution The axial force that causes yield, $P_{\text{yield}} = F_y A = 4 \times 10^8 \times \frac{\pi}{4} \times 0.01^2 = 31.42 \times 10^3$ N

The axial elongation at yield:

$$\Delta_{\text{yield}} = \frac{P_{\text{yield}}L}{AE} = \frac{31.42 \times 10^3 \times 0.4}{\frac{\pi}{4} \times 0.01^2 \times 200 \times 10^9} = 0.0008 \text{ m} = 0.8 \text{ mm}$$

Therefore, the final elongation (2 mm) is beyond yield and includes a plastic deformation of 1.2 mm. Unloading from 31.42 to 5 kN is a stress change = 336.4 MPa and the strain recovery, at a slope of E, is

$$\Delta\varepsilon = \frac{\Delta\sigma}{E} = \frac{336.4 \times 10^6}{200 \times 10^9} = 0.00168$$

The recovery of elongation is 0.00168 × 40 cm = 0.0673 cm = 0.673 mm
Final elongation = 2 − 0.673 = 1.327 mm.

Poisson's Ratio

As load is applied in one direction, an elastic material experiences a corresponding deformation (strain). We will refer to this direct effect as *longitudinal strain.* In addition, because of its elasticity, it also experiences an indirect and proportional deformation in the other two orthogonal (lateral) directions. We will refer to this as *lateral strain.* Poisson's ratio (ν) is the constant of proportionality between the longitudinal strain and the lateral strain. Thus

$$\varepsilon_{\text{lateral}} = -\nu\varepsilon_{\text{longitudinal}} \tag{8.12}$$

The minus sign in Eq. (8.12) signifies that the lateral strain is opposed in nature to the longitudinal strain. For example, if a prismatic rod is subject to axial tension, the longitudinal strain is positive (i.e., the rod elongates) while the lateral strain is negative, which causes the cross section to reduce. Poisson's ratio dictates that these two deformations are not independent of each other, but are in fact directly proportional to each other.

Shear Stress

Within the proportional limit, the relationship between shear stress τ and shear strain γ is given by

$$\tau = G\gamma \tag{8.13}$$

where the constant of proportionality is named the shear modulus or the modulus of rigidity (G).

For elastic materials, the two moduli (E and G) are related according to

$$G = \frac{E}{2(1+\nu)} \tag{8.14}$$

where ν is the Poisson ratio.

Example 8.6

A block of elastic material is subjected to a vertical load $P = 200$ kN, as shown below. The left face of the block is fused to a rigid surface, while the load P is distributed evenly over the right face as shown. The elastic properties of the material are modulus of elasticity = 4 GPa, Poisson's ratio = 0.28. What are the minimum values of dimensions "a" and "b" if the maximum allowable shear stress is 30 MPa and the maximum downward deflection of the loaded face is 1 mm?

A. $a = 45$ mm; $b = 70$ mm

B. $a = 35$ mm; $b = 70$ mm

C. $a = 55$ mm; $b = 85$ mm

D. $a = 35$ mm; $b = 85$ mm

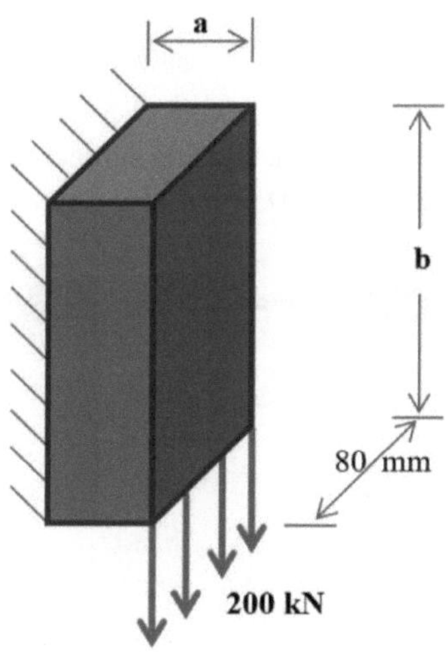

Solution The modulus of rigidity (shear modulus), G is given by $G = \dfrac{E}{2(1+\nu)} = \dfrac{4}{2(1+0.28)} = 1.56$ GPa

Shear stress: $\tau = \dfrac{P}{A} = \dfrac{200 \times 10^3}{0.08 \times b} \leq 30 \times 10^6 \Rightarrow b \geq 0.084 \text{ m} = 84 \text{ mm}$

Shear strain: $\gamma = \dfrac{\delta}{a}$

Modulus of rigidity:

$$G = \frac{\tau}{\gamma} = \frac{P/0.08b}{\delta/a} = \frac{Pa}{0.08b\delta} \Rightarrow \delta = \frac{Pa}{0.08bG} \leq 0.001$$

$$\Rightarrow a \leq \frac{0.001 \times 0.08 \times b \times 1.56 \times 10^9}{200 \times 10^3} \leq 0.624b$$

Thus, edge "*b*" must be at least 84 mm and thickness "*a*" must be at most 0.624 times *b*

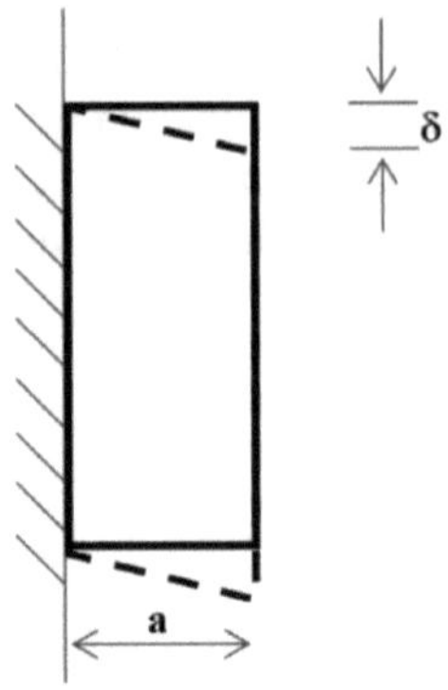

Answer is D (only one to meet *both* criteria).

Example 8.7

A steel tensile coupon cut from 5-mm-thick sheet is subject to a tensile force as shown below. What is the axial strain in the narrowed (20 mm wide) test section?

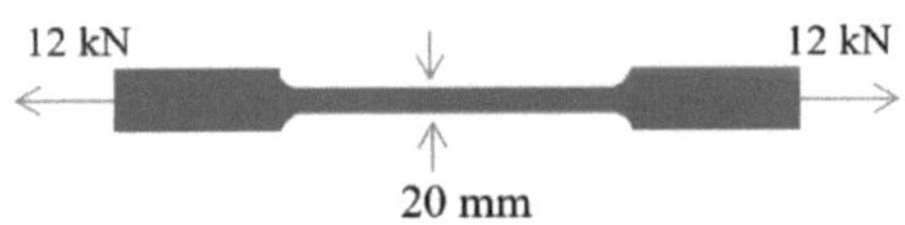

A. 100 $\mu\varepsilon$

B. 200 $\mu\varepsilon$

C. 600 $\mu\varepsilon$

D. 1200 $\mu\varepsilon$

Solution Cross-sectional area $A = 20 \times 5 = 100 \text{ mm}^2 = 1 \times 10^{-4} \text{ m}^2$

Axial stress: $\sigma = \dfrac{P}{A} = \dfrac{12{,}000}{1\times10^{-4}} = 1.2\times10^{8}\,\text{Pa} = 120 \text{ MPa}$

Axial (longitudinal) strain: $\varepsilon = \dfrac{\sigma}{E} = \dfrac{1.2\times10^{8}}{2\times10^{11}} = 6\times10^{-4} = 600\ \mu\varepsilon$

Answer is C.

Example 8.8

A steel tensile coupon cut from 5-mm-thick sheet is subject to a tensile force as shown below. What is the altered width (20 mm dimension)?

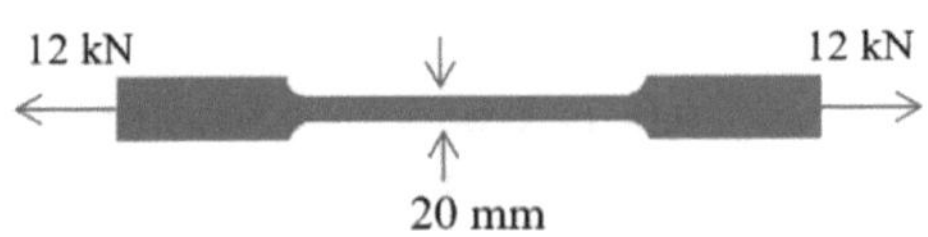

A. 20.0036 mm

B. 19.9964 mm

C. 20.0120 mm

D. 19.9880 mm

Solution Cross-sectional area $A = 20 \times 5 = 100\ \text{mm}^2 = 1 \times 10^{-4}\ \text{m}^2$

Axial stress: $\sigma = \dfrac{P}{A} = \dfrac{12{,}000}{1\times 10^{-4}} = 1.2\times 10^{8}\ \text{Pa} = 120\ \text{MPa}$

Axial (longitudinal) strain: $\varepsilon = \dfrac{\sigma}{E} = \dfrac{1.2\times 10^{8}}{2\times 10^{11}} = 6\times 10^{-4} = 600\ \mu\varepsilon$

Poisson's ratio $\nu = 0.3$ (for steel). Therefore, transverse strain in the other two directions is given by

$$\varepsilon_{\text{lateral}} = -0.3 \times 600\ \mu\varepsilon = -180\ \mu\varepsilon = -0.00018$$

Therefore, the new width $= 20 \times (1 - 0.00018) = 19.9964$ mm
Answer is B.

Generalized Hooke's Law

The discussion so far has been to relate a single component of normal stress to the corresponding component of normal strain. Equation (8.7) is therefore a statement of uniaxial Hooke's law.

When an element of an elastic material is subject to all nine (three normal + six shear) stress components as shown in Fig. 8.5, the statement of Hooke's law needs to be generalized to three dimensions as follows

$$\varepsilon_{xx} = \frac{\sigma_{xx}}{E} - \nu\frac{\sigma_{yy}}{E} - \nu\frac{\sigma_{zz}}{E} \tag{8.15a}$$

$$\varepsilon_{yy} = \frac{\sigma_{yy}}{E} - \nu\frac{\sigma_{xx}}{E} - \nu\frac{\sigma_{zz}}{E} \tag{8.15b}$$

$$\varepsilon_{zz} = \frac{\sigma_{zz}}{E} - \nu\frac{\sigma_{xx}}{E} - \nu\frac{\sigma_{yy}}{E} \tag{8.15c}$$

$$\gamma_{xy} = \frac{\tau_{xy}}{G} \tag{8.15d}$$

$$\gamma_{yz} = \frac{\tau_{yz}}{G} \tag{8.15e}$$

$$\gamma_{zx} = \frac{\tau_{zx}}{G} \tag{8.15f}$$

Each expression for normal strain has one longitudinal strain component and two transverse strain components.

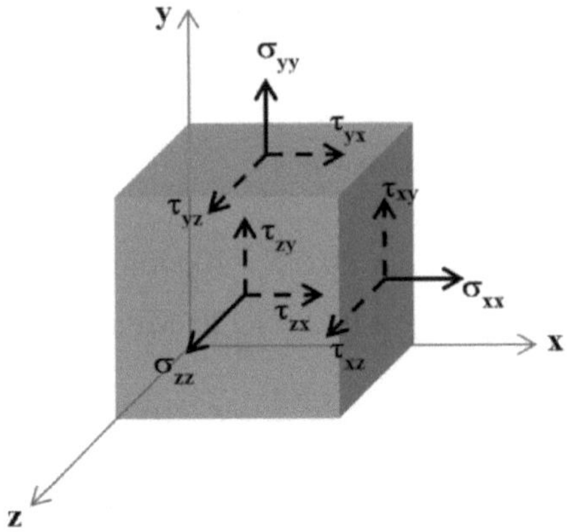

Figure 8.5 Stress components in three dimensions.

Example 8.9

The 3D steel element shown below is subject to concentric normal force pairs on all six faces. Assume that all loads are uniformly distributed on the faces of the element. What is the deformed length in the *z* direction?

A. 75.15 mm

B. 74.92 mm

C. 75.08 mm

D. 75.04 mm

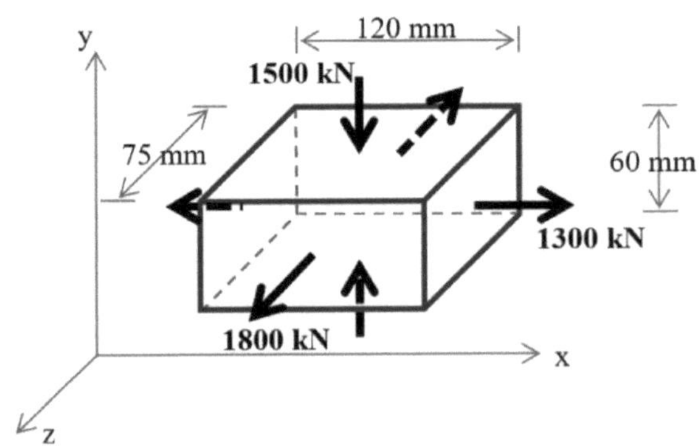

Solution The following elastic properties of steel are listed in Table 1 on page 84 of the *FE Reference Handbook*, version 9.4: $E = 200$ GPa; $\nu = 0.3$.

The normal stresses in the three orthogonal directions are

$$\sigma_{xx} = +\frac{1300\times10^3}{0.075\times0.06} = +289 \text{ MPa}$$

$$\sigma_{yy} = -\frac{1500\times10^3}{0.075\times0.12} = -167 \text{ MPa}$$

$$\sigma_{zz} = +\frac{1800\times10^3}{0.12\times0.06} = +250 \text{ MPa}$$

The normal strain in the *z* direction is

$$\varepsilon_{zz} = +\frac{\sigma_{zz}}{E} - \nu\left(\frac{\sigma_{xx}}{E}+\frac{\sigma_{yy}}{E}\right) = \frac{250\times10^6 - 0.3\times(289-167)\times10^6}{200\times10^9} = 1.067\times10^{-3}$$

Deformed length of the 75 mm edge

$$L' = 75(1+1.067\times10^{-3}) = 75.08 \text{ mm}$$

Answer is C.

Typical Stress-Strain Curve for Mild Steel

A material such as mild steel exhibits a wide range of behavior upon being subject to increasing stress. A typical plot of stress versus strain for a mild steel sample in tension is shown in Fig. 8.6. When stresses are low, the material remains elastic and exhibits

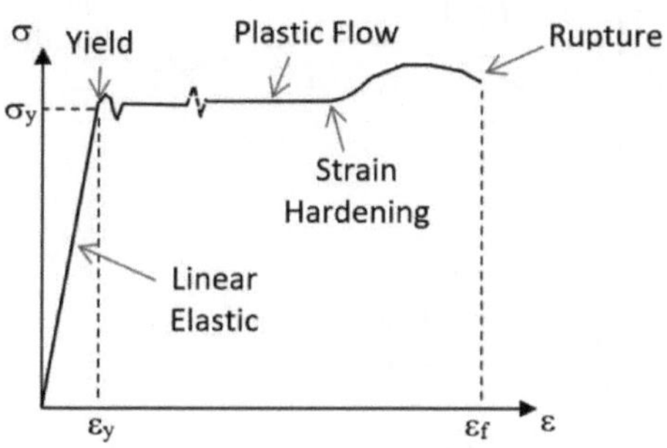

Figure 8.6 Stress-strain diagram for mild steel with well-defined yield point.

linear behavior (strain proportional to stress). In this regime, the slope of the stress-strain diagram is equal to the nominal value of the modulus of elasticity ($E_{steel} = 29 \times 10^6$ psi = 200 GPa). This continues up to the proportional limit, beyond which Eq. (8.8) is no longer valid. Often, the curve indicates an upper and a lower yield point, but for all practical purposes, these are essentially identical. Beyond the yield point, the material undergoes "plastic flow," where it does not take further increase of stress to cause additional strain. If the stress is simply held at the yield value, the strain increases until the strain hardening point is reached.

Beyond the strain hardening point, it again takes an increase in stress to cause additional strain. However, the stiffness (quantified by the slope of the stress-strain diagram) is less than the nominal value of the modulus of elasticity. The sample fails (fracture or rupture) at a failure strain ε_f.

For a stress-strain diagram (such as in Fig. 8.7) that does not show a well-defined yield point, an empirical rule called the "0.2% offset method" is used to identify the yield point. A line is drawn through the point $\varepsilon = 0.002$ and parallel to the initial straight portion of the σ-ε curve. The point where this line (shown dashed on the figure) intersects the σ-ε curve is taken to be the yield point and the x and y coordinates of this point are taken to be the yield strain ε_y and the yield stress σ_y, respectively.

Ductility

The ductility of a material is defined as its ability to deform under tensile stress. On the other hand, malleability is the material's ability to deform under compressive stress. A representative measure of a material's ductility is the ratio of the final strain (at failure) to the yield strain

$$\text{Ductility} = \frac{\varepsilon_f}{\varepsilon_y} \tag{8.16}$$

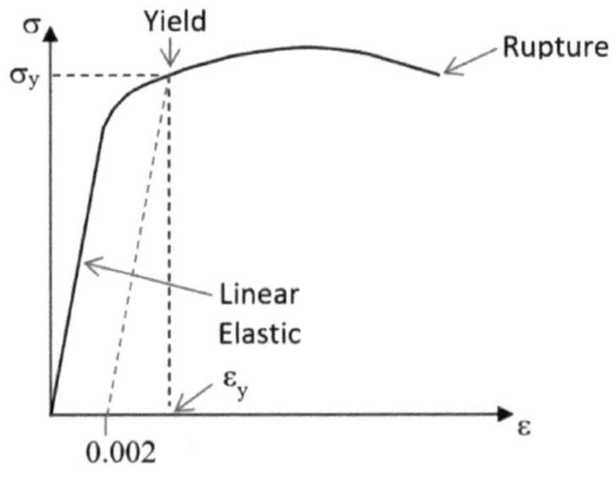

Figure 8.7 Stress-strain diagram for steel without well-defined yield point.

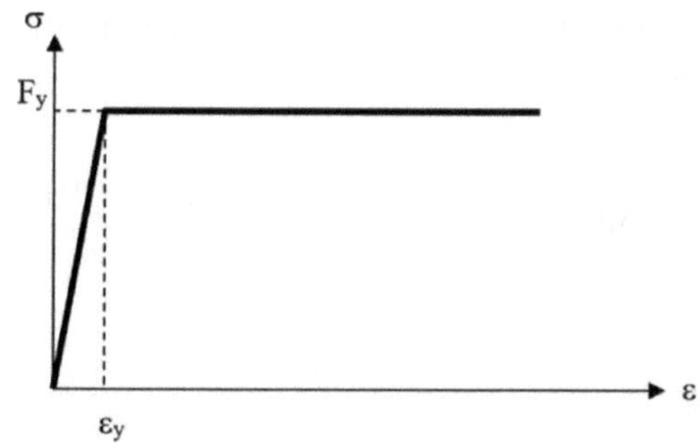

Figure 8.8 Elasto-platic model for steel.

Elasto-Plastic Model

The actual stress-strain diagrams, as shown in Figs. 8.6 and 8.7, are too complex for any kind of mathematical modeling. A simplified model that is adequate for many kinds of analyses is the elasto-plastic model, shown in Fig. 8.8. In this model, the stress-strain relationship for steel is assumed to be linear and elastic up to the yield point, and then perfectly plastic ($E = 0$) thereafter. According to this model, the stress remains constant at the yield stress F_y, once the yield strain is exceeded. This model does not account for the limit state of rupture.

Example 8.10

The results of a tensile test of a prismatic metal specimen are summarized in the curve below. What is most nearly the modulus of elasticity of the specimen?

A. 190 GPa

B. 230 GPa

C. 250 GPa

D. 300 GPa

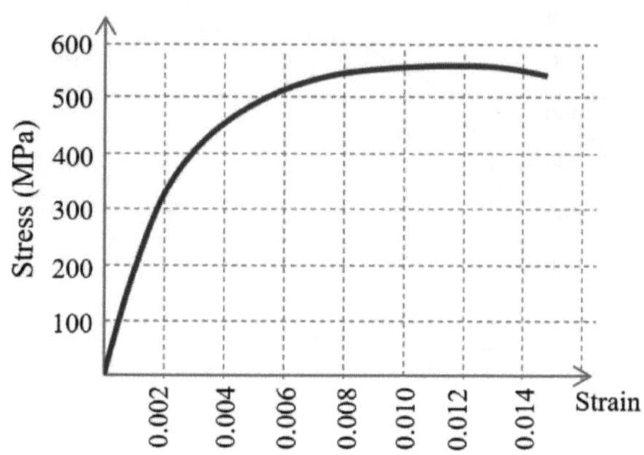

Solution The modulus of elasticity is calculated as the slope of the initial tangent of the curve. This is approximately

$$E = \frac{\Delta\sigma}{\Delta\varepsilon} = \frac{380 \times 10^6}{0.002} = 1.9 \times 10^{11} = 190 \text{ GPa}$$

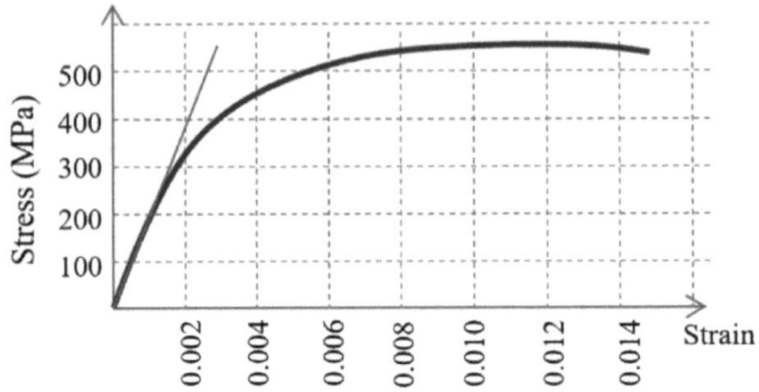

Answer is A.

Example 8.11

The results of a tensile test of a prismatic metal specimen are summarized in the curve below. What is most nearly the yield stress of the specimen?

A. 325 MPa

B. 375 MPa

C. 425 MPa

D. 475 MPa

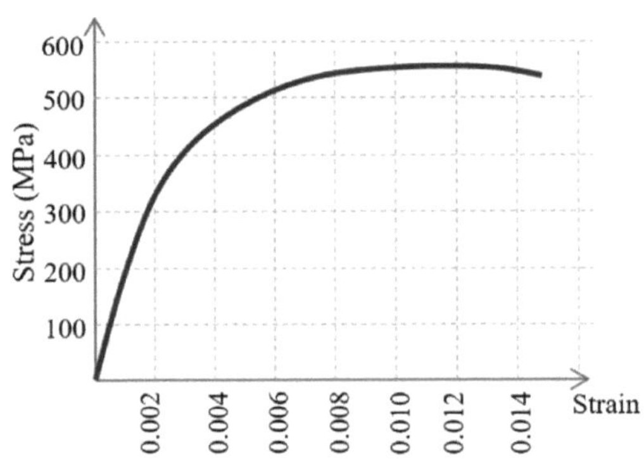

Solution The yield stress is estimated using the 0.2% offset rule. From a strain = 0.002, a line is drawn parallel to the initial tangent of the curve. This is shown on the figure below. The intersection of this line with the curve is nominally taken to be the yield point.

Yield stress: $\sigma_y = 470$ MPa

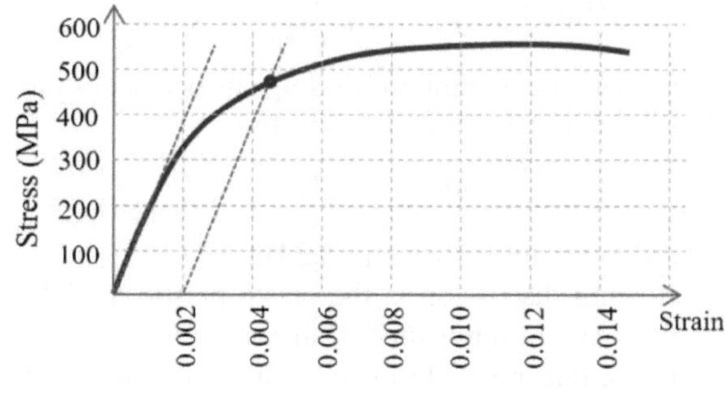

Answer is D.

Bending Stress

If the material behavior is linear and elastic, and subject to the assumption that *plane sections remain plane after bending*, the bending stress at a plane located at a distance y from the neutral plane (see Fig. 8.9) is given by

$$\sigma_b = -\frac{My}{I} \tag{8.17}$$

where M = bending moment on cross section
y = distance from the neutral axis to the location where the stress is being calculated
I = moment of inertia of the cross section about the neutral axis

NOTE In this discussion, it has been assumed that the bending moment occurs about the *x*-axis and therefore—the relevant properties of the section are I_x, S_x, etc. Similarly, the variable y represents the distance of a plane of interest from the x axis. The same concept can be extended to bending about the y axis.

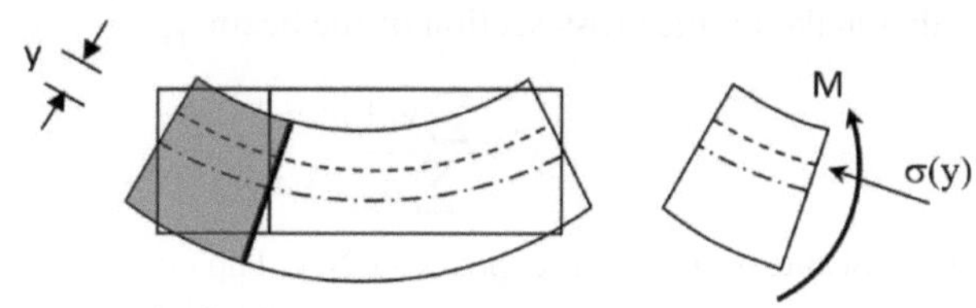

Figure 8.9 Bending deformation.

This stress distribution is linear, zero at the neutral plane and has maximum value at the plane furthest from the neutral plane ($y = y_{max}$).
Therefore, the maximum bending stress is given by

$$\sigma_{max} = \frac{M_x y_{max}}{I_x} = \frac{M_x}{I_x / y_{max}} = \frac{M_x}{S_x} \tag{8.18}$$

As the moment is increased, the maximum bending stress reaches the yield stress (F_y). The moment that causes first yield in the section is given by

$$M_{yield} = S_x F_y \tag{8.19}$$

where S_x is the elastic section modulus [see Eq. (8.23)].

The basis for the following discussion is the elasto-plastic model—commonly used for modeling structural steel behavior. In this model, shown in Fig. 8.8, the stress-strain relationship for steel is assumed to be linear and elastic up to the yield point, and then perfectly plastic ($E = 0$) thereafter. According to this model, the stress remains constant at the yield stress F_y, once the yield strain is exceeded.

Therefore, as the moment is increased beyond M_{yield} [Eq. (8.17)], the outer fibers begin to yield and the "plastic behavior zone" gradually progresses inward. The ultimate limit of such a stress distribution is when the plastic zones progress all the way to the neutral axis (all planes at or beyond yield strain). According to the elasto-plastic model, there is no more capacity for increased internal stress within the section and therefore, the internal moment capacity is at its maximum. Any increase in load beyond this point will cause the formation of an instability which exhibits itself as a "hinge-like" collapse mechanism. This is referred to as "plastic hinge formation." The moment that causes plastic hinge formation is the ultimate moment capacity of the section and is often referred to as the plastic moment capacity. This moment M_p can be calculated as

$$M_P = Z_x F_y \tag{8.20}$$

where Z_x is the plastic section modulus.

Elastic Section Modulus

The following steps must be followed to calculate the elastic section modulus:

1. Locate elastic neutral axis (ENA) which is the same as geometric centroid. The location of the ENA is the weighted arithmetic average of the centroid locations of the sub-areas that make up the cross section of the beam

$$\bar{y} = \frac{\sum y_i A_i}{\sum A_i} \tag{8.21}$$

2. Find moment of inertia I_{NA} with respect to ENA. The contribution of individual parts of the cross section to the overall value of I_{NA} is calculated according to the parallel axis theorem (also known as the moment of inertia transfer theorem)

$$I_{NA} = I_c + Ad^2 \tag{8.22}$$

where I_c = moment of inertia with respect to the local centroid
A = area of the section
d = distance between the centroidal axis of the shape and the overall centroid of the compound section

3. The elastic section modulus S is calculated by

$$S = \frac{I_{NA}}{y_{\max}} \tag{8.23}$$

where $y_{\max}$ is the distance of fiber farthest from ENA.

Plastic Section Modulus

The following steps must be followed to calculate the plastic section modulus:

1. Locate plastic neutral axis (PNA) which is the plane that divides the cross section into two equal halves ($A_{\text{top}} = A_{\text{bottom}}$).
2. The plastic section modulus Z is calculated as $Z = \sum A_i y_i$ which is the first moment of all areas about PNA. These first moments are all taken as positive.

Combined Axial and Bending Stress

For a section subject to an eccentric axial load, the eccentric load can be replaced by the combination of a concentric load P and a moment Pe, where e is the eccentricity of the load. The total stress is given by a superposition of the normal stress P/A and the bending stress My/I (Fig. 8.10).

$$\sigma_{\max} = \frac{P}{A} \pm \frac{My}{I} = \frac{P}{A} \pm \frac{Pey}{Ar^2} = \frac{P}{A}\left[1 \pm \frac{ey}{r^2}\right] \tag{8.24}$$

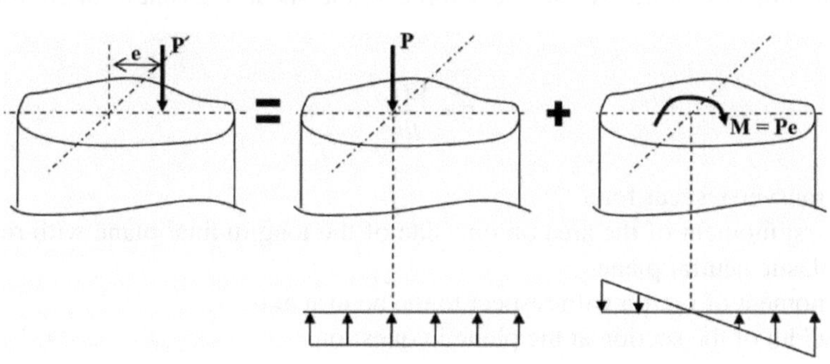

Figure 8.10 Superposition of axial and bending stress.

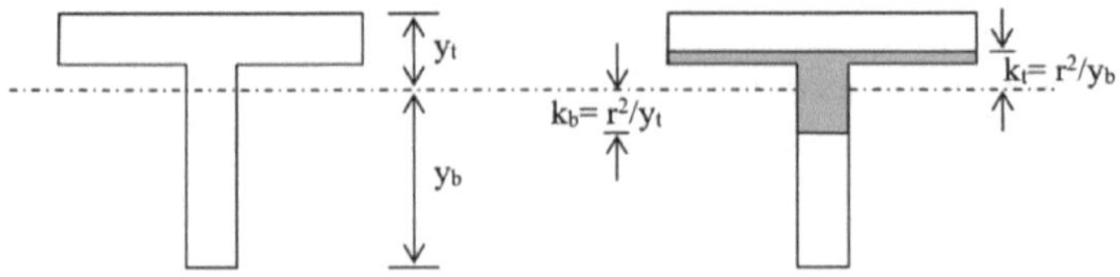

Figure 8.11 Kern limits.

For elements such as prestressed concrete beams, the underlying design philosophy is to avoid (or minimize) tensile stress in the cross section. This may be accomplished by keeping the eccentricity of the prestressing force within the "kern" of the cross section. The geometric limits of the kern may be calculated by setting the minimum stress equal to zero

$$1 \pm \frac{ey}{r^2} = 0 \Rightarrow e_{\max} = \frac{r^2}{y_{\max}} \tag{8.25}$$

For example, consider the T-section shown in Fig. 8.11. Let us say that the moment of inertia and area have been calculated and yield a radius of gyration equal to r. The neutral axis is shown, and this defines the distance of the extreme top fiber as y_t and the distance of the extreme bottom fiber as y_b. The outer limits of the kern of the section, shown shaded in the figure on the right, are calculated using

$$k_t = \frac{r^2}{y_b} \quad \text{and} \quad k_b = \frac{r^2}{y_t} \tag{8.26}$$

If the load is applied within the shaded area, the stresses are of the same type everywhere in the section (all compression or all tension).

For a rectangular section, $k_t = k_b = h/6$, leading to the common rule of thumb that stress can be of one type (all tension or all compression) if the load is *within the middle third* of the section width.

Shear Stress due to Transverse Load

For a section subject to a transverse shear force V, the shear stress at a longitudinal plane is given by

$$\tau = \frac{VQ}{Ib} \tag{8.27}$$

where V = transverse shear force
Q = first moment of the area on one side of the longitudinal plane with respect to the elastic neutral plane
I = moment of inertia with respect to the neutral axis
b = width of the section at the plane in question

These are illustrated in Fig. 8.12.

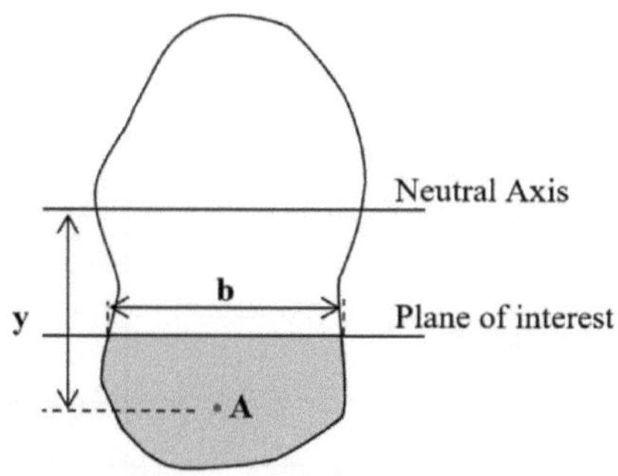

Figure 8.12 Parameters for shear stress calculation.

The first moment of the area A is calculated as

$$Q = Ay \tag{8.28}$$

It can be proved that the maximum shear stress occurs at the elastic neutral axis.

Example 8.12

The I-shaped section shown below is subject to transverse shear force $V = 90$ kips. Calculate the maximum shear stress (at the elastic neutral axis).

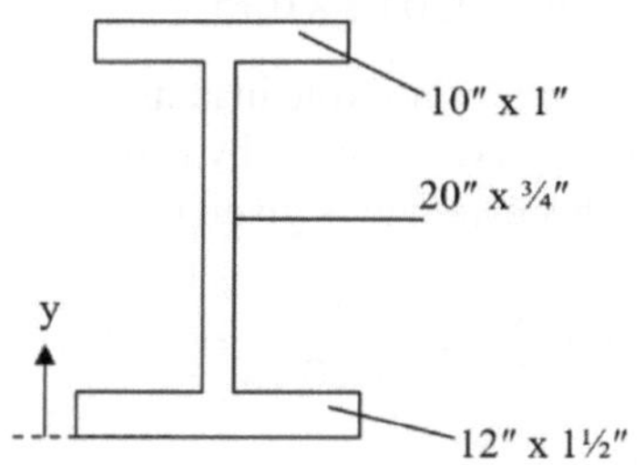

Solution The y coordinate of the centroid or the ENA is calculated with respect to (an arbitrarily chosen) datum located at the bottom of the bottom flange. This is designated as $y = 0$

$$\bar{y} = \frac{\sum A_i y_i}{\sum A_i} = \frac{(12\times1.5)\times0.75+(20\times0.75)\times11.5+(10\times1)\times22}{12\times1.5+20\times0.75+10\times1} = \frac{406}{43} = 9.44$$

The moment of inertia is calculated as

$$I_{NA} = \frac{1}{12}\times12\times1.5^3 + 18\times(0.75-9.44)^2$$
$$+\frac{1}{12}\times\frac{3}{4}\times20^3 + 15\times(11.5-9.44)^2$$
$$+\frac{1}{12}\times10\times1^3 + 10\times(22-9.44)^2 = 3504.7 \text{ in}^4$$

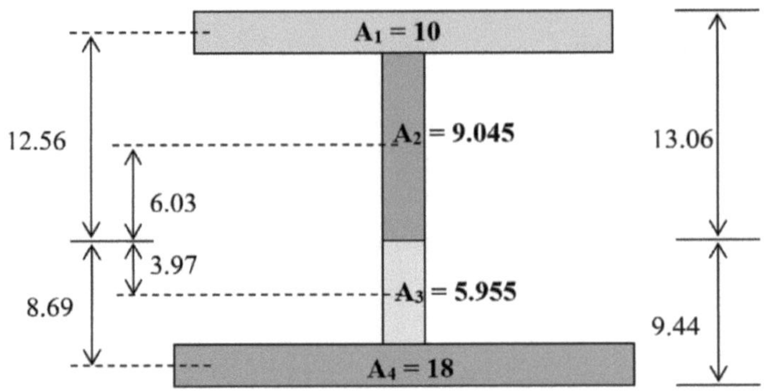

As shown in the figure above, the neutral axis subdivides the section into two "halves"—either the area $A_1 + A_2$ or the area $A_3 + A_4$. Either one can be used to calculate the first moment Q.

Using the top half, we have $Q = 10 \times 12.56 + 9.045 \times 6.03 = 180.1 \text{ in}^3$.

Using the bottom half, we have $Q = 5.955 \times 3.97 + 18 \times 8.69 = 180.1 \text{ in}^3$.

If this section is subject to a transverse shear force $V = 90$ kips, the shear stress can then be calculated as below. Note that the moment of inertia was previously calculated as $I = 3504.7 \text{ in}^4$

$$\tau = \frac{VQ}{Ib} = \frac{90 \times 180.1}{3504.7 \times 0.75} = 6.17 \text{ ksi}$$

For I-sections, the flanges play a very minor role in resisting shear. Thus, an approximate estimate of the maximum shear stress may be made by distributing the shear force "uniformly" over the web only. In this case, this estimate is given by

$$\tau \approx \frac{V}{A_w} = \frac{90}{20 \times 0.75} = 6.0 \text{ ksi}$$

This estimate has an error of 3%.

Typical distributions of shear stress in flanged sections are shown in Fig. 8.13. It can be seen that for I-shaped sections, the approximate estimate obtained by distributing the shear force equally over the area of the web is fairly close to the exact value of the maximum stress at the neutral axis. On the other hand, for a T-section, such an approximation does not yield a good estimate.

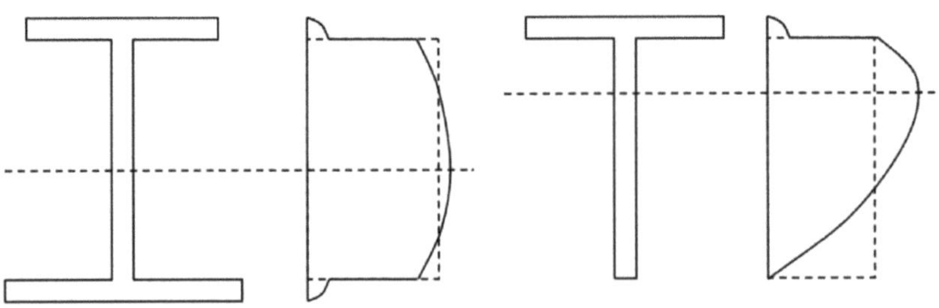

Figure 8.13 Approximations for longitudinal shear stress.

For a rectangular section (dimensions $b \times d$) subject to transverse shear V (such as is common for timber beams), the shear stress distribution is parabolic and the maximum shear stress (at mid-height) is given by

$$\tau = \frac{3V}{2A} = \frac{3V}{2bd} \tag{8.29}$$

For a circular section (radius r) subject to transverse shear V, the maximum shear stress is given by

$$\tau = \frac{4V}{3A} = \frac{4V}{3\pi r^2} \tag{8.30}$$

Beam Deflection—The Elastic Curve

If stresses are relatively low (well below the yield stress), the bending behavior of a beam is likely to be linear and elastic. Moreover, if shear deformations are ignored, the bending deformation is given by the moment-curvature relationship [Eq. (8.31)]. This relates the curvature at every spanwise location of the beam to the bending moment at that location. The exact form of the moment-curvature relation is given by

Curvature
$$\kappa = \frac{1}{\rho} = \frac{M}{EI} \tag{8.31}$$

If the deflected shape of the beam is described by the function $y(x)$, the curvature κ is given by

Curvature
$$\kappa = \frac{1}{\rho} = \frac{1}{\text{radius of curvature}} = \frac{y''}{\left[1+(y')^2\right]^{3/2}} \tag{8.32}$$

where y' = the first derivative of the deflection function = $\dfrac{dy}{dx}$

y'' = the second derivative = $\dfrac{d^2y}{dx^2}$

For small deformations, assuming slope (y') is small, $(y')^2 \approx 0$, then curvature $\kappa \approx \dfrac{d^2y}{dx^2}$ and the linearized version of the moment curvature equation becomes

$$\frac{d^2y}{dx^2} = \frac{M}{EI} \tag{8.33}$$

Solution of the moment curvature equation yields the deflection function $y(x)$. This solution may be obtained one of several ways. One of these methods is by successive integration of the second-order differential equation [Eq. (8.33)] to yield the deflection function $y(x)$.

Direct Integration Method

The second-order (linear) differential equation can be integrated twice to generate the function $y(x)$. This method can work well if the bending moment diagram is composed of a single function over the entire beam, thereby necessitating a single series of integrations. The disadvantage of this method is that the two integration constants must be evaluated from the boundary conditions to complete the function $y(x)$. This makes the process tedious if the bending moment diagram is composed of more than one function defined over several subdomains of the beam. Integration of Eq. (8.33) leads to

$$y' = \int \frac{M(x)}{EI} dx + c_1 \tag{8.34}$$

$$y = \int \left(\int \frac{M(x)}{EI} dx + c_1 \right) dx = \iint \frac{M(x)}{EI} dx\, dx + c_1 x + c_2 \tag{8.35}$$

Example 8.13

For the simply supported beam shown below, answer the following questions:

(i) What is the equation of the elastic curve?

(ii) Where does the maximum deflection occur?

(iii) What is the maximum deflection in inches?

(iv) What is the rotation at left and right ends (in degrees)?

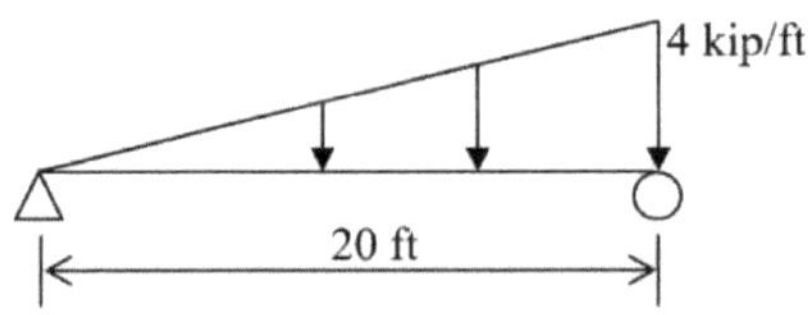

Solution We adopt the following units: kip (force) and inch (length)
Therefore, $L = 240$ in, $w = 0.333$ kip/in

Step 1. Calculate the support reactions.
Resultant load = ½ × 0.333 × 240 = 40 kips
Vertical reactions at left and right ends = 13.333 kips and 26.667 kips

Step 2. Make a section at distance x from the left end and solve for the bending moment function $M(x)$.

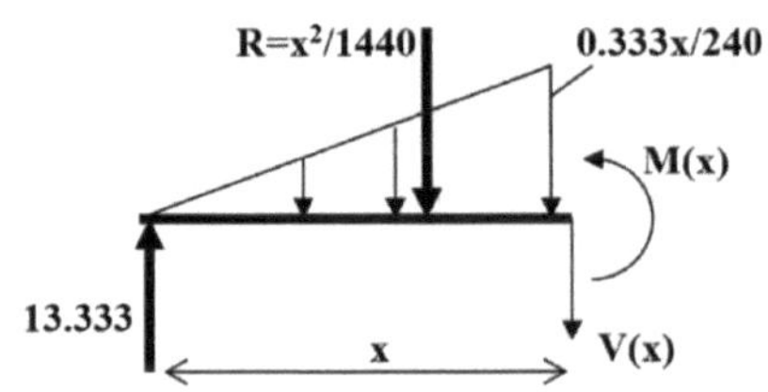

$$\sum M_{\text{cut}} = -13.333x + \frac{x^2}{1440}\frac{x}{3} + M(x) = 0 \Rightarrow M(x) = 13.333x - \frac{x^3}{4320}$$

Step 3. Moment curvature relationship

$$y'' = \frac{M}{EI} \Rightarrow EIy'' = M(x) = 13.333x - \frac{x^3}{4320}$$

$$EIy' = 6.6667x^2 - \frac{x^4}{17,280} + c_1$$

$$EIy = 2.2222x^3 - \frac{x^5}{86,400} + c_1 x + c_2$$

Step 4. Boundary conditions
(i) At $x = 0, \quad y = 0 \quad \rightarrow \quad c_2 = 0$
(ii) At $x = 240, \quad y = 0 \quad \rightarrow \quad c_1 = -89,600$

Step 5. Equation of the elastic curve

$$y = \frac{1}{EI}\left(2.2222x^3 - \frac{x^5}{86,400} - 89,600x\right)$$

ANSWER (i)

Step 6. Slope function (used to determine point of zero slope, thus maximum deflection)

$$y' = \frac{1}{EI}\left(6.6667x^2 - \frac{x^4}{17,280} - 89,600\right)$$

Step 7. Solving the equation $6.6667x^2 - \frac{x^4}{17,280} - 89,600 = 0$

By making the substitution $u = x^2$, we get the quadratic equation

$$5.787 \times 10^{-5} u^2 - 6.6667u + 89,600 = 0$$

which has the solutions

$$u = 99,669,\ 15,532 \qquad \text{leading to } x = 315.7,\ 124.7$$

The first solution is inadmissible (outside the range 0 to 240). Therefore, deflection is maximum at $x = 124.6$ in **ANSWER (ii)**

$$y_{\max} = \frac{1}{29,000 \times 700}\left[2.222 \times 124.6^3 - \frac{124.6^5}{86,400} - 89,600 \times 124.6\right] = -0.355 \text{ in}$$

ANSWER (iii)

Rotation (slope) at left end is $y'\big|_{x=0} = -0.0044$ rad $= -0.25°$ (clockwise)

Rotation (slope) at right end is $y'\big|_{x=240} = +0.005$ rad $= +0.29°$ (counterclockwise)

ANSWER (iv)

Maximum bending moment is obtained by setting $\frac{dM}{dx} = 13.3333 - \frac{x^2}{1440} = 0 \Rightarrow x = 138.6$

Therefore $M_{max} = M\big|_{x=138.6} = 1231.68 \text{ kip} \cdot \text{in.}$

Unit Load Method

The unit load method is derived from the principle of virtual work. Stated in terms of the deformation strain energy, it can be stated as: The deflection corresponding to a particular degree of freedom is equal to the partial derivative of the strain energy with respect to the (work) conjugate of the deflection. Thus, if a particular deflection Δ is desired and its work conjugate is the virtual load P, the bending strain energy is given by $U = \int \frac{M^2(x)dx}{2EI}$ and the desired deflection Δ is given by the partial derivative $\partial U/\partial P$

$$\Delta_P = \frac{\partial U}{\partial P} = \int \frac{M(x)\frac{\partial M}{\partial P}dx}{EI} = \int \frac{M(x)m(x)dx}{EI} \tag{8.36}$$

where the function $M(x)$ describes the bending moment diagram due to the external loads, and the function $m(x)$ describes the bending moment diagram due to the unit load. The unit load is chosen according to the desired displacement.

Example 8.14

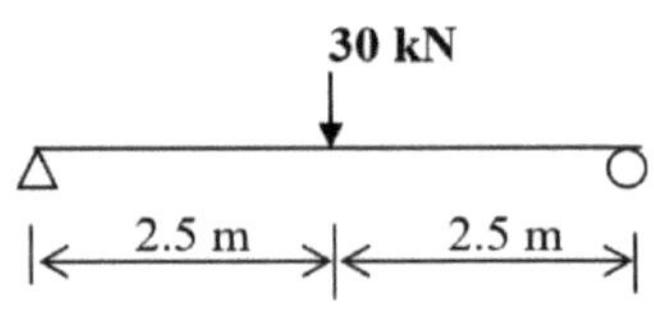

A steel beam supports a load P = 30 kN at the center of a 5-m span as shown. What is the maximum deflection using the unit load method? Use E = 200 GPa. $I = 3 \times 10^8$ mm^4.

Solution The virtual load corresponding to the question "What is the vertical deflection at midspan?" is a unit load (vertical) applied at midspan.

The function $M(x)$ is the bending moment function due to the actual load (P), while the function $m(x)$ is the bending moment function due to the virtual load (1). These are shown below.

Since the strain energy in the beam is predominantly from bending, the midspan deflection is given by

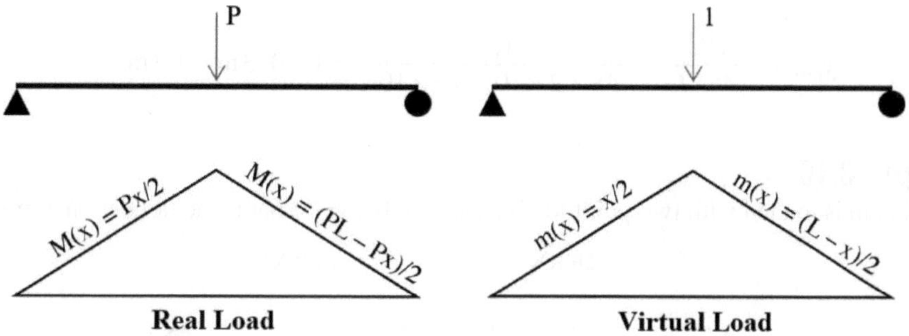

$$\Delta = \int_0^L \frac{M(x)m(x)}{EI}dx = \int_0^{L/2} \frac{\left(\frac{Px}{2}\right)\left(\frac{x}{2}\right)}{EI}dx + \int_{L/2}^{L} \frac{\frac{(PL-Px)}{2}\frac{(L-x)}{2}}{EI}dx = \frac{Px^3}{12EI}\Bigg|_0^{L/2}$$

$$+\frac{P(L^2x - Lx^2 + x^3/3)}{4EI}\Bigg|_{L/2}^{L} = \frac{PL^3}{96EI} + \frac{PL^3}{96EI} = \frac{PL^3}{48EI}$$

The positive sign indicates that the deflection is in the same direction as the virtual load, that is, downward.

Beam Deflection Equations

Pages 85 and 86 of the *Handbook* have 12 standard cases for beam deflections. Page 85 has six load cases for simply supported beams and page 86 has six more for cantilever beams. In each case, the origin is located at the left end and the vertical deflection is designated $v(x)$ and is considered positive upwards. The tables give the equation of the elastic curve $v(x)$, the rotation (radians) at either end ($x = 0$ and $x = L$) and location and value of the maximum deflection v_{max}. Rotation is considered counterclockwise positive.

Example 8.15

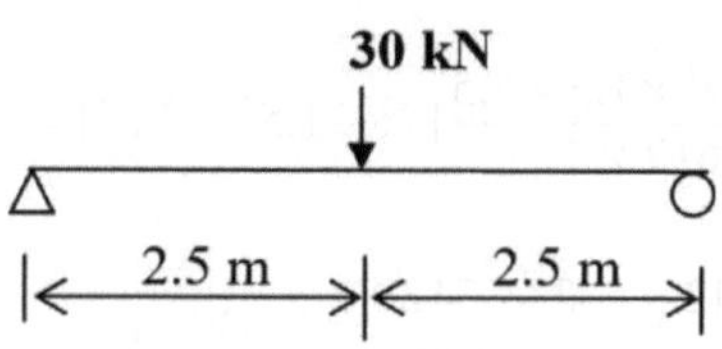

A steel beam supports a load $P = 30$ kN at the center of a 5-m span as shown. What is the maximum deflection using the standard load case tables in the *Handbook*? Use $E = 200$ GPa. $I = 3 \times 10^8$ mm^4.

Solution This is standard case no. 1 on page 85 of the *Handbook*.
Maximum deflection

$$y_{max} = -\frac{PL^3}{48EI} = -\frac{3\times10^4\times5^3}{48\times2\times10^{11}\times3\times10^{-4}} = 0.0013\,\text{m} = 1.3\,\text{mm}$$

Example 8.16

A steel beam is loaded with two point loads as shown below. What is the deflection at midspan?

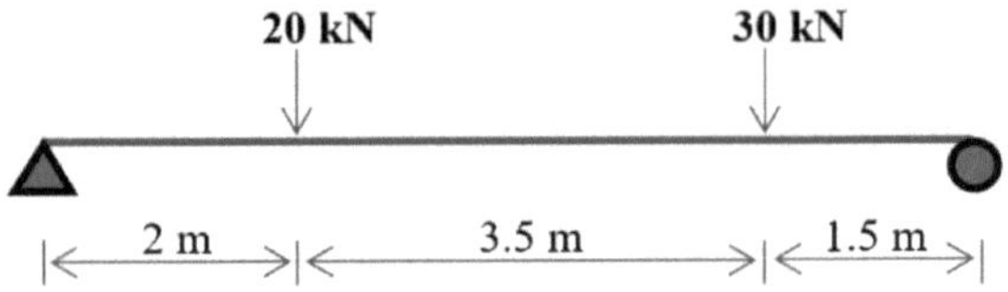

A. 40 mm

B. 34 mm

C. 22 mm

D. 18 mm

Solution For a simple span carrying a point load, the following equation is given in the *Handbook*:

$$\delta = -\frac{Pbx}{6EIL}[L^2 - b^2 - x^2] \text{ for } x < a$$

For the 20 kN load, using the origin at the right support, use $P = 20{,}000$, $a = 5$, $b = 2$, $L = 7$, $x = 3.5$

$$\delta_{20} = -\frac{2\times10^4\times2\times3.5}{6EI\times7}[7^2 - 2^2 - 3.5^2] = -\frac{109{,}167}{EI}$$

For the 30 kN load, using the origin at the right support, $P = 30{,}000$, $a = 5.5$, $b = 1.5$, $L = 7$, $x = 3.5$

$$\delta_{30} = -\frac{3\times10^4\times1.5\times3.5}{6EI\times7}[7^2 - 1.5^2 - 3.5^2] = -\frac{129{,}375}{EI}$$

Total deflection: $\delta = -\frac{238{,}542}{EI} = -\frac{238{,}542}{6\times10^6} = -0.0398\text{ m} = -39.8\text{ mm}$

Answer is A.

Torsion

Shear Stress due to Torsion—Circular Sections

When a solid circular section (radius R) is subjected to torsion, the shear stress between adjacent planes is given by

$$\tau = \frac{Tr}{J} \tag{8.37}$$

where J is the polar moment of inertia of the section, given by

$$J = \frac{\pi R^4}{2} \tag{8.38}$$

Thus, the maximum shear stress in a solid circular shaft due to a torque T is given by

$$\tau_{\max} = \frac{TR}{\pi R^4/2} = \frac{2T}{\pi R^3} \tag{8.39}$$

For a hollow cylindrical shaft (inner radius R_1 and outer radius R_2), the polar moment of inertia is given by

$$J = \frac{\pi(R_2^4 - R_1^4)}{2} \tag{8.40}$$

and the maximum shear stress is given by

$$\tau_{\max} = \frac{2TR_2}{\pi(R_2^4 - R_1^4)} \tag{8.41}$$

Shear Stress due to Torsion—Thin-Walled Sections

Economy of weight leads to hollow thin-walled sections being commonly used for structural members subject to significant torsional moments. The thin-walled assumption is considered approximately valid when the wall thickness is less than about one-tenth of the lateral dimension. The maximum shear stress in a thin-walled hollow shaft is given by

$$\tau_{\max} = \frac{T}{2A_m t} \tag{8.42}$$

where T = applied torque
A_m = area enclosed by the closed wall centerline of the section (Fig. 8.14)
t = thickness of the thinnest part of the wall

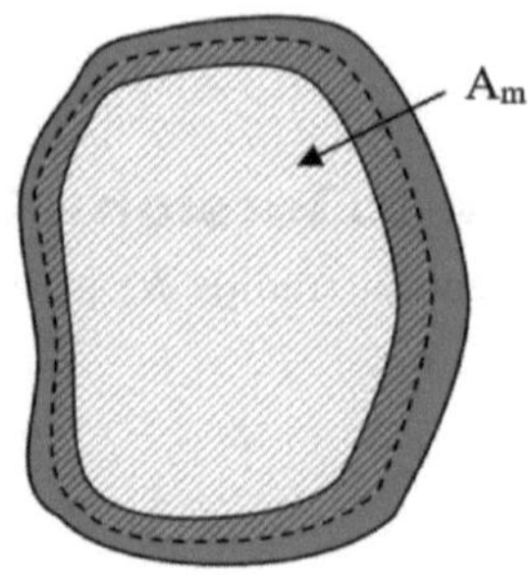

Figure 8.14 Thin-walled tube in torsion.

Example 8.17

A hollow circular steel shaft (O.D. = 40 mm; I.D. = 32 mm) is subject to a torque $T = 400$ Nm. What is the twist angle per unit length?

A. 2.4°/m

B. 1.9°/m

C. 2.8°/m

D. 3.2°/m

Solution Twist angle for a circular shaft is given by

$$\phi = \frac{TL}{GJ}$$

Polar moment of inertia: $J = \dfrac{\pi a^4}{2} = \dfrac{\pi(0.02^4 - 0.016^4)}{2} = 1.484 \times 10^{-7}\ \text{m}^4$

Twist angle per unit length

$$\frac{\phi}{L} = \frac{T}{GJ} = \frac{400}{80 \times 10^9 \times 1.484 \times 10^{-7}} = 0.0337 \frac{\text{rad}}{\text{m}}$$

Twist angle per unit length = 1.93°/m.

Answer is B.

Example 8.18

A hollow steel shaft (I.D. = 20 mm; O.D. = 30 mm) is subjected to a torsional moment M = 200 Nm. If the length of the shaft is 50 cm, what is the twist angle?

A. 1.12°

B. 0.85°

C. 0.62°

D. 0.32°

Solution For steel, modulus of rigidity $G = 80$ GPa
The polar moment of inertia of the shaft is

$$J = \frac{\pi}{32}(0.03^4 - 0.02^4) = 6.38 \times 10^{-8}$$

$$\phi = \frac{TL}{GJ} = \frac{200 \times 0.5}{80 \times 10^9 \times 6.38 \times 10^{-8}} = 0.0196 \text{ rad} = 1.12°$$

Answer is A.

Example 8.19

A hollow steel shaft has the cross section shown below. If the maximum torsional moment experienced by the section is 100 Nm, what is the maximum shear stress?

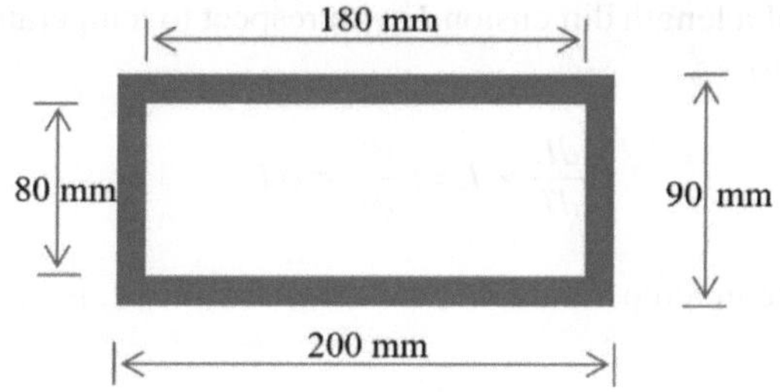

A. 0.62 MPa

B. 1.24 MPa

C. 2.78 MPa

D. 5.56 MPa

Solution For a thin-walled section, A_m is the total area enclosed by the wall centerline. This area is shown shaded below.

$$A_m = 0.19 \times 0.085 = 0.01615 \text{ m}^2$$

Wall thickness $t = 5$ mm for the longer wall segments (thinnest wall develops largest shear stress).

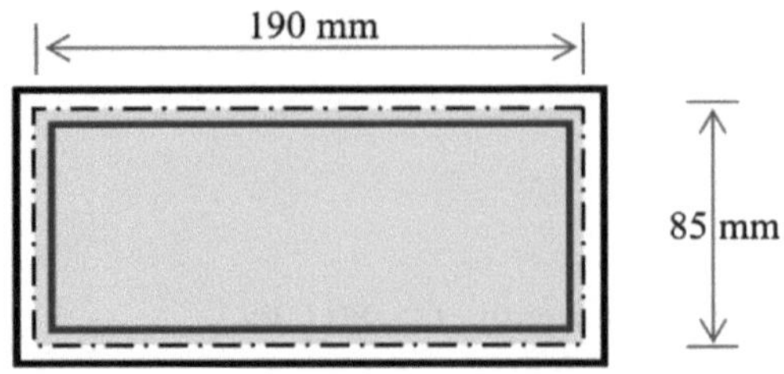

$$\tau = \frac{T}{2A_m t} = \frac{100}{2\times 0.01615\times 0.005} = 1.24\times 10^{6}\ \text{Pa}$$

Answer is B.

Thermal Stress and Strain

Most materials increase in size when heated, due to an increase in molecular kinetic energy and greater molecular separation. The rate of increase in characteristic length with temperature is called the coefficient of thermal expansion α.

The rate of change of a length dimension L with respect to temperature T is proportional to the length L according to

$$\frac{dL}{dT} \propto L \Rightarrow \frac{dL}{dT} = \alpha L \tag{8.43}$$

Thus, for a finite change in temperature ΔT, the thermal strain ε_T is given by

$$\varepsilon_T = \frac{\Delta L}{L} = \alpha \Delta T \tag{8.44}$$

The equation works well as long as the coefficient α remains relatively constant over the range of temperature ΔT and the thermal strain ε_T is small.

Example 8.20

A composite rod is constructed by fusing a steel sleeve onto a brass core as shown. Mechanical and thermal properties of brass and steel are shown in the table below:

	Steel	Brass
E (GPa)	205	110
ν	0.3	0.32
α (/°C)	11.7×10^{-6}	20.5×10^{-6}

If the assembly is stress free at 20°C and then heated to 150°C, what is the stress in the brass core?

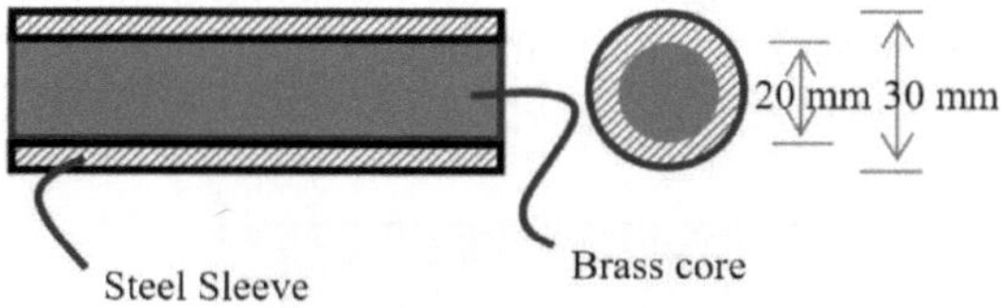

A. 103 MPa

B. 89 MPa

C. 75 MPa

D. 66 MPa

Solution Since the coefficient of thermal expansion for brass is more than that of the steel, the brass core is under compression and the steel sleeve is under tension. These two forces must be equal and opposite. Also, since the materials are fused, they must deform equally

$$A_{\text{brass}} = \frac{\pi}{4} 0.02^2 = 3.14 \times 10^{-4}$$

$$A_{\text{steel}} = \frac{\pi}{4}(0.03^2 - 0.02^2) = 3.93 \times 10^{-4}$$

For brass, total elongation = sum of thermal expansion + compression shortening is

$$\Delta L_B = +\alpha L \Delta T - \frac{PL}{AE} = 20.5 \times 10^{-6} \times L \times 130 - \frac{PL}{3.14 \times 10^{-4} \times 110 \times 10^9}$$

$$= L(2.67 \times 10^{-3} - 2.895 \times 10^{-8} P)$$

For steel, total elongation = sum of thermal expansion + tension lengthening is:

$$\Delta L_S = +\alpha L \Delta T + \frac{PL}{AE} = 11.7 \times 10^{-6} \times L \times 130 + \frac{PL}{3.93 \times 10^{-4} \times 205 \times 10^9}$$

$$= L(1.52 \times 10^{-3} + 1.241 \times 10^{-8} P)$$

Equating $\Delta L_B = \Delta L_S$
$P = 27,804$ N

Stress in brass core: $\sigma_B = \frac{P}{A} = \frac{27,805}{3.14 \times 10^{-4}} = 88.6$ MPa

Answer is B.

Example 8.21

A steel rod of diameter 20 mm is fixed at the left end as shown. At a temperature of 20°C, a gap of 0.1 mm exists between the right end of the rod and a rigid wall. If the temperature of the rod is raised to 120°C, what is the stress in the rod?

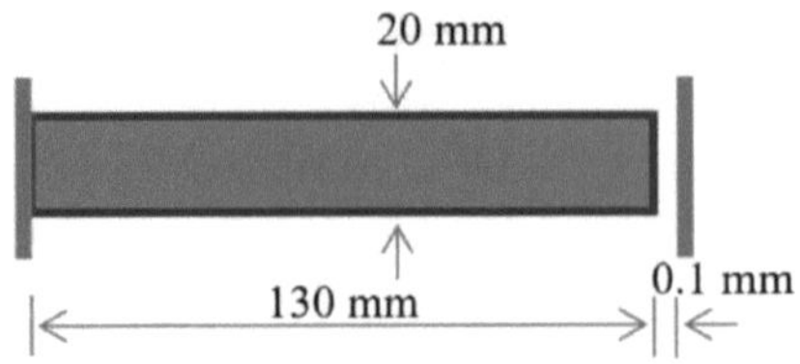

A. $\sigma = 20$ MPa

B. $\sigma = 40$ MPa

C. $\sigma = 80$ MPa

D. $\sigma = 160$ MPa

Solution The (unrestricted) thermal elongation of the rod would be

$$\Delta L = L\alpha\Delta T = 130 \times 11.7 \times 10^{-6} \times 100 = +0.152 \text{ mm}$$

This implies that the 0.1 mm gap will close. As a result, the rod will experience a compression reaction at each wall. Shortening of rod due to compression load

$$\Delta L = -\frac{PL}{AE}$$

The cross-sectional area $A = \pi r^2 = \pi \times 0.01^2 = 3.14 \times 10^{-4}$

The net length change of the rod is

$$\Delta L = +1.52 \times 10^{-4} - \frac{P \times 0.1301}{A \times 200 \times 10^9} = 1 \times 10^{-4}$$

$$0.52 \times 10^{-4} = \frac{P}{A} \times 6.5 \times 10^{-13}$$

Stress: $\sigma = \frac{P}{A} = \frac{0.52 \times 10^{-4}}{6.5 \times 10^{-13}} = 7.99 \times 10^7 \text{ Pa} = 79.9 \text{ MPa}$

Answer is C.

Mohr's Circle: Normal (σ) and Shear (τ) Stress Combination

Figure 8.15(a) shows an element in a state of plane stress. The normal stresses are σ_x and σ_y and the shear stresses are τ_{xy} and τ_{yx}. The stresses are given in a framework where the axes are the "traditional" *x*-*y* axes—a horizontal *x* axis and a vertical *y* axis. Sometimes, the engineer

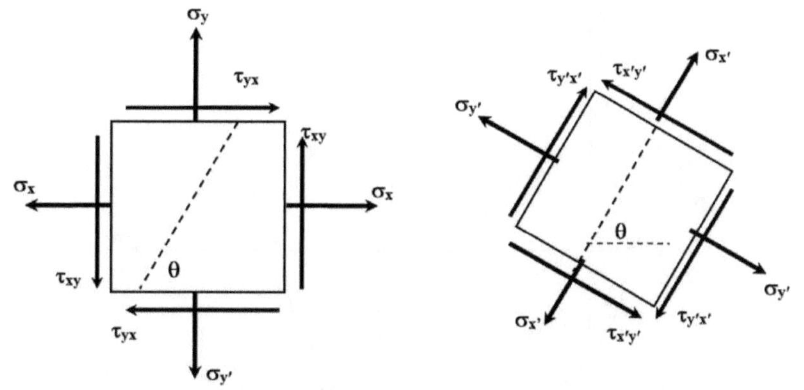

Figure 8.15 Stresses on 2D element (a) original; (b) transformed.

might be more interested in stresses oriented along a different set of axes. A "rotated" element (rotated counterclockwise though angle θ) is shown in Fig. 8.15(b). This element has "new" stress components—$\sigma_{x'}$, $\sigma_{y'}$, $\tau_{x'y'}$, and $\tau_{y'x'}$.

This "new" stress-state is not independent of the original stress-state, but rather a function of it (transformed). The transformation relations are given by Eq. (8.45). However, note that these transformation equations are *not* given in the *Handbook*. This means that you must be able to solve Mohr's circle problems without using these transformation equations

$$\begin{aligned}
\sigma_{x'} &= \frac{\sigma_x + \sigma_y}{2} + \frac{\sigma_x - \sigma_y}{2}\cos 2\theta + \tau_{xy}\sin 2\theta \\
\sigma_{y'} &= \frac{\sigma_x + \sigma_y}{2} - \frac{\sigma_x - \sigma_y}{2}\cos 2\theta - \tau_{xy}\sin 2\theta \\
\tau_{x'y'} &= -\frac{\sigma_x - \sigma_y}{2}\sin 2\theta + \tau_{xy}\cos 2\theta
\end{aligned} \tag{8.45}$$

A very convenient way of (graphically) expressing these results is through Mohr's circle. By eliminating θ from Eq. (8.43), we can write

$$\left(\sigma_{x'} - \frac{\sigma_x + \sigma_y}{2}\right)^2 + (\tau_{x'y'} - 0)^2 = \left(\frac{\sigma_x - \sigma_y}{2}\right)^2 + \tau_{xy}^2 \tag{8.46}$$

This represents the equation of a circle with center at $\left(\frac{\sigma_x + \sigma_y}{2}, 0\right)$ and radius R given by

$$R = \sqrt{\left(\frac{\sigma_x - \sigma_y}{2}\right)^2 + \tau_{xy}^2} \tag{8.47}$$

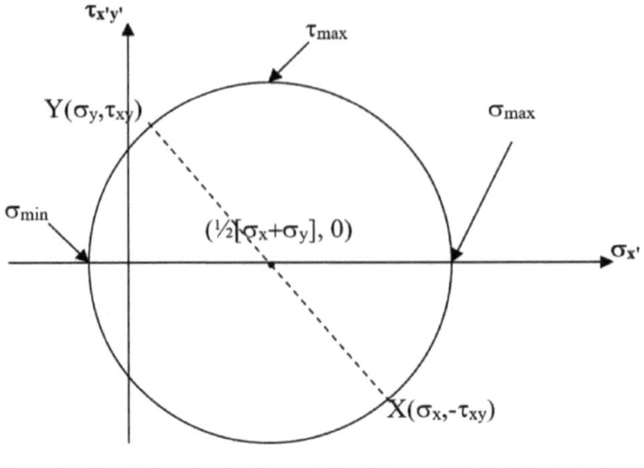

Figure 8.16 Mohr's circle parameters.

The Mohr circle with these parameters is shown in Fig. 8.16. Each point on the circle represents a particular direction (θ) and the coordinates (σ, τ) of the point represent the stress state along that plane. A plane is identified by the angle θ made by its normal. Thus, a vertical plane (whose normal is along the x axis) would be identified as a "x plane" and the angle θ associated with it would be 0°. Similarly, a horizontal plane (whose normal is along the y axis) would be identified as a "y plane" and the angle θ associated with it would be 90°.

Principal Stresses

Maximum and minimum normal stresses are designated σ_{max} and σ_{min}, respectively. These are also called the *principal stresses*. Note that these principal directions have only normal stress, that is, the principal directions are free of shear stress

$$\sigma_{\max}, \sigma_{\min} = \sigma_{\text{ave}} \pm R = \frac{\sigma_x + \sigma_y}{2} \pm \sqrt{\left(\frac{\sigma_x - \sigma_y}{2}\right)^2 + \tau_{xy}^2} \tag{8.48}$$

The maximum and minimum values of $\sigma_{x'}$ are calculated by setting $\dfrac{d\sigma'_x}{d\theta} = 0.$

These normal stresses (principal stresses) occur at two specific values of θ, given by

$$\tan 2\theta_p = \frac{2\tau_{xy}}{\sigma_x - \sigma_y} \tag{8.49}$$

Example 8.22

Consider a 2D element with the stress components as shown below. Calculate the (a) values of the principal stresses, and (b) the orientation of the principal planes.

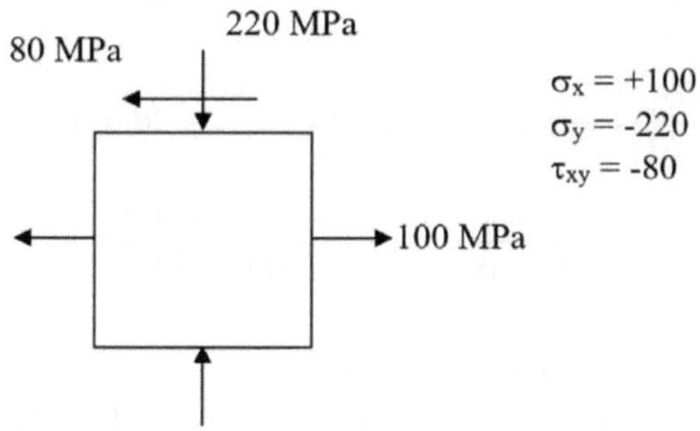

Solution From the given stresses, we can calculate the average normal stress σ_{ave} and the radius R of the Mohr circle

$$\sigma_{ave} = \frac{\sigma_x + \sigma_y}{2} = \frac{100 - 220}{2} = -60$$

$$R = \sqrt{\left(\frac{\sigma_x - \sigma_y}{2}\right)^2 + \tau_{xy}^2} = \sqrt{\left(\frac{100 + 220}{2}\right)^2 + (-80)^2} = 179$$

Therefore, $\sigma_{max}, \sigma_{min} = -60 \pm 179 = 119, -239$.

The resulting Mohr circle is shown below. Each tick mark on the axes is 60 MPa. Point X is located at coordinates (100, 80) and Y is located at (–220, –80).

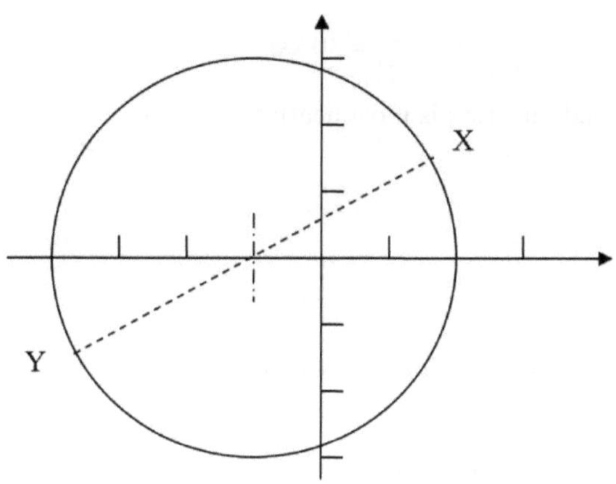

The values of maximum and minimum normal stress can be also calculated by taking the derivative of the $\sigma_{x'}$ with respect to θ

$$\sigma_{x'} = \frac{\sigma_x + \sigma_y}{2} + \frac{\sigma_x - \sigma_y}{2}\cos 2\theta + \tau_{xy}\sin 2\theta$$

$$\frac{d\sigma_{x'}}{d\theta} = -(\sigma_x - \sigma_y)\sin 2\theta + 2\tau_{xy}\cos 2\theta = 0$$

$$\tan 2\theta = \frac{2\tau_{xy}}{(\sigma_x - \sigma_y)} = \frac{2\times -80}{100+220} = \frac{-160}{320} = -0.5$$

This yields $2\theta = \tan^{-1}(-0.5) = -26.6°$ or $153.4°$.
Substituting $2\theta = -26.6°$, we get $\sigma_{x'} = -60 + 160 \times 0.894 + 80 \times 0.447 = 119$.
Substituting $2\theta = 153.4°$, we get $\sigma_{x'} = -60 - 160 \times 0.894 - 80 \times 0.447 = -239$.

Note that the point with normal stress $\sigma = 119$ is located at a position which is 26.6° clockwise (thus negative) with respect to the point X and $\sigma = -239$ is located at a position which is 153.4° counterclockwise (thus positive) with respect to X.

Example 8.23

At a particular point on a beam subject to transverse loads, the following stresses have been measured.

Bending stress $\sigma_{xx} = 28.5$ ksi (tension)

$\sigma_{yy} = 0$

$\tau_{xy} = 14.8$ ksi

The steel used for the beam has the following ultimate stresses:

$$\sigma_{\text{ult}} = 45 \text{ ksi (tension)}$$

$$\tau_{\text{ult}} = 30 \text{ ksi}$$

The factor of safety based on stress is most nearly:

A. $FS = 2.75$

B. $FS = 1.46$

C. $FS = 1.29$

D. $FS = 1.18$

Solution The maximum bending stress is given by

$$\sigma_{\max} = \frac{\sigma_{xx} + \sigma_{yy}}{2} + \sqrt{\left(\frac{\sigma_{xx} - \sigma_{yy}}{2}\right)^2 + \tau_{xy}^2}$$

With $\sigma_{yy} = 0$, this becomes

$$\sigma_{\max} = \frac{\sigma_{xx}}{2} + \sqrt{\frac{\sigma_{xx}^2}{4} + \tau_{xy}^2} = \frac{28.5}{2} + \sqrt{\frac{28.5^2}{4} + 14.8^2} = 34.8$$

The maximum shear stress is given by

$$\tau_{\max} = \sqrt{\frac{\sigma_{xx}^2}{4} + \tau_{xy}^2} = \sqrt{\frac{28.5^2}{4} + 14.8^2} = 20.6$$

Factor of safety based on normal stress

$$FS = \frac{\sigma_{\text{ult}}}{\sigma_{\max}} = \frac{45}{34.8} = 1.29$$

Factor of safety based on shear stress

$$FS = \frac{\tau_{\text{ult}}}{\tau_{\max}} = \frac{30}{20.6} = 1.46$$

Overall factor of safety = 1.29.

Answer is C.

Composite Materials

When a manufactured material is considered a composite of two or more distinct materials, present in fractions f_i (by volume), then certain properties of the composite material (such as density ρ, heat capacity c) can be calculated by using a weighted arithmetic average of the properties of component materials. For example, the density ρ_c and the heat capacity c_c of the composite are given by

$$\rho_c = \sum f_i \rho_i \tag{8.50a}$$

$$c_c = \sum f_i c_i \tag{8.50b}$$

For certain properties such as modulus of elasticity, the composite property E_c depends on the manner in which the composite is manufactured. Lower and upper bounds for E_c are given by the weighted harmonic mean and the weighted arithmetic mean, respectively

$$\frac{1}{\sum \frac{f_i}{E_i}} \le E_c \le \sum f_i E_i \tag{8.51}$$

Composite Section Undergoing Axial Force

For axially oriented, long, fiber-reinforced composites (composed of two materials 1 and 2), equating the axial strains results in

$$\left(\frac{\Delta L}{L}\right)_1 = \left(\frac{\Delta L}{L}\right)_2 \tag{8.52}$$

The applied load (and therefore, the stress) is distributed to the component materials such that the deformation (and therefore, the strain) is equal. This is illustrated in the following example.

Example 8.24

A composite rod is constructed by fusing a steel sleeve onto a brass core as shown. Mechanical properties of brass and steel are shown in the table below:

	Steel	Brass
E (GPa)	205	110
v	0.3	0.32

If the assembly is initially stress free and has a length = 400 mm, then an axial compression force $P = 60$ kN is applied uniformly at both ends of the rod, what is the shortening of the composite rod?

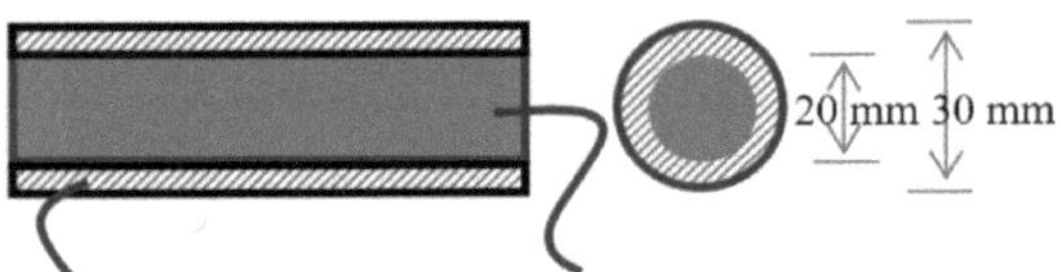

A. 0.1 mm

B. 0.2 mm

C. 0.3 mm

D. 0.4 mm

Solution Since the materials are fused, they must deform equally.
However, the load $P = 60$ kN will be shared between the two materials. Let us call the load carried by the brass core P. This means that the load carried by the steel sleeve is $(60{,}000 - P)$

$$A_{\text{brass}} = \frac{\pi}{4}0.02^2 = 3.14 \times 10^{-4}$$

$$A_{\text{steel}} = \frac{\pi}{4}(0.03^2 - 0.02^2) = 3.93 \times 10^{-4}$$

$$\text{Shortening of the brass core} = \Delta L_B = \frac{PL}{AE} = \frac{PL}{3.14 \times 10^{-4} \times 110 \times 10^9} = 2.895 \times 10^{-8}\,PL$$

Shortening of the steel sleeve

$$\Delta L_S = \frac{PL}{AE} = \frac{(60{,}000 - P)L}{3.93 \times 10^{-4} \times 205 \times 10^9} = 1.241 \times 10^{-8} L(60{,}000 - P)$$

Equating $\Delta L_B = \Delta L_S \Rightarrow 2.895P = 1.241(60{,}000 - P) \Rightarrow P = 18{,}003$ N
Shortening of the composite rod, calculated using the brass core (either material could have been used)

$$\Delta L_B = \frac{PL}{AE} = 2.895 \times 10^{-8} \times 18{,}003 \times 0.4 = 2.08 \times 10^{-4}\,\text{m} = 0.21\text{ mm}$$

Answer is B.

Composite Section Undergoing Flexure

A reinforced concrete beam (concrete-steel composite), a timber beam reinforced with a steel plate (wood-steel composite) are common examples of composite sections subject to flexure. In order to calculate stresses and strains, it is customary to convert both materials to one common basis. For example, for the steel girder-concrete deck system (as shown in Fig. 8.17), it is traditional to convert the concrete deck to an equivalent amount of steel.

The ratio of the moduli of elasticity of the two materials is called the modular ratio n

$$n = E_1/E_2 \tag{8.53}$$

The modular ratio n is then used to convert one of the materials (labeled 1) into an equivalent amount of the other material (labeled 2). The moment of inertia of the transformed section is then calculated as I_t.
The bending stress in the (base) material (2) is given by the standard formula

$$\sigma_2 = -\frac{My}{I_t} \tag{8.54}$$

whereas the bending stress in the other material (1) is given by the formula

$$\sigma_1 = -\frac{nMy}{I_t} \tag{8.55}$$

Equivalent Section

Figure 8.17 shows a very commonly encountered example of composite sections—steel girders topped with a concrete slab. In this instance, the concrete slab has a simpler geometry and therefore is easier to convert (to an equivalent amount of steel).

If the "base material" is chosen to be steel, then the concrete slab is converted to an "equivalent steel slab" by multiplying the width of the slab by the modular ratio $n = E_c/E_s$.

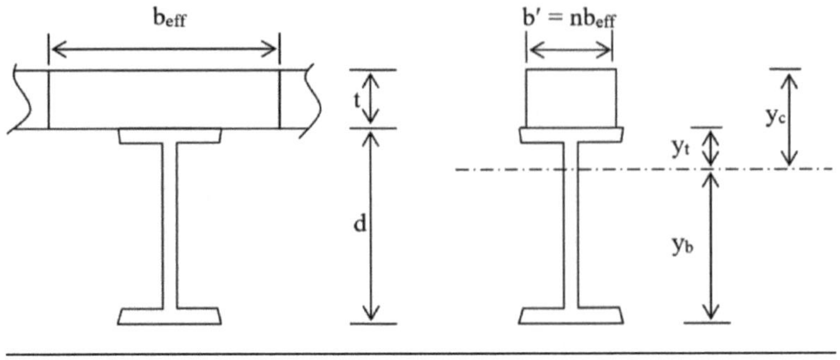

Figure 8.17 Equivalent (steel) section of a composite beam.

Elastic Neutral Axis of Composite Section

The converted slab width is shown as b' in Fig. 8.17. The ENA of this "equivalent steel" section can then be found as (using the bottom edge of the steel beam as the datum)

$$\bar{y} = \frac{A_s \dfrac{d}{2} + b't\left(d + \dfrac{t}{2}\right)}{A_s + b't} = y_b \tag{8.56}$$

Having established the location of the ENA, the moment of inertia of this equivalent section can be found as

$$I_{NA} = I_s + A_s\left(\bar{y} - \frac{d}{2}\right)^2 + \frac{1}{12}b't^3 + b't\left(d + \frac{t}{2} - \bar{y}\right)^2 \tag{8.57}$$

where I_s = moment of inertia of the steel beam
A_s = cross section area of the steel beam

The elastic bending stress at the top of the steel is given by

$$\sigma_{ts} = \frac{My_t}{I_t} \tag{8.58}$$

The bending stress at the bottom of the steel is given by

$$\sigma_{bs} = \frac{My_b}{I_t} \tag{8.59}$$

The stress at the top of the concrete (modified by modular ratio n) is given by

$$\sigma_c = \frac{nMy_c}{I_t} \tag{8.60}$$

Example 8.25

A composite steel girder is shown below. The effective width of the concrete slab = 0.2 m. The modulus of elasticity of concrete = 30 GPa and modulus of elasticity of steel = 205 GPa. Geometric properties of the girder are shown.

Find **(a)** The height of the centroid measured from the bottom edge of the girder
(b) The moment of inertia of the equivalent steel section

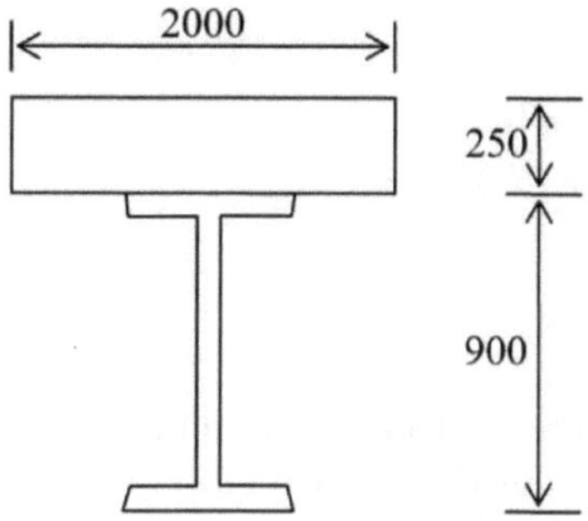

Steel Girder:
Depth = 900 mm
Area = 28,000 mm²
$I_x = 1.25\text{x}10^9$ mm⁴

All dimensions mm

Solution Modular ratio: $n = \frac{E_c}{E_s} = \frac{30}{205} = 0.1463$

Effective width of slab (equivalent steel): $b_{\text{eff}} = 0.1463 \times 2000 = 292.7$ mm

Area of equivalent slab = $292.7 \times 250 = 73{,}175$ mm^2

Height of centroid ($y = 0$ at bottom of girder): $y = \dfrac{28{,}000 \times 450 + 73{,}175 \times 1025}{28{,}000 + 73{,}175} = 866$ mm

Moment of inertia (of equivalent steel section):

$$I_s = 1.25 \times 10^9 + 28{,}000 \times (866 - 450)^2 + \frac{1}{12} \times 292.7 \times 250^3 + 73{,}175 \times (866 - 1025)^2$$

$$= 8.33 \times 10^9 \text{ mm}^4$$

Stability of Axially Loaded Columns—Euler Buckling

Euler's theory is based upon the assumption of elastic behavior, which is acceptable only for very slender columns which buckle at very low loads (i.e., very low stress levels). For a *concentrically loaded column with both ends pinned*, Euler's stability analysis yields the following expression for the critical buckling load P_e

$$P_e = \frac{\pi^2 EI}{L^2} \tag{8.61}$$

where I is the moment of inertia of the column cross section with respect to the axis about which buckling takes place. For columns with different end conditions, the theory may be extended by replacing the physical length of the column (L) with the effective length (KL)

$$P_e = \frac{\pi^2 EI}{(KL)^2} \tag{8.62}$$

The K-value depends on end conditions. Euler's K-values are listed on page 162 in Table C-A-7.1 of the *Handbook.*

Euler's critical buckling stress given by

$$F_e = \frac{\pi^2 E}{(KL/r)^2} \tag{8.63}$$

where r is the radius of gyration = $\sqrt{(I/A)}$.

Example 8.26

A rectangular timber column (E = 10 GPa) has a 100 × 200-mm cross section and length L = 4 m. The bottom of the column is restrained against rotation and lateral translation, while the top is only restrained against lateral translation. What is the critical buckling load of the column?

A. 210 kN

B. 420 kN

C. 630 kN

D. 840 kN

Solution Buckling occurs about the weaker axis (i.e., the one with the smaller moment of inertia).

The weak-axis moment of inertia of the cross-section is calculated as $I = 1/12 \times 0.2 \times 0.1^3 = 1.667 \times 10^{-5}\,\text{m}^4$.

From the Civil Engineering section, the K-value for the given condition (case b) is 0.7.

Euler's buckling load: $P_E = \dfrac{\pi^2 EI}{(KL)^2} = \dfrac{\pi^2 \times 10 \times 10^9 \times 1.667 \times 10^{-5}}{(0.7 \times 4)^2} = 209{,}812\ \text{N}$

Answer is A.

CHAPTER 9

Materials

Concrete Mix Design

Absolute Volume Method

Design of concrete mixes using the *absolute volume method* as outlined in the document *Design and Control of Concrete Mixtures* (Portland Cement Association) uses a series of specific steps. The method ends up with calculated volumes of cement, water, coarse and fine aggregate and for air-entrained concrete, air. Typically all results are stated for a unit volume of 1 cubic yard (yd^3) or 1 m^3.

The absolute volume method has the following steps:

1. From the specified compressive strength f_c', establish a required average compressive strength f_{cr}'.
2. Based on f_{cr}', the target compressive strength of the design mix, choose the desirable water-cement ratio. The aforementioned document dictates a maximum water-cement ratio for both air-entrained and non-air-entrained concrete.
3. The air content is determined based on the anticipated exposure of the concrete.
4. Based on aggregate size and target air content, choose the approximate mixing water (lb of water per yd^3 of concrete).
   a. Using the unit weight of water, the weight of water may be converted into the volume of water (ft^3 of water per yd^3 of concrete).
5. Knowing the water-cement ratio and the weight of water, the weight of cement (lb of cement per yd^3 of concrete) is calculated.
   a. Using the unit weight of cement, the weight of cement may be converted into the volume of cement (ft^3 of cement per yd^3 of concrete).
6. The bulk volume of coarse aggregate is multiplied by the bulk density to obtain the weight of coarse aggregate (lb of aggregate per yd^3 of concrete).
   a. Using the unit weight of coarse aggregate, the weight of aggregate may be converted into the volume of aggregate (ft^3 of coarse aggregate per yd^3 of concrete).

7. The volume of air in the mix is obtained from the air content determined in step 3.
8. The volume of fine aggregate is obtained as the volume remaining by subtracting air, water, cement and coarse aggregate volumes from 1 yd^3 (27 ft^3).
   a. The weight of fine aggregate is calculated (lb of aggregate per yd^3 of concrete) based on the known specific gravity.
9. Corrections are made to the mixing water for moisture contained within fine and coarse aggregates according to Eq. (9.1).

Table 9.1 is reproduced from the PCA document. It can be used to calculate steps 3 and 4 listed above.

Figure 9.1 (*FE Reference Handbook*, version 9.4, page 65) shows that with both air-entrained and non-air-entrained concrete, strength decreases with an increase in water content. For a particular water content, non-air-entrained concrete is stronger than air-entrained concrete, while the latter has greater durability (more resistant to cracking caused by freeze-thaw).

Figure 9.2 (*Handbook*, page 65) shows the relationship between compressive strength of concrete and age, as the hydration reactions of the cementitious materials continue. This process is called *curing*. Much of the strength gain occurs during the first 7 days of cure, beyond which the strength gain continues, but at a slower pace.

Table 9.1 Approximate Mixing Water and Target Air Content for Various Target Slump, Exposure and Aggregate Size

	Water, lb/yd³ of concrete, for indicated sizes of aggregate							
Slump, in	**⅜ in**	**½ in**	**¾ in**	**1 in**	**1½ in**	**2 in**	**3 in**	**6 in**
	Non-air-entrained concrete							
1 to 2	350	335	315	300	275	260	220	190
3 to 4	385	365	340	325	300	285	245	210
6 to 7	410	385	360	340	315	300	270	—
Approximate amount of entrapped air in non-air-entrained concrete, percent	3	2.5	2	1.5	1	0.5	0.3	0.2
	Air-entrained concrete							
1 to 2	305	295	280	270	250	240	205	180
3 to 4	340	325	305	295	275	265	225	200
6 to 7	365	345	325	310	290	280	260	—
Recommended average total air content, percent, for level of exposure:								
Mild exposure	4.5	4.0	3.5	3.0	2.5	2.0	1.5	1.0
Moderate exposure	6.0	5.5	5.0	4.5	4.5	3.5	3.5	3.0
Severe exposure	7.5	7.0	6.0	6.0	5.5	5.0	4.5	4.0

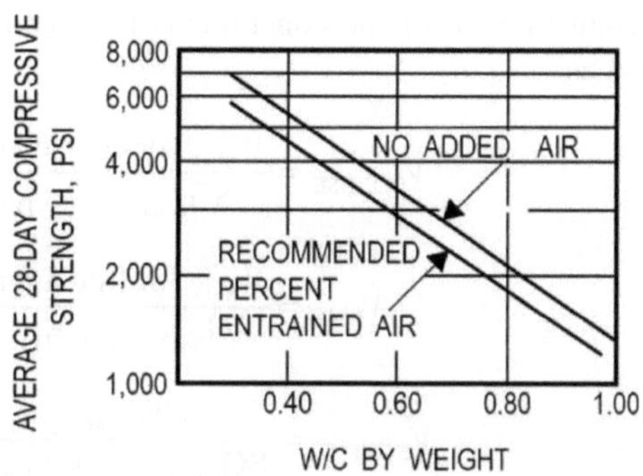

Figure 9.1 Effect of water and air content on concrete strength.

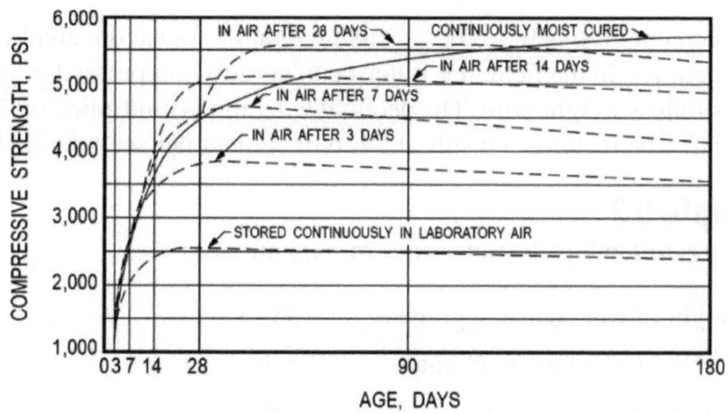

Figure 9.2 Concrete strength gain with age.

Surface Saturated Dry

Aggregate properties are typically given for the surface saturated dry (SSD) condition. This means that the aggregate is not oven dry but has a small amount of moisture adsorbed onto the surface. When these aggregates are used in a wet condition, the excess moisture (exceeding the SSD moisture content) becomes available as hydration water.

$$W_{\text{excess water}} = W_{\text{wet aggr}}\left(\frac{\text{MC}-\text{MC}_{\text{SSD}}}{100+\text{MC}}\right) \tag{9.1}$$

where MC = moisture content (%) of wet aggregate
MC_SSD = moisture content (%) of surface saturated dry aggregate

Example 9.1

Regular sand is used as the fine aggregate for a concrete mix. The SSD condition of the sand has a moisture content of 0.8%. If 1400 lb of wet sand (MC = 5%) is used, how much excess (available) water does the aggregate contain?

Solution 105 lb of wet sand contains the following constituents –100 lb of dry sand +0.8 lb of surface adsorbed water +4.2 lb of "free" water.

Therefore, free water contained in 1400 lb of wet sand $= \frac{4.2}{105} \times 1400 = 56$ lb

SSD sand = 1400 – 56 = 1344 lb

Once the weight of each of the constituents has been found, the corresponding volumes can be calculated as

$$V_{\text{COARSE}} = \frac{W_{\text{SSD}} \text{ for coarse aggr}}{\text{SG}_{\text{SSD}} \times \gamma_{\text{WATER}}} \tag{9.2}$$

$$V_{\text{FINE}} = \frac{W_{\text{SSD}} \text{ for fine aggr}}{\text{SG}_{\text{SSD}} \times \gamma_{\text{WATER}}} \tag{9.3}$$

$$V_{\text{CEMENT}} = \frac{W_{\text{CEMENT}}}{\text{SG}_{\text{CEDMENT}} \times \gamma_{\text{WATER}}} \tag{9.4}$$

Water cement ratio = $W_{\text{water}}/W_{\text{cement}}$ (using total or adjusted weight of mixing water)

Sometimes, in the field, the water-cement ratio is expressed as gallons per sack instead of a dimensionless weight ratio. The decimal w/c ratio is multiplied by 11.27 to convert to "gallons per sack." The factor is a result of 94 lb of cement per sack divided by 8.34 lb/gal of water.

Example 9.2

Given the following characteristics of the bulk mix:

Weight of wet coarse aggregate	135 lb
Weight of wet fine aggregate	95 lb
Weight of added water	21 lb
Weight of cement	48 lb
Volume of concrete	2.14 ft^3

Calculate (1) unit weight of the concrete, (2) volume of coarse aggregate, (3) volume of fine aggregate, and (4) water cement ratio.

Use the following data:

Coarse aggregate:

Moisture content of wet aggregate	2.8%
Moisture content of SSD aggregate	0.6%
Specific gravity of SSD aggregate	2.65

Fine aggregate:

Moisture content of wet aggregate	6.0%
Moisture content of SSD aggregate	1.4%
Specific gravity of SSD aggregate	2.60

Cement:

Specific gravity	3.15

Solution

1. Unit weight $\gamma = \dfrac{135+95+21+48}{2.14} = 139.7\ \text{lb/ft}^3$.
2. For each 100 lb of dry coarse aggregate, there is 0.6 lb of adsorbed moisture and an additional 2.2 lb of unattached water. This makes the weight of wet aggregate 102.8 lb. Therefore, 135 lb of wet coarse aggregate yields $100.6/102.8 \times 135 = 132.11$ lb of SSD aggregate and 2.89 lb of excess water.

   Volume of SSD coarse aggregate $= 132.11 \div (2.65 \times 62.4) = 0.799\ \text{ft}^3$.
3. Similarly, weight of SSD fine aggregate $= 101.4/106 \times 95 = 90.88$ lb of SSD sand and 4.12 lb of excess water.

   Volume of SSD fine aggregate $= 90.88 \div (2.60 \times 62.4) = 0.560\ \text{ft}^3$.
4. Actual water weight: water $= 21.0 + 2.89 + 4.12 = 28.01$ lb.

   Actual water cement ratio $= 28.01/48 = 0.584 = 6.58$ gal/sack.

Example 9.3

A concrete mix has proportions 1:1.6:2.6 (cement-sand-coarse aggregate) by weight. The following specifications are given

Cement	specific gravity = 3.15
	94 lb/sack
Sand	specific gravity = 2.62 (SSD)
Coarse aggregate	specific gravity = 2.65 (SSD)
Added water	5.8 gal/sack cement
Air	3% (by volume)

The quantity of cement (lb/yd^3) is most nearly

A. 630

B. 650

C. 670

D. 690

Solution

Assume 1 sack cement: $W = 94$ lb.	$V = 94/(3.15 \times 62.4) = 0.478\ \text{ft}^3$
Sand: $W = 1.6 \times 94 = 150.4$ lb.	$V = 150.4/(2.62 \times 62.4) = 0.920\ \text{ft}^3$
Coarse: $W = 2.6 \times 94 = 244.4$ lb.	$V = 244.4/(2.65 \times 62.4) = 1.478\ \text{ft}^3$
Added water:	$V = 5.8\ \text{gal} = 5.8/7.48 = 0.775\ \text{ft}^3$

Total volume of components $= 3.651\ \text{ft}^3$

The components above represent 97% of total volume, since air = 3% by volume.

Therefore, total volume = 3.651 ÷ 0.97 = 3.764 ft^3

The quantity of cement = 94 lb/3.764 ft^3 = 24.97 lb/ft^3 = 674 lb/yd^3

Answer is C.

Asphalt Mix Design: Volumetric Relationships

Hot mix asphalt (HMA) is a mix of the following constituents: aggregate (typically angular broken rock), asphalt binder, and air. The binder exists in two different states—some of it is absorbed onto the surface of the aggregate and the rest is free. The bulk volume of the aggregate includes the absorbed binder whereas the effective volume of the aggregate does not.

The various constituents of total mass and total volume of an asphalt mix are shown in Fig. 9.3.

Unit Volume Approach to Calculating Asphalt Properties

For a total volume = 1.0, the following relationships are useful in calculating all components of the asphalt mix:

Volume of bulk aggregate = mass of aggregate ÷ bulk SG of aggregate

Effective volume of aggregate = mass of aggregate ÷ effective SG of aggregate

Volume of absorbed asphalt = volume of bulk aggregate − effective volume of aggregate

Effective volume of asphalt = total volume of asphalt − volume of absorbed asphalt

Volume of air = total volume − total volume of asphalt − effective volume of aggregate

= total volume − total volume of asphalt + volume of absorbed asphalt − volume of bulk aggregate

= total volume − (effective volume of asphalt + volume of bulk aggregate)

Figure 9.3 Composition of an asphalt mix.

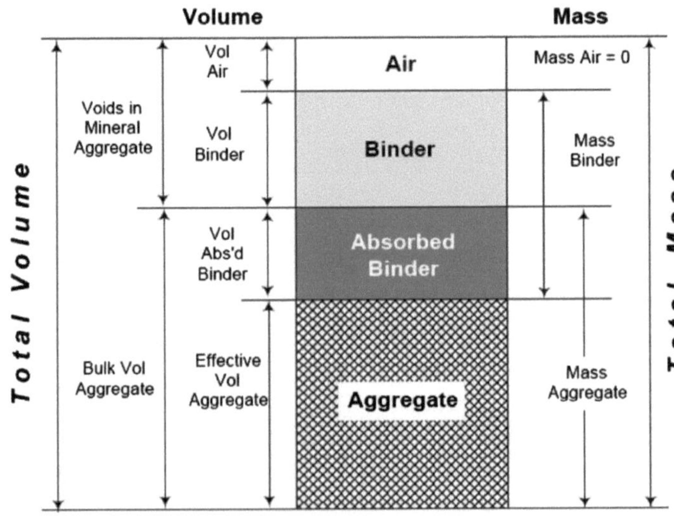

VMA = volume of air + effective volume of asphalt = total volume − volume of bulk aggregate

VFA = effective volume of asphalt ÷ VMA

Example 9.4

A sample of compacted hot mix asphalt is known to have the following properties at 25°C:

Mix bulk specific gravity = 2.329

Bulk specific gravity of aggregate = 2.705

Effective specific gravity of aggregate = 2.731

Asphalt binder specific gravity = 1.015

Asphalt content = 5% by weight

What are the (a) effective volume of aggregate per ft^3 total volume, (b) effective volume of asphalt per ft^3 total volume, and (c) air content?

Solution

Assume a total volume = 1.0 ft^3

Weight of asphalt mix = 2.329 × 62.4 × 1.0 (using bulk specific gravity)
= 145.33 lb

Total weight of asphalt = 0.05 × 145.33 (using asphalt content)
= 7.27 lb

Total volume of asphalt = 7.27 ÷ (1.015 × 62.4) (using asphalt specific gravity)
= 0.115 ft^3

Total weight of aggregate = 145.33 − 7.27 [using bulk specific gravity (AGGR)]
= 138.06 lb

Volume of bulk aggregate = 138.06 ÷ (2.705 × 62.4) [using bulk specific gravity (AGGR)]
= 0.818 ft^3

(a) Effective volume of aggregate = 138.06 ÷ (2.731 × 62.4) = 0.810 ft^3
Volume of absorbed asphalt = 0.818 − 0.810 = 0.008 ft^3

(b) Effective volume of asphalt = 0.115 − 0.008 = 0.107 ft^3

(c) Volume of air = 1.0 − 0.115 − 0.810 = 0.075 ft^3
Air content = 7.5%

Common Material Tests

Asphalt: Particle Size Analysis (AASHTO T30/ASTM D5444)

This test is typically run on asphalt mix samples after the binder has been extracted or burned off. The remaining sample is dried and then washed over a nest of sieves. The cumulative percent passing for each sieve is calculated and plotted as a particle size distribution curve.

The actual percentages are then compared to the corresponding percentages in the design mix. Potential ramifications of deviating from the design mix are segregation, harsh/tender mix, and out of balance volumetrics. The maximum aggregate size influences allowable pavement thickness, susceptibility to segregation, asphalt content, and volumetric requirements. The percent passing the no. 200 sieve influences compaction, asphalt content, and volumetric requirements. A low number for the percent passing no. 200 sieve may result in high air voids, permeability, and a lower asphalt concrete (AC) demand. A high value may result in low air voids and VMA, a higher AC demand and a tender mix. High dust proportion is typically accompanied by low binder content, which reduces pavement durability.

Asphalt: Moisture Susceptibility Test (AASHTO T283/ASTM D4867)

The purpose of this test is to show the susceptibility of the HMA to moisture damage. In order to conduct this test, six cylindrical samples are divided into two groups of three. One group of samples is tested under dry conditions while the other samples are subject to freezing under saturated conditions. The result of the test is reported as the tensile strength ratio (TSR) which is defined as

$$\text{TSR} = \frac{\text{Average wet tensile strength}}{\text{Average dry tensile strength}} \tag{9.5}$$

Typically, a minimum value of 0.75 is required. A failing TSR indicates that the mix is susceptible to stripping and raveling.

Asphalt: Specific Gravity and Absorption of Coarse Aggregate (AASHTO T85/ASTM C127)

After washing and drying the aggregate sample, the following measurements are made:

A = mass of oven-dry sample in air

B = mass of SSD sample in air

C = mass of SSD sample in water

The following can then be calculated:

$$\text{Bulk specific gravity of aggregate: } G_{sb} = \frac{A}{B-C} \tag{9.6}$$

$$\text{Bulk specific gravity of SSD aggregate} = \frac{B}{B-C} \tag{9.7}$$

$$\text{Apparent specific gravity of aggregate: } G_{sa} = \frac{A}{A-C} \tag{9.8}$$

$$\text{Absorption} = \frac{B-A}{A} \times 100\% \tag{9.9}$$

Asphalt Binder Content Test (AASHTO T308/ASTM D6307)

This test, which is performed on pretty much every HMA field sample, shows the percent of asphalt binder (by weight) in a HMA mix sample. A low binder content can lead to premature aging, tripping, or raveling. A high binder content can lead to flushing or bleeding in the pavement.

Voids in Mineral Aggregate (VMA) (AASHTO T166 and T209/ASTM D2726 and D2041)

VMA consists of the volume of air voids plus the volume of binder not absorbed into the aggregate. The mix needs a minimum VMA to have enough volume to hold the proper amount of air voids and the proper amount of binder.

Figure 9.4 shows the typical relationship between air content and the stability of an asphalt pavement.

Some common flexible pavement distresses are listed below:

Bleeding/flushing: Excess bituminous binder on the pavement surface, creating a shiny surface that may be tacky to the touch.

Stripping: The loss of the adhesive bond between asphalt cement and aggregate, most often caused by the presence of water in asphalt concrete, resulting in raveling, loss of stability and load carrying capacity of the pavement.

Raveling: Wearing away of the pavement surface in high-quality HMA concrete that may be caused by the dislodging of aggregate particles and loss of asphalt binder.

Segregation: The separation of coarse aggregate from fine aggregate as a result of mishandling of the mix during production, hauling, and placement. Segregation leads to nonuniform surface texture and nonuniform density.

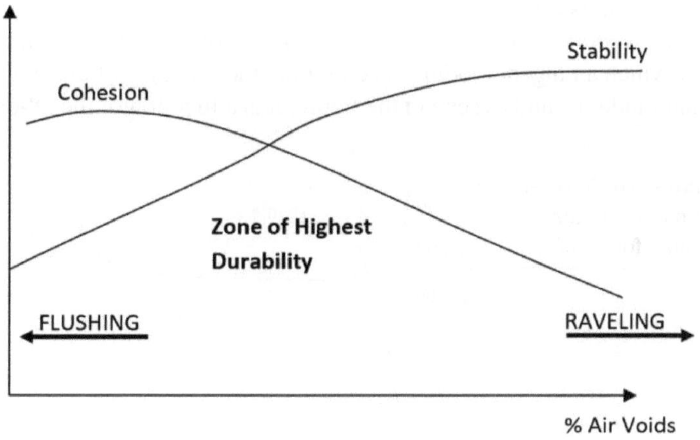

Figure 9.4 Effect of air voids on pavement characteristics.

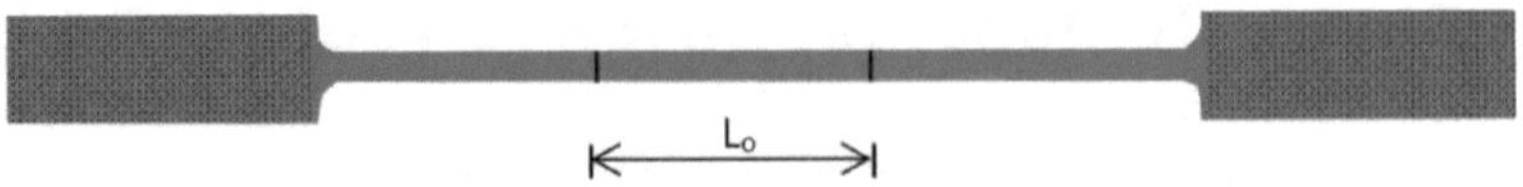

Figure 9.5 Tensile test coupon.

Steel: Tensile Test

The most common test for steel and other metals is the tensile test, where a sample with an instrumented test section (often called the "gage length") is subjected to axial tension. See Fig. 9.5. Because of the narrowing of the test section, most of the deformation (strain) is concentrated in the test section because the stress is much higher than in the end (grip) section. The gage length L_o is recorded at the beginning of the test (at zero load), and the extensometer records the elongation (Δ) of the gage length as the load progresses. These measurements (P, Δ) can then be converted to axial stress ($\sigma = P/A$) and strain ($\varepsilon = \Delta/L$).

The typical stress-strain diagrams for ductile and brittle steels are shown in Fig. 9.6. Most grades of steel exhibit a common initial tangent section. Thus, the slope of this tangent section—the modulus of elasticity is more or less constant across various grades of steel (E_s = 29,000 ksi or 200 GPa). If a distinct point of transition (yield point) is not apparent, then the 0.2% offset (line parallel to the initial tangent and passing through a strain intercept of 0.2% or $\varepsilon = 0.002$) is used to identify the yield point. The ratio of the final (fracture) strain to the yield strain is called the ductility.

Concrete: Compression Test

During a compression test of a concrete cylinder (test cylinders may be 6 in diameter × 12 in height or 4 in diameter × 8 in height or 3 in diameter × 6 in height), the cylinder is loaded in axial compression. Deformation of the cylinder (Δ) is measured by the travel of the loading crosshead and the resisting load (P) is measured by a load cell. These measurements (P, Δ) can be converted to axial stress ($\sigma = P/A$) and strain ($\varepsilon = \Delta/L$). See Fig. 9.7.

The typical stress-strain diagrams for various grades of concrete are shown in Fig. 9.8. Unlike steel, for which a tangent modulus is common, the modulus of elasticity of concrete is typically a secant modulus, and as seen in the figure, it is a function of the 28-day compressive

Figure 9.6 Typical stress-strain diagrams for steel.

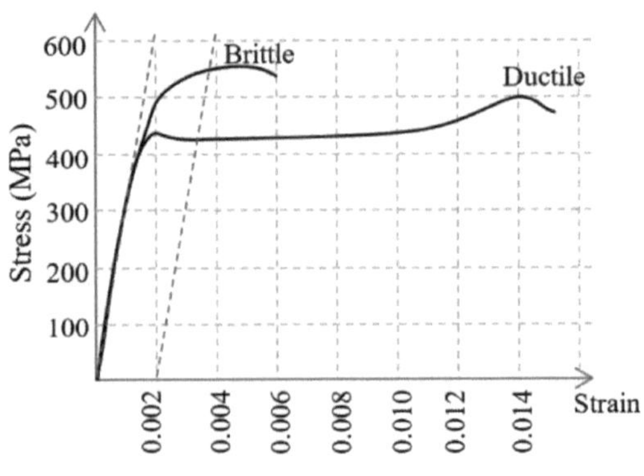

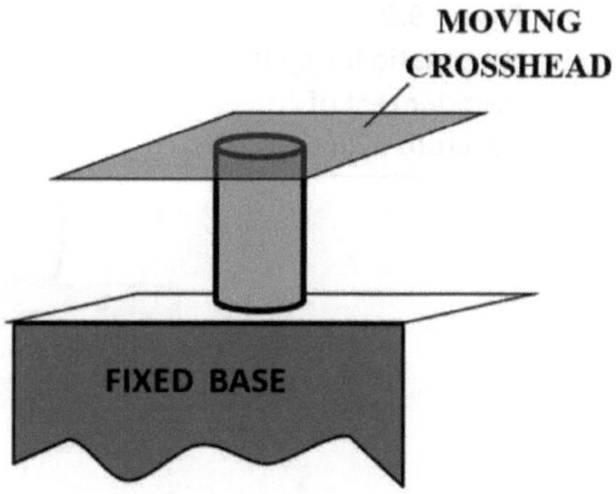

Figure 9.7 Schematic for compression test of concrete cylinders.

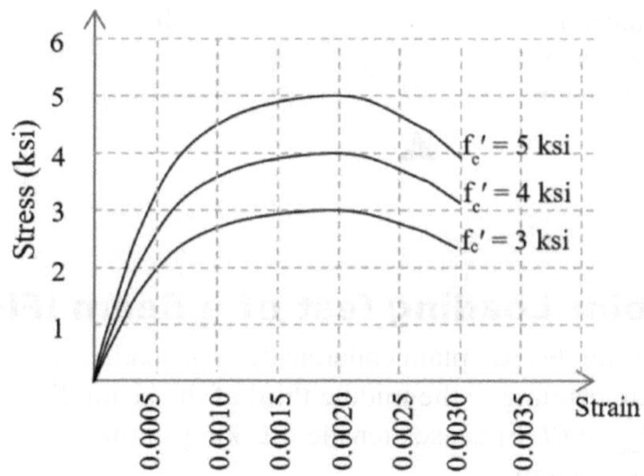

Figure 9.8 Typical stress-strain diagrams for concrete.

strength f_c'. The provisions of ACI 318 say that the modulus of elasticity of concrete E_c (psi) is given by

$$E_c = 33w_c^{1.5}\sqrt{f_c'} \tag{9.10}$$

where w_c is the unit weight of the concrete (lb/ft^3). Typically, the maximum compressive strain at which concrete is considered to have failed by *crushing* is $\varepsilon_c = 0.003$.

Concrete: Split Cylinder Test

In the split cylinder test, a concrete cylinder is placed on its side and loaded transversely in compression, as shown in Fig. 9.9. The load is applied diametrically and uniformly along the length of the cylinder until failure occurs along the vertical diameter. This produces a state of tension on the vertical plane and causes the cylinder to "split." The tensile stress on the vertical plane is given by

$$f_t = \frac{2P}{\pi DL} \tag{9.11}$$

Figure 9.9
Schematic for split-cylinder test of concrete cylinders.

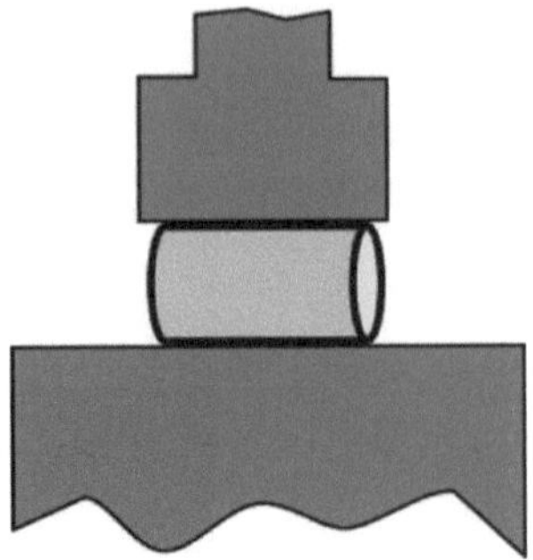

Figure 9.10
Schematic of third point loading (flexure) test of concrete beams.

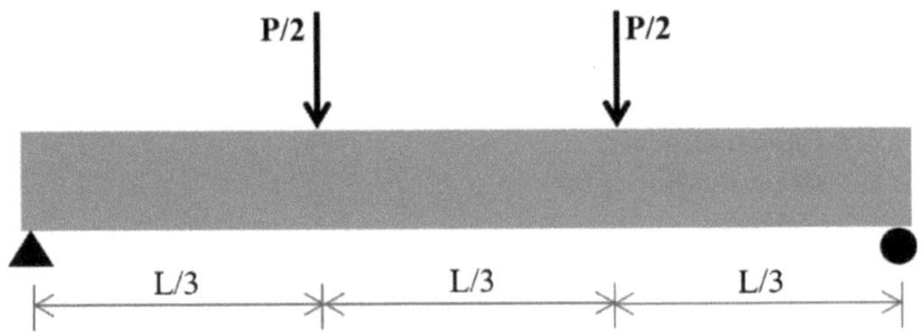

Third Point Loading Test of a Beam (Flexure)

In this test, a simple span plain concrete beam is loaded at one-third span points. This produces a state of pure bending in the middle third of the beam. See Fig. 9.10. The maximum bending moment ($M_{max} = PL/6$) causes tensile cracking in this zone. The tensile stress corresponding to a load P is given by

$$f_{ct} = \frac{\frac{PL}{6}}{\frac{bd^2}{6}} = \frac{PL}{bd^2} \tag{9.12}$$

where the rectangular cross section of the beam has width b and depth d.

At the load that causes failure, the value of this tensile stress is called the *modulus of rupture*. In ACI 318, the modulus of rupture is represented by f_r. For normal weight concrete, the ACI 318 recommends that the modulus of rupture be taken as

$$f_r = 7.5\sqrt{f_c'} \tag{9.13}$$

where f_r is modulus of rupture (tensile strength) of concrete (psi) and f_c' is 28-day compression strength (psi).

Example 9.5

A plain concrete cylinder beam (cross section: 6 in width × 8 in depth) is tested as shown. If the load P that causes cracking in the beam is 1600 lb, what is the tensile strength of the concrete?

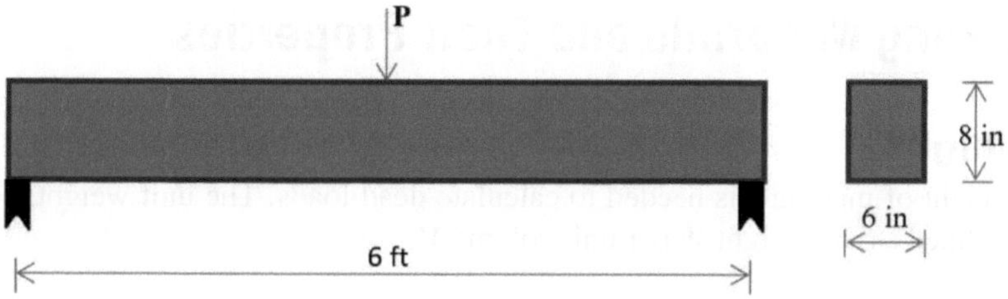

A. 350 psi

B. 450 psi

C. 550 psi

D. 650 psi

Solution The maximum moment due to load $P = 1600$ lb occurs at the bottom edge of the beam at midspan.

$$M = \frac{PL}{4} = \frac{1600 \times 6}{4} = 2400 \text{ lb} \cdot \text{ft} = 28{,}800 \text{ lb} \cdot \text{in}$$

Section modulus of the beam: $S = \frac{bh^2}{6} = \frac{6 \times 8^2}{6} = 64 \text{ in}^3$

Maximum bending stress: $\sigma = \frac{M}{S} = \frac{28{,}800}{64} = 450 \text{ psi}$

Answer is B.

Example 9.6

Concrete test cylinders (6 × 12 in) are tested as part of quality control measures for a project. Results from compression tests are summarized in the table below. What is the mean compressive strength predicted by this sample?

Sample	1	2	3	4	5	6	7	8	9
Load at failure (lb)	88,500	91,235	90,886	91,214	89,055	90,765	90,882	89,561	91,238

A. 3200 psi

B. 3300 psi

C. 3400 psi

D. 3500 psi

Solution The average failure load = $\Sigma P/N = 722{,}571/9 = 90{,}321$ lb

Compressive strength (average): $\sigma = \frac{P}{A} = \frac{90321}{\pi(3)^2} = 3194.5 \text{ psi}$

Answer is A.

Engineering Materials and Their Properties

Unit Weight

The unit weight of materials is needed to calculate dead loads. The unit weight γ or specific weight is defined as the weight W per unit volume V.

$$\gamma = \frac{W}{V} = \rho g \tag{9.14}$$

Steel: Tensile Strength

Materials such as steel exhibit significant strength when subject to tensile loads. Some materials exhibit brittle characteristics while others are more ductile. A ductile material undergoes significant deformation (strain) before failure. The longitudinal deformation increases rapidly as the material approaches failure. The longitudinal deformation (stretching) is accompanied by a significant reduction of area ("necking"), causing the stresses to increase rapidly until failure. The tensile strength of a material is calculated as the failure load divided by the cross-sectional area at the location of fracture.

Concrete: Compressive Strength

Contrary to steel, concrete has tensile strength which is negligible compared to its compressive strength. Typically, for commonly used concrete mixes, the tensile strength is between 10% and 15% of the compressive strength and is therefore ignored. In ACI 318 (Building Code Requirements for Structural Concrete) the 28-day compressive strength is designated f_c'.

Modulus of Elasticity

When an internal force N acts normal to a plane (as shown below), it produces a normal stress $\sigma = N/A$. As a result of the applied stress, the material undergoes a deformation, which is quantified by the strain, $\varepsilon = \delta l / l_o$. Most materials, when only moderately loaded, will exhibit a linear relationship between stress and strain. The constant of proportionality is called the modulus of elasticity E.

$$\sigma = E\varepsilon \tag{9.15}$$

Poisson's Ratio

As the bar shown (Fig. 9.11) elongates axially (longitudinally), its elasticity causes it to contract in the two directions perpendicular (transverse) to the longitudinal direction. The linear relationship between the longitudinal strain and the lateral strain is given by the Poisson ratio ν

$$\varepsilon_{\text{lateral}} = -\nu\varepsilon_{\text{longitudinal}} \tag{9.16}$$

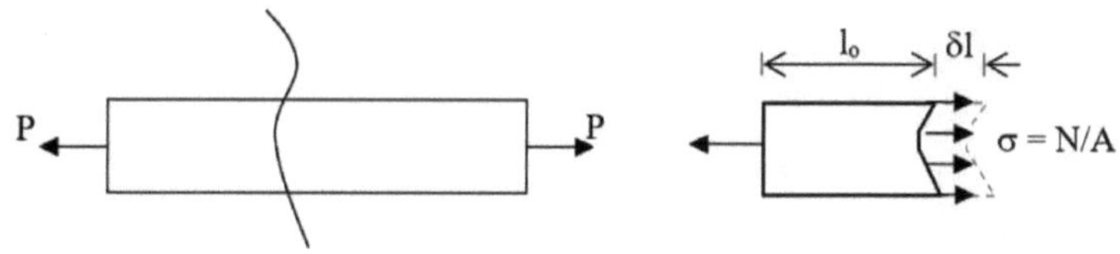

Figure 9.11 Elongation due to normal stress.

Thermal Expansion

As the temperature of an object is increased, it is accompanied by a corresponding increase in molecular energy, which causes an increase in size. The coefficient of thermal expansion measures the fractional change in size per degree change in temperature at constant pressure. Thermal expansion coefficient may be developed for volumetric, area or linear expansion. For liquids, the coefficient of volumetric thermal expansion is the most relevant, whereas for solids, one might only be concerned about the expansion in length for a specific area. The area and volumetric thermal expansion coefficients are approximately twice and three times the linear coefficient of expansion. Substances that expand at the same rate in all directions are called isotropic. For more on thermal stress and strain, see Chap. 8.

Composites

For composite materials, see Chap. 8.

CHAPTER 10

Fluid Mechanics

Fluid Properties

The table on page 114 of the *FE Reference Handbook*, version 9.4, summarizes some of the fundamental properties (specific weight, density, viscosity, kinematic viscosity, vapor pressure) of water as a function of temperature. Both SI and USCS units are listed.

Density

Density of a fluid (ρ) is defined as the mass per unit volume. The SI unit for density is kg/m^3. In USCS units, it is $slug/ft^3$, which can also be expressed as $lb \cdot s^2/ft^4$.

Specific volume, on the other hand, is defined as volume per unit mass. It is the inverse of density. Specific volume has units of m^3/kg (SI).

Specific weight γ is defined as weight per unit volume. It is calculated as the product of density ρ and acceleration due to gravity g. The SI unit for specific weight is N/m^3. In USCS units, specific weight can be measured in lb/ft^3.

Specific gravity is the ratio of the density of the fluid to the density of water. It can also be expressed as the ratio of the unit weights. In SI units, since the density of water (at normal temperatures) is about 1.0 g/cm^3, the specific gravity is numerically equal to the density.

Surface Tension

At a liquid–air interface, the liquid molecules experience a greater attraction to other liquid molecules than to the molecules in the air. As a result of these unbalanced forces, there is a net force at the liquid–air interface. This is in contrast to liquid molecules within the bulk of the liquid, where the force exerted by adjacent liquid molecules cancels out.

The result is an inward (directed from the interface toward the fluid) force that tends to minimize the surface area of the interface. This makes the liquid behave as if its surface were covered with a stretched membrane.

Surface tension is measured in force per unit length

$$\sigma = \frac{F}{L} \tag{10.1}$$

where F = surface force at the interface
L = length of the interface

Usually, surface tension decreases with an increase in temperature because the cohesive forces decrease with an increase in molecular thermal activity.

Example 10.1

A glass slide is in contact with a water surface along one of its sides as shown below. What is the magnitude of the force required to overcome surface tension as the slide is lifted from the water surface? The surface tension of water at 20°C is 0.073 N/m. The dimensions of the plate are $12 \times 5 \times 0.02$ cm.

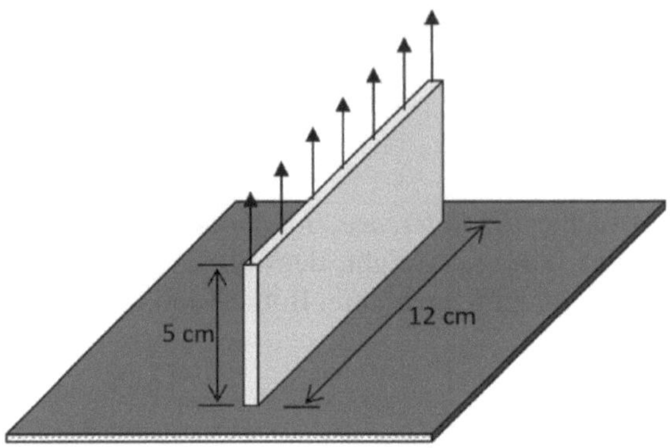

Solution The surface tension forces will act along the *perimeter* (not area) of the bottom face of the slide (which is in contact with the water surface). This perimeter is 24.04 cm = 0.2404 m.

The required force $= 0.073 \times 0.2404 = 0.0175$ N (equivalent of the weight of a 1.8-g mass).

Capillary Action

Capillary motion of a liquid occurs when cohesion (within the liquid) and adhesion (between the liquid and container walls) forces are strong enough to overcome gravitational forces. In narrow spaces such as in a thin tube or intercellular spaces, the forces are strong enough for the liquid to rise to a certain height within the space. A meniscus, or a crescent-shaped surface, forms on the upper surface of the liquid. A meniscus can be either concave or convex, depending on the characteristics of the liquid–wall interface. A concave meniscus forms when the liquid particles are more strongly attracted to the wall surface (adhesion forces) than to each other (cohesion), causing the liquid to "climb" the walls of the container. A concave meniscus forms between glass and water. See Fig 10.1(a). On the other hand, a convex meniscus forms when the cohesion forces are greater than the adhesion forces. A convex meniscus forms between glass and mercury, as in thermometers and barometers. See Fig. 10.1(b).

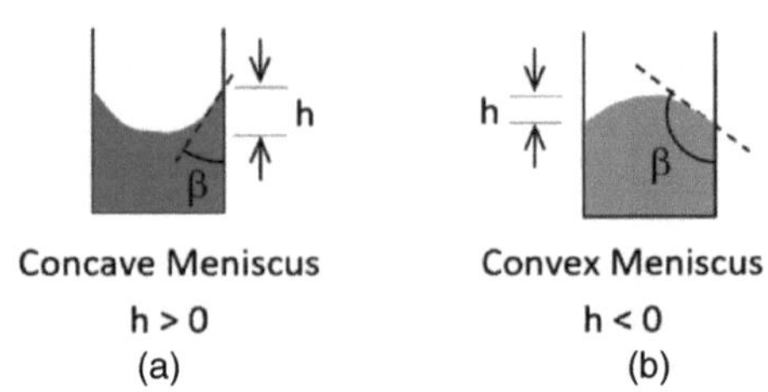

Figure 10.1 (a) Concave, (b) convex meniscus due to capillary action.

The height of rise (or fall) of a liquid in a tube due to capillarity is given by

$$h = \frac{4\sigma \cos\beta}{\rho g d} \tag{10.2}$$

where h = height of liquid rising ($h > 0$) or falling ($h < 0$) due to capillarity
d = diameter of fine tube
ρ = density of liquid
σ = surface tension
β = angle between wetted wall surface and tangent to fluid surface

Example 10.2

A glass capillary tube (diameter 1 mm) is inserted into an open reservoir of a liquid. The surface tension of the liquid is 85 dynes/cm. The inset diagram shows a magnified view of the liquid inside the tube.

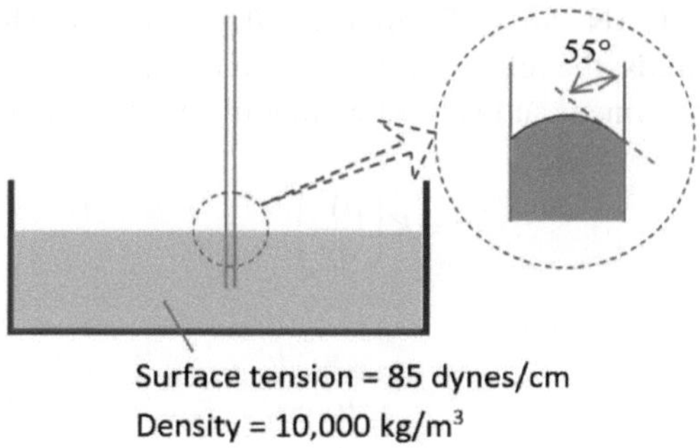

Which of the following is true?

A. Liquid rises 2 mm in the tube.

B. Liquid falls 2 mm in the tube.

C. Liquid rises 4 mm in the tube.

D. Liquid falls 4 mm in the tube.

Solution The angle β is the angle between the wetted tube and the tangent to the surface. Therefore, $\beta = 125°$.

σ = 85 dynes/cm = 85×10^{-5} N/cm = 85 x 10^{-3} N/m

$$h = \frac{4\sigma \cos\beta}{\rho g d} = \frac{4 \times 0.085 \times \cos 125}{10{,}000 \times 9.81 \times 1 \times 10^{-3}} = -0.002$$

Thus, the liquid experiences a capillary *drop* of 2 mm.

Viscosity

The viscosity of a fluid is a measure of its ability to resist shear stresses. When a shear stress, no matter how small, is applied to any fluid, a velocity gradient (dV/dy) develops transverse to the plane on which the shear stress is applied. See Fig. 10.2. For Newtonian fluids, the applied

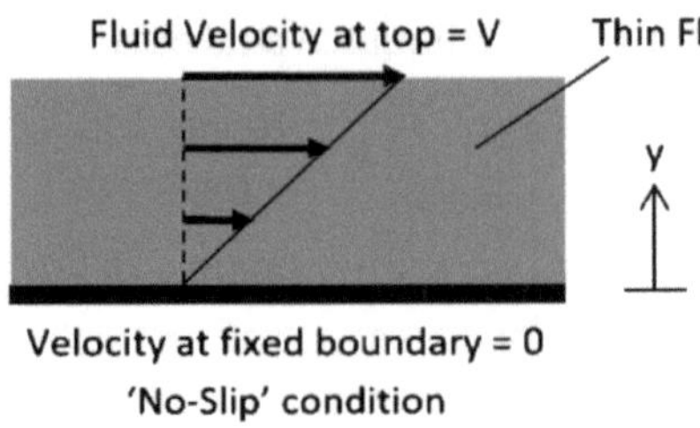

Figure 10.2 Velocity gradient in thin fluid film due to shear stress.

shear stress causes a proportional velocity gradient. The constant of proportionality is termed the viscosity, or dynamic viscosity, μ

$$\tau = \mu \frac{dV}{dy} \tag{10.3}$$

Viscosity has units of FT/L^2 ($N \cdot s/m^2$ or $lb \cdot s/ft^2$) or M/LT ($kg/m \cdot s$ or $slug/ft \cdot s$).

For non-Newtonian fluids, the relationship between the applied shear stress and the resulting velocity gradient is nonlinear and is most commonly expressed as a power law

$$\tau = K\left(\frac{dV}{dy}\right)^n \tag{10.4}$$

The kinematic viscosity ν is given by

$$\nu = \frac{\mu}{\rho} \tag{10.5}$$

For a Newtonian fluid, the Reynolds number, which is a dimensionless parameter expressing the ratio between the inertial forces and the viscous forces acting on the fluid, is given by

$$Re = \frac{\rho VD}{\mu} = \frac{VD}{\nu} \tag{10.6}$$

A non-Newtonian fluid, where the relationship between the applied shear stress and the resulting velocity gradient is given by Eq. (10.4), can be further subclassified into pseudo-plastic ($n < 1$) and dilatant ($n > 1$) fluids. For non-Newtonian fluids, the Reynolds number is given by

$$Re' = \frac{\rho V^{2-n} D^n}{K\left(\frac{3n+1}{4n}\right)^n 8^{n-1}} \tag{10.7}$$

NOTE For $n = 1$, Eq. (10.7) degenerates to Eq. (10.6).

Example 10.3

A Newtonian fluid with viscosity $\mu = 0.01$ lb·s/ft^2 fills the annular space between a fixed inner cylinder diameter = 2.0 in and an outer rotating drum (inner diameter = 2.04 in). The length of the assembly is 35 in. The drum is rotated at a constant speed of 2000 rev/min. What is the torque required to maintain the rotation of the drum?

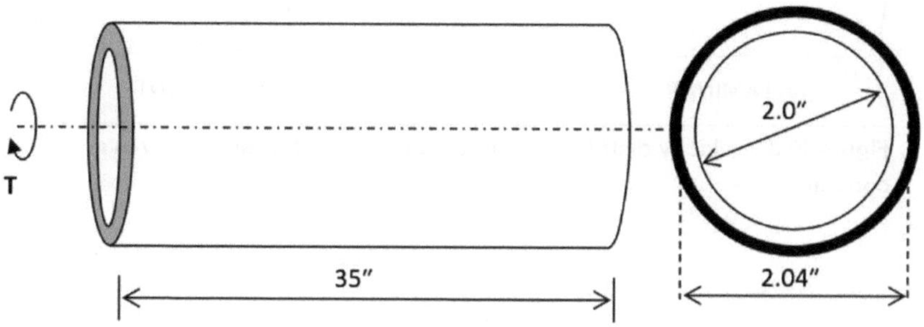

Solution Rotational speed $\omega = 2000 \text{ rev/min} = \dfrac{2000 \times 2\pi}{60} = 209.44$ rad/s

The tangential velocity of the drum $v = r\omega = 1.02 \times 209.44 = 213.6$ in/s

Thickness of fluid film = $0.04 \div 2 = 0.02$ in

Velocity gradient in the thin fluid film (in the annular space) $\dfrac{dV}{dy} = \dfrac{213.6}{0.02} = 10{,}681 \text{ s}^{-1}$

Shear stress at the surface of the rotating drum $\tau = \mu \dfrac{dV}{dy} = 0.01 \times 10{,}681 = 106.8 \text{ lb/ft}^2$

The (cylindrical) surface on which this shear stress acts is

$$A = 2\pi rL = 2\pi \times 1.0 \times 35 = 220 \text{ in}^2 = 1.527 \text{ ft}^2$$

Therefore, the torque needed is the moment of the resultant surface force about the axis

$$T = \tau \times A \times r = 106.8 \times 1.527 \times 1.0 \text{ lb} \cdot \text{in} = 14 \text{ lb} \cdot \text{ft}$$

Laminar versus Turbulent Flow

Reynolds number is the flow parameter used to distinguish laminar and turbulent flows. The term *laminar flow* indicates that individual streamlines remain parallel and distinct and mixing does not occur. For turbulent flows, slight perturbations introduced upstream will cause adjacent streamlines to cross, thereby inducing mixing. For flow through closed conduits, the *Handbook* states that laminar flow occurs for $Re < 2100$ and a transition to turbulence occurs between $2100 < Re < 10{,}000$. For $Re > 10{,}000$, the flow is considered fully turbulent.

Figure 10.3 shows laminar and turbulent flow profiles in circular pipes and between parallel plates. For the turbulent flow profile shown in Fig. 10.3(b), the boundary layer, within which the flow velocity grows from the "no slip" zero velocity at the boundary to the maximum velocity, is very thin. As a result, for fully turbulent flow, the average velocity is a good indicator of the true maximum velocity, which occurs at the center streamline. On the other hand, for laminar flow, a parabolic velocity profile develops, given by

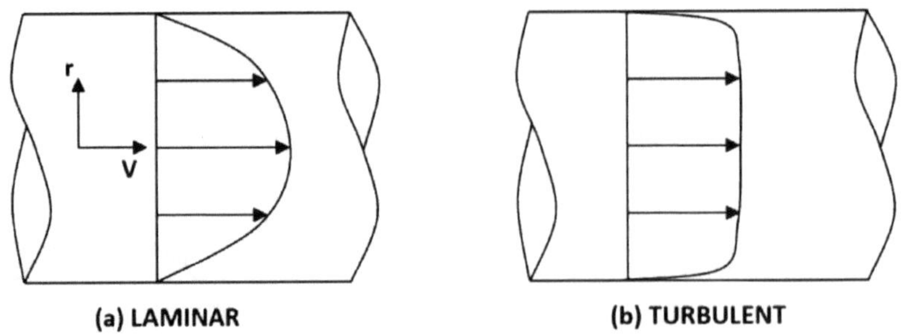

Figure 10.3 Velocity distribution for (a) laminar; (b) turbulent flow in a conduit.

$$V(r) = V_{max}\left[1-\left(\frac{r}{R}\right)^2\right] \tag{10.8}$$

where r = radial distance from centerline
R = radius (circular pipe) or half the distance between parallel plates

Relationship between Average Velocity and Maximum Velocity

The nominal velocity of flow in a conduit is often calculated as the average velocity which is given by dividing the flow rate by the cross-sectional area. A direct result of the parabolic profile is that for laminar flow in a circular pipe, the maximum velocity is twice the average velocity and for laminar flow between parallel plates, the maximum velocity is 1.5 times the average velocity. For fully turbulent flow in a circular conduit, the maximum velocity is approximately equal to 1.18 times the average velocity.

Example 10.4

The velocity is measured along the centerline of a circular pipe (internal diameter 2 in) experiencing laminar flow as $V = 1.5$ in/s. What is the flow rate (cfs)?

Solution Since the flow is laminar flow in a circular pipe, the theoretical velocity distribution is parabolic and the average velocity is half of the maximum velocity (along centerline). Thus, average velocity, $V_{ave} = 0.75$ in/s

Flow rate $Q = V_{ave}A = 0.75 \times \frac{\pi}{4} \times 2^2 = 2.36\ \text{in}^3/\text{s} = 0.0014\ \text{ft}^3/\text{s}$

Static Pressure on Submerged Surfaces

A submerged static object is subject to hydrostatic pressure acting on the various surfaces defining the boundary of the object. The resultant of these static pressures is the buoyancy force acting on the object. For an object of negligible volume, these forces are in static equilibrium and the buoyancy is zero. See Fig. 10.4.

Figure 10.4 Hydrostatic forces acting on a submerged body.

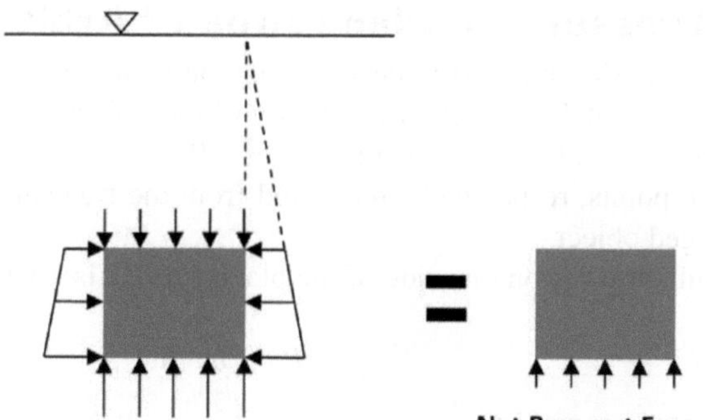

Archimedes' principle on buoyancy may be stated as

1. The buoyant force on a submerged object is equal to the weight of the displaced fluid.
2. An object in neutral equilibrium (floating) displaces a volume of fluid whose weight equals the weight of the object.

Thus, an ideal submerged plate (zero thickness, zero volume) develops equal and opposite pressure profiles on either side, and therefore has no buoyancy.

Example 10.5

A prismatic object is neutrally buoyant 20 cm below the free surface as shown. The object has plan dimensions 1 × 1 m and is 50 cm deep. Fluids 1 and 2 are immiscible. What is the weight of the object?

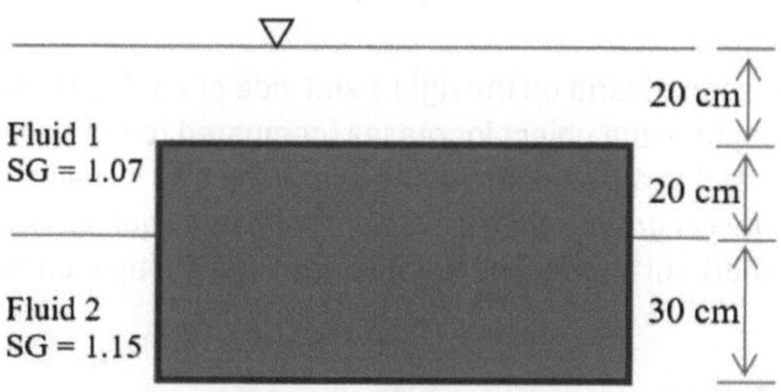

A. 2812 N

B. 3625 N

C. 4350 N

D. 5500 N

Solution According to Archimedes' principle, an object floats when the buoyant force (which is the weight of the displaced liquid) is equal to the weight of the object

$$W = \rho_1 V_1 g + \rho_2 V_1 g = (1070 \times 0.2 + 1150 \times 0.3) \times 9.81 = 5484 \text{ N}$$

Answer is D.

Static Pressure on Plane Area of Arbitrary Shape

Figure 10.5 shows a plane surface which is completely submerged and inclined at an angle θ with the horizontal. The center of gravity of the object is marked as CG and the "effective location" of the resultant force is marked as CP (center of pressure). Z_C and Z_R are distances to these two points, respectively, measured from the free surface and parallel to the plane of the submerged object.

The resultant force on one side of the plane surface is given by

$$R = \gamma h_c \cdot A \tag{10.9}$$

The depth of the center of gravity (measured vertically) is

$$h_c = Z_c \sin\theta \tag{10.10}$$

The distance Z_R is given by

$$Z_R = Z_C + \frac{I}{AZ_C} \tag{10.11}$$

where γ = unit weight of the fluid = ρg
h_c = depth of centroid (vertical)
Z_c = depth of centroid (inclined)
Z_R = depth of center of pressure (inclined)
I = second moment of area (moment of inertia) of the submerged plane about a horizontal axis passing through its center of gravity
A = cross-sectional area of the submerged plane

NOTE The second term on the right-hand side of Eq. (10.11) gets smaller as the immersion depth of the object increases (compared to the size of the object). For an object immersed to a significant depth below the free surface, we can assume that *while Z_R is always greater than Z_C, the difference in such cases is negligible.* This may allow for a short cut to be taken, particularly for a multiple choice exam.

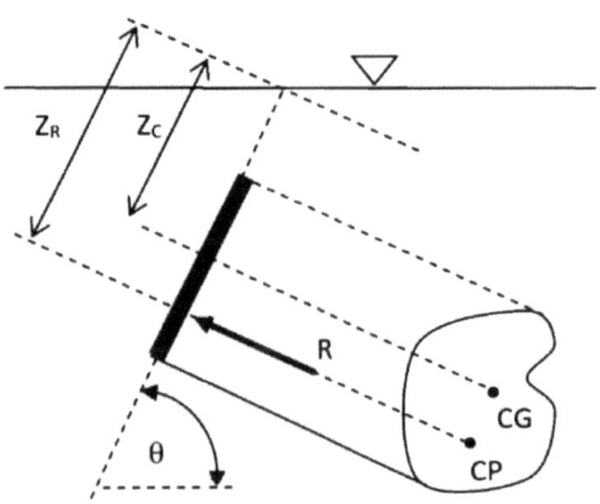

Figure 10.5 Static fluid pressure on a submerged plane surface.

Example 10.6

A triangular plate serves as a gate for a water storage tank as shown. The gate is hinged at the top (point A in the tank cross section or line AA in the side view of the gate) and at the bottom (point B). Calculate (i) the resultant force (lb) acting on the gate, (ii) the effective depth of the resultant force (ft), and (iii) the force exerted at the hinge B.

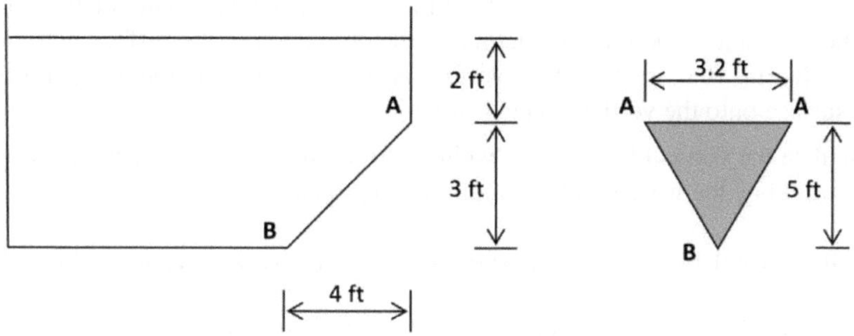

Solution Length of the triangle (perpendicular distance from the vertex B to the base AA) = 5 ft.

Moment of inertia (centroidal axis) of a triangle $I = \dfrac{bh^3}{36} = \dfrac{3.2 \times 5^3}{36} = 11.11\ \text{ft}^4$

Area of triangle $A = \dfrac{bh}{2} = \dfrac{3.2 \times 5}{2} = 8\ \text{ft}^2$

The depth of the centroid is $h_c = 2 + \dfrac{1}{3} \times 3 = 3$ ft

The inclined distance to the centroid is $Z_c = h_c \times \dfrac{5}{3} = 5$ ft

The inclined distance to the center of pressure is $Z_R = Z_c + \dfrac{I}{AZ_c} = 5 + \dfrac{11.11}{8 \times 5} = 5.278$ ft

(i) The resultant force is $R = \gamma h_c \cdot A = 62.4 \times 3 \times 8 = 1497.6$ lb

(ii) The (vertical) depth to the center of pressure is $h_R = Z_R \times \dfrac{3}{5} = 3.167$ ft

(iii) Since the top of the gate is at depth 2 ft, the bottom is at depth 5 ft, and the depth of the center of pressure is at 3.167 ft, the reaction at B, using the lever rule, is given by

$$F_B = \frac{3.167 - 2}{5 - 2} \times 1497.6 = 582.6\ \text{lb}$$

Static Pressure on Compound Area (Curved or Multiple Linear Segments)

Calculating the resultant force on a nonplanar surface can be very complex if performed as an integration of the pressure profile acting normal to infinitesimal elements. An easier and equivalent procedure is to perform the calculation in two steps as follows:

1. Draw a vertical plane through the toe of the surface and calculate the horizontal force as the resultant of the pressure acting on the projection of the surface onto this vertical plane. In Fig. 10.6, the toe of the surface is at the point *d* and the vertical projection of the surface onto the vertical plane is *de*.
2. Calculate the vertical force as the weight of the fluid above the surface. In Fig. 10.6, this would be the weight of the fluid in the region *abcde*.

To calculate total force on the curved surface *cd*, draw a vertical plane *de* through the toe.
The horizontal force H = resultant of the hydrostatic pressure on *de*.
The vertical force V = weight of fluid in area *abcde*.
The resultant force acting on the surface can then be calculated from

$$R = \sqrt{H^2 + V^2}$$

Manometers

Manometers are used to measure pressure by converting pressure to an equivalent height of a manometer fluid. In a simple manometer, the U-shaped manometer tube has one end attached to a pressure tap and the other end either open or at a reference pressure, typically atmospheric pressure. Such a configuration is shown in Fig. 10.7.

Considering the horizontal plane AB as the datum, the pressure at A (created by the pressure in the conduit + a column of fluid 1) must equal the pressure at B (created by the atmospheric pressure + a column of fluid 2)

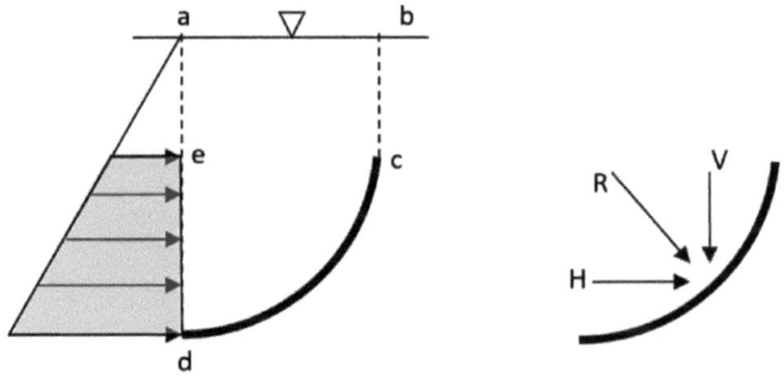

Figure 10.6 Static fluid pressure on a submerged curved surface.

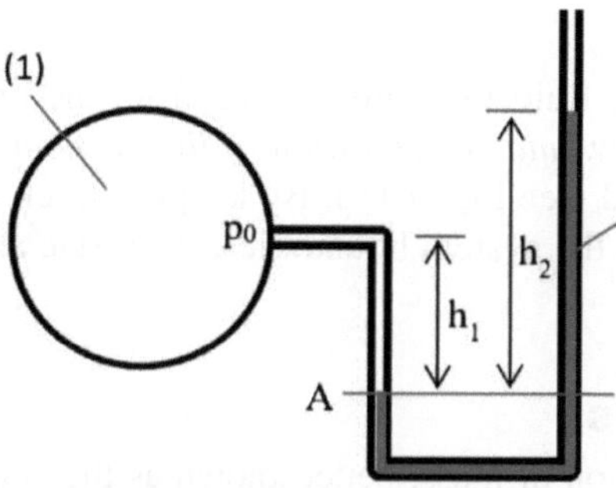

Figure 10.7 Manometer.

$$p_o + \rho_1 g h_1 = p_{atm} + \rho_2 g h_2 \quad (10.12)$$

$$p_o = p_{atm} + \rho_2 g h_2 - \rho_1 g h_1 \quad (10.13)$$

On the other hand, a differential manometer can be used between two pressure taps. For an illustration of such use, see the section on Venturi meter and orifice meter.

Barometer

Another device that works on the same equilibrium principle is the barometer. In this device, the atmospheric pressure is indirectly measured by observing the height of the barometric fluid in the closed tube. At the top of the tube is a closed pocket where the pressure is equal to the vapor pressure of the barometric fluid. See Fig. 10.8. Based on equilibrium of pressures

$$p_{atm} = p_v + \rho g h \quad (10.14)$$

where ρ = density of the barometric fluid
p_v = vapor pressure of the barometric fluid

Figure 10.8 Barometer.

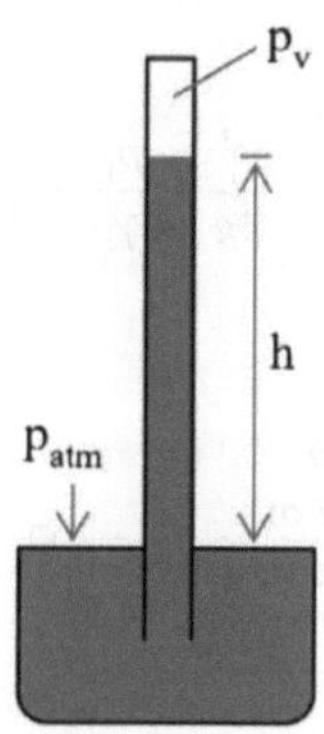

Fluid Dynamics

The basic equations governing fluid flow are derived from the principles *of conservation of mass, conservation of energy, and conservation of linear momentum.* These are derived using a control volume of fluid (see Fig. 10.9). By keeping track of the transport of mass, momentum, and energy across the system boundaries, we arrive at the following conservation principles.

Conservation of Mass

The statement of the conservation of mass, better known as the *continuity equation*, may be stated as—the rate of mass of fluid entering the system must equal that exiting the system. In other words, mass is neither created nor destroyed within the control volume. The mass flow rate through a closed control volume is given by

$$\dot{m} = \rho VA = \text{constant} \tag{10.15}$$

For incompressible fluids (such as water under normal pressures) the density ρ is constant, leading to $Q = VA = \text{constant}$, where Q is the volumetric flow rate.

Conservation of Energy

The statement of the conservation of energy for incompressible fluids, better known as *Bernoulli's equation*, may be stated as—for steady, ideal flow in an incompressible fluid, the energy per unit weight (specific energy) is conserved between two locations, as long as there are no energy loss mechanisms between these locations. The specific energy can be written as the sum of the kinetic energy, pressure energy, and potential energy of the fluid. These terms are shown (in order) in Eq. (10.16)

$$E = \frac{v^2}{2g} + \frac{p}{\gamma} + z \tag{10.16}$$

If energy losses between sections 1 and 2 are ignored, this leads to the ideal form of Bernoulli's equation

$$\frac{v_1^2}{2g} + \frac{p_1}{\gamma} + z_1 = \frac{v_2^2}{2g} + \frac{p_2}{\gamma} + z_2 \tag{10.17}$$

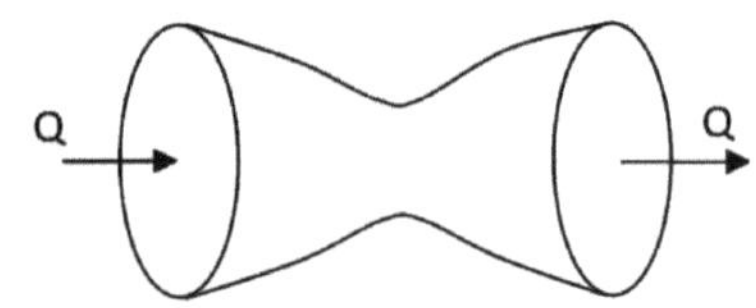

Figure 10.9 Continuity of volumetric flow rate.

If friction and form losses between sections 1 and 2 are written as h_f and a pump or turbine is located between 1 and 2, this may be extended to

$$\frac{-\eta \dot{W}_S}{\gamma Q} = \frac{v_2^2 - v_2^2}{2g} + \frac{p_2 - p_1}{\gamma} + z_2 - z_1 + h_f \tag{10.18}$$

where $\dot{W}_S$ = shaft power (positive for turbines, negative for pumps)
η = efficiency of pump or turbine
h_f = head loss between locations 1 and 2 (for details about calculating h_f, see Chap. 11)

NOTE If the equation is applied as shown above, pressures may be expressed as absolute pressure or gage pressure, without introducing error.

Example 10.7

Flow occurs in a rectangular open channel (6 ft wide) with velocity of 8 ft/s and depth of 5.8 ft. The elevation of the channel floor is 568.5 ft above sea level. At a downstream location, the channel floor elevation is 564.3 ft. The depth at the downstream end is 6 ft. Determine the head loss due to friction.

Solution Using the continuity equation, we solve for the velocity at the downstream section

$$V_1 A_1 = V_2 A_2 \Rightarrow V_1 d_1 = V_2 d_2 \Rightarrow V_2 = \frac{V_1 d_1}{d_2} = \frac{8 \times 5.8}{6} = 7.73 \text{ ft/s}$$

The energy grade line (EGL) slopes downward due to the head loss. Thus, the head loss is equal to the decrease of total energy from locations 1 to 2. The surface streamline is at atmospheric pressure, gage pressure = 0).

The elevation of the channel surface at the upstream location is 568.5 + 5.8 = 574.3 ft and at the downstream location is 564.3 + 6.0 = 570.3 ft

$$h_f = E_1 - E_2 = \left(\frac{v_1^2}{2g} + \frac{p_1}{\gamma} + z_1\right) - \left(\frac{v_2^2}{2g} + \frac{p_2}{\gamma} + z_2\right)$$

$$= \left(\frac{8^2}{2 \times 32.2} + 0 + 574.3\right) - \left(\frac{7.73^2}{2 \times 32.2} + 0 + 570.3\right) = 4.07\text{ft}$$

Example 10.8

Water at temperature of 10°C flows through a 150-mm-diameter galvanized iron pipe. The average flow velocity is 4 m/s. What is the pressure loss due to friction across a 100 m length of this pipe?

A. 3 m

B. 9 m

C. 11 m

D. 16 m

Solution The kinematic viscosity of water at 10°C is $\nu = 1.306 \times 10^{-6}$ m²/s.

Reynolds no $Re = \dfrac{VD}{\nu} = \dfrac{4 \times 0.15}{1.306 \times 10^{-6}} = 4.6 \times 10^5$.

For galvanized iron pipe, absolute roughness e = 0.15 mm. Relative roughness e/D = 0.15/150 = 0.0010.

For $Re = 4.6 \times 10^5$ and $e/D = 0.001$, $f = 0.02$

Head loss: $h_f = f \dfrac{L}{D} \dfrac{V^2}{2g} = \dfrac{0.02 \times 100 \times 4^2}{0.15 \times 2 \times 9.81} = 10.87$ m

Answer is C.

Conservation of Momentum

The statement of conservation of momentum is derived from Newton's second law which states that the sum of external forces is equal to the rate of change of linear momentum. This principle may be used to calculate the net force acting on a node where a change in momentum occurs. Examples are bends in pipes where the flow changes direction and pipe section enlargements (or contractions) where the flow velocity decreases (or increases).

$$\sum F = \dot{m}(v_2 - v_1) = \rho Q(v_2 - v_1) \tag{10.19}$$

The computed net force acts on the fluid. Thus, if we are interested in computing the force imparted *by* the fluid, we should reverse the sign of the computed force.

With constant flow rate, force on a rigid surface due to impinging flow = $\rho Q \Delta V$, where ΔV is the change in velocity effected by the static surface.

Example 10.9

A pipeline has a horizontal bend + a reducer as shown below. The flow rate through the pipe is 0.2 m³/s. The gage pressure at point A is 500 kPa and at point B is 490 kPa. The bend is supported by an anchor. What is the force on the anchor?

A. 12 kN in the northeast direction

B. 39 kN in the southeast direction

C. 68 kN in the northwest direction

D. 90 kN in the southwest direction

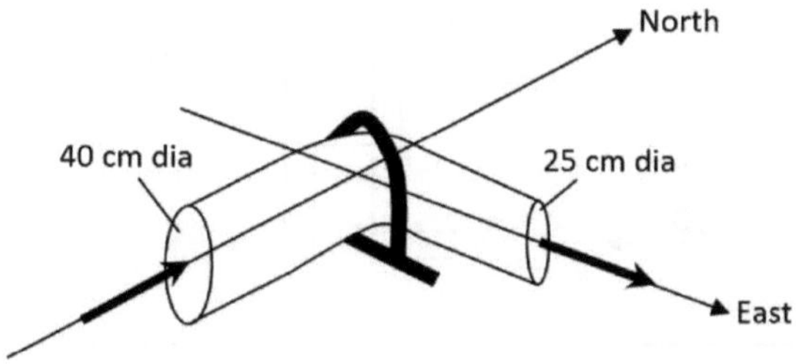

Solution From the continuity equation, the velocities at locations 1 and 2 are

$$V_1 = \frac{Q}{A_1} = \frac{0.2}{\frac{\pi}{4}\times 0.4^2} = 1.592\frac{\text{m}}{\text{s}}$$

$$V_2 = \frac{Q}{A_2} = \frac{0.2}{\frac{\pi}{4}\times 0.25^2} = 4.074\frac{\text{m}}{\text{s}}$$

If the force components acting on the fluid are Q_x and Q_y, then the statement of conservation of momentum, for each Cartesian direction, is

$$Q_x - 490\times 10^3 \times \frac{\pi}{4}\times 0.25^2 = 1000\times 0.2\times(4.074-0) \Rightarrow Q_x = 24{,}868\ \text{N}$$

$$Q_y + 500\times 10^3 \times \frac{\pi}{4}\times 0.4^2 = 1000\times 0.2\times(0-1.592) \Rightarrow Q_y = -63{,}150\ \text{N}$$

Q_x and Q_y are forces acting on the fluid. So, the forces acting on the bracket are equal and opposite:

R_x and R_y are forces acting on the bracket:

$R_x = -24{,}868$ N and $R_y = +63{,}150$ N. The resultant of this is 67,870 N acting in the N-W direction.

Answer is C.

Energy Grade Line and Hydraulic Grade Line

With respect to a horizontal datum, the EGL and the hydraulic grade line (HGL) may be plotted as the functions

$$\text{EGL} \qquad z + \frac{p}{\gamma} + \frac{v^2}{2g}$$

$$\text{HGL} \qquad z + \frac{p}{\gamma} \qquad (10.20)$$

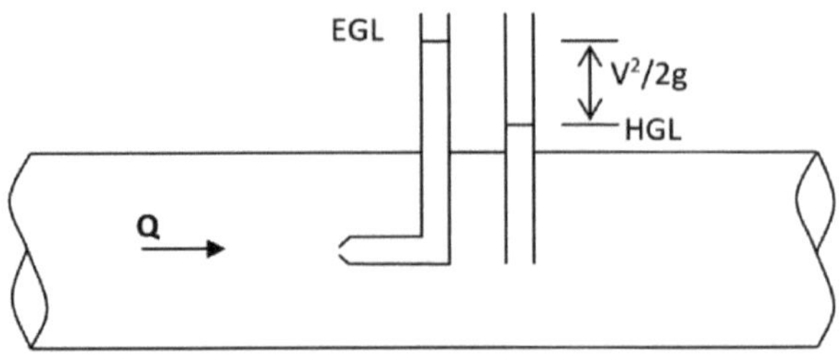

Figure 10.10 Elevation of HGL and EGL in a pitot tube.

For example, if a horizontal pipe is tapped with a pitot tube as shown in Fig. 10.10, the stagnation pressure at the head of the tube causes the liquid column to rise to the EGL elevation, whereas a tap into the same central streamline (but which does not stagnate the flow) causes the liquid column to rise to the elevation of the HGL. If these two pressure taps were connected to a differential manometer, the height difference recorded on the manometer is proportional to the quantity $V^2/2g$, where V = flow velocity.

These lines are plotted for the simple system shown in Fig. 10.11. A system of pipes carries flow from the reservoir on the left to the one on the right. The flow is aided by a pump and regulated by a valve. In the figure, the dark solid line plots the EGL, while the dashed line plots the HGL.

The following observations may be made about Fig. 10.11:

1. At free surfaces of large reservoirs, since the gage pressure is zero and the velocity is effectively zero, both EGL and HGL coincide with the reservoir free surface.
2. The slope of the EGL and HGL along each pipe segment represents friction loss per unit length along that segment.

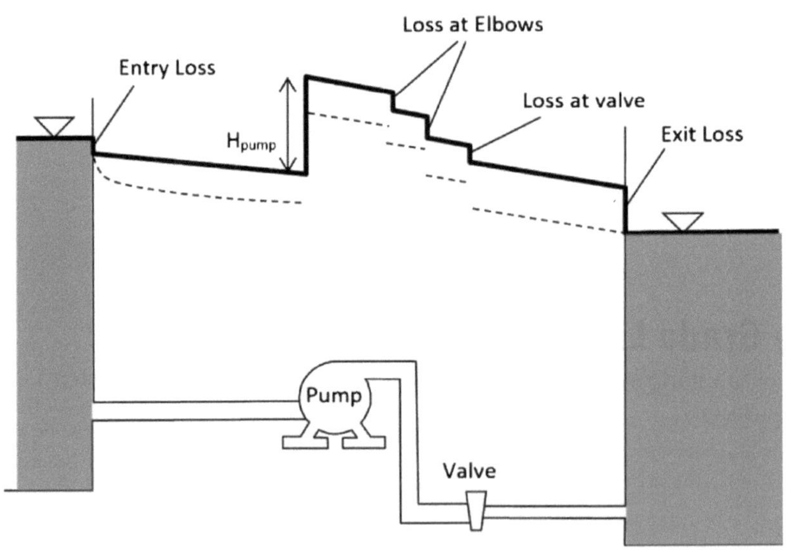

Figure 10.11 Elevation of HGL and EGL in a water delivery system.

3. The vertical separation between EGL and HGL is proportional to the square of the velocity. Thus, with the same flow rate through various pipes, this separation will be greater for smaller pipe diameters.
4. At the location of the pump, there is a discontinuous jump in both HGL and EGL.
5. At all locations where form losses occur (such as entry into the pipe system, exit from the pipe system, pipe bends, etc.), there is a discontinuous jump in both HGL and EGL.

Example 10.10

Five manholes (A to E) with characteristics shown in the table exist along a storm sewer as shown. During extreme rainfall events, the pipe flows full with the pressures given in the fourth column. At which manholes does the storm water overflow through the manhole onto the street above?

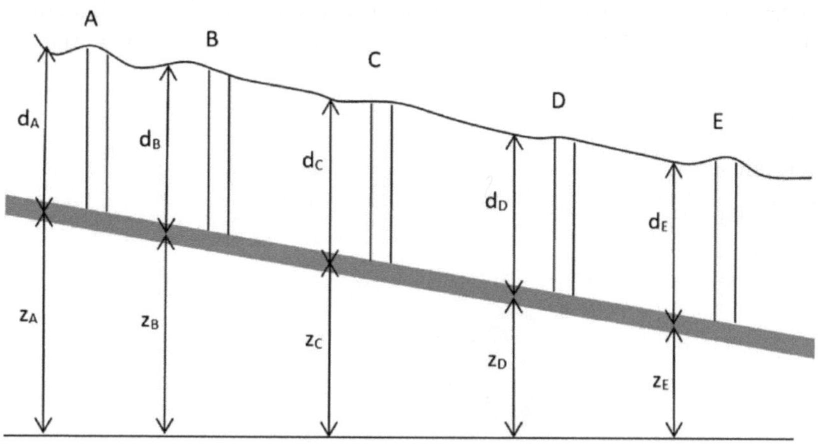

Location	Pipe C.L Depth *d* (ft)	Pipe C.L Elevation *z* (ft)	Pressure lb/in²
A	8.2	100	0
B	10.5	86	4.2
C	12.4	71	6.3
D	6.8	54	3.1
E	6.2	43	1.2

Solution For each location, the pressure is converted to pressure head (listed in the third column in the table below)

For example, $4.2 \text{ psi} = 4.2 \times 144 \text{ psf} = \dfrac{4.2 \times 144}{62.4} = 9.7 \text{ ft head}$

The elevation of the HGL is given by $\text{HGL} = \dfrac{p}{\gamma} + z$ (listed in the fourth column of the table).

Location	Pressure lb/in²	Pressure Head p/γ(ft)	HGL Elevation $p/\gamma + z$ (ft)	Ground Elevation (ft)
A	0	0	100	108.2
B	4.2	9.7	95.7	96.5
C	6.3	15.7	86.7	83.4
D	3.1	7.2	61.2	60.8
E	1.2	2.8	45.8	49.2

Comparing HGL elevation to surface elevation, manholes C and D will overflow. In other words, if the pressure head at a certain location exceeds the burial depth of the sewer, the manhole will overflow.

Example 10.11

A tank contains water to a fixed elevation as shown. A rounded orifice with diameter = 1 cm exists with its centerline 2.1 m below the water surface. What is the flow rate through the nozzle?

A. 0.1 L/min

B. 0.5 L/min

C. 7.2 L/min

D. 30 L/min

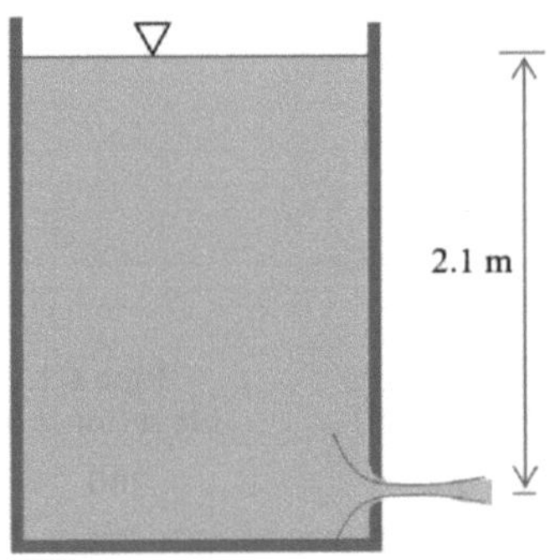

Solution For a jet discharging freely into the atmosphere

$$Q = CA_o\sqrt{2gh}$$

For a rounded orifice, the discharge coefficient $C = 0.98$ (see page 112 of the *Handbook*).
Orifice area: $A_o = \pi r^2 = \pi \times 0.005^2 = 7.854 \times 10^{-5}\,\text{m}^2$

$$Q = 0.98 \times 7.854 \times 10^{-5}\sqrt{2 \times 9.81 \times 2.1} = 4.94 \times 10^{-4}\,\frac{\text{m}^3}{\text{s}} = 0.494\,\frac{\text{L}}{\text{s}} = 29.6\text{ L/min}$$

Answer is D.

Fluid Power

The fluid power (power needed to transmit a flow rate Q of a fluid with unit weight γ and provide it with a lift H) is given by $P = \gamma QH$. In order that a pump operating at an efficiency η provide this fluid power, it must be rated at a higher power given by

$$\dot{W}_p = \frac{\gamma QH}{\eta} \tag{10.21}$$

This is called the rate of shaft work. Also, see Eq. (10.18).

Example 10.12

Water is being delivered from a reservoir by a centrifugal pump operating at 85% efficiency. The total dynamic head for the pump is 75 m. If the flow rate = 0.1 m³/s, what is the required brake horsepower of the pump?

Solution The pump power is calculated as $W = \dfrac{\rho g QH}{\eta} = \dfrac{1000 \times 9.81 \times 0.1 \times 75}{0.85} =$ $86{,}559 \text{ W} = 116 \text{ hp}$

Dynamic Similarity

In conducting model tests on fluid behavior, certain dimensionless groups are used to demonstrate that dynamic similarity (or dissimilarity) exists between different fluid flow fields. For example, if model tests are carried out in a laboratory experiment (such as in a wind tunnel or water tunnel) to study a full-scale phenomenon, one must ensure that the model and prototype flow regimes have similar characteristics. This may be ensured by maintaining the same Reynolds number (Re) or Froude number (Fr) or Mach number (M) in both flow fields.

There are two aspects to maintaining similarity between model and prototype. The first is *geometric similarity*, which is to maintain the same geometric proportions (model is true to scale in length, area, and volume). The second is *dynamic similarity*, which is to maintain equal ratios of all types of forces for the model and prototype.

There are various types of forces at work in a fluid field. Some of these are inertial forces, viscous forces, surface tension forces, gravitational forces. The relative importance of a particular fluid property is expressed in the form of a dimensionless number. A list of some of these numbers is on page 112 of the *Handbook*. For the field of civil engineering, the two most important of these dimensionless numbers are the Reynolds number and the Froude number.

Case 1—Inertial and Viscous Forces Dominate (Reynolds Number Similarity)

For systems where inertial and viscous forces dominate, the relative effect of these forces is expressed in terms of the Reynolds number (Re). Examples of systems where Reynolds number similarity is desirable are subsonic aircraft, closed pipe-flow, pumps, submarines, turbines, and drainage through tank orifices

$$\text{Reynolds number } Re = \frac{VD}{\nu} = \frac{\rho VD}{\mu} \tag{10.22}$$

Thus, if the model (subscript m) and prototype (subscript p) are to have equal Reynolds number

$$Re_m = \frac{\rho_m \cdot V_m \cdot L_m}{\mu_m} = \frac{\rho_p \cdot V_p \cdot L_p}{\mu_p} = Re_p \tag{10.23}$$

This can be rearranged to relate the velocity ratio (or scale) to the length scale as given below. The length scale, velocity scale, and viscosity scale are expressed as λ_L, λ_V, and λ_μ, respectively

$$\lambda_V = \frac{V_m}{V_p} = \frac{\rho_p}{\rho_m}\frac{L_p}{L_m}\frac{\mu_m}{\mu_p} = \frac{1}{\lambda_\rho}\frac{1}{\lambda_L}\lambda_\mu \tag{10.24}$$

Thus, if the fluid surrounding the prototype structure and the model are the same (density and viscosity are the same), this would reduce to the simpler relation

$$\lambda_V = \frac{V_m}{V_p} = \frac{1}{\lambda_L} \tag{10.25}$$

Example 10.13

A 1:50 scale model of a suspension bridge is tested in a water tunnel. The model is completely immersed in the fluid. If the temperature of the water is 70°F, what is the velocity of the water needed to replicate wind effects (air temp 60°F) due to a 60 mph wind?

Viscosity of water at 70°F = 1.059×10^{-5} ft²/s

Viscosity of air at 60°F = 1.58×10^{-4} ft²/s

Solution This is a case of submerged flow. Therefore, Reynolds number similarity must be maintained

$$Re_m = \frac{V_m \cdot L_m}{\nu_m} = \frac{V_p \cdot L_p}{\nu_p} = Re_p \Rightarrow V_m = V_p \times \frac{L_p}{L_m} \times \frac{\nu_m}{\nu_p}$$

$$V_m = 60\text{ mph} \times \frac{50}{1} \times \frac{\nu_{\text{water } 70°\text{F}}}{\nu_{\text{air } 60°\text{F}}} = 60 \times 50 \times \frac{1.059 \times 10^{-5}}{15.8 \times 10^{-5}} = 201\text{ mph}$$

Case 2—Inertial and Gravitational Forces Dominate (Froude Number Similarity)

For systems where inertial and gravitational forces dominate, the relative effect of these forces is expressed in terms of the Froude number (*Fr*). Examples of systems where Froude number similarity is desirable are flows over spillways, weirs, open channel flow with varying surface levels, surface ships, surge and flood waves

Froude number

$$Fr = \frac{V^2}{Lg} \tag{10.26}$$

Thus, if the model (subscript *m*) and prototype (subscript *p*) are to have equal Froude number

$$Fr_m = \frac{V_m^2}{L_m g_m} = \frac{V_p^2}{L_p g_p} = Fr_p \tag{10.27}$$

This can be rearranged to relate the velocity ratio (or scale) to the length scale as given below. The length scale, velocity scale, and gravity scale are expressed as λ_L, λ_V, and λ_g, respectively

$$\lambda_V = \frac{V_m}{V_p} = \sqrt{\frac{L_m}{L_p}\frac{g_m}{g_p}} = \sqrt{\lambda_L}\sqrt{\lambda_g} \tag{10.28}$$

Thus, if the gravity field strength for the prototype structure and the model are the same, this would reduce to the simpler relation

$$\lambda_V = \sqrt{\lambda_V} \tag{10.29}$$

Example 10.14

A 1:50 scale model of a spillway is constructed to replicate flow phenomena for a flow rate of 500 cfs. What must be the flow rate in the model?

Solution This is a case of free surface flow. Therefore, Froude number similarity must be maintained

$Fr_m = Fr_p \Rightarrow \dfrac{V_m^2}{L_m g_m} = \dfrac{V_p^2}{L_p g_p}$, which can be used to solve for the velocity scale

$$\frac{V_m}{V_p} = \sqrt{\frac{L_m}{L_p}\frac{g_m}{g_p}} = \sqrt{\frac{1}{50}\frac{1}{1}} = \sqrt{0.02} = 0.1414$$

Flow rate is given by $Q = VA$. From a point of view of dimensional analysis, $Q = VL^2$. Therefore, the flow rate scale is given by

$$\frac{Q_m}{Q_p} = \frac{V_m}{V_p}\frac{A_m}{A_p} = \frac{V_m}{V_p}\left(\frac{L_m}{L_p}\right)^2 = 0.1414 \times 0.02^2 = 0.0000566$$

Therefore $Q_m = 0.0000566 \times 500 = 0.0283$ cfs.

Flow Measurement Devices

Flow measurement techniques can be broadly classified as either direct or indirect. Direct methods involve the actual measurement of the volume of fluid passing a point in a specific length of time. Indirect methods usually involve the measurement of pressure changes which are related to the flow rate. For example, an orifice meter restricts the flow by forcing it through an orifice. The resulting pressure drop is measured and since it is related to the velocity of flow, it can also be related to the flow rate.

Pitot Tube

Possibly the simplest flow measurement device is the pitot tube. It is an open-ended tube of very fine bore that is placed parallel to the flow such that the flow impinges on the open end. See Fig. 10.12. The flow stagnates at the tip of the tube, thereby converting the kinetic energy to pressure energy. If the pressures within the free stream and at the tip of the tube are compared, they show a pressure differential. This pressure rise can then be recorded on a manometer. The fundamental physics involved in flow measuring devices of this philosophy is the conversion of velocity head ($V^2/2g$) to pressure energy (p/γ). Thus

$$\frac{V^2}{2g} = \frac{p}{\gamma} \Rightarrow V = \sqrt{\frac{2gp}{\gamma}} = \sqrt{2gH} \tag{10.30}$$

However, if the monitored fluid and the manometer fluid are different, the pressure head recorded (H) would change by a factor dependent on the densities of the two fluids. This results in the following expression for the stream velocity V

$$V = \sqrt{2gH\left(\frac{\rho_m - \rho}{\rho}\right)} \tag{10.31}$$

where ρ = density of the fluid in the conduit
ρ_m = density of the manometer fluid
H = manometer differential column height

Figure 10.12 Pitot tube.

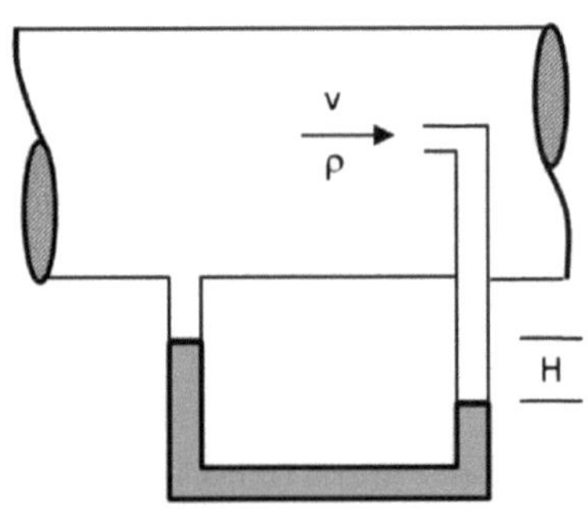

Note that this term reappears in the expression for flow rate for the orifice meter and the venturi meter as well.

Orifice Meter

The orifice meter uses a sharp-edged orifice (diameter D_o) to constrict the flow in a pipe. See Fig. 10.13. Pressure taps are installed at distances D and ½ D upstream and downstream of the orifice, respectively. The flow narrows to its smallest diameter (*vena contracta*) downstream of the orifice. The area of the contracted jet A_c is smaller than the area of the orifice A_o by a constant called C_c (coefficient of contraction) given by

$$C_c = \frac{A_c}{A_o} = \left(\frac{D_c}{D_o}\right)^2 \tag{10.32}$$

The flow rate in the conduit is then given by

$$Q = C_f A_o \sqrt{\frac{2gH(\rho_m - \rho)}{\rho}} \tag{10.33}$$

where the discharge coefficient is given by

$$C_f = \frac{C_v C_c}{\sqrt{1 - C_c^2 A_o^2 / A_1^2}} \tag{10.34}$$

where A_o and A_1 are the areas of the orifice and the upstream section, respectively. C_v is a velocity coefficient which is less than 1.0 due to significant viscous effects at low to moderate Reynolds numbers. At high Reynolds numbers, $C_v \approx 1.0$.

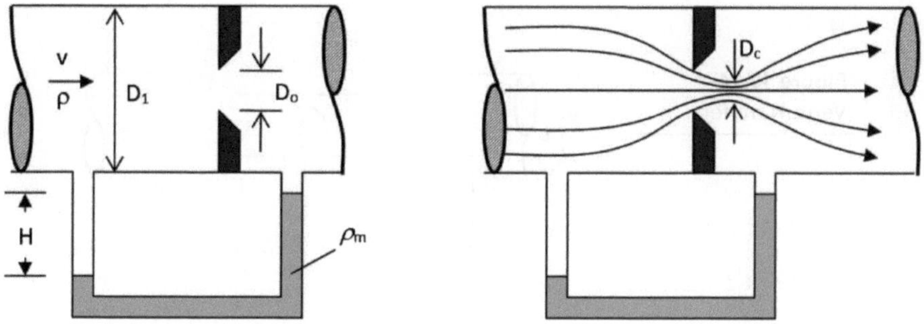

Figure 10.13 Orifice meter.

Venturi Meter

The Venturi meter constricts the flow in a conduit through a small diameter "throat" before allowing the flow to expand to the original diameter. The advantage of the Venturi meter is that while very similar to the orifice meter in principle, it imposes a much smaller head loss because of the more streamlined flow. The two pressure taps (marked 1 and 2 in Fig. 10.14) are located upstream of and in the throat section, respectively.

The flow rate for a Venturi meter is given by Eq. (10.35)

$$Q = \frac{A_2 C_d}{\sqrt{1-\beta^4}} \sqrt{\frac{2gH(\rho_m - \rho)}{\rho}} \tag{10.35}$$

where $\beta = \dfrac{D_2}{D_1}$

C_d = discharge coefficient (approaches 1.0 for large Reynolds numbers and small β)

Example 10.15

An orifice meter (orifice diameter = 50 mm) is inserted in a 10-cm-diameter cast iron pipe carrying water at 15°C. The pressure difference across the meter is 1.5 atm. The orifice coefficient is 0.72. The discharge (L/min) through the pipe is most nearly?

A. 650 L/min

B. 1050 L/min

C. 1500 L/min

D. 2350 L/min

Solution The flow rate of an orifice meter is given by

$$Q = CA_o \sqrt{2g\left(\frac{p_1}{\gamma} + z_1 - \frac{p_2}{\gamma} - z_2\right)}$$

Figure 10.14
Venturi meter.

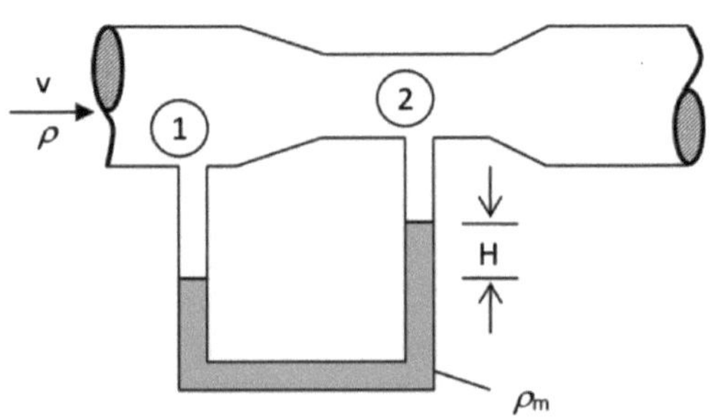

Ignoring the elevation difference between the two pressure taps at 1 and 2, we get

$$Q = CA_o\sqrt{2g\left(\frac{p_1 - p_2}{\gamma}\right)} = CA_o\sqrt{2\left(\frac{p_1 - p_2}{\rho}\right)}$$

1.0 atm = 1.013×10^5 Pa

$$Q = 0.72 \times \pi \times 0.025^2\sqrt{2\left(\frac{1.5 \times 1.013 \times 10^5}{1000}\right)}$$

$$Q = 0.02464 \text{ m}^3/\text{s} = 24.64 \text{ L/s} = 1478.7 \text{ L/min}$$

Answer is C.

Flow Measurement with Weirs

Sharp-Crested Weirs

Measurement weirs are vertical, sharp-crested plates obstructing flow. See Fig. 10.15. If the weir opening does not span the entire width of the channel, the flow width is contracted and the effective weir length is less than the actual weir length. In such a case, the weir is said to be *contracted*. If a weir has no end walls, or if contractions are suppressed by flaring the walls, it is called a *suppressed* weir. For example, the Cipoletti weir, which is commonly used in irrigation channels, uses a trapezoidal section with the side walls inclined at an angle of 28° to the vertical to accomplish suppression of end contractions.

The discharge equation for a sharp-crested rectangular weir, with end contractions suppressed, is given by

$$Q = CLH^{3/2} \tag{10.36}$$

where C is the weir coefficient and is given as 3.33 (USCS) or 1.84 (SI).

For a rectangular weir with two contracted ends, the physical width of the weir L is replaced by the effective width L_{eff} given by $L_{eff} \approx L - 0.2H$.

Figure 10.15 Discharge over a sharp-crested weir.

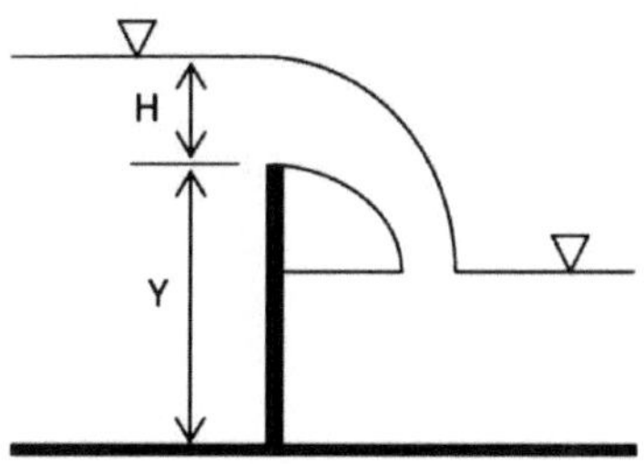

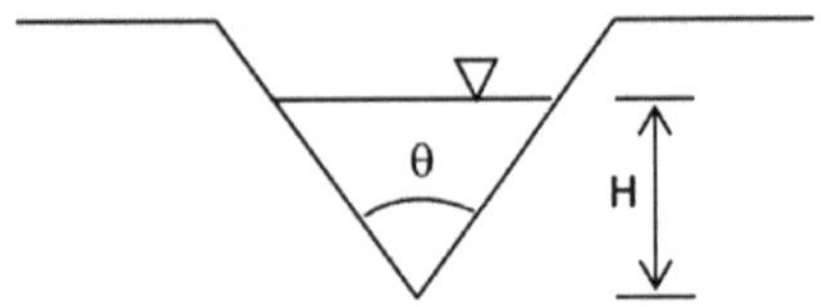

Figure 10.16 Discharge through a triangular weir.

Triangular Weirs

For low flow rates, triangular weirs are commonly used. The discharge equation for a triangular weir (Fig. 10.16) is given by

$$Q = CH^{5/2} \tag{10.37}$$

where C is the weir coefficient. The *Handbook* gives the following weir coefficients:

For 90° V-notch weirs: $C = 2.54$ (USCS units), 1.40 (SI units).

Example 10.16

A sharp-crested rectangular contracted weir (width = 8 m) discharges water flowing in a stream. The elevation of the weir crest is 132.54 m above sea level and the elevation of the water surface behind the weir is 134.88 m. What is the flow rate (m^3/s)?

Solution The weir head $H = 134.88 - 132.54 = 2.34$ m.

For a contracted weir, flow rate $Q = 1.84 L_{eff} H^{3/2} = 1.84 \times (8 - 0.2 \times 2.34) \times 2.34^{3/2} =$ 49.61 m^3/s.

CHAPTER 11

Hydraulics and Hydrologic Systems

Hydrologic Balance

Hydrology is the study of the distribution and movement of water throughout the earth (hydrologic cycle). The hydrologic balance for a watershed involves positive flows into the system (such as from precipitation) and negative flows out of the system (such as through evaporation). The overall water balance (hydrologic budget) may be written as

$$R = P - ET - \Delta S \tag{11.1}$$

where R = stream flow
P = precipitation (gross rainfall)
ET = evapotranspiration
ΔS = change in storage

Precipitation

The average precipitation on a watershed may be written as

$$\text{Average precipitation} = \frac{\text{total rainfall volume}}{\text{watershed area}} \tag{11.2}$$

The total rainfall volume may be estimated from an examination of isohyetal contours. An isohyet is a line drawn through geographical points recording equal amounts of precipitation during a specific period. These lines are labeled using precipitation depth (such as 0.5, 1.0 in rainfall, etc.). If one considers a specific region for study, the area enclosed by an isohyet within the region may be estimated using standard techniques. Figure 11.1 shows isohyets for a watershed.

Typical precipitation data may be as shown in Example 11.1.

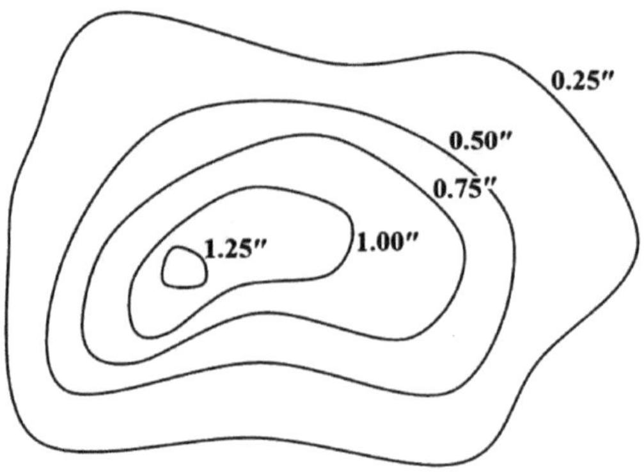

Figure 11.1
Isohyetal contours.

Example 11.1

The table below shows rainfall data for a region. Lines 1 to 5 represent isohyets and the table shows precipitation depth and enclosed area for each isohyet. Calculate the average precipitation depth (inches) for the region:

Line	Precipitation (in)	Enclosed Area (ac)
1	0.25	321
2	0.50	126
3	0.75	85
4	1.00	47
5	1.25	12

Solution Using the data in the table, one may calculate the incremental areas (region between two adjacent isohyets) as shown in the table below. Each incremental area is then represented by the average precipitation between the two adjacent isohyets. The average precipitation over the entire watershed can be calculated as the weighted average for all these incremental areas that make up the watershed:

Area	Precipitation Range	Precipitation (in)	Enclosed Area (ac)	Volume (ac·in)
1	<0.25			
2	0.25–0.50	0.375	195	73.125
3	0.50–0.75	0.625	41	25.625
4	0.75–1.00	0.875	38	33.250
5	1.00–1.25	1.125	35	39.375
6	>1.25	1.375	12	16.500
Total				187.875

Average precipitation (gross rainfall depth) over the entire watershed is then given by

$$P_{\text{ave}} = \frac{187.875}{321} = 0.59 \text{ in}$$

Rainfall Intensity

Rainfall intensity, on the other hand, is calculated from the temporal (rather than spatial) distribution of precipitation. Typical units of rainfall intensity are cm/h or in/h. Rain gage data may be used to determine rainfall intensity during specific time bands of a precipitation event. A rain gage basically records cumulative precipitation depth over a period of time. Typical data collected in a rain gage may appear as in the following example.

Example 11.2

The following data are recorded from a rain gage. Calculate (a) the average intensity over the entire 40 min and (b) the peak intensity during the rainfall event.

Time (min)	Precipitation (in)
0.00	0.0
10.00	0.4
20.00	0.6
30.00	1.2
40.00	1.5

Solution The storm hyetograph is the bar graph of instantaneous rainfall versus time. For the data shown above, the hyetograph is shown in the figure below.

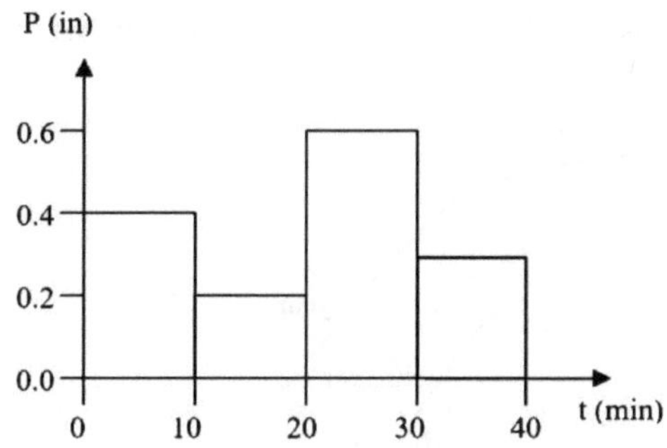

(a) The average rainfall intensity during the entire 40 min duration is 1.5 in/40 min = 2.25 in/h.

(b) The rainfall intensity during the time interval 0–10 min is 0.4 in/10 min = 2.4 in/h, the intensity during the time interval 10 to 20 min is 0.2 in/10 min = 1.2 in/h, etc.

The peak intensity occurs during the third interval and is equal to 3.6 in/h.

Intensity-Duration-Frequency Curves

Estimation of peak flow can be separated into two categories: (1) sites with measured stream gage data, and (2) sites without gaged data. When reliable gaged data are available, statistical analysis of the flow record can be used to estimate flood peaks for various return periods. When gaged data are not available, estimates are made by empirical equations such as the rational method or by regional regression equations. Regional regression equations are typically appropriate for larger drainage areas, and methods such as the rational method are commonly used for smaller areas, less than about 200 ac.

The recurrence interval is the period during which, on an average, a given event (e.g., rainfall or runoff) is equaled or exceeded. The (annual) exceedance probability is the reciprocal of recurrence interval. Thus, the "100-year flood" is defined as the event which has an annual probability of exceedance = 1/100 = 0.01.

If the recurrence interval is too short, the design event is of low intensity, resulting in a low initial cost, but the life-cycle cost is high because of frequently occurring damage. On the other hand, if the recurrence interval is too long, the initial cost is high, but annual maintenance costs are low. Somewhere between these limits lies the design frequency which will produce a reasonable balance of construction cost, annual maintenance cost, and risk of flooding.

Based on regional precipitation data recorded over a long period of time, intensity-duration-frequency (IDF) curves are plotted. A typical example of a set of IDF curves is shown in Fig. 11.2.

In order to select the governing rainfall intensity for a watershed, the duration is set equal to the time of concentration and the appropriate curve (based on return period) is used to predict the intensity.

Time of Concentration

The time of concentration for an inlet is the time required for water to flow from the most remote (in terms of time of flow) point of the area to the inlet once the soil has become saturated and minor depressions have been filled. It is assumed that when the duration of the storm

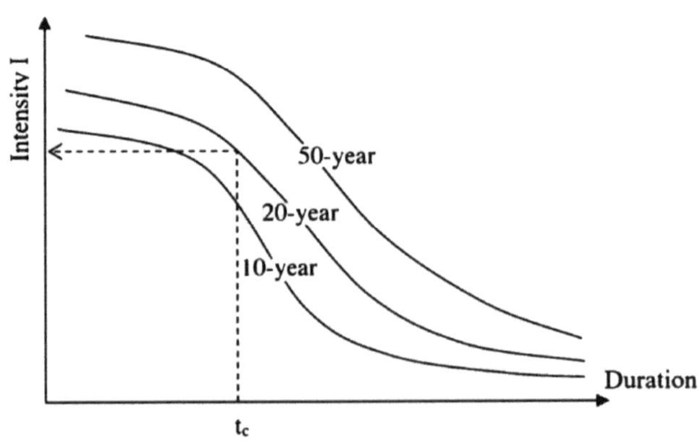

Figure 11.2 Typical intensity-duration-frequency curves.

equals the time of concentration, all parts of the watershed are contributing simultaneously to the discharge at the inlet. For this reason, the time of concentration is chosen as the duration while using the IDF curves.

Time of concentration is calculated as the sum of overland flow, swale or ditch flow, and channel flow

$$t_c = \underset{\text{flow}}{t_{\text{sheet}}} + \underset{\text{flow}}{t_{\text{ditch}}} + \underset{\text{flow}}{t_{\text{channel}}} \tag{11.3}$$

Several empirical models (Kirpich's equation, FAA formula, Manning's kinematic wave formula) are available for estimating the time of concentration for a watershed. All of these models employ the following parameters—the length of the hydraulically longest path, the average slope along the path, and some parameter that quantifies the land use of the terrain (such as rational coefficient C, or curve number CN).

Hydrograph

Flow rate in a stream is often measured by measuring river stage. Previously established rating curves that relate stage (surface elevation) to flow rate can then be used to estimate the flow rate. A stream hydrograph is a plot of stream flow rate versus time. When interception and infiltration of incident rainfall (precipitation) is complete, the runoff (excess precipitation) begins to occur and the stream carries more discharge than the previously existing flow in the stream (baseflow). Thus, there is a time lag between the start of precipitation and the start of excess precipitation.

Once the baseflow is separated from the stream flow, the flow that can be ascribed to excess precipitation (runoff) is plotted against time. This represents the direct runoff due to the rainfall event. Once the total depth of runoff is calculated, the unit hydrograph for the rainfall event can be derived by dividing the direct runoff discharge values by the runoff depth.

Thus, a unit hydrograph of time T represents a hypothetical case of 1 unit (1 mm, 1 cm, 1 in) of runoff occurring from the catchment area during a time interval T. Therefore, ordinates of a unit hydrograph have units such as $ft^3/s/in$ or $m^3/s/cm$.

Hydrograph Separation

Net runoff lags rainfall since the soil has a finite initial abstraction, based upon its infiltration rate. Runoff will only occur when the rain falls on a very wet watershed that is unable to absorb more moisture.

Stream monitoring stations record the discharge in a stream that collects runoff flow from a contributing watershed. The recorded discharge includes the pre-existing flow (baseflow) that must be separated to identify the discharge produced by the storm runoff. The most common of baseflow separation methods is the *constant discharge method*, which assumes that baseflow is constant during the storm hydrograph. The minimum streamflow immediately prior to the rising limb is used as the constant value. This is illustrated in Fig. 11.3.

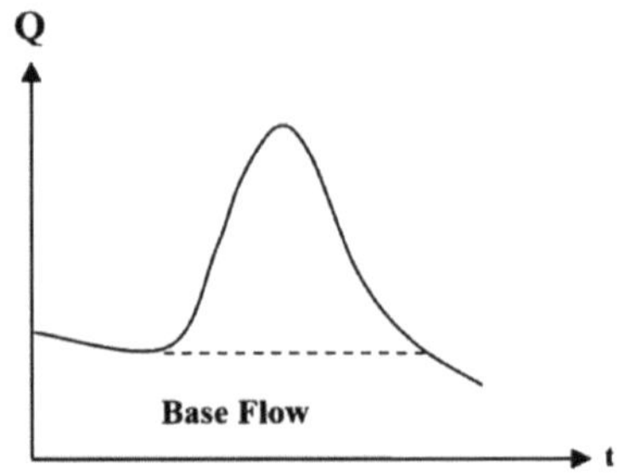

Figure 11.3 Constant discharge method of baseflow separation.

The excess precipitation due to the storm is calculated by subtracting the baseflow from the stream discharge, resulting in a plot of Q_{net} versus time. The volume of excess rainfall from the watershed is calculated as the area under the Q_{net} versus time curve. If this volume is spread evenly over the watershed area, the average depth of excess precipitation is given by

$$P_{av} = \frac{V}{A_d} \tag{11.4}$$

where A_d = drainage area
V = total volume of excess precipitation = area under the Q_{net} versus time plot

Example 11.3

A stream monitoring station has a watershed area = 225 ac. The table below shows time-discharge data recorded in the 5-h period following a rainfall event:

Time (h)	0.0	1.0	2.0	3.0	4.0	5.0
Discharge (ft³/s)	25.0	95.0	176.0	132.0	66.0	27.0

What is the average depth of excess precipitation produced by the rain event?

Solution The baseflow in the stream appears to be 25.0 cfs. Subtracting it from all discharge values, we obtain the net discharge (produced by the runoff from the rainfall event):

Time (h)	0.0	1.0	2.0	3.0	4.0	5.0
Discharge (ft³/s)	0.0	70.0	151.0	107.0	41.0	2.0

The total volume of excess precipitation can be calculated as

$$V = \sum (Q\Delta t) = \Delta t \sum Q = 3600 \text{ s} \times 371 \text{ cfs} = 1{,}335{,}600 \text{ ft}^3$$

Depth of excess precipitation: $d = \frac{Q}{A} = \frac{1{,}335{,}600}{225 \times 43{,}560} = 0.136 \text{ ft} = 1.64 \text{ in.}$

Unit Hydrograph

If the overland flow hydrograph is divided everywhere by the average precipitation P_{av}, we get the unit hydrograph of excess precipitation. Typical units of Q_{unit} are ft³/s/in or m³/s/cm.

For example, a 1-h unit hydrograph for a watershed is the pattern of discharge produced by a 1-h period of uniform excess precipitation equal to 1 unit (1 in or 1 cm, for example).

A unit hydrograph developed from one storm can be used as design basis for other storms. This is subject to the following assumptions:

- All storms in the watershed have same duration.
- Shape of the rainfall curve is the same.
- Only the total rainfall amount varies.

Hydrograph Synthesis

The unit hydrograph can be used to predict runoff from a storm only if the duration of the storm is the same as that of the unit hydrograph.

Hydrograph synthesis techniques can be used to construct the hydrograph of a longer storm from the unit hydrograph of a shorter storm.

Example 11.4

The 1-h unit hydrograph for a watershed is given below:

Time (h)	0.0	1.0	2.0	3.0	4.0	5.0
Discharge (cfs/in)	0.0	0.7	3.3	1.9	0.5	0.0

These data are shown below graphically. The time $t = 0$ marks the beginning of excess precipitation from the storm (after interception and infiltration are complete).

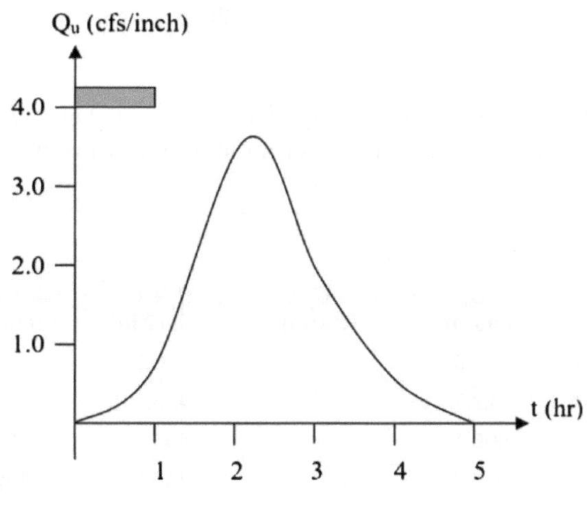

What is the runoff at $t = 2$ h due to a storm that is described as an hour of uniform excess precipitation (average intensity $i = 0.6$ in/h) followed by a second hour of uniform excess precipitation (average intensity $i = 0.3$ in/h)?

Solution Thus, at time $t = 2$ h, the contribution of the two periods of uniform precipitation (0.6 in. in the first hour followed by 0.3 in during the second hour) can be calculated as

$$Q = 0.6 \times 3.3 + 0.3 \times 0.7 = 2.19\,\text{cfs}$$

This represents the contribution of the first hour of excess precipitation due to a lag time of 2 h plus the contribution of the second hour of excess precipitation due to a lag time of 1 h.

Similarly, at time $t = 3$ h, the contribution of the two periods can be calculated as

$$Q = 0.6 \times 1.9 + 0.3 \times 3.3 = 2.13\,\text{cfs}$$

This represents the contribution of the first hour of excess precipitation due to a lag time of 3 h plus the contribution of the second hour of excess precipitation due to a lag time of 2 h.

Example 11.5

Ordinates for a unit hydrograph for a 2-h storm are shown below. Create the unit hydrograph for a 4-h storm synthesized from this storm:

Time t (h)	Discharge Q (cfs/in)
0.0	0.0
1.0	423.7
2.0	805.1
3.0	1016.9
4.0	296.6
5.0	42.4

Solution The table below shows the application of the lagging storm method. This method is applicable since the synthesized storm duration (4 h) is an integer multiple of the basis storm (2 h):

Time t (h)	Q_1 (cfs/in)	Q_2 (cfs/in)	$Q_1 + Q_2$ (cfs/2 in)	$(Q_1 + Q_2)/2$ (cfs/in)
0.0	0.0		0.0	0.0
1.0	423.7		423.7	211.9
2.0	805.1	0.0	805.1	402.6
3.0	1016.9	423.7	1440.6	720.3
4.0	296.6	805.1	1101.7	550.9
5.0	42.4	1016.9	1059.3	529.7
6.0		296.6	296.6	148.3
7.0		42.4	42.4	21.2

NOTE Since the original hydrograph was a unit hydrograph and the 4-h storm was formed by combining two such storms, the sum is then divided by two to obtain the new unit hydrograph.

Runoff Estimation by Natural Resources Conservation Service Curve Number

The Natural Resources Conservation Service (NRCS), formerly the Soil Conservation Service (SCS), specifies an approach for calculating storm runoff as specified in the document TR55. This method is also referred to as the soil cover complex method. In this approach, a curve number (CN) is used as the basis for calculating the runoff. It is applicable to any size homogeneous watershed with a known percentage of imperviousness. SCS method assumes a type II storm. This is not representative of storms that drop most of their precipitation early as in Hawaii, Alaska, and coastal regions in the Gulf of Mexico.

The CN for a watershed is a single number representing the aggregate effect of land use and soil type. The following soil types are used:

Type A Deep sands, deep loess, and aggregated silts are classified as type A soils. These soils exhibit high infiltration rate (>0.3 in/h) and therefore low runoff potential.

Type B Shallow loess and sandy loam are classified as type B soils. These soils exhibit moderate infiltration rates (0.15 to 0.3 in/h) if thoroughly wet.

Type C Clay loams, shallow sandy loam, and soils low in organic content are classified as type C soils. These soils exhibit low infiltration rates (0.05 to 0.15 in/h) if thoroughly wet.

Type D Soils that swell significantly when wet. These soils exhibit low infiltration rates (<0.05 in/h) and therefore have a high runoff potential. Examples are heavy plastic clays and saline soils.

Paved regions within the watershed are assigned CN = 98.

For a composite area that can be partitioned into separable sections, each qualifying as a homogeneous watershed, a composite CN may be used as in Eq. (11.5)

$$\overline{\text{CN}} = \frac{\sum (A_i \times \text{CN}_i)}{\sum A_i} \tag{11.5}$$

Procedure for the NRCS Method

The methodology for estimating runoff depth according to NRCS is outlined below:

1. Choose soil type and land use for each region.
2. Use the TR-55 tables to look up CN_{II} for each region.

3. **3.** Correct to CN_I or CN_{III} if necessary (AMC not "normal" conditions). Antecedent moisture condition (AMC) represents the moisture condition of the soil just before the storm, ranging from dry (AMC I) to saturated (AMC III).
4. Calculate composite CN for entire watershed.
5. Calculate storage capacity of soil $S = \dfrac{1000}{\text{CN}} - 10$.
6. Estimate time of concentration for the watershed.
7. Using the time of concentration as the storm duration and a recurrence interval, determine total rainfall P_g.
8. Modify the gross rainfall for large areas.
9. Total runoff (net rain or excess rain) is calculated as

$$Q_{\text{inches}} = \frac{(P_g - I_a)^2}{P_g - I_a + S} = \frac{(P_g - 0.2S)^2}{P_g + 0.8S} \tag{11.6}$$

Example 11.6

A watershed (area = 370 ac) is subdivided into five distinct land use classifications, as shown in the table below. Using the NRCS method, the net runoff (inches) from a 20-year storm with gross rainfall = 5.6 in is most nearly:

A. 1.6 in

B. 2.2 in

C. 2.9 in

D. 3.6 in

Region	Area (ac)	Land Use	Soil Type	Time for Overland Flow (min)	Curve Number	Rational Runoff Coefficient
A	80	Lawns: fair condition	B	30	69	0.4
B	80	Forest	C	45	45	0.2
C	50	Paved	B	15	98	0.9
D	90	Residential: 4 lots/ac	D	25	87	0.6
E	70	Forest	A	45	35	0.2

Solution Using the NRCS approach, we can calculate the runoff depth based on the composite CN for the watershed, which is given by

$$\overline{\text{CN}} = \frac{\sum \text{CN}_i A_i}{\sum A_i} = \frac{69 \times 80 + 45 \times 80 + 98 \times 50 + 87 \times 90 + 35 \times 70}{370} = 66$$

For CN = 66 and a gross rainfall $P_g = 5.6$ in

$$S = \frac{1000}{\text{CN}} - 10 = \frac{1000}{66} - 10 = 5.15$$

$$Q = \frac{(P_g - 0.2S)^2}{P_g + 0.8S} = \frac{(5.6 - 0.2 \times 5.15)^2}{5.6 + 0.8 \times 5.15} = 2.15 \text{ in}$$

Runoff depth = 2.15 in

Answer is B.

Rational Method for Predicting Runoff

The error in the runoff estimate increases as the size of the drainage area increases. For these reasons, the rational method should not be used to determine the rate of runoff from large drainage areas. For the design of highway drainage structures, the use of the rational method should be restricted to drainage areas less than 80 ha (200 ac).

The assumptions involved in using the rational method are

1. The peak flow occurs when the entire watershed is contributing.
2. The rainfall intensity is uniform over a duration equal to the time of concentration.
3. The frequency of the computed peak flow is equal to the frequency of the rainfall intensity. In other words, the 10-year rainfall intensity is assumed to produce the 10-year flood.

Most urban watersheds served by storm drains contribute flow to these large drainage channels well before they reach significant size and therefore the rational method can be applied to these tributary areas. The rational method uses Eq. (11.7) to predict peak runoff flow rate Q_p from a watershed of area A_d due to a storm with intensity of rainfall I

$$Q_p = CIA_d \tag{11.7}$$

The intensity (typical units in/h) may be estimated from historical data presented in the form of IDF curves.

The rational coefficient C is a single number representing land use and land cover for the watershed.

Example 11.7

Land use classification for a watershed in the western United States is summarized in the table below. What is the flow to the storm sewer resulting from a 10-year storm? Use a design intensity $I = 2.4$ in/h.

Land Use	Area (ac)	Rational Coefficient
Shingle roof	0.8	0.85
Concrete surface	1.1	0.88
Asphalt surface	2.6	0.83
Poorly drained lawn	10.0	0.15

Solution The runoff coefficient for the watershed is calculated as the weighted average

$$\overline{C} = \frac{\sum C_i A_i}{\sum A_i} = \frac{0.85 \times 0.8 + 0.88 \times 1.1 + 0.83 \times 2.6 + 0.15 \times 10}{0.8 + 1.1 + 2.6 + 10} = 0.366$$

Runoff discharge is given by

$$Q = CIA_d = 0.366 \times 2.4 \times 14.5 = 12.74 \text{ ac-in/h} = 12.85 \text{ cfs}$$

Example 11.8

A small development has 45 houses, with average occupancy of four persons per house. Each roof has dimensions 65 × 40 ft. Assuming an average sewage production of 115 gpcd, what is the peak flow rate for which the sewer must be designed? Include domestic flow rate plus drainage from the roofs from a storm with rainfall intensity = 1.5 in/h.

Solution Domestic sewage flow = 45 × 4 × 115 = 20,700 gpd = 14.38 gpm = 0.032 cfs (average)

Assuming the maximum hour to carry three times the average, $Q = 0.1$ cfs

Roof area = 45 × 65 × 40 = 117,000 ft^2 = 2.69 ac

Assuming a rational $C = 1.0$ (impervious roof), the roof runoff is

$$Q = CIA = 1.0 \times 1.5 \times 2.69 = 4.04 \text{ ac-in/h} = 4.04 \text{ cfs}$$

The combined flow rate is 4.14 cfs.

Water Distribution Systems

Reservoirs must be located such that gravity flow can be utilized for much of the water distribution to a community. In establishing a proper reservoir elevation, the rule of thumb is that every foot of head produces approximately 0.434 psi. Therefore, to generate about 60 psi in the water distribution system, the storage reservoir needs to be located approximately 150 ft above the service area. For certain low-lying areas, regulators must be installed to protect systems from overpressure.

Pumps should be installed at locations where gravity flow does not result in adequate pressures. These "booster" locations and the required pumping power should be chosen such that they have optimal effect on the entire network.

Reservoirs should have adequate capacity to provide continuous domestic flow and anticipated fire flows for a reasonable duration. A reasonable rule of thumb is that storage should be sufficient to provide at least 2 days of peak domestic consumption plus required fire flows. For example, in a residential area, if fire codes call for 2 h of fire flow at 1000 gpm, then additional fire flow storage is approximately 1.44 million gallons. For industrial areas, fire flow of 4000 to 5000 gpm for 3 h is more typical.

A water distribution system must have some redundancy built into it so that adequate flow and pressure can be maintained in spite of closure of one or more branches of the network.

Each point of the distribution system should be supplied by primary feeders which are typically 16 in. in diameter. Secondary feeders (typically 12-in diameter) are typically used to supply specific neighborhoods. Distributor mains of 8-in diameter are typically used to supply water to individual streets.

Valves should be provided at every junction for mains branching from those junctions. Operation of these valves should allow individual sections of the main to be isolated for service or repair, so that "failure" of one section does not result in loss of water delivery to parts of the network.

Inflow and Infiltration

Inflow and infiltration (I&I) are terms used to describe the ways that groundwater and stormwater enter into dedicated wastewater or sanitary sewer systems.

Inflow is stormwater that enters into sanitary sewer systems at points of direct connection to the systems from various sources.

Infiltration is groundwater that enters sanitary sewer systems through cracks and/or leaks in the sanitary sewer pipes.

I&I add clear water to sewer systems increasing the load on the systems. Once the sanitary sewer system reaches capacity or becomes overloaded, wastewater flows at much higher water level than normal and creates the potential for backflow into households and/or overflows through manholes into streets. This poses a public health hazard and also creates the potential for environmental fines and litigation costs. Pumps that transport wastewater operate round the clock and must work harder as the sewer system's water level load increases, putting additional strain on the pumps and shortening their life expectancy. Failure on the part of a municipality to mitigate these overflow events can compromise their rating and reduce property values in affected areas.

To quantify the I&I into a sanitary sewer system, information relating to a variety of issues must be collected and carefully evaluated. The corrected data should be tabulated and analyzed to make comparisons between the measured I&I and the corresponding rainfall intensity. Data under surcharge conditions should be avoided for analysis purposes. The analysis allows the prioritization of drainage basins for future studies and potential I&I reduction.

Typical units used for inflow an infiltration rates are gallons/day/foot, mgd/1000 ft, gpd/in-mile of pipe, mgd/ac, etc.

Example 11.9

The design infiltration rate for a sewer pipeline is 12 gal/day per in-mile of pipe. A section of the sewer is a 16-in-diameter main that has a length of 850 ft between two designated nodes A and B. What is the added flow between nodes A and B that can be ascribed to inflow and infiltration?

Solution For the pipe segment, diameter = 16 in and length = 850 ft = 0.161 mi. Therefore, in-mile product = $16 \times 0.161 = 2.58$ in-mi.

Therefore, I&I contribute a flow rate of $12 \times 2.58 = 31$ gal/day.

Closed Conduit Hydraulics

Pipes conveying flow under the influence of a pressure gradient experience friction losses which are dependent on pipe roughness, flow regime (laminar or turbulent), and flow velocity. The two most prevalent models for quantifying friction loss in pipes are those due to Darcy and Weisbach and to Hazen and Williams. The Darcy-Weisbach model gives a dimensionally consistent formula which requires inputs about Reynolds number and pipe roughness. It is applicable to all fluids. The Hazen-Williams model leads to an empirical equation which yields good results for water at a temperature of 60°F. It is not applicable to other fluids.

For conduits or channels transporting water containing fine sediment (silt), a minimum velocity of 2 to 3 ft/s is recommended. Factors such as channel erosion and hydraulic surge limit maximum velocities to about 10 to 20 ft/s.

Darcy-Weisbach Equation

According to the Darcy-Weisbach model, the head loss due to friction h_f is given by

$$h_f = f \cdot \frac{L}{D} \cdot \frac{V^2}{2g} \tag{11.8}$$

where f = friction factor
L = length of pipe
D = pipe diameter
V = average flow velocity in the pipe

The Darcy friction factor f is a function of relative roughness ε/D of the pipe and Reynolds number *Re*. The Moody (Stanton) diagram for the Darcy friction factor is on page 115 of the *FE Reference Handbook*, version 9.4, and is reproduced in Fig. 11.4.

In order to navigate this diagram, we need to calculate the *relative* roughness of the conduit. The inset box on the lower left of Fig. 11.4 gives range of values of *absolute* roughness for various pipe materials (for the multiple-choice FE exam, a specific value of absolute roughness will be given). Calculating the relative roughness ε/D will allow us to choose one of the many curves. In order to locate a point on the selected curve, we can either (1) calculate the Reynolds number (lower horizontal scale) or (2) calculate the product *VD* where velocity V is in feet per second and diameter D is in inches. The upper scale is to be used only for water at 60°F.

If it appears that answer choices are well separated, then a shortcut may be used by assuming that the flow is "fully turbulent." Each of the curves becomes horizontal beyond a certain value of Reynolds number. In other words, in this regime, the friction factor is no longer dependent on the Reynolds number and the flow is deemed to have become fully turbulent. If this assumption is made, one only needs to calculate the relative roughness, choose the correct curve, and read the friction factor for the "flat" portion of the curve. For example, if the relative roughness is 0.002, friction factor for fully turbulent flow is about 0.024.

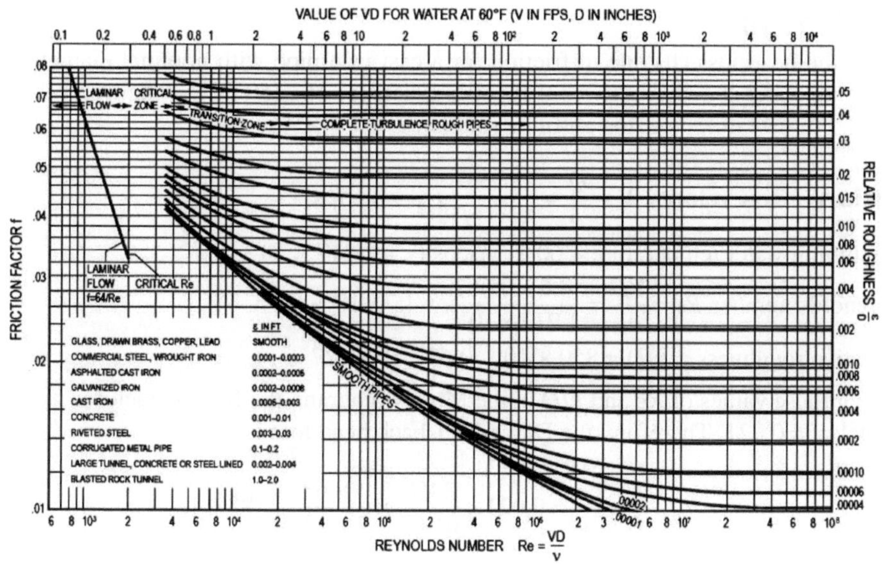

Figure 11.4 Moody (Stanton) diagram.

Laminar Flow

For Reynolds numbers less than about 2000, laminar flow typically occurs in pipes. The friction factor for laminar flow can be approximated by

$$f_{\text{Laminar}} = \frac{64}{Re} = \frac{64\nu}{VD} = \frac{64\mu}{\rho VD} \tag{11.9}$$

where ν is the kinematic viscosity and μ is the absolute viscosity of the fluid. The properties of water are tabulated on page 114 of the *Handbook.*

Thus, for laminar flow, the friction head loss in a circular pipe is given by the Hagen-Poiseuille equation

$$h_f = \frac{64}{\text{Re}} \cdot \frac{L}{D} \cdot \frac{V^2}{2g} = \frac{32\mu LV}{\rho g D^2} = \frac{128\mu LQ}{\pi\rho g D^4} \tag{11.10}$$

This can be used to relate the flow rate Q to the corresponding pressure loss over a given length of pipe L

$$Q = \frac{\pi D^4 \Delta P_f}{128\mu L} \tag{11.11}$$

where ΔP_f is the pressure loss that occurs over a pipe length L. Equations (11.10) and (11.11) apply to laminar flow only.

Example 11.10

Estimate the head loss (ft) due to friction for a cast iron pipe with nominal diameter = 8 in transporting water at a flow rate = 3.0 cfs over a distance of 2450 ft. Use the Darcy-Weisbach equation. For cast iron pipe, the absolute roughness is $\varepsilon = 850 \times 10^{-6}$ ft.

Solution Flow velocity $V = \dfrac{Q}{A} = \dfrac{3.0}{\pi \times (4/12)^2} = 8.6$ fps

Assuming water temperature = 60°F, viscosity = 1.217×10^{-5} ft²/s

Reynolds number $Re = \dfrac{VD}{\nu} = \dfrac{8.6 \times 0.667}{1.217 \times 10^{-5}} = 4.7 \times 10^5$

Relative roughness = $\varepsilon/D = 850 \times 10^{-6}/0.67 = 0.00127$

Using these values of Re and ε/D in the Moody diagram, the friction factor is approximately $f = 0.021$. Therefore, the Darcy-Weisbach head loss is

$$h_f = f \cdot \frac{L}{D} \cdot \frac{V^2}{2g} = 0.021 \times \frac{2450}{0.667} \times \frac{8.6^2}{2 \times 32.2} = 88.6 \text{ ft}$$

Hydraulic Radius and Hydraulic Diameter

The hydraulic radius R_H reflects a combined effect of the flow area and the wetted perimeter and is given by

$$R_H = \frac{\text{area of flow}}{\text{wetted perimeter}} \tag{11.12}$$

The hydraulic diameter D_H is given by four times the hydraulic radius

$$D_H = 4R_H \tag{11.13}$$

NOTE For a circular pipe flowing full, the wetted perimeter is the circumference $2\pi R$ and the flow area is πR^2. Therefore, the hydraulic radius [according to Eq. (11.12)] is $R/2 = D/4$.

Hazen-Williams Equation

The Hazen-Williams equation should be used only for turbulent flow. It yields good results for liquids having kinematic viscosity around 1.2×10^{-5} ft²/s. Thus, it yields good results for water around 60°F.

The flow velocity according to the Hazen-Williams equation is given by Eq. (11.14) (in the Fluid Mechanics section of the *Handbook*)

$$V = 1.318 C R_H^{0.63} S^{0.54} \quad \text{(US)} \tag{11.14a}$$

$$V = 0.849 C R_H^{0.63} S^{0.54} \quad \text{(SI)} \tag{11.14b}$$

where S = (decimal) slope of the energy grade line and is equal to h_f/L
R = hydraulic radius (ft or m)
C = Hazen-Williams roughness coefficient. Some typical values of C are given on page 168 of the *Handbook*

In order to apply the Hazen-Williams equation to non-circular ducts, the hydraulic diameter D_h must be used in place of the variable D. The hydraulic diameter is calculated as four times the hydraulic radius.

Head Loss in Circular Conduits

For a circular conduit flowing full, substituting $R_H = D/4$ leads to

$$h_f = \frac{4.73L_{\text{ft}}}{C^{1.852}D_{\text{ft}}^{4.87}}Q_{\text{cfs}}^{1.852} \text{ (where } h_f, L, \text{ and } D \text{ are in ft; } Q \text{ is in ft}^3\text{/s)} \qquad (11.15)$$

The pressure loss in a pipe P is given by

$$P = \frac{4.52Q^{1.85}}{C^{1.85}D^{4.87}} \text{ (US units)} \qquad (11.16)$$

where P = pressure loss (psi/ft)
Q = flow rate (gal/min)
D = pipe diameter (in)
C = Hazen-Williams coefficient

$$P = \frac{6.05\times10^5Q^{1.85}}{C^{1.85}D^{4.87}} \text{ (SI units)} \qquad (11.17)$$

where P = pressure loss (bars per meter). 1 bar = 1×10^5 Pa
Q = flow rate (L/min)
D = pipe diameter (mm)
C = Hazen-Williams coefficient

Example 11.11

Estimate the head loss (ft) and pressure loss (psi) due to friction for a cast iron pipe with nominal diameter = 8 in transporting water at a flow rate = 3.0 cfs over a distance of 2450 ft. Use the Hazen-Williams equation. Use Hazen-Williams roughness coefficient C = 100.

Solution The head loss is calculated as

$$h_f = \frac{4.73L_{\text{ft}}}{C^{1.852}D_{\text{ft}}^{4.87}}Q_{\text{cfs}}^{1.852} = \frac{4.73\times2450}{100^{1.852}\times0.667^{4.87}}\times3^{1.852} = 126 \text{ ft}$$

The pressure loss (psi/ft) is calculated as (note the changes in units: $Q = 3\times448.8 = 1346.4$ gpm, diameter D = 8 in)

$$P = \frac{4.52Q^{1.85}}{C^{1.85}D^{4.87}} = \frac{4.52\times1346.4^{1.85}}{100^{1.85}\times8^{4.87}} = 0.0222 \text{ psi/ft}$$

Total pressure loss (psi) = 54.4 psi. (This is equivalent to a head loss of 126 ft.)

Minor Losses

When flow passes through regulating devices such as valves and gates, through path changes such as bends or through change in flow area such as pipe entrances and exits, there is a loss of flow energy at that location. For typical systems, the magnitude of these losses is small compared to the losses that occur due to friction from the pipe walls. For this reason, these losses are called minor losses. However, like pipe friction loss, these losses are also proportional to the kinetic head or velocity head $V^2/2g$. Thus, minor losses are expressed in a manner very similar to friction losses, except that the source of energy loss is discrete (located at the specific locations of these devices) rather than occurring continuously along the length of the pipe. Each location is assigned a standardized value or coefficient C and the total system minor loss may be expressed in terms of the C values for all devices

$$h_{\text{minor}} = C\frac{V^2}{2g} \tag{11.18}$$

Equivalent Length

Alternatively, by analogy with the Darcy-Weisbach friction loss formula, the loss coefficient C may also be expressed in terms of an equivalent length L_e. This is the length of pipe (of specific material and diameter) which would cause friction loss equivalent to that caused by the device

$$f\frac{L}{D}\cdot\frac{V^2}{2g} = C\frac{V^2}{2g} \Rightarrow C = \frac{f \cdot L_e}{D} \quad \text{or} \quad L_e = \frac{CD}{f} \tag{11.19}$$

Thus, the critical parameters for the pipeline are f (friction factor), L (length), and D (diameter), whereas the devices (valves, etc.) and geometry changes (enlargements, bends, etc.) are assigned C values and the system is represented by the *total* minor loss coefficient ΣC.

Some typical minor loss coefficients are given on page 106 of the *Handbook*.

Example 11.12

A fully open globe valve ($C = 10$) is installed on a 4-in diameter cast iron pipe. If the nominal friction factor for the pipe is 0.022, what is the equivalent length (ft) of the valve?

Solution The equivalent length is given by

$$L_{\text{eq}} = \frac{CD}{f} = \frac{10\times 4}{0.022} = 1818 \text{ in} = 152 \text{ ft}.$$

Pipe Networks

Pipe networks such as those for municipal water distribution systems are composed of thousands of nodes (connection or branching points) and links (pipe length between two nodes). Additionally, these systems have one or more sources (such as water flowing into the system from a distribution station) and a large number of loads (such as household water connections). The final solution of such a system must satisfy the following conditions:

1. The sum of flow rates at each node must be zero. In other words, total flow rate into each node must be equal to the total flow rate out of that node.
2. The sum of head losses in all links in any closed loop around the network must be zero.

Obtaining such a solution requires an iterative approach. One of the most prominent methods for such a problem is the Hardy-Cross Method.

Two-Node Network

While the process described above may be "computationally tedious," a simple two node, two link "network" such as the one in Example 11.10 can be solved without iteration.

Example 11.13

A 6-in-diameter steel pipe is in place to carry a flow rate of 2.4 cfs. The pipe is deemed inadequate and a second 4-in-diameter steel pipe is laid in parallel to the first pipe. The length of both pipes is 1200 ft. What is the velocity of flow in the original 6-in pipe after connecting the new 4-in pipe?

Solution The flow will split in the pipes such that the head loss h_f in each branch is equal.

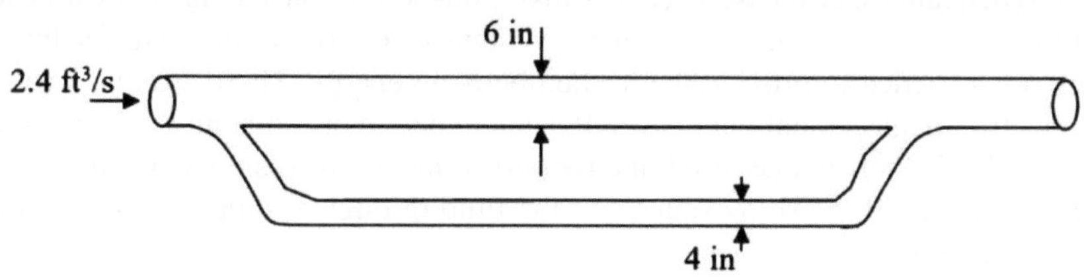

Using the Darcy-Weisbach approach for friction loss in pipes, we can write

$$h_f = f\frac{L}{D}\frac{V^2}{2g} = f\frac{L}{D}\frac{Q^2}{2gA^2} = f\frac{L}{D}\frac{Q^2}{2g(\pi D^2/4)^2} \Rightarrow h_f \propto \frac{fLQ^2}{D^5}$$

For an approximate solution, let us assume that the friction factor f is the same for both pipes. Since the lengths are also equal

$$\therefore \frac{Q_1^2}{D_1^5} = \frac{Q_2^2}{D_2^5} \Rightarrow \frac{Q_1}{Q_2} = \left(\frac{D_1}{D_2}\right)^{5/2} = \left(\frac{6}{4}\right)^{5/2} = 2.8$$

Split the total flow of $Q = 2.4$ cfs into fractions 2.8/3.8 and 1.0/3.8.

Therefore

$$Q_1 = \frac{2.8}{3.8} \times 2.4 = 1.77 \text{ cfs} \qquad v_1 = \frac{Q_1}{\pi D^2/4} = 9.0 \text{ fps}$$

NOTE Since the relative roughness of pipe 2 is greater than that of pipe 1 (due to smaller diameter of pipe 2), pipe 1 will actually carry slightly higher flow than 1.77 fps. Thus the velocity in pipe 1 will be slightly greater than 9 fps.

Also, note that if the Hazen-Williams friction loss equation were used

$$h_f = \frac{4.725LQ^{1.85}}{C^{1.85}D^{4.865}}$$

And from $h_{f1} = h_{f2}$, we get

$$\frac{Q_1^{1.85}}{D_1^{4.865}} = \frac{Q_2^{1.85}}{D_2^{4.865}} \Rightarrow \frac{Q_1}{Q_2} = \left(\frac{D_1}{D_2}\right)^{4.865/1.85} = \left(\frac{6}{4}\right)^{2.63} = 2.9$$

These two solutions are fairly equivalent.

Pumps

When water needs to be conveyed to a community through several miles of pipelines while maintaining adequate water pressure, energy losses due to friction and due to flow obstructions and modifications (valves, bends, etc.) may be significant. To compensate for these losses, pumps are often needed to provide the needed boost in energy. There are two basic types of pumps—positive displacement pumps (PDPs) and dynamic or momentum change pumps.

PDPs may be further subclassified into reciprocating pumps (using a piston or diaphragm) and rotary pumps. PDPs work by squeezing the fluid through volume changes, adding pressure head to the fluid.

Dynamic pumps add momentum to the fluid using moving blades or vanes. Rotary dynamic pumps may have centrifugal flow or axial flow. The fluid's increased velocity converts to an increased pressure when the fluid exits through the diffuser.

When a pump is used to deliver a volumetric flow rate Q from locations 1 to 2, the modified Bernoulli equation for energy balance is given by

$$H_P = \frac{\eta \dot{W}_s}{\gamma Q} = \frac{p_2 - p_1}{\gamma} + z_2 - z_1 + \frac{v_2^2 - v_1^2}{2g} + h_f \tag{11.20}$$

where H_p = pump head
η = pump efficiency
γ = fluid unit weight
Q = volumetric flow rate

The rated power of the pump (also called *the rate of doing shaft work* $\dot{W}_s$) is given by

$$\dot{W}_s = \frac{\gamma Q H_P}{\eta} = \frac{\dot{m} g H_P}{\eta} \tag{11.21}$$

The mass flow rate (typical units kg/s or slugs/s or lb·s/ft) is given by

$$\dot{m} = \rho Q = \frac{\gamma Q}{g} \tag{11.22}$$

In US units, typical units of pump power are lb·ft/s, which may be converted to horsepower (hp) or kilowatts (kW) using

$$1\ \text{hp} = \frac{Q_{\text{gpm}} \times H_{\text{ft}} \times \text{SG}}{3960} = 550\ \text{lb}\cdot\text{ft/s} = 0.746\ \text{kW} = 746\ \text{Nm/s}$$

where SG is the specific gravity of the fluid.

System Curve

Consider a pump which is used to deliver a rate of flow Q from reservoir 1 (surface elevation Z_1) to reservoir 2 (surface elevation Z_2), as shown in Fig. 11.5. The head loss in the piping system is taken as the sum of friction losses and minor losses.

The friction loss may be calculated using either the Darcy-Weisbach equation

$$h_f = f\frac{L}{D}\frac{V^2}{2g} \text{ or the Hazen-Williams equation } h_f = \frac{LV^{1.85}}{C^{1.85}D^{1.165}}$$

Since both locations 1 and 2 are at reservoir surface, they are under equal pressure ($p_1 = p_2 = p_{\text{atm}}$) and since the reservoirs are of large extent compared the pipe, the continuity equation dictates that velocity is negligible ($V_1 = V_2 \approx 0$). The equation then simplifies to

$$H_P = \left(\frac{p_2 - p_1}{\gamma}\right) + (Z_2 - Z_1) + \left(\frac{V_2^2 - V_1^2}{2g}\right) + h_f + h_M = (Z_2 - Z_1) + f\frac{L}{D}\frac{V_{\text{pipe}}^2}{2g} + \Sigma C\frac{v_{\text{pipe}}^2}{2g}$$

Since $V_{\text{pipe}} = Q/A_{\text{pipe}}$, this can also be written in terms of the flow rate Q

$$H_P = (Z_2 - Z_1) + \left(\frac{fL}{D2gA^2} + \frac{\Sigma C}{2gA^2}\right)Q^2 \tag{11.23}$$

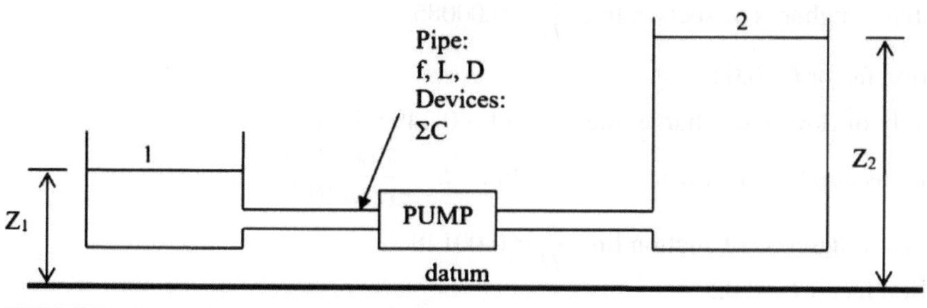

Figure 11.5 Pump delivering flow between two reservoirs.

The elevation difference to be overcome by the pump is called the static head

$$H_s = Z_2 - Z_1$$

The second term in H_p from Eq. (11.23) is due to the flow conveyed by the pipe and is called the dynamic head

$$H_d = bQ^2$$

where b is a system constant.

Thus, the total dynamic head TDH = static head + dynamic head.

Note that if the Hazen-Williams equation for head loss were used, the dynamic head H_d would be of the form

$$H_d = bQ^{1.85}$$

Example 11.14

The pump shown in the figure below operates at efficiency of 85%. The suction line is 12 in. in diameter and 500 ft long. The discharge line is 8 in. in diameter and 2200 ft long. In order to pump water at a discharge of 1200 gpm, what is the required power rating of the pump (hp)? Assume all pipes to be cast iron. Assume minor losses totaling 15 ft of headloss.

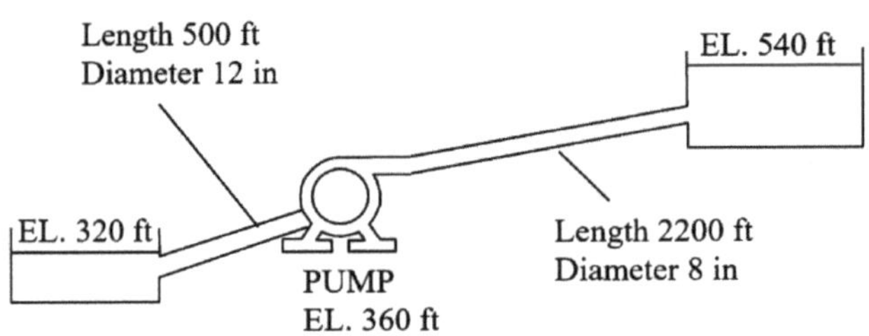

Solution Flow rate Q = 1200 gpm = 2.674 cfs

Velocity of flow in suction line = 2.674 ÷ 0.785 = 3.4 fps

Reynolds number of flow in suction line $Re = \dfrac{3.4 \times 1.0}{1.2 \times 10^{-5}} = 2.8 \times 10^5$

Relative roughness of suction line $\dfrac{\varepsilon}{D} = 0.00085$

Friction factor $f = 0.02$

Velocity of flow in discharge line = 2.674 ÷ 0.349 =7.66 fps

Reynolds number of flow in discharge line $Re = \dfrac{7.66 \times 1.0}{1.2 \times 10^{-5}} = 6.4 \times 10^5$

Relative roughness of suction line $\dfrac{\varepsilon}{D} = 0.00128$

Friction factor $f = 0.02$

Note that, depending on the relative spread between answer choices, an assumption of $f = 0.02$ for both pipes would have been a very good time saving assumption.

$$\text{Head loss (friction) } h_f = \sum f \frac{L}{D}\frac{V^2}{2g} = \frac{0.02 \times 500 \times 3.4^2}{1 \times 2 \times 32.2} + \frac{0.02 \times 2200 \times 7.66^2}{0.667 \times 2 \times 32.2} = 62 \text{ ft}$$

Total dynamic head = static head + head loss = (540 – 320) + 15 + 62 = 297 ft

$$\text{Rated power of pump } \dot{W}_s = \frac{\gamma Q H_P}{\eta} = \frac{62.4 \times 2.674 \times 297}{0.85} = 58{,}302 \text{ lb} \cdot \text{ft/s} = 106 \text{ hp}$$

Pump Curves

The pump manufacturer provides calibrated "pump curves" which plot various pump parameters (efficiency, required suction head, operating head) against pump discharge. These curves may be developed at a factory or in the field. When the system curve is overlaid with the pump curves, the intersection of the *H*-*Q* plots yields the operating point for the pump-system combination. A typical set of pump curves is shown in Fig. 11.6. The operating head (*H*), pump efficiency (η), and net positive suction head required (NPSHR) are shown plotted versus discharge (*Q*) in this figure.

Every pump has a shutoff head, which is the maximum operating head for the pump. When the head exceeds this value, the discharge becomes zero.

When a particular pump is used in a specific system, the intersection of the pump's *H* versus *Q* curve and the system's *H* versus *Q* curve gives us the operating point for the pump (see Example 11.12). Once this operating point is identified, parameters such as operating discharge, efficiency, and NPSHR can be estimated from the corresponding manufacturer's curves.

Example 11.15

The manufacturer's curves (*H* versus *Q* and η versus *Q*) are given below. The pump is used to deliver water from reservoir A (surface elevation 100 ft above sea level) to reservoir B

Figure 11.6 Typical pump manufacturer's curves.

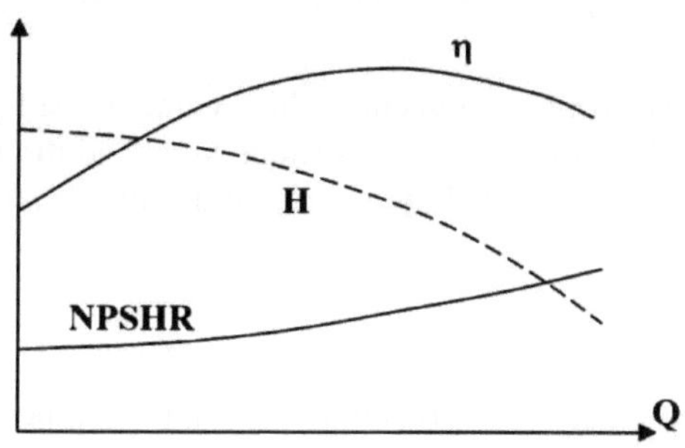

(surface elevation 140 ft above sea level) through 800 ft of 8-in-diameter cast iron pipe (C = 100). Determine the flow rate and the efficiency at the operating point of the pump.

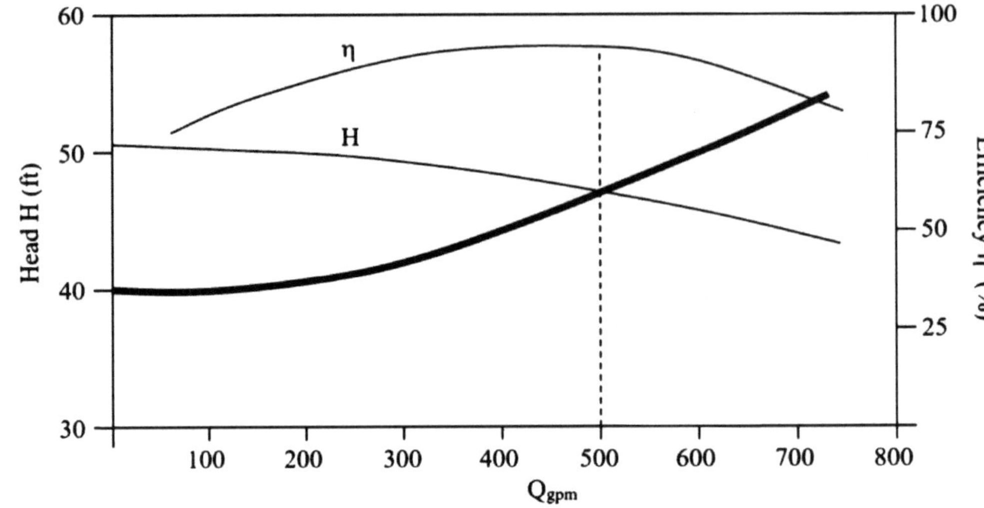

Solution Using the Hazen-Williams equation to determine the frictional head loss

$$h_f = \frac{4.73\,LQ^{1.852}}{C^{1.852}D^{4.87}} = \frac{4.73 \times 800 \times Q^{1.852}}{100^{1.852}0.667^{4.87}} = 5.376Q_{\text{cfs}}^{1.852} = 6.59 \times 10^{-5}Q_{\text{gpm}}^{1.852}$$

NOTE To convert flow rate in ft³/s to gal/min, multiply by 448.83 ($Q_{\text{gpm}} = 448.83 \times Q_{\text{cfs}}$) To convert flow rate in gal/min to ft³/s, multiply by 0.002228 ($Q_{\text{cfs}} = 0.002228 \times Q_{\text{gpm}}$).

The system curve equation, obtained by applying Bernoulli's equation to reservoir surfaces, is shown below

$$H_P = h_f + \left(\frac{p_2 - p_1}{\gamma}\right) + (z_2 - z_1) + \left(\frac{V_2^2 - V_1^2}{2g}\right) = h_f + (z_2 - z_1) = 40 + 6.59 \times 10^{-5}Q_{\text{gpm}}^{1.852}$$

This is used to plot the system curve (heavy line in the figure above).

The point of intersection of the system curve and the H-Q pump curve is at approximately Q = 510 gpm. This is called the operating discharge. At this discharge, the efficiency of the pump is $\eta_{\text{op}} = 90\%$.

Cavitation

Liquids can be made to boil by either raising the temperature to the boiling point while pressure remains constant, or by lowering the pressure to the saturation pressure at a particular temperature. For hydraulic machines such as pumps and turbines, the fluid can reach very high velocities. As a result, as predicted by Bernoulli's law of energy conservation, pressures can fall to very low values. When the hydrodynamic pressure is less than the vapor pressure (temperature dependent), small pockets of vapor (cavities) can form spontaneously. These vapor pockets collapse, causing noise, vibration, and pitting corrosion of surfaces. This phenomenon is called cavitation.

Cavitation can occur when the available head is less than the head required for satisfactory operation. The minimum fluid energy required at the pump inlet is the NPSHR. This is a specification of the pump manufacturer.

Net Positive Suction Head Available

The actual fluid energy (head) at the pump inlet is called the positive suction head available (PSHA). This is a function of the specific parameters of the system. For example, consider flow from a reservoir (free surface elevation Z_1) through a pipe to a pump whose suction line inlet is at elevation Z_2. A schematic of this system is shown in Fig. 11.7. The head losses in the suction line (friction losses plus minor losses) may be written as h_f and the static head $h_s = Z_1 - Z_2$.

The PSHA at the pump inlet is the sum of all pressure sources contributing to positive pressure minus all the sources of pressure (or head) loss

$$\text{PSHA} = h_{\text{atm}} + h_s - h_f \tag{11.24}$$

where h_{atm} = the atmospheric pressure head (approximately 33.9 ft)
h_s = elevation head at the pump inlet = $Z_1 - Z_2$
h_f = friction loss in the suction line

Since cavitation becomes possible when the pressure falls below the vapor pressure, the net positive suction head available (NPSHA) is given by

$$\text{NPSHA} = h_{\text{atm}} + h_s - h_f - h_{\text{vp}} \tag{11.25}$$

where h_{vp} is the vapor pressure (temperature dependent). The vapor pressure increases with water temperature, reaching 1 atm (101.3 kPa) at 100°C (boiling point of water at standard

Figure 11.7 Positive suction head for a pump.

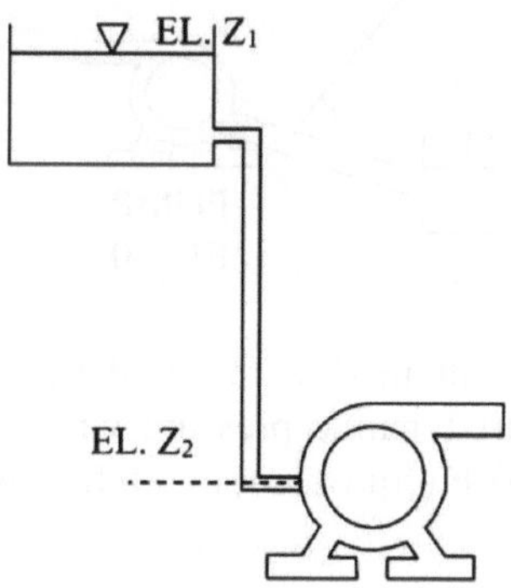

atmospheric pressure). Values of vapor pressure of water at various temperatures are given in tables on page 114 of the *Handbook*.

In order to increase the NPSHA, the following steps may be used:

1. Increase h_s by increasing the height of the fluid source (usually difficult, if not impossible).
2. Increase h_s by lowering the pump (not difficult).
3. Reduce h_f by shortening suction line length.
4. Reduce h_f by increasing suction line diameter (thereby reducing velocity).
5. Reduce the pump speed.
6. Locate the pump in the coldest part of the system, so that the vapor pressure is low (negligible effect).

Where cavitation cannot be eliminated, its effects can be mitigated by

1. The boundaries may be designed so that cavities form and collapse out in the flow field away from the boundaries.
2. Introducing air into the fluid changes cavitation from vaporous to gaseous. This eliminates the explosive growth and complete collapse of cavities, which reduces the potential for damage.
3. The damaging effects of cavitation can be reduced by using materials such as stainless steels.

Example 11.16

A pump is used to pump water from reservoir A (surface elevation 50.0 ft) to reservoir B (surface elevation 220.0 ft). The pump is located at elevation 60 ft as shown. All pipes are 3-in-diameter steel pipe (assume friction factor 0.02). The water temperature is 60°F. What is the maximum flow rate permissible in the system (based on cavitation potential)?

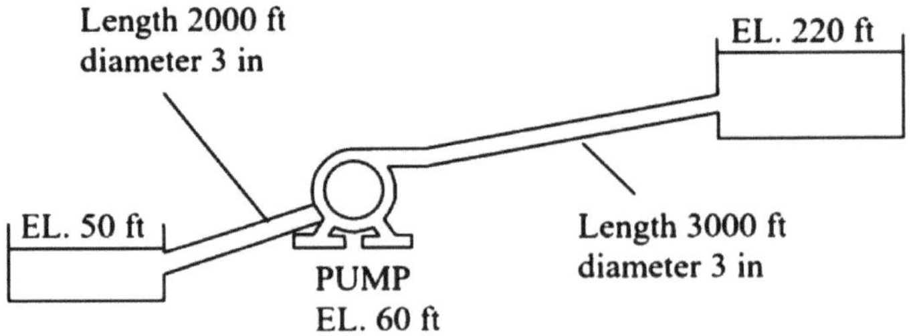

Solution The maximum flow rate is determined by the potential for cavitation, which occurs when the hydrodynamic pressure within the pump falls below the vapor pressure, which is 0.26 psi at 60°F (equivalent to 0.6 ft of water).

Therefore, setting the (absolute) pressure at pump location to be 0.6 ft, we can obtain the maximum flow velocity in the system. Note that the parameters of the system downstream of the pump do not affect this calculation

$$\frac{p_{\text{atm}}}{\gamma} + 50 + \frac{0^2}{2g} - f\frac{L}{D}\frac{V^2}{2g} = h_{\text{vp}} + 60 + \frac{V^2}{2g}$$

$$33.9 + 50 + 0 - 0.02\frac{2000}{0.25}\frac{V^2}{2\times 32.2} = 0.6 + 60 + \frac{V^2}{2\times 32.2}$$

$$2.5V^2 = 33.9 + 50 + 0 - 0.6 - 60 = 0 \Rightarrow V = 3.05 \text{ fps}$$

Flow rate $Q = \left(\frac{\pi D^2}{4}\right)V = 0.15 \text{ cfs} = 67.2 \text{ gpm}.$

Circular Pipe Head Loss

For a circular pipe of diameter D, the head loss can be expressed as

$$h_f = \frac{4.73LQ^{1.852}}{C^{1.852}D^{4.87}} \quad (11.26)$$

where h_f = head loss (ft)
L = pipe length (ft)
D = pipe diameter (ft)
C = Hazen-Williams coefficient
Q = flow rate (ft^3/s)

Expressed in terms of pressure loss, this is written

$$\Delta P = \frac{4.52Q^{1.852}}{C^{1.852}D^{4.87}} \quad (11.27)$$

where ΔP = pressure loss (psi per ft of pipe)
D = pipe diameter (in)
Q = flow rate (gal/min)

The SI version of Eq. (11.27) is given by Eq. (11.28)

$$\Delta P = \frac{6.05Q^{1.852}}{C^{1.852}D^{4.87}} \times 10^5 \quad (11.28)$$

where ΔP = pressure loss (bars per meter of pipe)
D = pipe diameter (mm)
Q = flow rate (L/min)
1 bar = 1×10^5 Pa

Open Channel Flow Fundamentals

NOTE In the *Handbook*, this topic is covered in both the Civil Engineering section (page 167) and the Fluid Mechanics section (page 106).

The primary driving force for flow in open channels is gravity, rather than pressure gradients (as in pipes flowing full). The longitudinal slope of the channel bottom is therefore a key parameter in determining flow rate and flow velocity in an open channel. The specific energy of the open channel (total energy per unit weight) is described as the energy of the free surface streamline with respect to the channel bottom

$$E = \alpha \frac{V^2}{2g} + y = \alpha \frac{Q^2}{2gA^2} + y \tag{11.29}$$

where E = energy per unit weight (specific energy)
y = depth of flow
Q = discharge
V = average flow velocity
g = acceleration due to gravity
A = cross-sectional area of flow
α = factor to account for non-uniform velocity distribution (usually taken as 1.0)

The key dimensionless parameter that defines flow characteristics in an open channel is the Froude number (Fr) given by

$$Fr = \frac{V}{\sqrt{g y_h}} \tag{11.30}$$

where y_h is the hydraulic depth, given by

$$y_h = \frac{\text{area of flow}}{\text{width at free surface}} \tag{11.31}$$

For a rectangular channel, the hydraulic depth is equal to the actual depth of flow.

The expression $\sqrt{g y_h}$ represents the speed of propagation of a small disturbance wave on the channel surface.

If the Froude number is greater than 1.0, the flow is termed supercritical (shallow depth, high velocity) while if the Froude number less than 1.0, the flow is termed subcritical (high depth, low velocity). When the Froude number is equal to 1.0, the channel experiences critical flow. At critical flow, the specific energy of the channel is minimized.

For uniform flow, the depth of flow remains constant and the free surface is parallel to the channel bottom.

Rectangular Open Channels

For a rectangular channel, if the flow rate is Q, and the (constant) width is B, then the flow rate per unit width may be written $q = Q/B$.

The average velocity of flow (at depth y) is given by

$$V = \frac{Q}{By} = \frac{q}{y} \tag{11.32}$$

Therefore, the specific energy of the channel can be written as

$$E = \frac{V^2}{2g} + y = \frac{q^2}{2gy^2} + y \tag{11.33}$$

The plot of this function is shown in Fig. 11.8. (Only the quadrant $E > 0$, $d > 0$ is shown.)

Note that the function is cubic in nature, exhibiting a minimum for energy E (called the critical point). The minimum value of E is called the critical specific energy E_c and the depth corresponding to it is called the critical depth y_c.

For constant discharge Q, the channel has minimum specific energy E when flowing at critical depth. Also, for constant specific energy, the channel conveys maximum discharge at the critical condition.

The critical depth for a rectangular open channel is given by minimizing E with respect to the depth y, leading to the following expressions for critical depth y_c and critical velocity V_c and the critical specific energy E_c

$$y_c = \sqrt[3]{\left(\frac{q^2}{g}\right)} = \sqrt[3]{\left(\frac{Q^2}{gB^2}\right)} \tag{11.34}$$

$$V_c = \sqrt[3]{qg} = \sqrt[3]{\left(g\frac{Q}{B}\right)} \tag{11.35}$$

$$E_c = \frac{3}{2}y_c \tag{11.36}$$

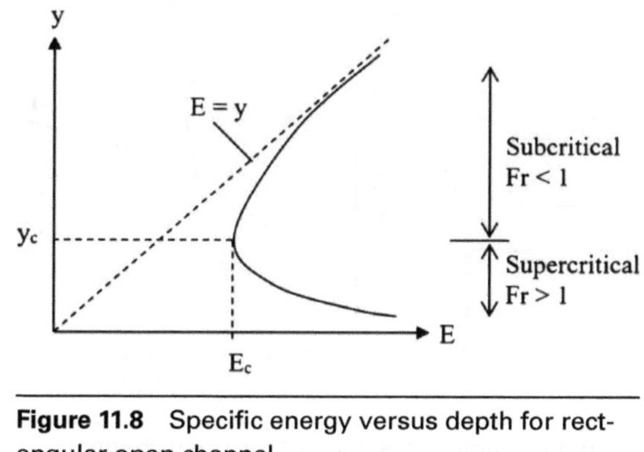

Figure 11.8 Specific energy versus depth for rectangular open channel.

Flows with depth less than the critical depth ($y < y_c$) are called supercritical flow (shallow, rapid flow) and flows with depth greater than the critical depth ($y > y_c$) are called subcritical flow (deep, tranquil flow).

Critical Depth of Flow in Open Channels

In general, for any open channel (any cross section), the following holds true when flow is occurring at critical depth (Froude number $Fr = 1$)

$$\frac{Q^2}{g} = \frac{A^3}{T} \tag{11.37}$$

where Q = volumetric flow rate
T = width of the channel at the free surface
A = cross-sectional flow area
g = acceleration due to gravity

For natural streams that do not have a regular cross section, bathymetry data may be used to represent the depth versus cross-section area and depth versus surface width relationships. These are shown for a typical stream in Fig. 11.9.

If the depth of flow is known, such curves can then be used to determine the corresponding values of T and A.

Example 11.17

The table below shows flow data for an open channel with an irregular cross section. At which flow rate is the channel most nearly conveying flow at critical depth? What is the critical velocity for this channel?

Depth of Flow (ft)	Width at Top (ft)	Area of Flow (ft^2)	Flow Rate (cfs)
2	3.52	4.12	52
4	4.95	6.75	108
6	7.10	24.90	265
8	10.34	36.22	430

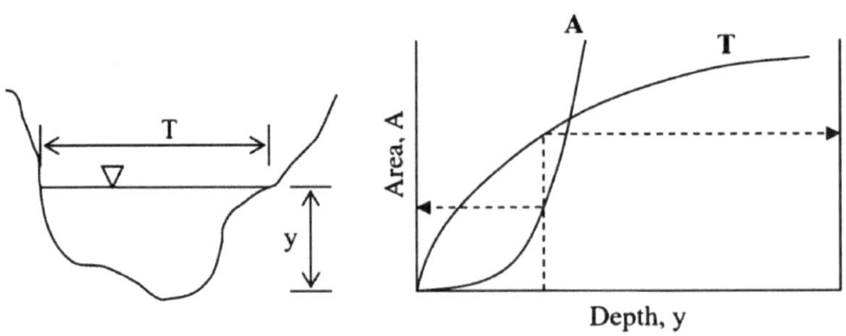

Figure 11.9 Hydraulic parameters of a natural open channel.

Solution Since at critical depth of flow, the relation $Q^2T = A^3g$ must hold, we calculate the expression Q^2T/A^3 for all four data points.

Point 1: Value = 136.1

Point 2: Value = 187.7

Point 3: Value = 32.3

Point 4: Value = 40.2

Of these, 32.3 is numerically closest to g = 32.2 ft/s². Thus, when the depth is 6 ft (flow rate Q = 265 cfs), the channel is most nearly conveying critical flow. Therefore, the critical velocity is

$$V_c = \frac{Q}{A} = \frac{265}{24.9} = 10.6 \text{ fps}$$

Alternate Depths

Two depths with the same value of specific energy E are called alternate depths. One of these is on the subcritical branch ($y_1 > y_c$) and the other on the supercritical branch ($y_2 < y_c$), as shown in Fig. 11.10.

Close to the critical point, flow in an open channel is very sensitive to channel conditions. When the specific energy is close to its minimum value, a very small perturbation to such flows (such as small changes in the channel bottom profile) can cause large changes to the type of flow. The effect of a localized perturbation, such as an abrupt change in the channel floor, can be illustrated in Example 11.15.

Example 11.18

The figure shows a rectangular (width = 8 ft) open channel with an approach flow velocity of 6.25 fps and a depth of 4 ft. At location A, there is an abrupt rise of the channel bottom by 4 in. What is the effect on the channel surface due to the "bump" on the channel bottom?

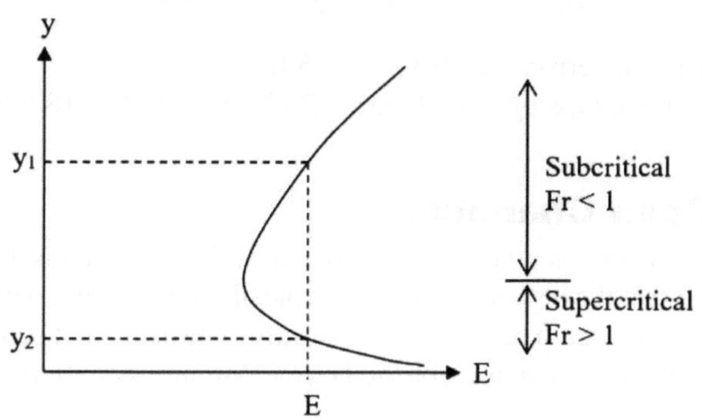

Figure 11.10 Alternate depths for rectangular open channel.

Solution Considering two planes 1 and 2 with negligible separation, we may assume that the absolute energy (with respect to a horizontal datum) remains constant. However, if the specific energy is used as defined earlier (i.e., measured with respect to the channel bottom), then the specific energy at location 2 (E_2) is less than that at location 1 (E_1) by the magnitude of the channel bottom "bump" (4 in = 0.333 ft)

$$E_2 = E_1 - 0.333$$

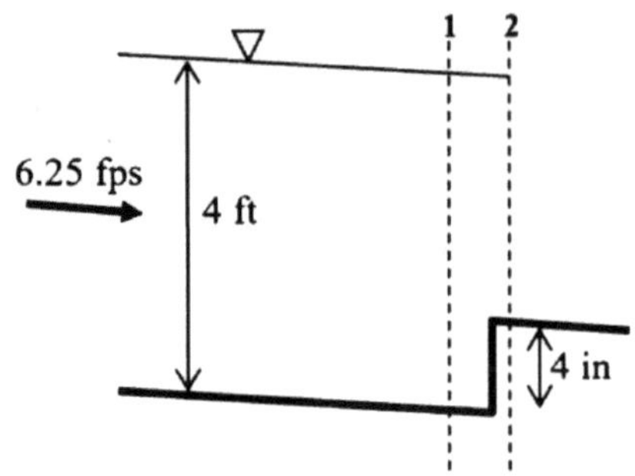

For a rectangular channel, the actual depth of flow is equal to the hydraulic depth.

The specific energy E_1 is given by

$$E_1 = y_1 + \frac{V_1^2}{2g} = 4 + \frac{6.25^2}{2 \times 32.2} = 4.607 \text{ ft}$$

Therefore, $E_2 = E_1 - 0.333 = 4.274 \text{ ft}$

Using the continuity equation, we have $V_1 y_1 = V_2 y_2 \Rightarrow V_2 = \frac{V_1 y_1}{y_2} = \frac{6.25 \times 4}{y_2} = \frac{25}{y_2}$

$$E_2 = y_2 + \frac{V_2^2}{2g} = y_2 + \frac{25^2}{2 \times 32.2 \times y_2^2} = y_2 + \frac{9.705}{y_2^2} = 4.274$$

Solving this by trial and error, we get $y_2 \approx 2.125$ ft.
Thus, the water surface dips by 4.0 – 0.333 – 2.125 = 1.542 ft = 18.5 in.

Velocity in Open Channels

The flow rate in the channel is given by the product of the average discharge velocity V and the flow area A. Note that the velocity of flow is typically not constant with depth, but reaches a maximum value at some depth below the free surface. The average discharge velocity (given by either the Chezy-Manning equation or the Hazen-Williams equation) does not reflect this variation

$$Q = VA \tag{11.38}$$

While the driving force for the flow is gravity, induced by the slope of the bottom of the channel S_o, the resisting force is provided by the frictional resistance from the sides of the channel. The side friction from the walls is quantified in terms of the Manning's constant n or Hazen-Williams constant C. The equilibrium depth flowing in an open channel is called the normal depth y_n. The average discharge velocity in the channel is given by either the Chezy-Manning empirical formula

$$V = \frac{K}{n} R_h^{2/3} \sqrt{S} \qquad K = 1 \text{ (SI units) or } K = 1.486 \text{ (US units)}$$

or, the Hazen-Williams empirical formula

$$V = kCR_h^{0.63} S^{0.54} \qquad k = 0.849 \text{ (SI units) or } k = 1.318 \text{ (US units)}$$

The discharge Q, according to the Chezy-Manning equation, is given by

$$Q = VA = \frac{K}{n} AR_h^{2/3} \sqrt{S} = K\sqrt{S} \tag{11.39}$$

where K is called the conveyance of the channel. The conveyance is a function of the flow cross section (which determines the area A and the hydraulic radius R_h) and the channel material (which determines the roughness coefficient n).

In both models, the hydraulic radius R_h reflects a combined effect of the flow area and the wetted perimeter and is given by

$$R_h = \frac{\text{area of flow}}{\text{wetted perimeter}} \tag{11.40}$$

The hydraulic diameter D_h is given by four times the hydraulic radius

$$D_h = 4R_h \tag{11.41}$$

Hydraulic Parameters of Straight-Sided Open Channels

An open channel with straight sides is shown in Fig. 11.11. If the side slope parameter $m = 0$, the channel is rectangular. If the bottom width $b = 0$, the channel is triangular. For $b \neq 0$ and $m \neq 0$, the channel is trapezoidal. The hydraulic parameters of a straight-sided channel are given by Eqs. (11.42) to (11.45).

Figure 11.11 Parameters of a straight-sided open channel.

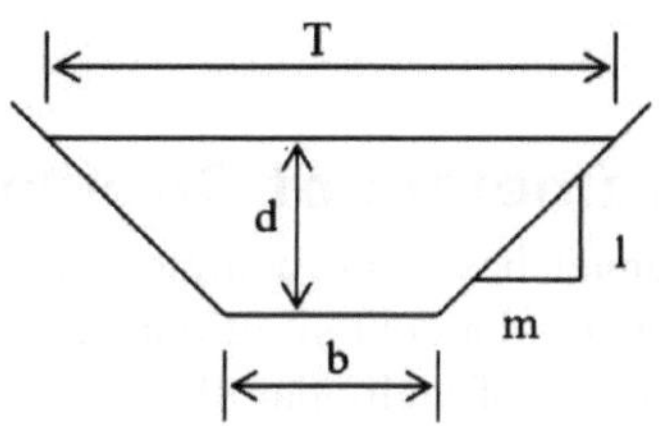

Width at the free surface is given by

$$T = b + 2md \tag{11.42}$$

The area of flow is given by

$$A = (b + md)d \tag{11.43}$$

The wetted perimeter is given by

$$P = b + 2d\sqrt{1 + m^2} \tag{11.44}$$

The hydraulic radius is given by

$$R_h = \frac{d(b + md)}{b + 2d\sqrt{1 + m^2}} \tag{11.45}$$

Example 11.19

A trapezoidal open channel with bottom width = 10 ft and side slopes $2H:1V$ conveys water at a uniform depth of 5 ft. If Manning's roughness coefficient = 0.014 and the longitudinal slope of the channel bottom is 0.4%, what is the flow rate?

A. 755 ft^3/s

B. 960 ft^3/s

C. 1420 ft^3/s

D. 1730 ft^3/s

Solution For depth = 5 ft and side slope = 2:1, the horizontal flare on each side = 10 ft, resulting in a top width = 30 ft.

The area of flow = 100 ft^2

Wetted perimeter: $P = 10 + 2 \times 5\sqrt{5} = 32.36$ ft

Hydraulic radius: $R_h = \dfrac{A}{P} = \dfrac{100}{32.36} = 3.09$ ft

Velocity: $V = \dfrac{1.486}{n} R^{2/3} S^{1/2} = \dfrac{1.486 \times 3.09^{2/3} \times 0.004^{1/2}}{0.014} = 14.24$ fps

Flow rate: $Q = VA = 1424$ cfs

Answer is C.

Hydraulic Parameters of Circular Open Channels

A circular open channel has more complex geometry than a channel with straight sides. Figure 11.12 shows a circular conduit of diameter D conveying flow at a partial depth $d < D$. The hydraulic parameters of a circular channel are best found using the hydraulic elements

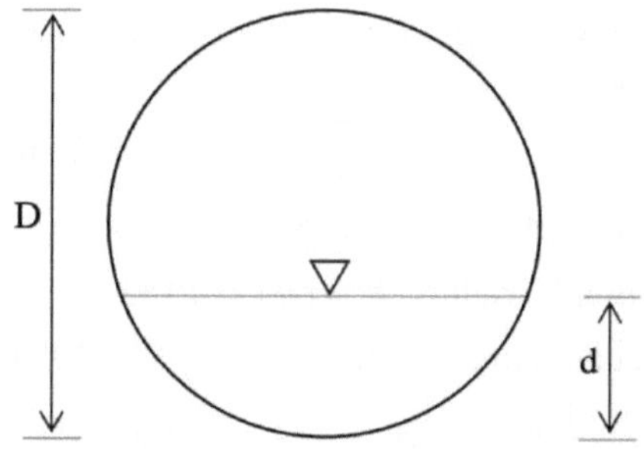

Figure 11.12 Geometry of circular conduit flowing partially full.

curves on page 166 of the *Handbook*. The most significant hydraulic parameters are area of flow A and hydraulic radius R_h.

Example 11.20

A circular conduit, diameter 600 mm, has Manning's $n = 0.015$ and a longitudinal slope $S = 1\%$. Normal depth of flow occurs in the conduit at a depth of 240 mm. What is the flow rate (m^3/s)?

Solution Depth ratio: $d/D = 240/600 = 0.4$

For $d/D = 0.4$, the diagram gives $A/A_f = 0.37$ and $R/R_f = 0.87$ (see the figure below).

Hydraulic-Elements Graph for Circular Sewers

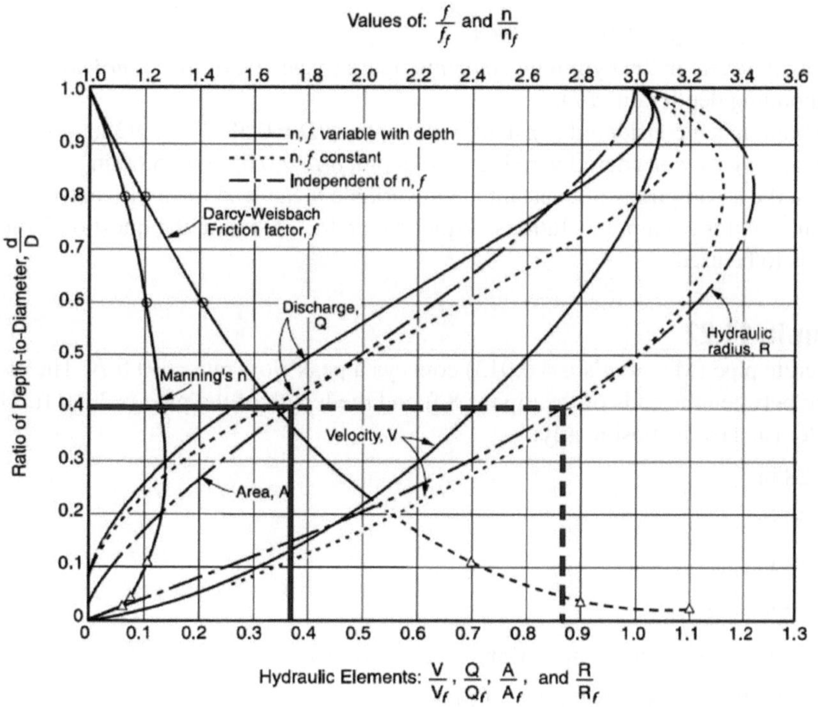

Recognizing that $A_f = \pi D^2/4$ and $R_f = D/4$, we can compute the hydraulic radius R and the area of flow A.

Full flow area: $A_f = \frac{\pi}{4}D^2 = \frac{\pi}{4} \times 0.6^2 = 0.2827 \text{ m}^2$

Actual flow area: $A = 0.37 \times 0.2827 = 0.1046 \text{ m}^2$

Full flow hydraulic radius: $R_f = D/4 = 0.15$ m

Actual hydraulic radius $R = 0.87 \times 0.15 = 0.1305$ m

Flow rate: $Q = \frac{1}{n} R^{2/3} S^{1/2} A = \frac{1}{0.015} \times 0.1305^{0.667} \times 0.01^{0.5} \times 0.1046 = 0.1793 \text{ m}^3/\text{s}$

Example 11.21 illustrates the use of the hydraulic elements curves to solve for unknown depth in a circular open channel, given the flow rate.

Example 11.21

What is the depth of flow if a flow rate of $Q = 6$ cfs is conveyed by a 24-in-diameter pipe? Bottom slope = 0.3% and nominal value of Manning's $n = 0.017$.

Solution This is an example of a pipe flowing partially full.
When the pipe flows full, hydraulic radius $R_h = D/4 = 0.5$ ft and area = $\pi D^2/4 = 3.14 \text{ ft}^2$

Flow rate for the circular pipe flowing full is given by

$$Q = \frac{1.486}{0.017} \times 0.5^{0.67} \times 0.003^{0.5} \times 3.14 = 9.45 \text{ cfs}$$

Flow rate ratio: $Q/Q_f = 6.0/9.45 = 0.64$

Using this ratio in the hydraulic elements figure (page 166 of the *Handbook*), we obtain the corresponding depth ratio d/D.

If Manning's n considered constant with depth, $d/D = 0.58$ $d = 13.92$ in

If Manning's n considered variable with depth, $d/D = 0.65$ $d = 15.60$ in

Thus, the assumptions of constant n and variable n can lead to significantly different (12% difference in this example) solutions. Typically, on the FE exam, the question indicates which option is to be used.

Example 11.22

A concrete pipe (Manning's $n = 0.013$) conveys a peak flow rate = 30 ft³/s. The elevation difference between the ends of the pipe is 8 ft and the length of the pipe is 2000 ft. The required diameter (inches) is most nearly:

A. 28 in

B. 30 in

C. 32 in

D. 34 in

Solution For concrete pipe, Manning's $n = 0.013$

Longitudinal slope $S = 8/2000 = 0.004$

Peak flow rate = 30 ft³/s

Flow rate conveyed by circular pipe flowing full (based on Chezy-Manning formula):

Flow area: $A = \dfrac{\pi D^2}{4}$

Wetted perimeter: $P = \pi D$

Hydraulic radius: $R_h = \dfrac{A}{P} = \dfrac{D}{4}$

Flow rate (flowing full): $Q_f = VA = \left(\dfrac{1.486}{n} R^{2/3} S^{1/2}\right)\left(\dfrac{\pi D^2}{4}\right) = \dfrac{0.4632}{n} D^{8/3} S^{1/2}$

Required diameter: $D = \left[\dfrac{nQ_f}{0.4632 \times \sqrt{S}}\right]^{3/8} = \left[\dfrac{0.013 \times 30}{0.4632 \times \sqrt{0.004}}\right]^{3/8} = 2.64 \text{ ft} = 31.7 \text{ in}$

Answer is C.

Momentum in Open Channels

In general, the principle of conservation of momentum can be stated as "Net force acting on a fluid domain between two boundaries is equal to the rate of change of momentum between those boundaries."

Thus, when there are no forces acting on the fluid between sections 1 and 2, the momentum is conserved. The generalized expression for specific force (force per unit weight acting on the water) is given by the sum of the pressure force and the momentum flow through the channel

$$M = \frac{Q^2}{gA} + Ah_c \tag{11.46}$$

where A = area of flow
h_c = vertical distance from centroid of flow area to water surface
Q = flow rate
g = acceleration due to gravity

Applied to a rectangular open channel, the principle of conservation of momentum can be written as

$$\frac{d_1^2}{2} + \frac{qV_1}{g} = \frac{d_2^2}{2} + \frac{qV_2}{g} \tag{11.47}$$

where the momentum per unit width m may be written in terms of the flow rate per unit width q as

$$m = \frac{d^2}{2} + \frac{qV}{g} \tag{11.48}$$

Thus, if the depth and velocity of flow are known at a location upstream of a channel transition, then the depth and/or velocity at a location downstream of the transition can be solved.

Example 11.23

A rectangular channel (width = 5 ft) conveys a flow rate $Q = 300$ cfs at a depth of 10 ft. The channel smoothly transitions to a 4-ft-wide channel with no change in elevation. What is the depth of flow after the width transition?

Solution For the approach flow:

Area: $A = 50 \text{ ft}^2$

Depth of centroid, $h_c = 5$ ft

Momentum: $M = \dfrac{Q^2}{gA} + Ah_c = \dfrac{300^2}{32.2 \times 50} + 50 \times 5 = 305.9$

For downstream section (unknown depth y_2)

Momentum: $M = \dfrac{Q^2}{gA} + Ah_c = \dfrac{300^2}{32.2 \times 4 \times y_2} + 4 \times y_2 \times \dfrac{y_2}{2} = \dfrac{698.8}{y_2} + 2y_2^2 = 305.9$

Solving this by trial and error, $y_2 = 11.0$.

Two depths with the same value of specific momentum M are called sequent (or conjugate) depths. One of these is on the supercritical branch ($y_1 < y_c$) and the other on the subcritical branch ($y_2 > y_c$), as shown in Fig. 11.13.

Hydraulic Jump

When a supercritical flow (Froude number >1) in an upstream section of the channel is forced to become subcritical, there may occur a sudden increase in depth, accompanied by a significant loss of energy. This phenomenon is known as a hydraulic jump and is often the preferred mechanism of dissipating energy at the bottom of spillways. By controlling the depth of the tailwater, or receiving reservoir, the jump may be forced to occur within the apron of the spillway, thereby reducing the erosion of the channel further downstream. All following results are derived for a horizontal channel, but are fairly applicable to channels of moderate slope (less than approximately 2%).

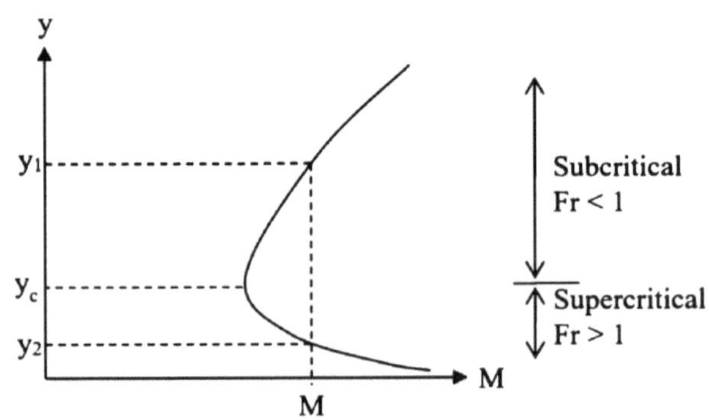

Figure 11.13 Conjugate depths for rectangular open channel.

The principal parameter that affects hydraulic jump performance is the upstream Froude number. A stable jump (that is insensitive to downstream conditions) forms when the Froude number is greater than 4.5 and causes significant energy dissipation (45% to 70%).

Depths on either side of a hydraulic jump are called conjugate depths. The momentum is conserved in a hydraulic jump. For a rectangular channel, the conjugate depths y_1 and y_2 are related by

$$y_2 = \frac{y_1}{2}\left[-1+\sqrt{1+8Fr_1^2}\right] \tag{11.49}$$

where y_1 = depth at upstream supercritical flow location
y_2 = depth at downstream subcritical flow location
Fr_1 = upstream Froude number is $Fr_1 = \frac{V_1}{\sqrt{gy_1}}$

Example 11.24

A spillway operates with 3 ft of head as shown. The top of the spillway is 40 ft above the toe. Assume the spillway discharge coefficient to be 3.45.

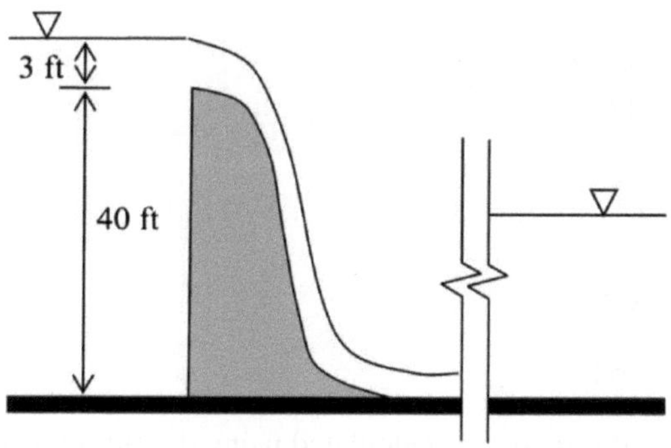

1. What is the discharge per foot width of the spillway?
2. What is the depth of flow at the toe?
3. What must be the tailwater depth in order to force a hydraulic jump to occur at the toe?
4. What is the energy loss (in hp) per foot width?

Solution

1. Discharge per foot width $Q = C_s bH^{3/2} = 3.45 \times 1 \times 3^{3/2} = 17.93$ cfs/ft
2. Equating total energy upstream of the spillway crest and at the toe (i.e., ignoring energy dissipation over the spillway)

$$43 + \frac{V^2}{2g} = d_{toe} + \frac{V_{toe}^2}{2g}$$

Neglecting upstream velocity (V) and depth at the toe (d_{toe}), we get

$$V_{toe} = \sqrt{2 \times 32.3 \times 43} = 52.62 \text{ ft / s}$$

Using continuity equation $V_1A_1 = V_2A_2$ to calculate d_{toe}, we get

$$d_{toe} = \frac{17.93}{52.62} = 0.34 \text{ ft}$$

NOTE This is small enough to justify neglecting it in the previous step.

---

3. $d_1 = d_{toe} = 0.34$ $V_1 = 52.62$ ft/s

$$d_2 = -\frac{1}{2}d_1 + \sqrt{\frac{2V_1^2 d_1}{g} + \frac{d_1^2}{4}} = 7.48 \text{ ft}$$

4. $V_2 = \dfrac{V_1 d_1}{d_2} = \dfrac{52.62 \times 0.34}{7.48} = 2.39$ fps

$$E_1 = 0.34 + \frac{52.62^2}{2 \times 32.2} = 43.33$$

$$E_2 = 7.48 + \frac{2.39^2}{2 \times 32.2} = 7.57$$

$$\Delta E = E_1 - E_2 = 35.76 \text{ ft}$$

The rate of energy dissipation is calculated using the fluid power equation

$$P = \gamma Q \Delta E = 62.4 \times 17.93 \times 35.76 = 40{,}009 \text{ lb} \cdot \text{ft/s} = 72.74 \text{ hp}$$

Example 11.25

A reservoir empties through a rectangular weir of width 3 m as shown. The contractions are suppressed. The elevation of the bottom of the weir is 120.5 m above sea level. If the elevation of the reservoir free surface immediately behind the weir is 124.7 m above sea level, what is the discharge through the weir?

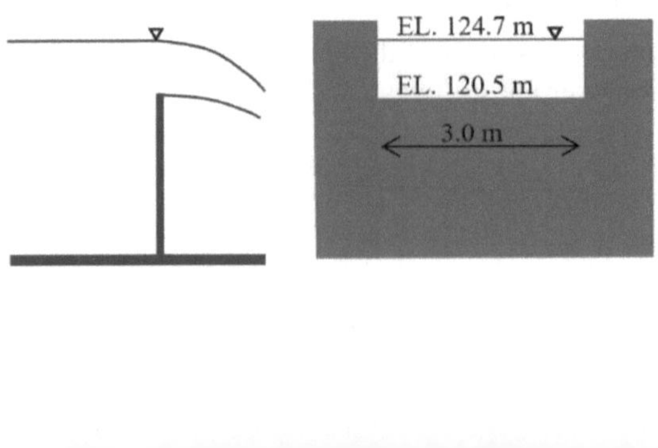

A. 86.0 m^3/s

B. 78.5 m^3/s

C. 63.2 m^3/s

D. 47.5 m^3/s

Solution Weir head $H = 124.7 - 120.5 = 4.2$ m
Free discharge (for suppressed weir):

$$Q = CLH^{3/2} = 1.84 \times 3 \times 4.2^{3/2} = 47.5 \frac{\text{m}^3}{\text{s}}$$

Answer is D.

For more on discharge through weirs, see Chap. 10.

Groundwater Distribution

Groundwater storage may be divided into the following zones:

1. The soil-water zone extends from the ground surface through the major root zone. The soil in this zone is unsaturated except during heavy infiltration.
2. The intermediate zone extends from the bottom of the soil-water zone down to the top of the capillary fringe.
3. The capillary zone extends from water table up to a height determined by the capillary rise that can be generated by the soil.
4. In the groundwater zone, soil voids are completely saturated with water. Not all of this water can be extracted from the soil because of molecular and surface tension forces. The water that can be drained from the soil by gravity is called the specific yield. The soil water that is held (retained) against gravity drainage is called the specific retention.

For more on the topic of groundwater—hydraulic conductivity, seepage in soils and draw-down in aquifers—see Chap. 14.

Groundwater Dewatering

Dewatering makes excavated material lighter and easier to handle, prevents the formation of a quick condition at the bottom of the excavation, and reduces lateral loads on sheeting and bracing. In most soils, the groundwater table must be at least 2 ft and preferably 5 ft below the bottom of the excavation to keep an excavation reasonably dry.

The most common method for dewatering excavations for large projects is the use of well points. Well points are metal well screens about 2 to 3 in. in diameter and about 4 ft long. To ensure good drainage in fine and dirty sands or layers of silt or clay, the well point and riser should be surrounded by a sand filter to just below the ground water table. The space above the filter should be sealed with silt or clay to keep air out of the well point. A pipe connects each well point to a header, from which water is pumped at a certain rate. Spacing of well

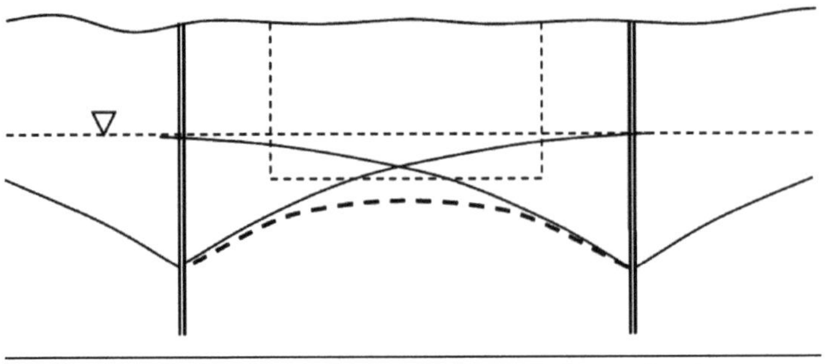

Figure 11.14 Dewatering an excavation site using well points.

points usually ranges from 3 to 12 ft. Well points are not suitable in very fine-grained soils and soils of low permeability. For dewatering very deep excavations or where the groundwater table must be lowered significantly (more than about 15 to 20 ft), other methods may become more economical.

Lowering the groundwater table below a construction site may be achieved using well points with plan locations chosen so that their combined field of action results in the desired lowering. For example, consider the dewatering scheme shown in Fig. 11.14. The horizontal dashed line represents the original groundwater table, which must be lowered to a certain minimum vertical clearance below the bottom of the proposed excavation. Dewatering of the site can be achieved by using well points on either side of the excavation. When these wells are pumped at a specified rate, at steady state, each well has a cone of influence as shown in the figure. The dashed line below the bottom of the excavation represents the superposition of these well fields.

CHAPTER 12

Structural Analysis

Structural Determinacy and Stability

Before embarking on analysis of any structural system, two things must be established. First, the overall stability of the structure must be investigated. If a structure is to be analyzed and then designed, the structural form must first be stable.

Once it has been established that the structure has no instabilities, the determinacy of the structure must be queried. A stable, statically determinate structure is one for which the equations of static equilibrium are sufficient in number to solve for the unknown quantities inherent in the structure. These unknowns can be external in nature, such as external reactions, or internal such as internal member forces. If the structure is stable but statically indeterminate, the number of unknown forces exceed the number of independent equilibrium equations. This means that simply using equilibrium equations is not sufficient to solve for the unknowns—equations broadly known as compatibility equations must also be employed.

For examples of the calculation of external reactions for beams, trusses, and frames, see Chap. 6.

Determinacy and Stability of Trusses, Frames, and Beams

Trusses

Determinacy of a plane [two dimensional (2D)] truss is determined using the rules given below

$m+r<2j$ Unstable

$m+r=2j$ Stable and statically determinate

$m+r>2j$ Stable and statically indeterminate. Degree of indeterminacy = $m + r - 2j$

where m = number of members
r = number of independent reactions
j = jumber of joints

For a space [three dimensional (3D)] truss, these rules become

$m+r<3j$	Unstable
$m+r=3j$	Stable and statically determinate
$m+r>3j$	Stable and statically indeterminate. Degree of indeterminacy = $m+r-3j$

Frames

For a plane frame, the rules to establish stability and determinacy are as follows:

$3m+r<3j+c$	Unstable
$3m+r=3j+c$	Stable and statically determinate
$3m+r>3j+c$	Stable and statically indeterminate. Degree of indeterminacy = $3m+r-3j-c$

In the last set, c is the number of condition equations based on known internal constraints, such as internal moment at a hinge = 0.

Beams

For beams, the geometry is by definition one-dimensional and three independent constraints are required at a minimum for stability. The only exceptions to this rule are (1) if all reactions are parallel and (2) if all reactions are concentric. These exceptions are shown in Fig. 12.1

Example 12.1

Which of the following statements are true?

1. Trusses consist of members that are assumed to be pinned together at their ends.
2. Truss members generally develop axial loads.
3. Truss members are rigidly connected at the joints.
4. The force system in a 2D truss is typically concurrent and coplanar.
5. Frame members develop bending moments, but not shear and axial forces.
6. A frame is generally a rigid structure.

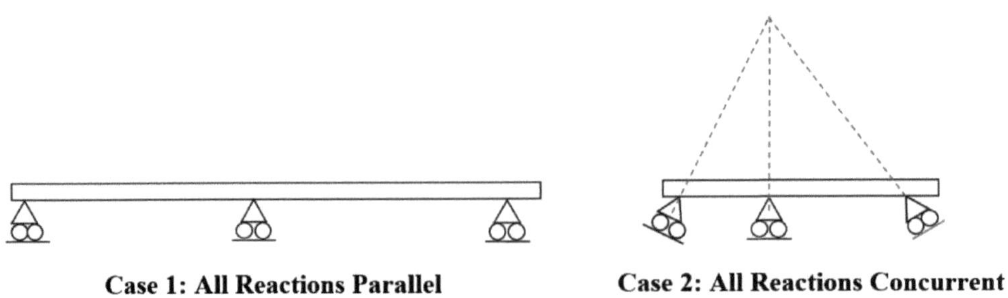

Figure 12.1 Beams unstable in spite of three constraints.

7. Moments cannot be transmitted from one member to another in a frame.

A. 1, 2, 5, and 6

B. 1, 3, 6, and 7

C. 1, 2, and 6

D. 1, 2, and 7

Solution Statements 3, 4, 5, and 7 are false as explained below:

3 is false because joints in an ideal truss are assumed to be pinned (moment-released) joints.

4 is false because the forces on a 2D truss are coplanar, but not necessarily concurrent (passing through a common point).

5 is false because frame members in general develop all three internal loads—axial force, shear force, and bending moment.

7 is false because frame members are generally rigidly connected to each other and therefore, moments ARE transmitted from one member to the next.

Answer is C.

Example 12.2

A 2D frame has eight joints, of which five are rigid and three are moment free. The frame has eight members and is supported by two fixed supports. Which of the following statements is true?

A. The frame is stable and statically determinate.

B. The frame is stable and statically indeterminate to the second degree.

C. The frame is stable and statically indeterminate to the third degree.

D. The frame is unstable.

Solution

Number of members, $m = 8$

Number of reactions, $r = 6$ (in 2D, each fixed support has three reactions)

Number of joints, $j = 8$

Number of condition equations (internal hinges), $c = 3$

$3m + r = 30$ represents the number of unknowns inherent in the structure

$3j + c = 27$ represents the number of equilibrium equations available

Structure is statically indeterminate to the third degree ($30 - 27 = 3$).

Answer is C.

Example 12.3

A crate weighing 900 lb is being lifted by four cables attached to the corners as shown. The attachment point for the cables (O) is 8 ft above the top surface of the crate and directly above its center of gravity. The tension in each cable is most nearly

A. 225 lb

B. 270 lb

C. 312 lb

D. 348 lb

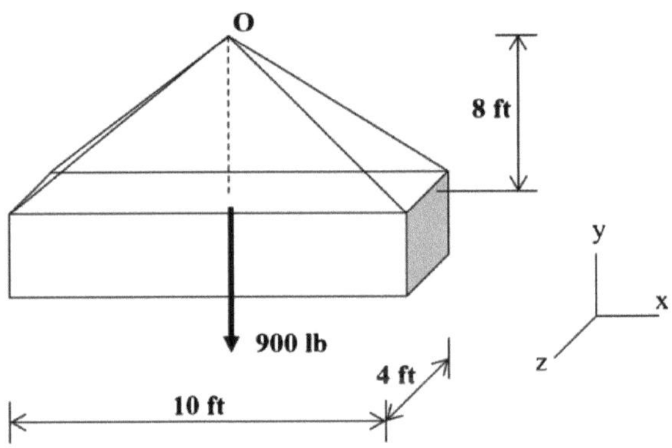

Solution Even though the problem is 3D, we can take advantage of the symmetry in the problem. All cables carry equal load. The length of each cable is given by

$$L = \sqrt{5^2 + 8^2 + 2^2} = 9.64 \text{ ft}$$

The vertical component of each cable tension can therefore be written:

$$T_y = T\frac{8}{9.64} = 0.83T$$

Therefore, the equation of vertical equilibrium is

$$4T_y - 900 = 0 \Rightarrow 4 \times 0.83T = 900 \Rightarrow T = 271 \text{ lb}$$

Answer is B.

Example 12.4

For the cable system shown, find the tension in cables AB and BC.

A. $T_{AB} = 12.8$ kips; $T_{BC} = 18.8$ kips

B. $T_{AB} = 15.8$ kips; $T_{BC} = 21.2$ kips

C. $T_{AB} = 21.2$ kips; $T_{BC} = 16.8$ kips

D. $T_{AB} = 15.8$ kips; $T_{BC} = 24.8$ kips

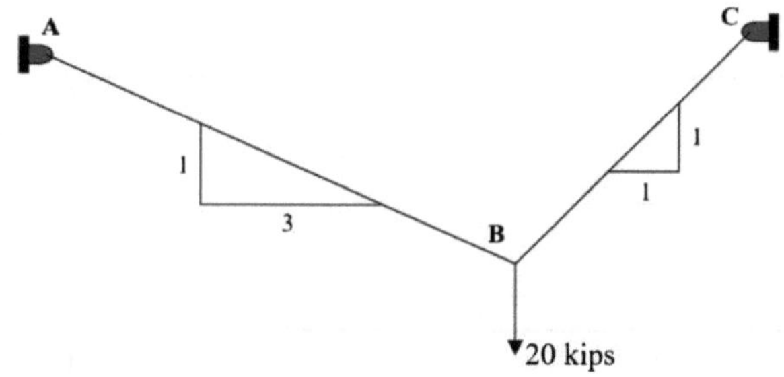

Solution The free body diagram of joint B is shown below. Since this FBD has two unknowns, the two equations of equilibrium may be used to solve for them.

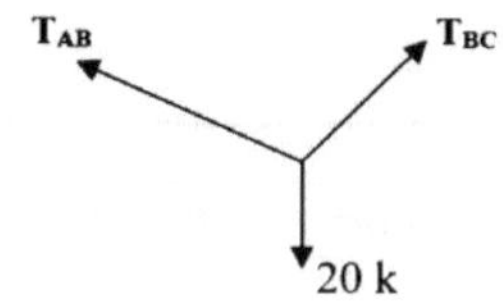

$$\sum F_x = -T_{\mathrm{AB}} \frac{3}{\sqrt{10}} + T_{\mathrm{BC}} \frac{1}{\sqrt{2}} = 0 \Rightarrow T_{\mathrm{AB}} = 0.745 T_{\mathrm{BC}}$$

$$\sum F_y = T_{\mathrm{AB}} \frac{1}{\sqrt{10}} + T_{\mathrm{BC}} \frac{1}{\sqrt{2}} = 20 \Rightarrow \left(0.745 \frac{1}{\sqrt{10}} + \frac{1}{\sqrt{2}}\right) T_{\mathrm{BC}}$$

$$= 20 \Rightarrow T_{\mathrm{BC}} = 21.21 \text{ kips; } T_{\mathrm{AB}} = 15.81 \text{ kips}$$

Answer is B.

Load Path and Tributary Area

The location of an element in a structure dictates the nature and magnitude of loads that will be incident to that element. The distribution of loads to the various elements of a typical residential floor system is described in the next section.

Floor System: Load Path

A typical floor framing system (shown in Fig. 12.2) has a load path, where the floor slab is supported by parallel floor beams, which are typically arranged with constant spacing. The aspect ratio of the slab panels is relatively large, resulting in "one-way" behavior. The tributary width of the slab (to each beam) is therefore the spacing (w) between parallel beams. The beams are supported by, and therefore transfer their loads to, girders. The columns and girders in one vertical plane form a frame. In addition to resisting the vertical loads transmitted to them, the frame also acts to resist lateral loads by either acting as a braced frame or a moment resisting frame.

The first structural component in the load path is the floor slab that is supported by the floor beams. Since the floor slab has support on two parallel edges only, the flexural behavior of the slab is predominantly one way, where the slab primarily bends in the short direction

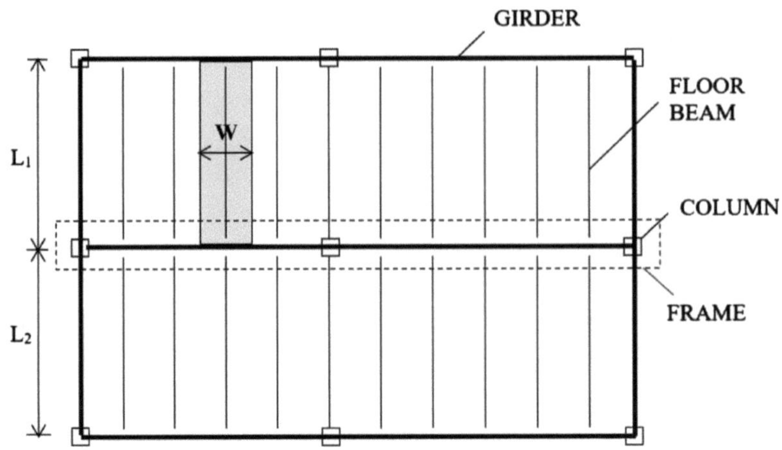

Figure 12.2 Floor system supported by floor beams, girders, and columns.

(i.e., between supporting beams). If the slab is continuous over three or more supports, it is reasonable to assume that the bending moment per unit width of the slab is given by

$$M_{\text{slab}} = \frac{qW^2}{10} \tag{12.1}$$

Assuming the floor beams are simply supported between girders and the uniformly distributed load on them is equal to the floor load q multiplied by tributary width W, the floor beam's free body diagram can be shown in Fig. 12.3.

The maximum bending moment in the beams is given by

$$M_{\text{beam}} = \frac{qWL_1^2}{8} \tag{12.2}$$

The end reactions of the beams are transmitted to the girders as equally spaced point loads as shown in Fig. 12.4. These loads are the end reactions from the beams. For exterior girders, reaction loads are transmitted from beams on one side only and the load P is given by

$$P = \frac{qWL_1}{2} \tag{12.3}$$

Figure 12.3 Floor beam with uniformly distributed load.

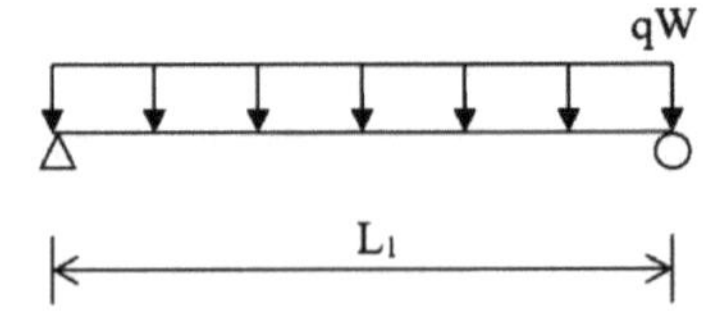

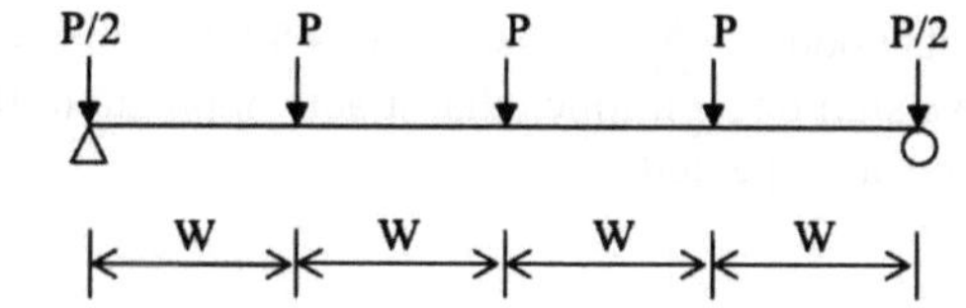

Figure 12.4 Girder supporting floor beams.

For interior girders, the load P is equal to

$$P = \frac{qW(L_1 + L_2)}{2} \tag{12.4}$$

Example 12.5

A steel angle is anchored to a wall as shown. Longitudinal spacing between anchor bolts is 18 in. The heel of the angle is continuously supported by timber blocking (height = 1 in, thickness = 0.5 in). The angle supports a wall panel centered at a horizontal distance of 4 in from the heel of the angle. The tension in each anchor bolt is most nearly

A. 45 lb

B. 56 lb

C. 98 lb

D. 125 lb

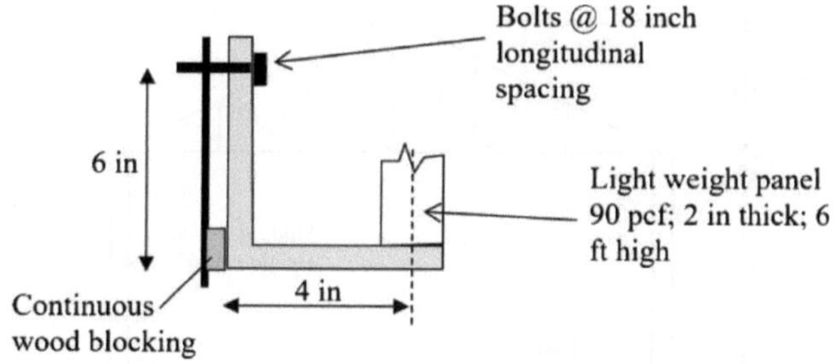

Solution Considering a 18-in-long strip of the angle (load incident on *one* bolt), the weight of the wall panel = $90 \times 2/12 \times 6 \times 18/12 = 135$ lb.

Shear in anchor bolts is represented by $A_y = 135$ lb.

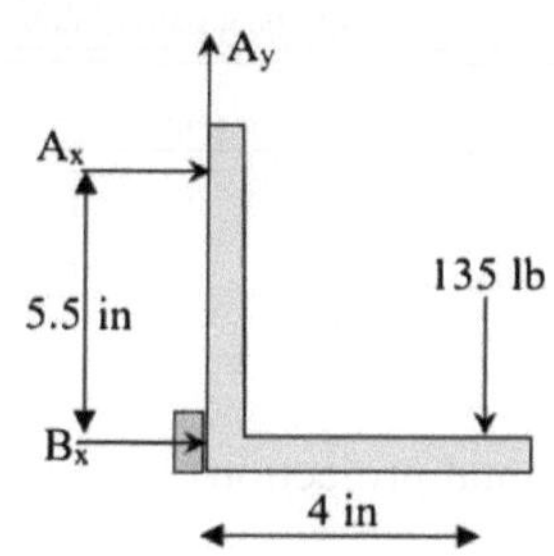

Taking moments about B: $\sum M_B = 0 \Rightarrow A_x \times 5.5 + 135 \times 4 = 0 \Rightarrow A_x = -98.2$ kips

The negative sign on A_x implies that it acts opposite to the direction shown in the FBD, or is a tensile force, as expected.

Answer is C.

Example 12.6

A steel beam supports a 6-in lightweight concrete masonry wall (unit weight = 85 lb/ft³). The wall is 10 ft high and 12 ft long. The load on the beam (lb/ft) is most nearly

A. 850 lb/ft

B. 425 lb/ft

C. 5100 lb/ft

D. 510 lb/ft

Solution The load on the beam = $85 \times 6/12 \times 10 = 425$ lb/ft

Answer is B.

Example 12.7

A two-story timber framed bearing wall building with a truss roof is shown below. The walls are framed with studs spaced at 16 in o.c. Roof trusses are spaced at 10 ft o.c.

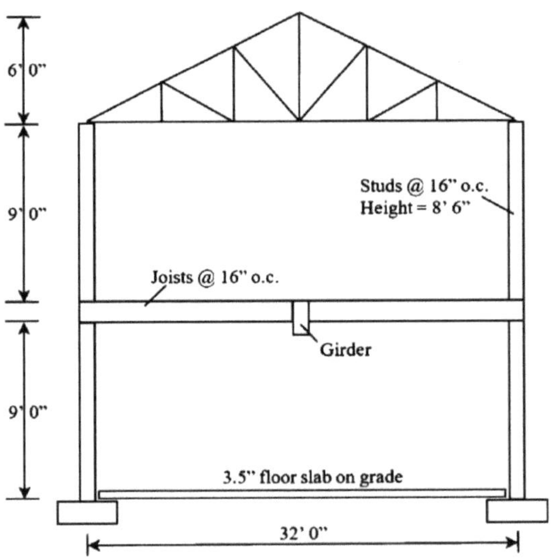

The following service loads apply

Floor $DL = 8$ lb/ft²

Roof $DL = 10$ lb/ft² (on horizontal projection)

Floor $LL = 40$ lb/ft²

Roof $LL = 20$ lb/ft^2 (on horizontal projection)

Wall $DL = 7$ lb/ft^2

Wind pressure on stud walls = 25 lb/ft^2

The service gravity load applied to each wall footing (lb/ft) is most nearly

A. 764

B. 934

C. 1000

D. 1400

Solution The roof load is incident on 32 ft (horizontal projection). Each wall carries half of that (16 ft tributary width).

Total load from roof × 16 ft = 30 × 16 = 480 lb/ft.

The floor load is incident on 32 ft (horizontal projection). Each 16-ft-wide half is carried equally by the girder and one wall. Therefore, the girder receives load from 16-ft-wide strip and each wall receives load from an 8-ft floor width.

Total load from floor × 8 ft = 48 × 8 = 384 lb/ft

Self-weight of wall = 7 × 18 = 126 lb/ft

Total load to footing = 990 lb/ft

Answer is C.

Analysis of Trusses

For a detailed discussion of method of joints and method of sections, see Chap. 6.

Example 12.8

For the simple truss shown below, the force in member AH is most nearly

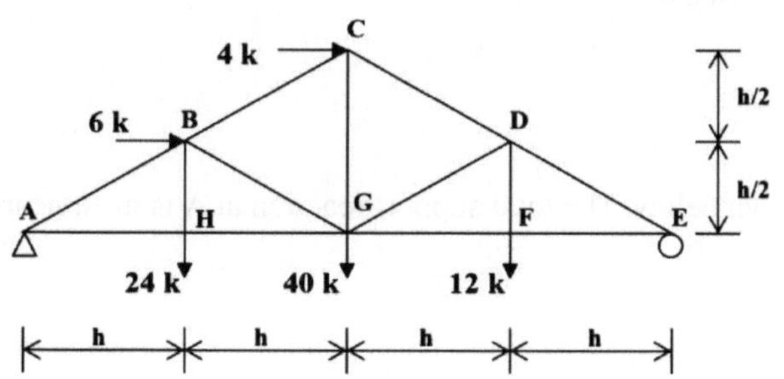

A. 89 kips (compression)

B. 89 kips (tension)

C. 79 kips (tension)

D. 79 kips (compression)

Solution The vertical reaction at the roller E can be calculated by taking moments about A (since E_y produces a *counterlockwise* moment about A, we find the *net clockwise moment about A* and divide it by the moment arm (distance between G and A_y)

$$E_y = \frac{6\frac{h}{2} + 4h + 24h + 40 \times 2h + 12 \times 3h}{4h} = 36.75 \text{ kips}$$

Reactions: $A_x = -10$ kips, $A_y = 39.25$ kips
Method of joints at A yields: $F_{AH} = 88.5$ kips (tension)

NOTE For a node such as A, where three forces intersect, the polygon of forces is a triangle. Therefore, the force triangle will be similar to the geometric triangle. The free body diagram at node A is shown below.

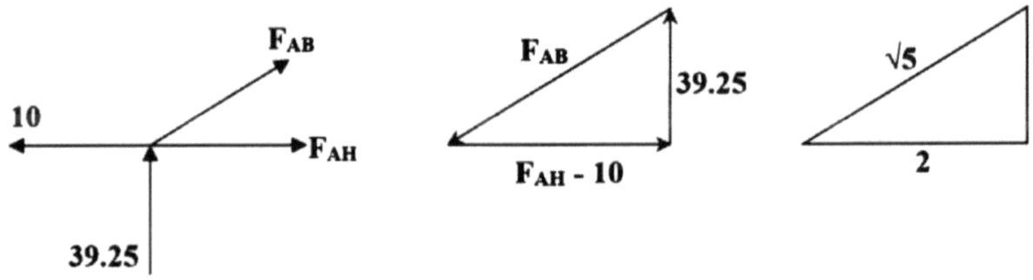

This leads to the following similarity relations:

$$\frac{F_{AH} - 10}{2} = \frac{39.25}{1} = \frac{F_{AB}}{\sqrt{5}}$$

leading to $F_{AH} = 88.5$ (tension).

Answer is B.

Example 12.9

A plane truss is shown below. The total support reaction at A is most nearly

A. 24.5 kips
B. 26.6 kips
C. 29.3 kips
D. 32.8 kips

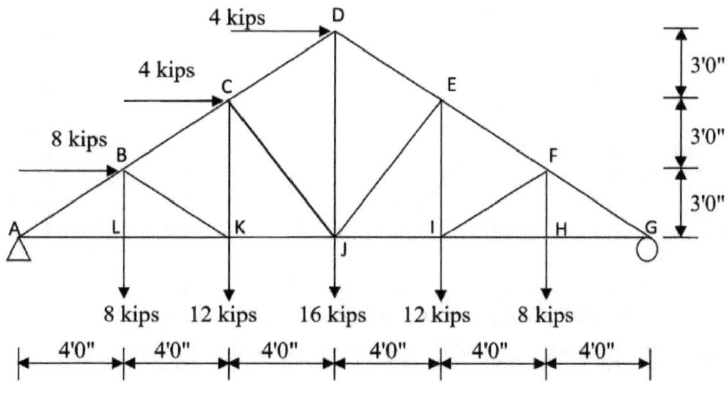

Solution The vertical reaction at A can be calculated by taking moments about G (since A_y produces a *clockwise* moment about G, we find the *net counterclockwise moment about G* and divide it by the moment arm (distance between G and A_y)

$$A_y = \frac{-8 \times 3 - 4 \times 6 - 4 \times 9 + 8 \times 4 + 12 \times 8 + 16 \times 12 + 12 \times 16 + 8 \times 20}{24} = 24.5 \text{ kips}$$

NOTE A shortcut, using the lever rule, may be described as: The vertical load is symmetric; therefore, it will create equal (upward) reactions at both supports $= 56/2 = 28$ kips. The clockwise overturning moment $(8 \times 3 + 4 \times 6 + 4 \times 9 = 84 \text{ kip} \cdot \text{ft})$ will be resisted by a couple (downward reaction at A and upward reaction at G) made up of equal and opposite forces $= 84 \div 24 = 3.5$ kips. Therefore, the vertical reaction at A $= 28 - 3.5 = 24.5$ kips.

The horizontal reaction at A = total horizontal force on truss = 16 kips

Therefore, total reaction at hinge A: $R_A = \sqrt{A_x^2 + A_y^2} = \sqrt{16^2 + 24.5^2} = 29.3$ kips

Answer is C.

Example 12.10

A truss is shown below. The axial force (kips) in the highlighted member (CJ) is most nearly

A. 31.2 kips (compression)

B. 31.2 kips (tension)

C. 17.6 kips (tension)

D. 17.6 kips (compression)

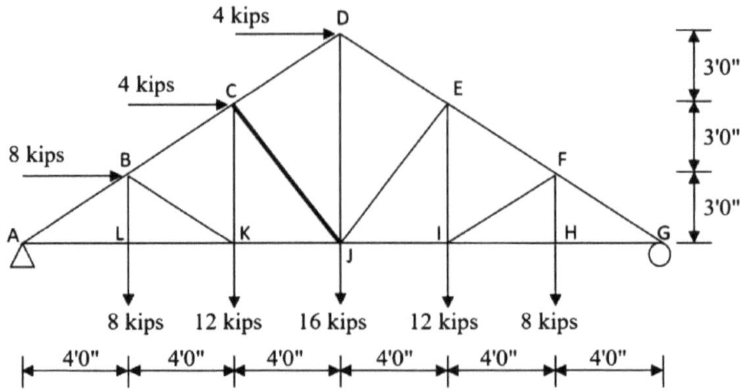

Solution Taking moments about A (so that the intersecting forces F_{CD} and F_{KJ} are not involved):

$$\frac{4}{\sqrt{52}}F_{CJ}\times 6+\frac{6}{\sqrt{52}}F_{CJ}\times 8+8\times 3+4\times 6+8\times 4+12\times 8=0\Rightarrow F_{CJ}=-17.63 \text{ kips}$$

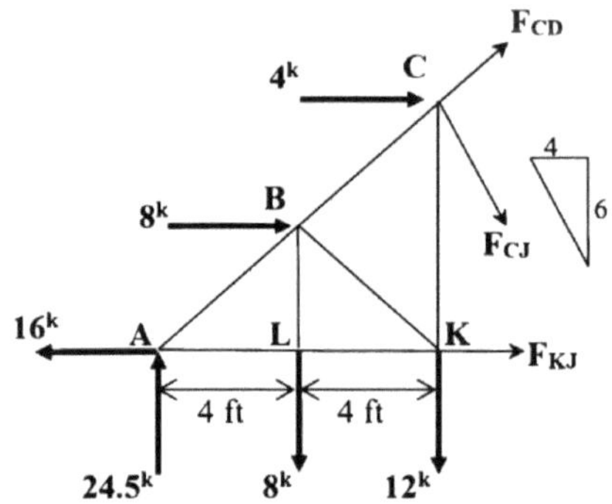

Note that if the strategy of taking moments about A is seen ahead of time, then it is obvious that the reaction at A will not be necessary and therefore one can avoid that calculation.

Answer is D.

Load Types

Loads acting on structures are broadly classified into vertical loads and lateral loads. Vertical, or gravity loads include dead loads, live loads, snow loads, etc., while lateral loads include wind and earthquake loads.

The most common loads that are encountered in structural systems are

1. Dead load, which comprises all loads that are fixed in location and magnitude.
2. Live loads, which arise out of the occupancy of the structure. Live loads may be considered to be partially applied, if that creates the most critical effect on the structure.
3. Environmental loads such as wind, seismic, snow, rain, etc. These loads depend on the structural form as well as site characteristics such as geographical location, site topography, etc.

In the United States, ASCE-7 (Minimum Design Loads for Buildings and Other Structures) and the International Building Code (IBC) prescribe procedures for determining appropriate values for loads acting on buildings, whereas AASHTO (AASHTO LRFD Bridge Design Specifications) prescribes the procedures for bridges.

Dead Loads

Dead loads include all loads that are fixed in magnitude, location, and extent. They are calculated based on materials specifications, particularly the unit weight of the material.

Example 12.11

A building floor consists of the following components:

Stay in place form	3.5 lb/ft^2
Concrete ($\gamma = 140$ pcf) topping	3.5 in
Carpeting	2.0 lb/ft^2

What is the total dead load (lb/ft^2)?

Solution The dead load of the concrete topping: $\gamma h = 140 \times \frac{3.5}{12} = 40.8$ lb/ft^2. Total dead load = 3.5 + 40.8 + 2.0 = 46.3 lb/ft^2.

Example 12.12

An elevated water tank is supported by four tower legs as shown. The empty tank weighs 56,000 lb and the full tank weighs 300 kips. The resultant wind force of 50 kips acts at a height of 65 ft as shown. Each tower leg is supported by an isolated square footing. The maximum design uplift force for designing anchor bolts for each footing is most nearly

A. 40

B. 54

C. 94

D. 108

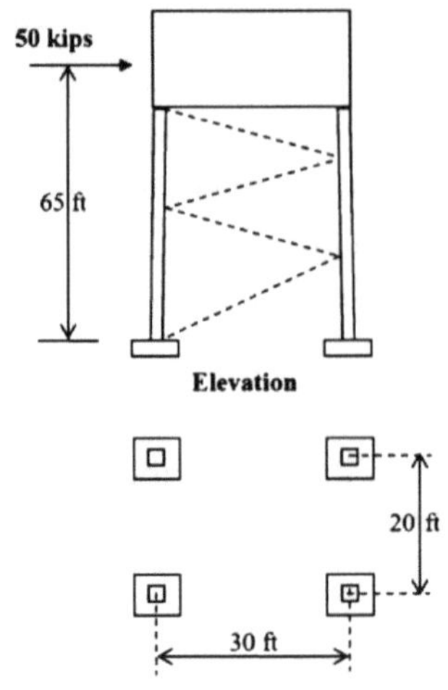

Solution Since gravity forces counteract uplift, we must consider the empty tank case here. The weight of 56 kips is carried equally by each leg. Therefore, the compression at each footing = 14 kips.

The lateral force creates an overturning moment = 50 kips × 65 ft = 3250 kip · ft on the horizontal plane at the top of the footings. This overturning moment is shared equally by two couples (four legs in two pairs). Therefore, each resisting couple = 1625 kip · ft.

Since the lever arm on each couple is 30 ft, the force at each footing = 1625 ÷ 30 = 54.2 kips. This is an added compression under the legs on the far side (right) of the tower and an uplift under the legs on the near side (left).

Therefore, the near side legs experience a "net" uplift of 54.2 – 14 = 40.2 kips.

Answer is A.

Example 12.13

A temporary warning sign is constructed at a worksite by using a 14-in diameter bucket filled with ballast so as to serve as a counterweight (see figure). If the maximum (3 s gust) wind pressure is 55 lb/ft^2 (ignore the wind pressure on the bucket), the minimum required weight (lbs) of the counterweight is most nearly:

A. 435

B. 565

C. 612

D. 659

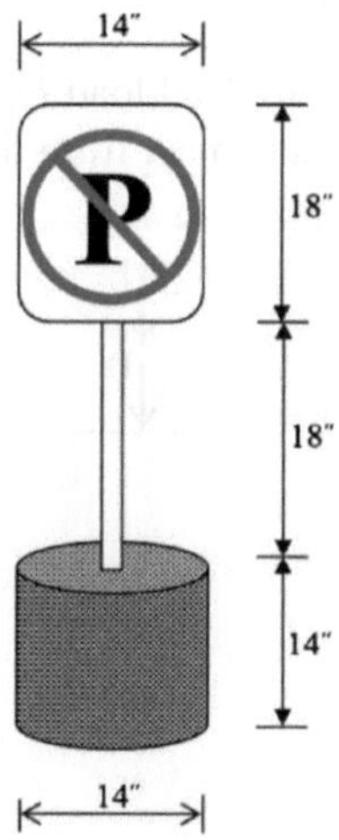

Solution The solution will assume that the critical condition is overturning due to wind gusts.

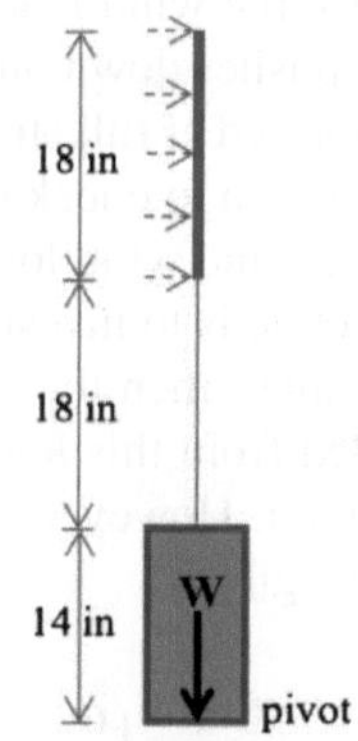

Resultant wind force on the sign = $14 \times 18 \times 55/144 = 96.25$ lb
Overturning moment about the tipping heel = $96.25 \times 41 = 3946.3$ lb·in
Stabilizing moment = $W_{bucket} \times 7$ in = 3946.3. Therefore, $W_{bucket} = 563.8$ lb.

Answer is B.

Live Loads

Unlike dead loads, live loads can be applied at the discretion of the designer, in a way that creates the worst effect on the component being designed. It must be understood that the live load placement for designing one element of a structure can be different from the load placement to create the critical effect in another.

Example 12.14

A single-story building frame is subject to dead load *D*, live load *L*, and wind load *W* as indicated. The wind load is reversible (i.e., it can act from the left or from the right).

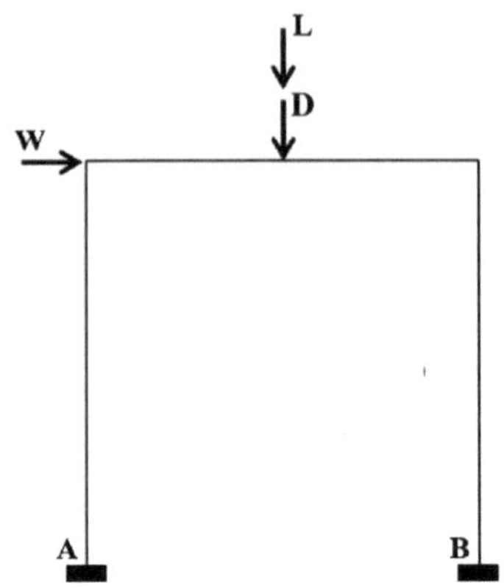

When designing the footings at A and B for bearing capacity of the soil, one must focus on the far side support (B as shown) because the wind load (as shown) causes a clockwise overturning moment on the structure, which pushes down on footing B and tends to cause uplift at footing A. In addition, the live load, if applied at full strength, will create additional compression at support B. Thus, the load combination to check the footing for bearing capacity of the soil is as shown above (with both live load and wind load acting at full strength).

On the other hand, if the design objective is to investigate the possibility of uplift at a support (nearside support A is most vulnerable), then the chosen load combination should be as shown below. The live load *L* is excluded from this load case, because it has the potential to counter the uplift caused by the wind load. However, we have no such discretion about the dead load, which must appear at full strength.

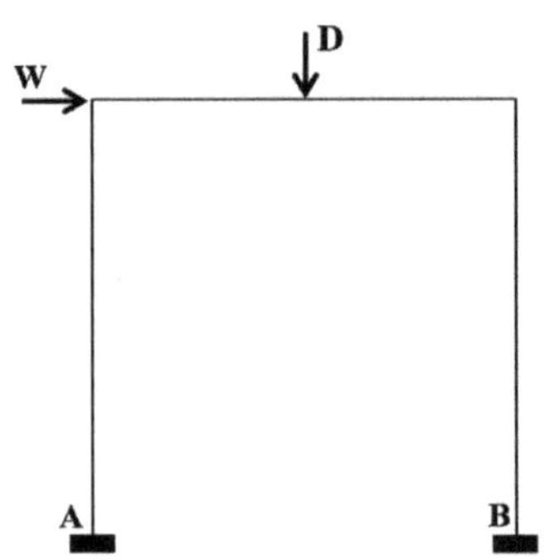

Live Load Reduction

Members for which a factored tributary area ($K_{LL}A_T$) is greater than 400 ft^2 are permitted to be designed for a reduced live load in accordance with Eq. (12.5)

$$L_{\text{reduced}} = L_{\text{nominal}}\left(0.25+\frac{15}{\sqrt{K_{LL}A_T}}\right) \geq 0.4L_{\text{nominal}} \tag{12.5}$$

where L_{reduced} = reduced design live load per ft^2 of area supported by the member
L_{nominal} = nominal live load given in ASCE 7
K_{LL} = live load element factor = 4 (typical columns) or 2 (typical beams and girders)
A_T = tributary area in (ft^2) supported by the member

The reduced live load L shall not be less than $0.40L_o$ for members supporting two or more floors.

Example 12.15

The nominal live load for a four-story commercial building is 80 lb/ft^2. The vertical load resisting system consists of flat plate floor slabs supported by a grid of columns spaced 25 ft in one plan direction and 20 ft in the other plan direction. What is the service live load for a first floor interior column, reduced according to ASCE-7 guidelines? Ignore roof live loads.

A. 48.0 kips

B. 53.3 kips

C. 60.0 kips

D. 120 kips

Solution For a first floor column, load accumulates from 3 stories (2nd – 4th floors). The total area tributary to an interior column on this level = 3 × 25 × 20 = 1500 ft^2.

According to ASCE-7 guidelines, the nominal live load can be reduced to

$$L_{\text{reduced}} = L_{\text{nominal}}\left(0.25+\frac{15}{\sqrt{K_{LL}A_T}}\right) = 80\left(0.25+\frac{15}{\sqrt{4\times1500}}\right) = 35.5 \text{ lb/ft}^2$$

This reduced live load is subject to a lower limit of $0.4L_{\text{nominal}}$ = 32 lb/ft^2.

Therefore, the service live load used for the design of this column should be 35.5 lb/ft^2.

Total design load for the column = 35.5 × 1500 = 53,250 lb = 53.3 kips.

Answer is B.

Snow Loads

Ground snow loads for the contiguous United States are shown in Fig. 7-1 of ASCE7 and for Alaska in Table 7-1. Section 7.3 of ASCE7 outlines the flat roof snow load for roofs sloped at or less than 5° (1 in/ft = 4.76°).

Flat Roof Snow Load

The flat roof snow load p_f is given by Eq. (12.6)

$$p_f = 0.7C_eC_tI_sp_g \tag{12.6}$$

where $p_g \le 20$ lb/ft², the design value of p_f is subject to a minimum value $p_f = I_sp_g$.

Where $p_g > 20$ lb/ft², the design value of p_f is subject to a minimum value $p_f = 20I_s$.

The exposure factor (C_e) is a function of terrain and roof exposure.

The thermal factor (C_t) is a function of the roof's thermal condition.

Sloped Roof Snow Load

Snow loads on a sloping surface shall be assumed to act on the horizontal projection of that surface. The balanced snow load on a sloped roof is calculated by multiplying the flat-roof snow load p_f by a roof slope factor C_s.

$$p_s = C_sp_f \tag{12.7}$$

Wind Loads

The basic formula for converting the design wind speed V (from maps) to the velocity pressure q_h is

$$q_h = 0.00256K_hK_{zt}K_dV^2 \tag{12.8}$$

where q_h = velocity pressure (lb/ft²)
K_h = exposure coefficient
K_{zt} = topographic factor
K_d = directional factor
V = design wind speed (mi/h)

Seismic Loads

Earthquake-induced loads are different from other load types in the sense that they are *inertial*. The load experience by a structure exists simply because the structure's mass gives it inertia, which tends to oppose imposed motion, such as due to ground accelerations.

The base shear V experienced by a structure due to ground accelerations is expressed as

$$V = C_sW \tag{12.9}$$

where C_s is a seismic coefficient and W is seismic weight of the structure, which includes the tributary dead load plus any other gravity loads that may be reasonably expected to be acting at the time of the design earthquake.

The seismic coefficient C_s is a function of the geographic location, site characteristics, structure type, and structure flexibility (expressed as the fundamental period T). A very stiff structure has a low value of T, whereas a very flexible structure has a high value of T.

The fundamental period of the structure is the most important characteristic that determines how much energy the structure "extracts" from the design earthquake.

Moving Loads: Influence Lines

An influence line is a plot of the internal force response of a beam (bending moment or shear force) as a function of the location of a single transverse point load on the beam.

Principle of Müller-Breslau

Derived from the principle of virtual work, the Müller-Breslau principle states: "The influence line of a given function (reaction, shear, or bending moment) is proportional to the deflected shape of the beam when the internal force in question is released and a unit displacement or rotation is applied in its place."

In short, the three steps to apply this principle are

1. Release the constraint corresponding to the quantity whose influence diagram is sought.
2. Apply a unit displacement corresponding to the released constraint.
3. Visualize the displaced shape of the beam. This is the shape of the influence diagram. In some cases, such as the influence diagram for bending moment, the diagram may require scaling.

NOTE For determinate structures, the influence diagram is always composed of straight lines. In other words, applying the unit displacement (step 2 above) is accommodated by the structure such that different parts of the structure deflect as rigid bodies (no bending). The same principle and the outlined procedure is also valid for indeterminate structures, the only difference being that the resulting influence diagram may be composed of nonlinear curves.

Example 12.16

For the beam shown below, what is the influence line diagram for the vertical reaction at the fixed support A?

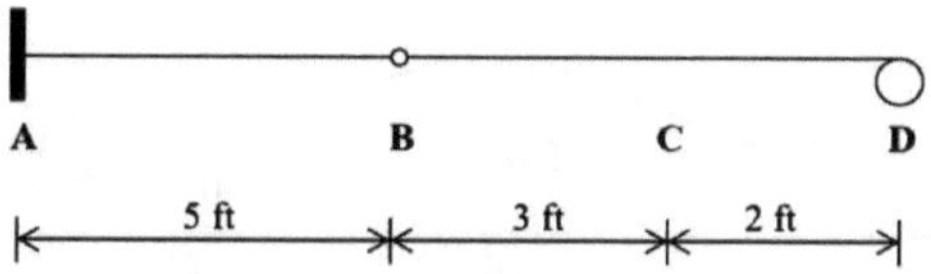

Solution The support at A provides three independent constraints to the beam—it restricts it from translating horizontally, it restricts it from translating vertically and it restricts it from rotating. By the Müller-Breslau principle, we will first remove the vertical constraint while maintaining the other two constraints, and then apply a unit upward deflection to the beam at A.

This deflection must not be accompanied by any horizontal translation or rotation of the beam at A (since we are supposed to release *only* the vertical restraint).

The segment AB has to remain horizontal so that there is no rotation at A. The beam accommodates such a deflected shape without any bending because of the presence of the internal hinge at B, which allows the segment AB to remain horizontal while rotating segment BC in a rigid body mode (without bending). The resulting deflected shape is given by the broken line in the figure below.

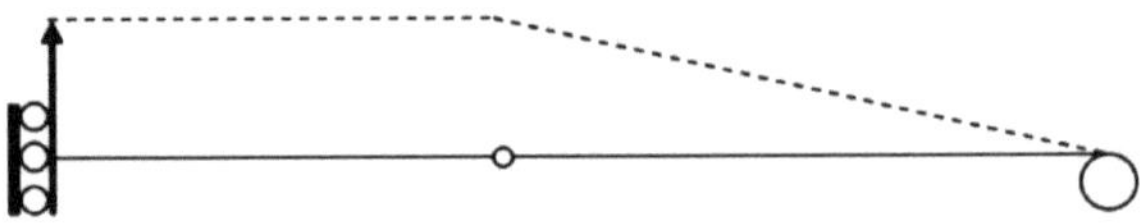

Thus, the *shape* of the influence line diagram for vertical reaction at A is as shown above. Noting that when the unit load is directly above the support A, the vertical reaction at A is equal to 1.0, the scaling factor for the above diagram is 1.0/1.0 = 1.0

Example 12.17

For the beam in Example 12.16, what is the influence line diagram for the reaction moment at the fixed support A?

Solution For this function (reaction moment at A), the beam must be *released* from the rotational constraint while maintaining the other constraints—horizontal and vertical translation. The way to achieve this is to create a hinge at A and apply a unit rotation to it. This action is shown in the figure below. The deflected shape of the beam due to this (rotational) displacement is shown as a broken line.

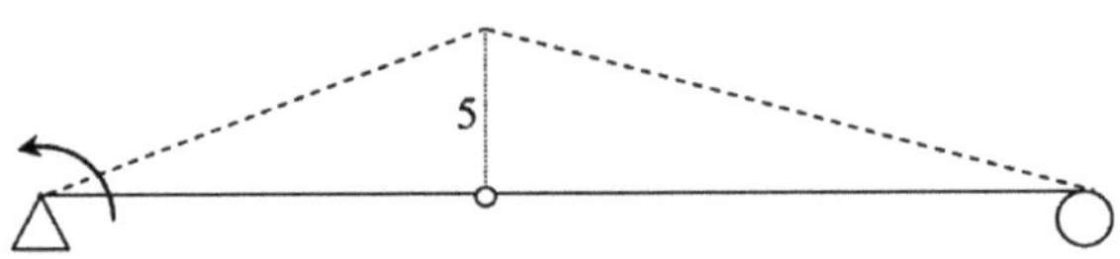

When the (unit downward) load is applied at hinge B on the original structure, the support at C does not participate in carrying this load and the beam behaves just like a cantilever AB. Therefore, the reaction moment at A is +5 (counterclockwise) when the unit load is at the hinge B. Thus, the scaling factor for the influence diagram is given by +5/+5 = 1.0. As a result, the scaled influence line diagram for the reaction moment at A is given by

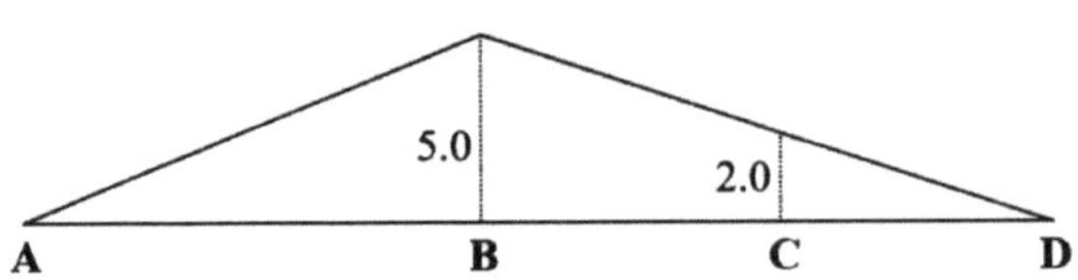

Example 12.18

For the beam in Example 12.16, what is the influence line diagram for the internal shear at point?

Solution For this function (shear at C), the beam must be *released* from the internal restraint against shear. This effect is shown in the following figure. A unit shear is then applied to point C (where the beam has been "split open" and therefore released from internal shear resistance).

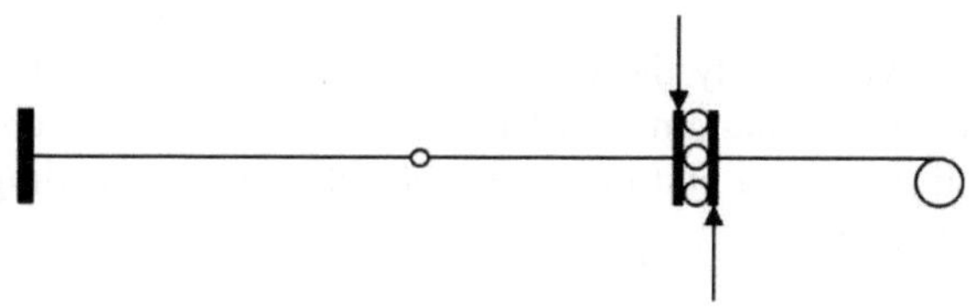

Under the combined action of the force pair, the beam will deflect down on the left of point C, and deflect up on the right of point C. The total relative deflection at C must be equal to 1.0. The theoretical derivation of the equation for the influence diagram for shear indicates that the slopes of the two sides of the influence line diagram are to be equal.

Thus, the total deformation of 1.0 is shared in proportion to the spans BC and CD. In the above case, BC:CD = 3:2. Thus, the deflections at C must be 0.6 and 0.4. Applying these deformations to point C and allowing the rest of the beam to deflect compatibly, we obtain the scaled influence line diagram for shear at C.

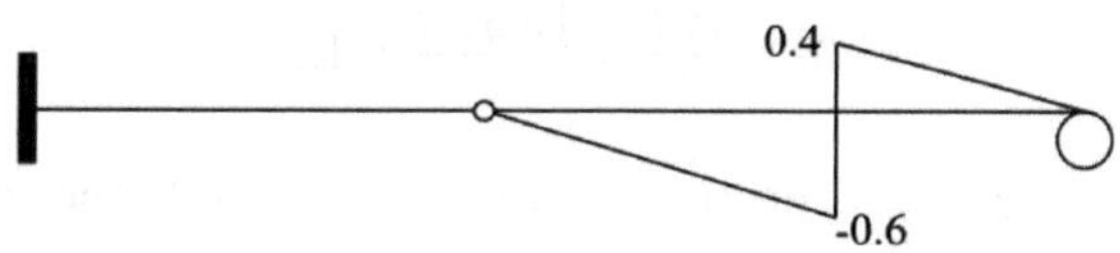

Note that because of the rotational resistance at support A and the ability of the hinge B to allow rotation, the segment AB does not deflect as a result of the deflection of segment BC. This implies that at point C, there is no internal shear produced if the load is on span AB (since the influence diagram has a zero ordinate over AB).

Example 12.19

For the beam in Example 12.16, what is the influence line diagram for the bending moment at C?

Solution In order to apply the Müller-Breslau principle for bending moment at C, the internal point C must be released from the internal restraint against bending without affecting the internal restraint against axial or shear deformation. This can be accomplished by inserting an internal hinge at C. The internal hinge has no resistance to bending. However, it can fully resist axial deformation as well as transverse (shear) deformation. Once the hinge has been inserted, a pair of rotations representing a unit positive bending moment should be applied on either side of C. This is shown below. The deflected shape of the beam due to this imposed displacement is shown using dashed lines.

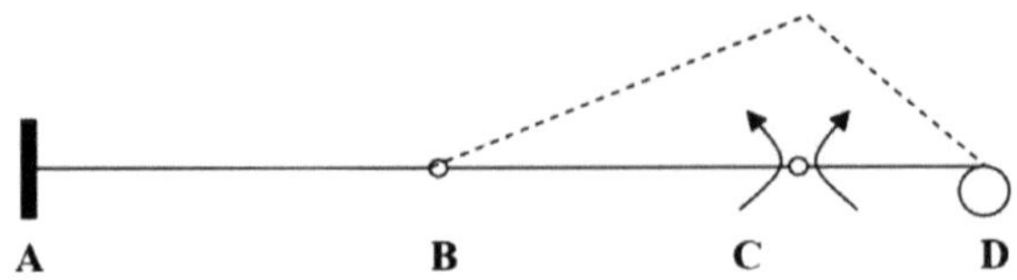

The principle above yields only the *shape* of the diagram. We still need to calculate the *value* of any one point on the diagram in order to *scale* it. The obvious choice for such a point above is point C. Thus, placing a unit load at C, we calculate the bending moment at C. This specific case is shown below.

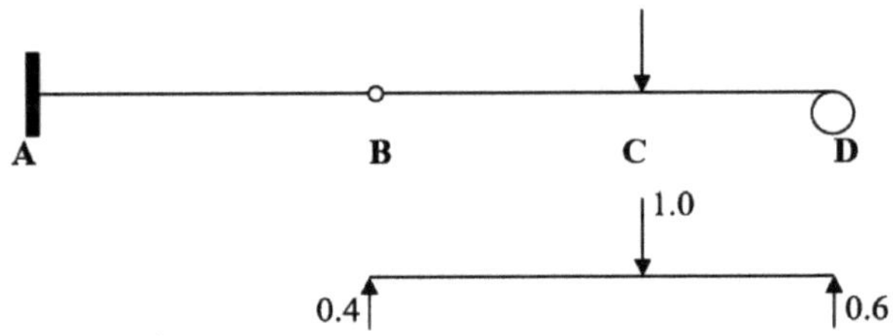

The bending moment under a load *P* on a simple span is given by *Pab*/*L*. This leads to

$$M = \frac{Pab}{L} = \frac{1 \times 3 \times 2}{5} = 1.2$$

Thus, the bending moment at C due to a unit load at C is 1.2. The influence line diagram, scaled accordingly, is shown below.

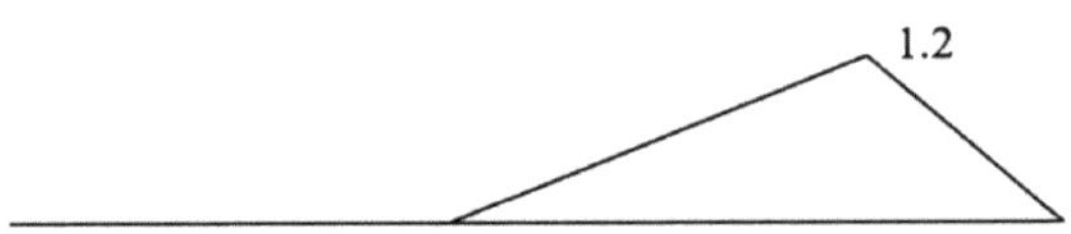

Example 12.20

The beam shown below is a propped cantilever with an internal hinge at B. The loads are

Uniformly distributed dead load = 2 kips/ft

Uniformly distributed live load = 1.4 kips/ft

Single concentrated live load = 10 kips

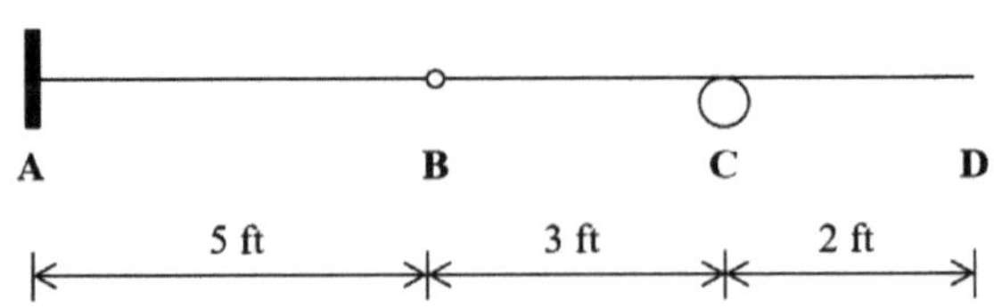

The maximum vertical reaction at A is most nearly

A. 21.8 kips

B. 28.6 kips

C. 30.8 kips

D. 32.0 kips

Solution By the Müller-Breslau principle, first remove the vertical constraint while maintaining the other two constraints, and then apply a unit upward deflection to the beam at A (not be accompanied by any horizontal translation or rotation of the beam at A). The resulting deflected shape is given by the broken line in the figure below.

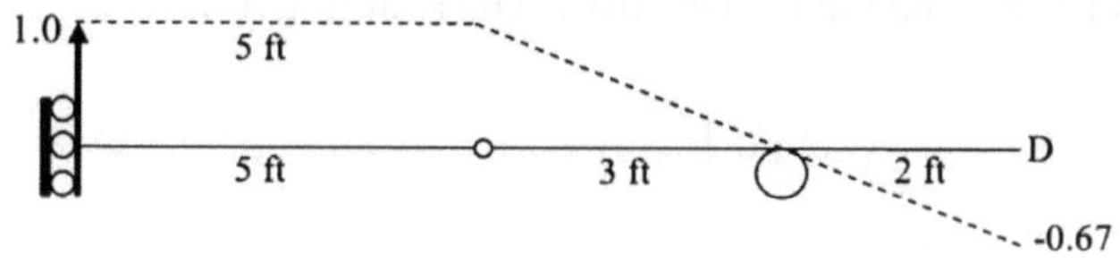

Thus, the *shape* of the influence line diagram for vertical reaction at A is as shown above. When the unit load is directly above the support A, the vertical reaction at A is equal to 1.0, the scaling factor for the above diagram is 1.0/1.0 = 1.0. By similar triangles, the value at $D = -0.67$.

Placing the dead load over the entire beam (no choice), live load over AC (to maximize positive contribution), and the point load anywhere along AB, the total vertical reaction A_y is

$$A_y = 2\times\left(1\times\frac{5+8}{2}-\frac{1}{2}\times2\times0.67\right)+1.4\times\left(1\times\frac{5+8}{2}\right)+10\times1=30.77 \text{ kips}$$

Example 12.21

Calculate the maximum possible uplift reaction due to live load at support A for the beam shown below. The live load is uniformly distributed $w = 4$ kips/ft. Partial live load placement is permitted. Internal hinges exist at B and D.

A. 7 kips

B. 12 kips

C. 14 kips

D. 22 kips

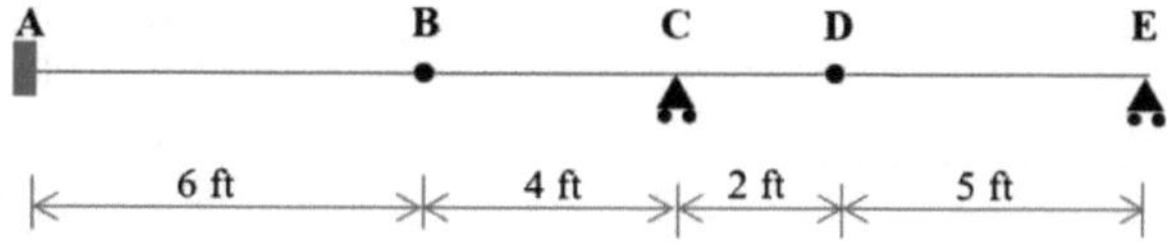

Solution Using the principle of Müller-Breslau (release vertical constraint at A, apply unit displacement, draw displaced shape), the influence diagram for vertical reaction at A (A_y) is shown the figure below.

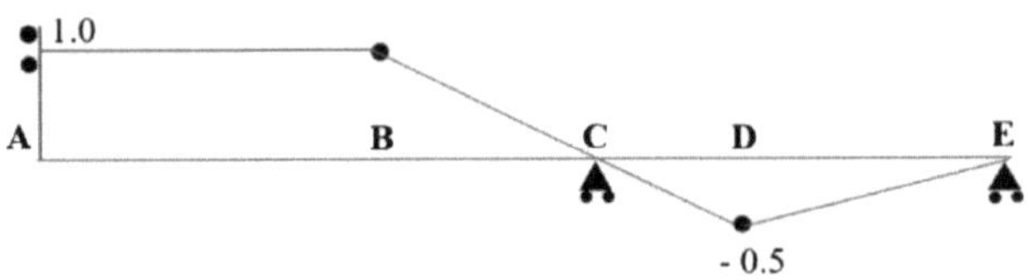

Therefore, to maximize uplift (negative reaction) at A, one must place the live load over CDE (where the influence diagram is negative). The resulting A_y is

$$A_y = 4 \times \frac{1}{2} \times -0.5 \times 7 = -7 \text{ kips}$$

Influence of a Series of Concentrated Loads

Highway and railway bridges are designed to carry train and truck loads. The design vehicle, which forms the basis for design, is a sequence of defined wheel loads with specific spacing between wheels. For example, the AASHTO HS-20 vehicle is shown in Fig. 12.5.

Absolute Maximum Shear

To determine the absolute maximum shear, one must figure out the location where the maximum shear is going to occur (for any load placement) and then determine the load placement that will create the maximum shear there. For complex structural systems, this is accomplished by allowing the load to travel to various locations on the structure, analyzing the structure for each load condition and then creating an "envelope" of analysis results. This, of course, is very time consuming and is facilitated by software.

However, for simpler structures, such as the simply supported beam shown in Example 12.22, the critical location can be determined fairly easily. This is then followed by placing the load according to the shape of the influence diagram at that location.

Example 12.22

A three axle truck with axle loads of 3, 8, and 10 kips traverses a simply supported bridge (span = 64 ft). What is the absolute maximum shear force (kips) in the bridge deck?

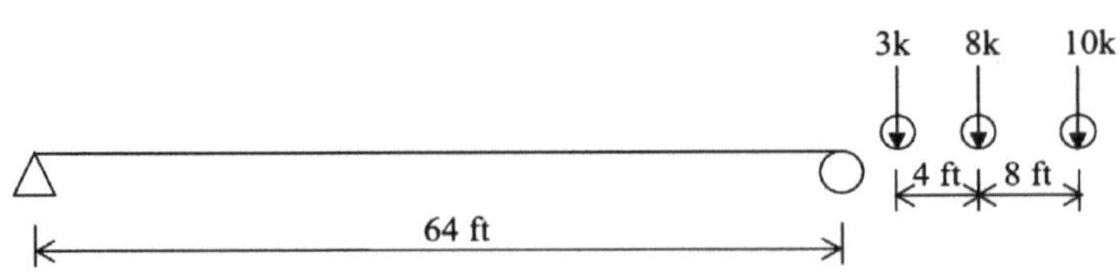

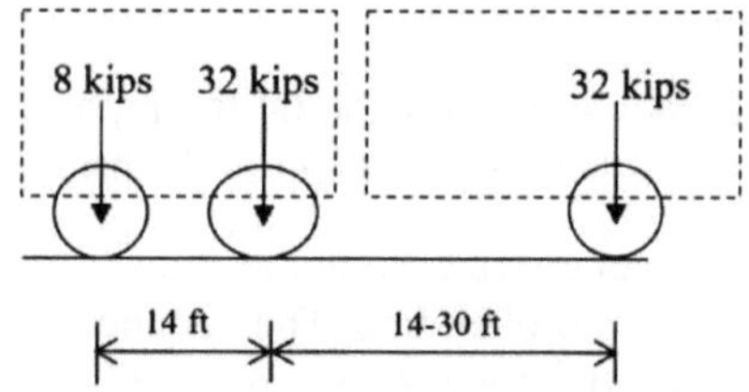

Figure 12.5 AASHTO standard vehicle (HS20).

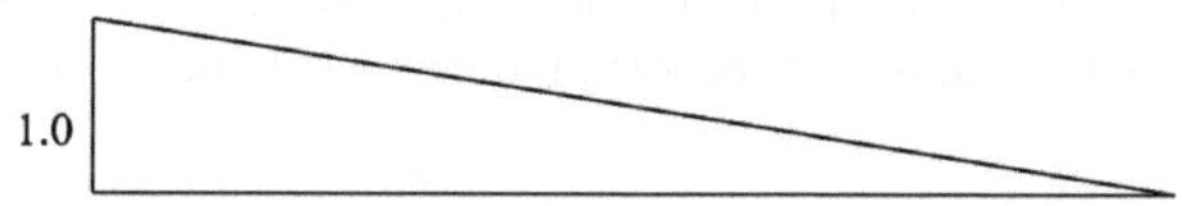

Figure 12.6 Influence line for vertical reaction at A.

Solution The absolute maximum shear in the beam will occur by maximizing the reaction at one of the supports. This can be accomplished by recognizing that the influence diagram for reaction (at A, for example) is shown in Fig. 12.6.

The reaction at A can therefore be maximized by placing the heaviest part of the vehicle (rear axle load 10 kips) over A and the rest of the vehicle ahead of it—in other words, a vehicle moving from left to right with the rear axle having just entered the bridge. This load position is shown below.

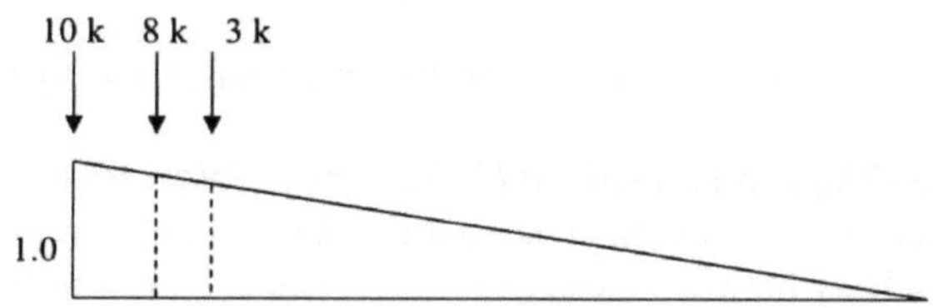

Once the ordinates have been found by similar triangles as 56/64 (under the 8-kip load) and 52/64 (under the 3-kip load), the absolute maximum shear can be found as

$$V_{max} = 1.0 \times 10 + \frac{56}{64} \times 8 + \frac{52}{64} \times 3 = 19.44 \text{ kip}$$

Absolute Maximum Bending Moment

Even for a simple structure such as a simply supported beam, the identification of the critical location for the absolute maximum bending moment is not trivial, as it was for the absolute maximum shear force (see above). The following "rule" can be established analytically.

RULE For a simply supported beam, the absolute maximum bending moment produced by a set of "*n*" moving loads occurs when the resultant of the load group and an adjacent load are equidistant from the midpoint of the beam. Since the resultant will be usually located between two of the loads, two possible load set positions must be considered, one for each adjacent load.

If the resultant is located exactly at one of the loads, then only one scenario needs to be considered, placing that load at the beam centerline.

Example 12.23

For the three-axle truck traversing the 64 ft simple span bridge (see Example 12.22), what is the absolute maximum bending moment in the bridge deck?

Solution The wheel group used in the above example is shown below. The resultant is the 21-kip force shown by a broken line. The distance of the resultant from the left (3-kip) wheel is 7.238 ft. In other words, if we were to place this wheel group on the simply supported beam (span = 64 ft) such that the midpoint of the beam ($x = 32$) were halfway between a wheel and the resultant, the maximum bending moment would occur under that wheel.

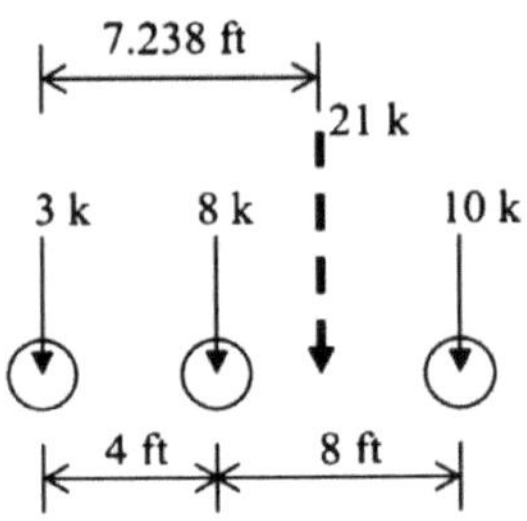

$$\bar{x} = \frac{3\times 0+8\times 4+10\times 12}{3+8+10} = \frac{152}{21} = 7.238$$

Since the resultant is between the 8-kip and the 10-kip wheels, let us try two different scenarios.

Scenario 1: Wheel 2 (8 kips) and resultant (21 kips) equidistant from beam midspan

The distance between wheel 2 and the resultant is 3.238 ft. Since we want to locate the beam midpoint ($x = 32$) midway between these two forces, wheel 2 must be located at $x = 32 - 0.5 \times 3.238 = 30.381$ ft and the resultant at $x = 32 + 0.5 \times 3.238 = 33.619$ ft. This loading is shown in the figure below. Under this loading, the maximum bending moment should occur directly under wheel 2 (at $x = 30.381$ ft).

Thus, the influence line for the bending moment at $x = 30.381$ is drawn below (Fig. 12.7) and the load placed such that wheel 2 is at $x = 30.381$. (This places wheel 1 at $x = 26.381$ ft and wheel 3 at $x = 38.381$ ft.)

Maximum bending moment (due to this wheel placement)

$$= 3 \times 13.86 + 8 \times 15.96 + 10 \times 12.16 = 290.86 \text{ kip} \cdot \text{ft}$$

Thus, the maximum bending moment of $M = 290.86$ kip · ft occurs at $x = 30.381$ when wheel no. 2 is located at that location.

Scenario 2: Wheel 3 (10 kips) and resultant (21 kips) equidistant from beam midspan

The distance between wheel 3 and the resultant is 4.762 ft. Since we want to locate the beam midpoint ($x = 32$) midway between these two forces, the resultant must be located at $x = 32 - 0.5 \times 4.762 = 29.619$ ft and wheel 3 at $x = 32 + 0.5 \times 4.762 = 34.381$ ft. This loading is shown in the figure below. Under this loading, the maximum bending moment should occur directly under wheel 3 (at $x = 34.381$ ft).

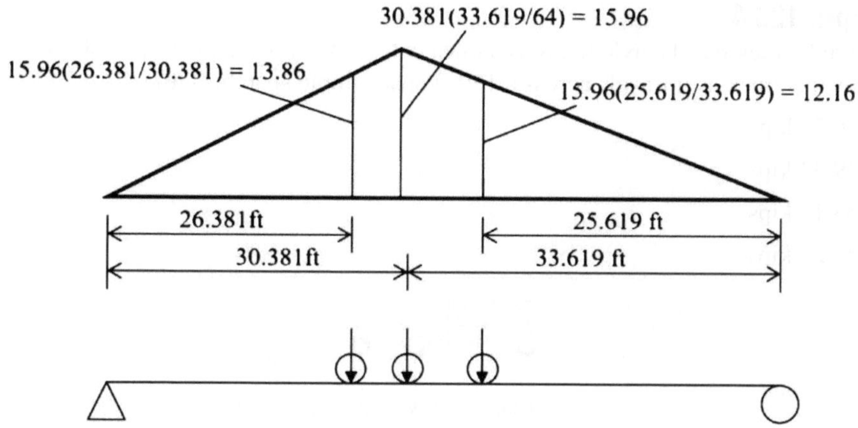

Figure 12.7 Influence line for bending moment under wheel 2.

Thus, the influence line for the bending moment at $x = 34.381$ is drawn below (Fig. 12.8) and the load placed such that wheel 3 is at $x = 34.381$. (This places wheel 1 at $x = 22.381$ ft and wheel 2 at $x = 26.381$ ft.)

Maximum bending moment (due to this wheel placement)

$$= 3 \times 10.36 + 8 \times 12.21 + 10 \times 15.91 = 287.86 \text{ kip} \cdot \text{ft}$$

Thus, the maximum bending moment of $M = 287.86$ kip · ft occurs at $x = 30.381$ when wheel no. 2 is located at that location. This is smaller than the maximum bending moment for scneario 1.

The absolute maximum bending moment in the beam is therefore $M_{max} = 290.86$ kip · ft at $x = 30.381$ ft.

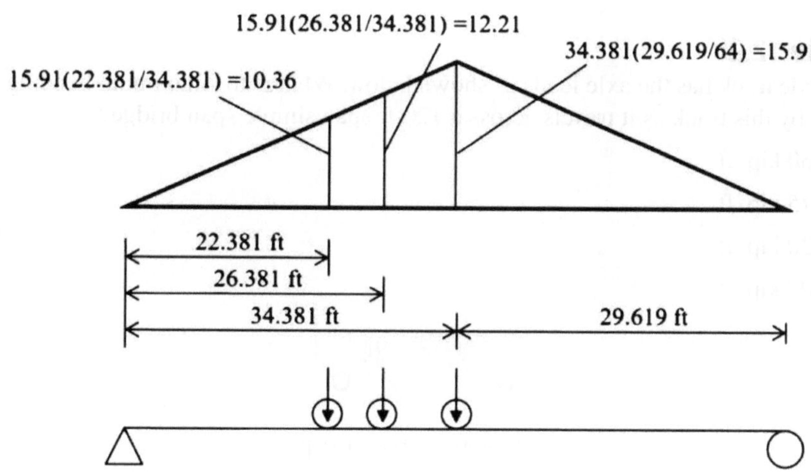

Figure 12.8 Influence line for bending moment under wheel 3.

Example 12.24

A three-axle truck has the axle loads as shown below. What is the maximum shear force produced by this truck as it travels across a 120-ft span simple span bridge?

A. 46.50 kips

B. 49.33 kips

C. 53.17 kips

D. 58.22 kips

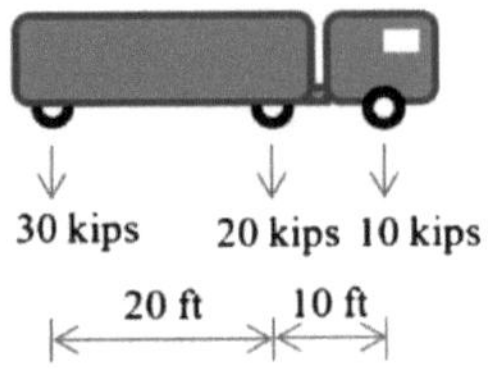

Solution Positioning the vehicle such that the heaviest wheel is immediately adjacent to a support creates the maximum reaction and therefore maximum shear force in the beam.

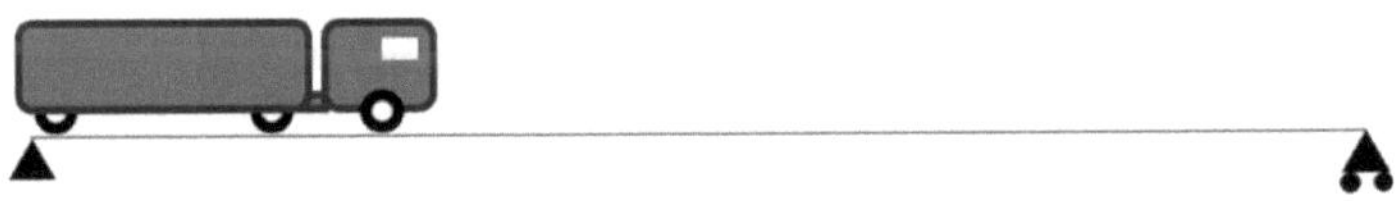

The reaction at the left support:

$$A_y = 30 + \frac{20 \times 100}{120} + \frac{10 \times 90}{120} = 54.17 \text{ kips}$$

Answer is C.

Example 12.25

A three-axle truck has the axle loads as shown below. What is the maximum bending moment produced by this truck as it travels across a 120-ft span simple span bridge?

A. 1460 kip · ft

B. 1575 kip · ft

C. 1720 kip · ft

D. 1800 kip · ft

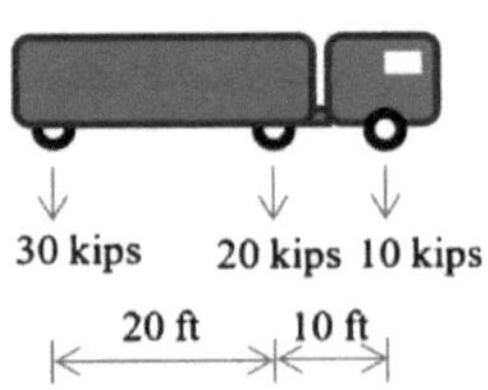

Solution The centroid of the wheel group is located at a distance x (from the rear axle) given by

$$\bar{x} = \frac{20 \times 20 + 10 \times 30}{60} = 11.67 \text{ ft}$$

Positioning the vehicle such that the distance (8.33 ft) between wheel 2 and this resultant is bisected by the midpoint ($x = 60$) of the beam, wheel 2 is located at $x = 64.17$ ft, wheel 3 at $x = 44.17$ ft and wheel 1 at $x = 74.17$ ft. This is shown below.

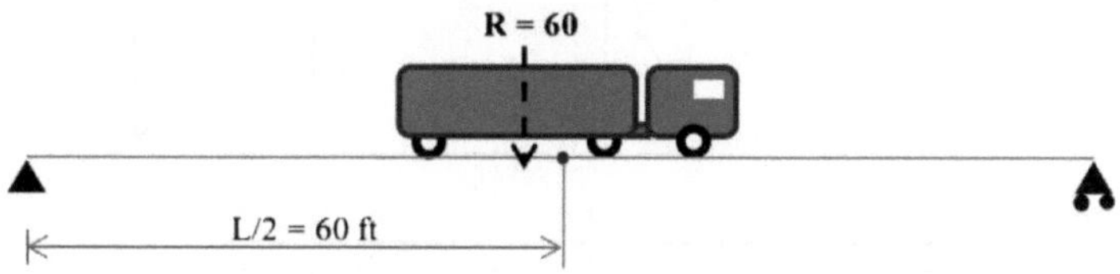

For this load placement, the reaction at left support is calculated as

$$R_1 = \frac{30 \times 75.83}{120} + \frac{20 \times 55.83}{120} + \frac{10 \times 45.83}{120} = 32.08 \text{ kips}$$

The shear diagram is shown below:

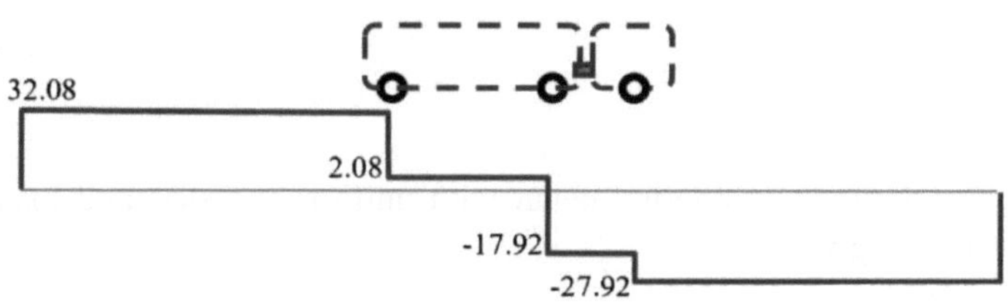

For this load placement, the maximum bending moment (under wheel 2) is calculated as

$$M_{max} = 32.08 \times 44.17 + 2.08 \times 20 = 1458.6 \text{ kip·ft}$$

Answer is A.

Deflection

The topic of deflection of beams is discussed in detail in Chap. 8. The methods outlined in that chapter are (1) integration of the elastic curve and the (2) unit load method.

Pages 85 to 86 of the *FE Reference Handbook,* version 9.4, list results (functions for slope and deflection as well as expressions for maximum slope and deflection) for six load cases for simply supported beams (page 85) and five standard load cases for a cantilever beam (page 86). In addition, on page 86, the results for the fixed-fixed beam are also listed.

Example 12.26

A system of regularly spaced rectangular timber beams supports a floor as shown. Assume modulus of elasticity of the timber = 1.5×10^6 psi. Beams are 2×10 in (1.5×9.25 in dressed size). Beams have a simple span = 18 ft.

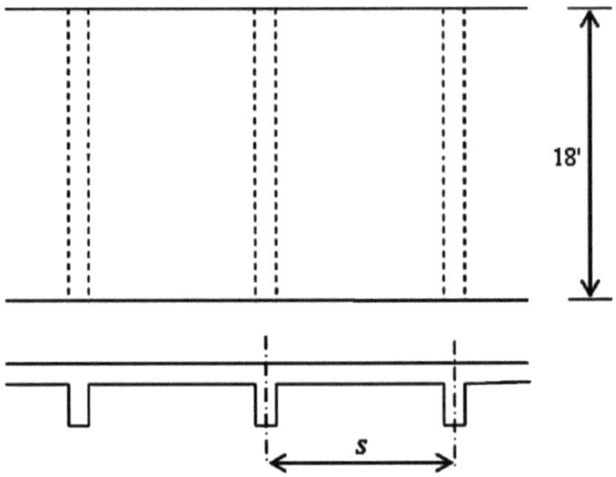

The floor load is 55 lb/ft². If the maximum deflection of the beams is L/240, the allowable spacing between beams is most nearly

A. 2.5 ft

B. 2.0 ft

C. 1.5 ft

D. 1.0 ft

Solution For a simply supported beam with uniformly distributed load, the maximum (elastic) deflection is given by

$$\Delta_{max} = \frac{5wL^4}{384EI}$$

Therefore,

$$\Delta_{max} = \frac{5wL^4}{384EI} \leq \frac{L}{240} \Rightarrow w \leq \frac{384EI}{1200L^3}$$

Moment of inertia of the dressed timber beam:

$$I = \frac{1}{12}\text{bh}^3 = \frac{1.5 \times 9.25^3}{12} = 98.93 \text{ in}^4$$

$$w \leq \frac{384EI}{1200L^3} = \frac{384 \times 1.5 \times 10^6 \times 98.93}{1200 \times 216^3} = 4.71\frac{\text{lb}}{\text{in}} = 56.54\frac{\text{lb}}{\text{ft}}$$

Maximum spacing $S = 56.54$ lb/ft ÷ 55 lb/ft² = 1.03 ft. Use joists at 12 in spacing.

Answer is D.

Truss Deflection

Applying the unit load method to trusses gives the following equation for the deflection of a joint of a truss subjected to external loading

$$\Delta = \sum f_i \Delta L_i = \sum f_i \left(\frac{FL}{AE}\right)_i = \sum \left(\frac{fFL}{AE}\right)_i \tag{12.10}$$

where f_i = force in member i due to the unit load
F_i = force in member i due to external loading
L_i, A_i, E_i = length, cross-sectional area, and modulus of elasticity of member i

For a truss subjected to thermal deformations, the term ΔL_i is replaced with $(\alpha L \Delta T)_i$, resulting in

$$\Delta = \sum f_i \Delta L_i = \sum f_i (\alpha L \Delta T)_i \tag{12.11}$$

where f_i is the force in member i due to the unit load and L_i, α_i, ΔT_i is the length, coefficient of thermal expansion, and temperature change of member i.

For a truss where one or more members are of imperfect length, the term ΔL_i is replaced with the misfit ε_i (positive if too long, negative if too short), resulting in

$$\Delta = \sum f_i \varepsilon_i \tag{12.12}$$

where f_i is the force in member i due to the unit load and ε_i is the lack of fit of member i.

Example 12.27

A two-member truss is loaded with a vertical load of 20 kips at node B as shown. The cross sections of the members AB and BC are 2 in^2 for AB, 3 in^2 for BC. Modulus of elasticity E = 29,000 ksi for both members. What is the vertical deflection at B?

A. 0.1 in

B. 0.15 in

C. 0.2 in

D. 0.25 in

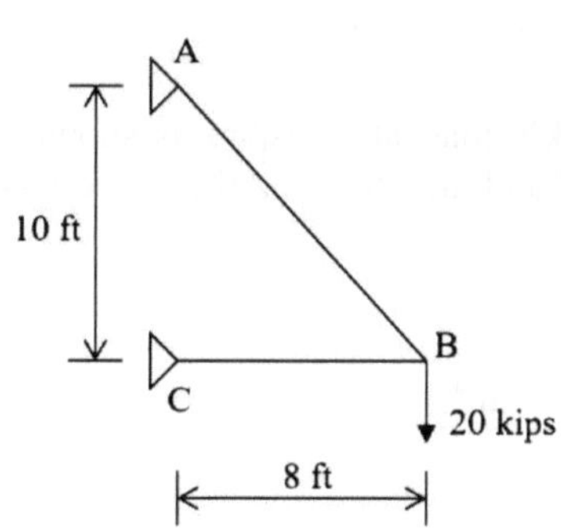

Solution Solve for truss member forces using equilibrium of joint B

$$F_{AB} = +25.6 \text{ kips (T)}$$
$$F_{BC} = -16.0 \text{ kips (C)}$$

Create the virtual load (consistent with desired deflection).

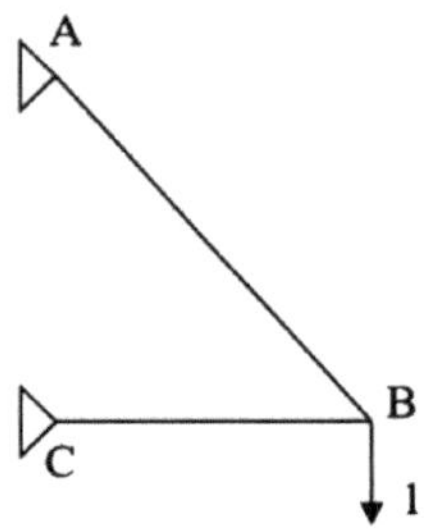

Calculate the member forces due to virtual load (note that in this case, the real load and the virtual load look similar, so we can use scaling)

$$f_{AB} = +1.28 \text{ (T)}$$
$$f_{BC} = -0.80 \text{ (C)}$$

$$\Delta = \sum \frac{FfL}{AE} = \frac{25.6 \times 1.28 \times (12.81 \times 12)}{2 \times 29{,}000} + \frac{-16 \times -0.8 \times (8 \times 12)}{3 \times 29{,}000} = 0.1 \text{ in}$$

NOTE The positive sign of the answer indicates that this deflection is in the same sense as the assumed unit load (downward).

Answer is A.

Example 12.28

A truss carries a vertical 50 kN load at midspan as shown. The member forces (tension positive) have been calculated and are shown. All members are made of steel and have the following properties:

Area = 3000 mm^2

Moment of inertia = 3×10^6 mm^4

What is the vertical deflection at the point of loading?

A. 5.7 mm

B. 8.3 mm

C. 11.1 mm

D. 15.8 mm

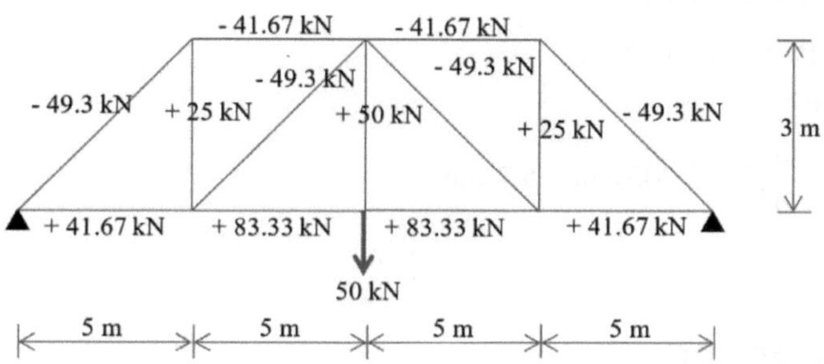

Solution According to the method of virtual work, the deflection at a point of truss is given by

$$\Delta = \sum \frac{FfL}{AE}$$

Since all the members have the same cross-sectional area *A* and modulus of elasticity *E*, this can be simplified to

$$\Delta = \frac{\sum FfL}{AE}$$

Since the question asks for vertical deflection at the load point, the unit load case a vertical load of 1 at the same point as the 50 kN load. Therefore, the force (f) produced by the unit load can be obtained by scaling.

There are six 5-m-long (horizontal) members, three 3-m-long (vertical) members, and four 5.83-m-long (diagonal) members.

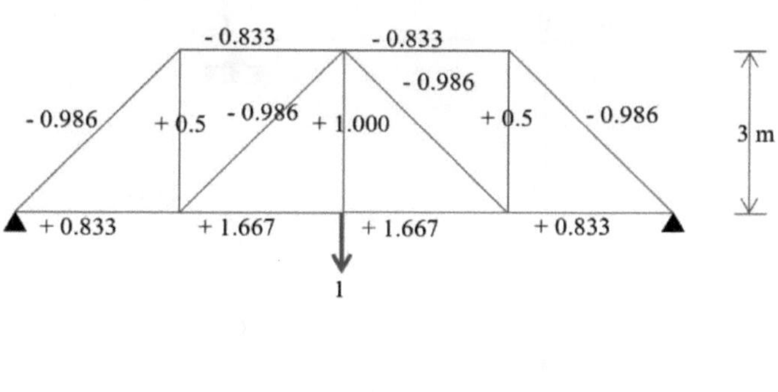

$$\Delta = \frac{\sum FfL}{AE}$$

$$= \frac{(-41.67 \times -0.833 \times 2 + 41.67 \times 0.833 \times 2 + 83.3 \times 1.667 \times 2) \times 5 + (25 \times 0.5 \times 2 + 50 \times 1) \times 3 + -49.3 \times -0.986 \times 4 \times 5.83}{AE}$$

$$= \frac{3442}{AE}$$

$$A = 3000 \text{ mm}^2 = 3 \times 10^{-6} \text{ m}^4$$

$$E = 200 \text{ GPa} = 2 \times 10^{11} \text{ N/m}^2$$

$$AE = 6 \times 10^5 \text{ Nm}^2$$

Deflection: $\Delta = \dfrac{3442}{AE} = 0.0057 \text{ m} = 5.7 \text{ mm}$

Answer is A.

Frame Deflection

For frame members, axial and bending deformations coexist. However, if axial deformations are considered to be small (strain energy dominated by bending strain energy), then applying the unit load method gives the following equation for the deflection of a joint of a frame subjected to external loading

$$\Delta = \sum \int_0^{L_i} \frac{m_i M_i}{EI_i} dx \qquad (12.13)$$

where m_i = moment function in member i due to the unit load
M_i = moment function in member i due to external loading
L_i, I_i = length and moment of inertia of member i

Example 12.29

For the two-member frame shown below, find the horizontal deflection at node B. The flexural rigidity $EI = 5.3 \times 10^7$ kip · in^2 for both members.

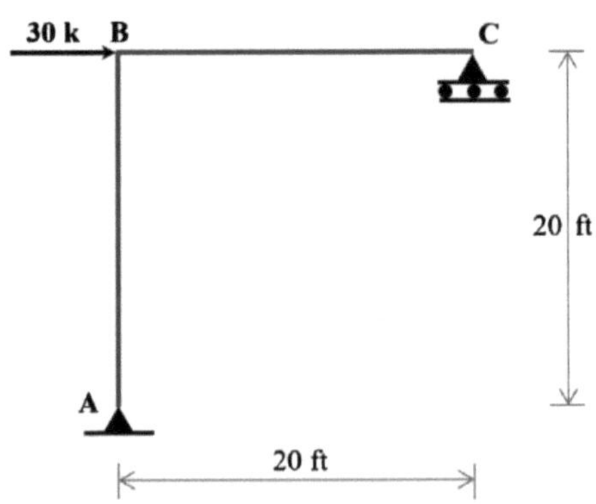

Solution Since the desired deflection is the horizontal deflection at B, the unit load is the conjugate horizontal load = 1 applied at node B. Since this looks "just like" the applied load, the bending moment for the unit load case will be a scaled down version of the bending moment for the external loads. The calculated reactions and shear and bending moment diagrams for the frame (under external loads) are shown below.

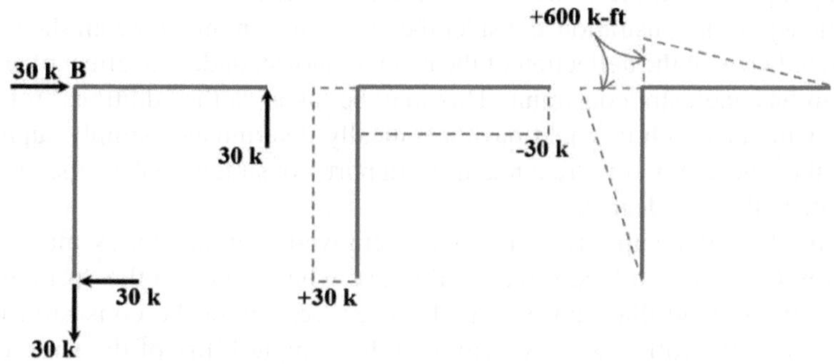

By scaling, the bending moment diagram for the unit load case is shown below. Note the difference in units (the unit load is a dimensionless load = 1, not 1 kip).

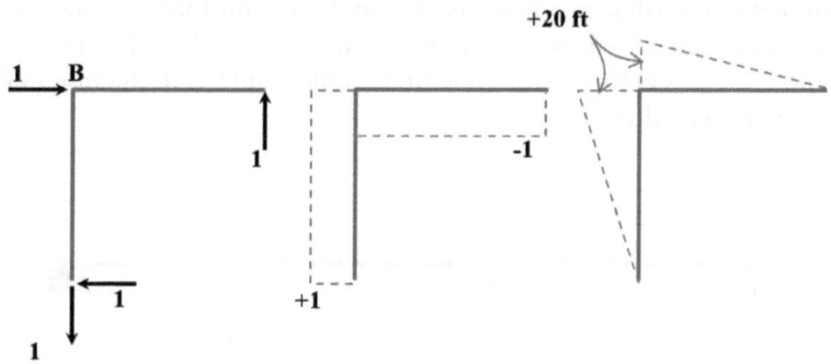

The product $\int_0^L mMdx$ of two triangular shapes is equal to 1/3 mML.

Therefore,

$$\Delta = \frac{1/3 \times 600 \text{ kip}\cdot\text{ft} \times 20 \text{ ft} \times 20 \text{ ft}}{EI} + \frac{1/3 \times 600 \text{ kip}\cdot\text{ft} \times 20 \text{ ft} \times 20 \text{ ft}}{EI}$$

$$= \frac{160{,}000}{EI} = \frac{1.6 \times 10^5}{\dfrac{5.3 \times 10^7}{144}} = 0.435 \text{ ft} = 5.22 \text{ in}$$

Introduction to Indeterminate Structural Analysis

Compatibility

Just as equilibrium equations express the idea of a *balance* of forces (and/or moments), compatibility equations express the idea of the *consistency* of deformations (displacements and/or rotations). As an illustration, consider the two-span continuous beam shown in Fig. 12.9. An exaggerated view of the deflection of the beam is shown, under the effect of arbitrary loads.

The beam has one extra constraint. This may be taken as the additional roller support in the middle, without which it would have a statically determinate "simply supported beam." Therefore, the one extra (or "redundant") support constraint makes the beam statically indeterminate to the first degree.

If the middle roller support were to be removed, but the loads maintained on the resulting simply supported beam, we would have what is termed the "primary structure," as shown in the second diagram of Fig. 12.9. By removing the constraint (vertical displacement = 0) at the roller, we have violated the compatibility of the structure, since the simply supported primary beam can have a nonzero displacement (Δ_p) at that point, unlike the original beam. The redundant force thus removed is then "put back" in the secondary structure, which is shown in the third diagram. According to the principle of superposition, the combination of the primary structure and the secondary structure must yield the original structure, as long as we use the correct vertical force B_y at the roller support. Since, we do not know what this force is, we apply a "unit load" to the secondary structure. The beam deflects (δ_s) under the action of the unit load. If the structure behaves in a linear fashion, the principle of superposition applies and the deflection under the actual redundant force B_y would be $B_y\delta_s$.

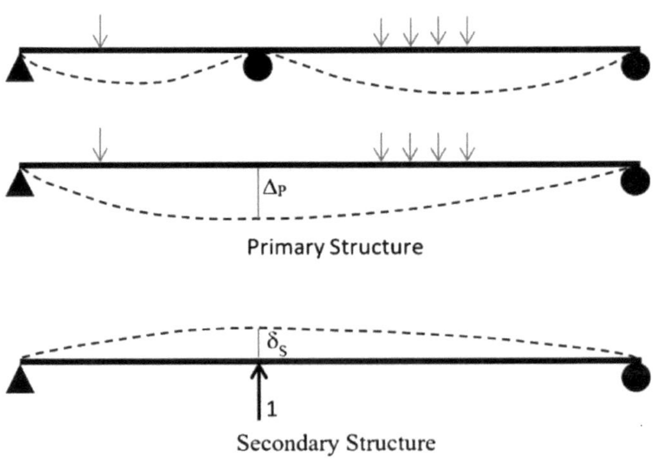

Figure 12.9 Method of consistent deformations.

Applying the compatibility relation at B,

$$\Delta_P + B_y\delta_S = 0 \Rightarrow B_y = -\frac{\Delta_P}{\delta_S} \qquad (12.14)$$

This approach is known as the principle of consistent deformations and is one of the approaches to solve indeterminate structures.

Indeterminate Trusses

For an indeterminate truss, the source of the indeterminacy can be external (due to too many external reactions) or internal (due to too many members). If the indeterminacy is external (too many reactions), then the solution is very similar to the method as applied to the structure in Fig. 12.9. The redundant supports are removed to obtain the primary structure and the redundant forces are applied one at a time. Enforcement of compatibility yields a sufficient number of equations to solve for the redundant forces.

On the other hand, if the indeterminacy is internal (due to too many members), then the redundant members are removed to obtain the primary structure. The redundant members forces are then inserted, one by one, in the secondary structures. Care must be taken to insert the member forces as a force pair, one acting at each end of, and in line with, the member removed.

Displacement Methods: Slope Deflection and Moment Distribution

A broad class of methods called displacement methods are based on the idea that connected members must deform in a compatible fashion and not independently. Each member has a bending stiffness which is a function of its flexural rigidity (*EI*) and its length (*L*). See Fig. 12.10.

In these methods, the first step is to "clamp" all joints. Once these joints are thus fixed, adjacent spans become independent from each other and develop "fixed-end-moments," which is simply a function of the span length and the nature and magnitude of the load. Page 152 of the *Handbook* lists fixed end moments for two load cases: (1) uniformly distributed load over the entire span and (2) a single concentrated load anywhere on the span.

In the second step, these joints (nodes) are then allowed to deform (bend). The moments developed at these nodes must be such that the resulting deformations (calculated using bending stiffnesses) are compatible.

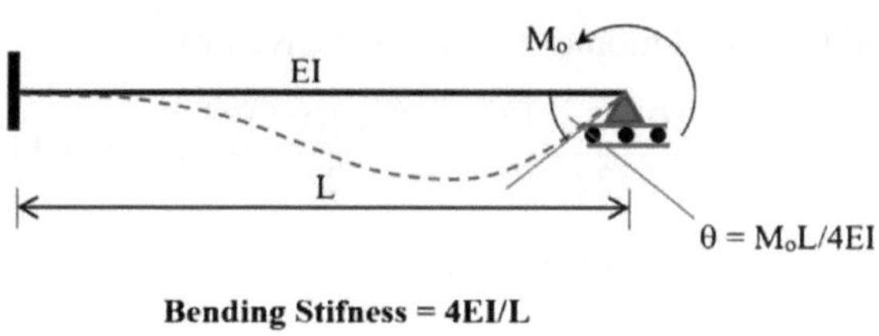

Figure 12.10 Bending stiffness of beams.

Example 12.30

A prismatic beam is fixed at one end and supported by a hinge at the other end, as shown below. What happens to the bending stiffness of the beam is the span if doubled and the moment of inertia is tripled?

A. Bending stiffness increases by a factor of 6.0.

B. Bending stiffness increases by a factor of 1.5.

C. Bending stiffness decreases by a factor of 3.0.

D. Bending stiffness decreases by a factor of 1.5.

Solution

Bending stiffness of the original beam is $4EI/L$.

Bending stiffness of the new beam is $4E(3I)/(2L) = 6EI/L$.

This means that the bending stiffness increases by a factor of 6/4 = 1.5. Answer is B.

Example 12.31

What is the vertical reaction at A for the beam shown below?

A. 37 kips

B. 39 kips

C. 41 kips

D. 43 kips

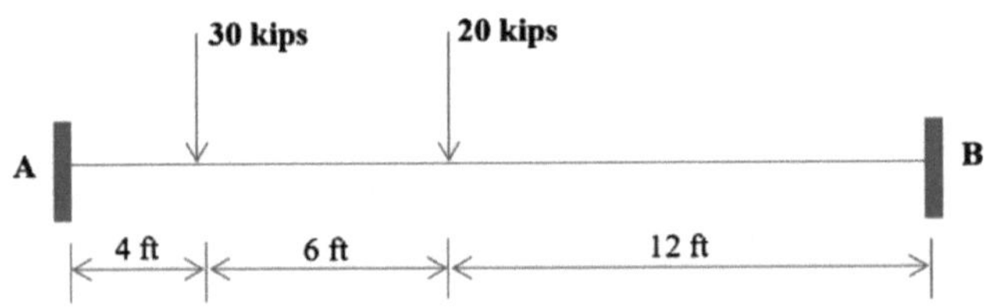

Solution The total fixed end moment at end A is given by

$$\text{FEM}_{AB} = \frac{30 \times 4 \times 18^2}{22^2} + \frac{20 \times 10 \times 12^2}{22^2} = 139.8 \text{ kip} \cdot \text{ft}$$

The total fixed end moment at end B is given by

$$\text{FEM}_{BA} = \frac{30 \times 18 \times 4^2}{22^2} + \frac{20 \times 12 \times 10^2}{22^2} = 67.4 \text{ kip} \cdot \text{ft}$$

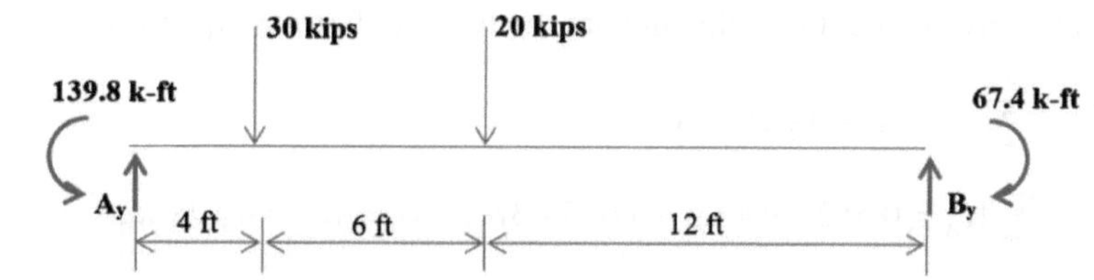

Taking moments about B: $M_B = 0 \Rightarrow 67.4 - 20\times 12 - 30\times 18 - 139.8 + 22A_y = 0$

Vertical reaction at A: $A_y = 38.75$ kips

Answer is B.

Example 12.32

Computer analysis of a beam ABC shown below results in the following joint moments:

$M_{AB} = 0.552$ kip · ft (counterclockwise)

$M_{BA} = 5.645$ kip · ft (clockwise)

$M_{BC} = 5.645$ kip · ft (counterclockwise)

$M_{CB} = 0$

The vertical reaction at B is most nearly

A. 3.67 kips

B. 6.20 kips

C. 8.12 kips

D. 9.86 kips

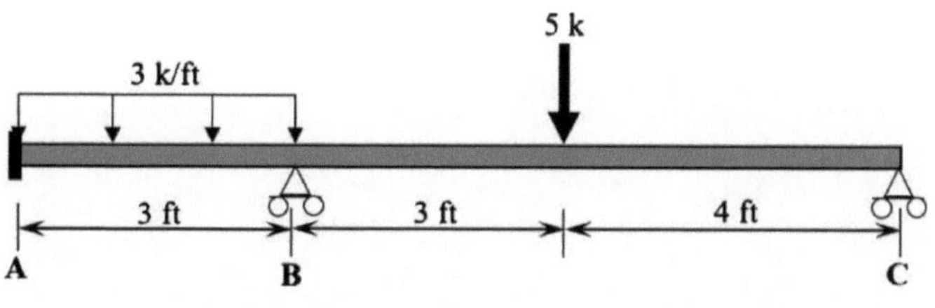

Solution Knowledge of the nodal moments M_{AB}, M_{BA}, M_{BC}, etc. allows the "decoupling" of the individual spans, as shown below.

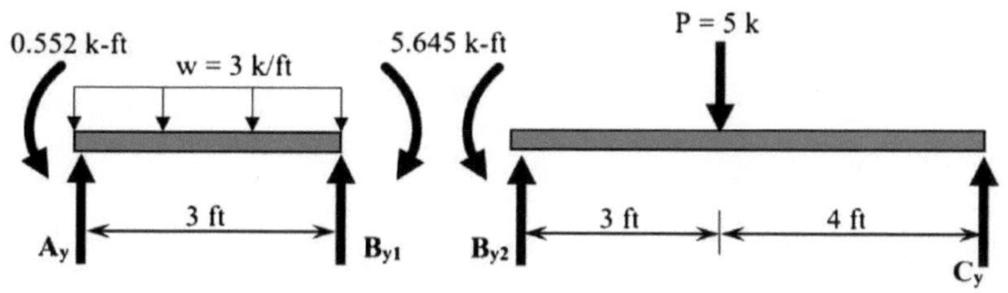

For the left substructure, the following are the two equilibrium equations:

$$\sum F_y = A_y - 9 + B_{y1} = 0$$

$$\sum M_A = 0.552 - 9 \times 1.5 - 5.645 + 3B_{y1} = 0 \Rightarrow B_{y1} = 6.198 \text{ kips}$$

For the right substructure, the following are the two equilibrium equations:

$$\sum F_y = B_{y2} - 5 + C_y = 0$$

$$\sum M_C = 5.645 + 5 \times 4 - 7B_{y2} = 0 \Rightarrow B_{y2} = 3.664 \text{ kips}$$

Therefore, the total vertical reaction at B is $B_y = B_{y1} + B_{y2} = 6.198 + 3.664 = 9.862$ kips

Answer is D.

CHAPTER 13

Structural Design

Reinforced Concrete Design

Basic Provisions of ACI 318-2011

In the United States, design of reinforced concrete structures and structural elements is performed according to the strength design specifications of the *ACI 318: Building Code Requirements for Structural Concrete* and is updated every 3 years. The version referenced in the *FE Reference Handbook,* version 9.4, is the 2011 edition of ACI 318.

In the strength design method, the strength of the concrete member under an ultimate limit state is compared to the demand created by the governing factored load combination.

Unless otherwise specified, the unit weight of normal weight concrete is taken to be $w_c = 150$ lb/ft^3. Lightweight structural concrete has unit weight 90 to 110 lb/ft^3.

According to the ACI code, the modulus of *elasticity of concrete* is given by

$$E_c = 33w_c^{1.5}\sqrt{f_c'} \tag{13.1}$$

where E_c = modulus of elasticity of concrete (psi)
w_c = unit weight of the concrete (pcf)
f_c' = 28-day compressive strength (psi)

In the United States, a standard bag of cement weighs 94 lb. Thus, a concrete mix referred to as a six-bag mix contains six standard (94 lb) bags of cement per cubic yard (564 lb cement per cubic yard concrete).

Portland cements are designated as the following types:

Type I	Normal
Type II	Modified for hot weather or large structures
Type III	High early strength
Type IV	Low heat (massive structures)
Type V	Sulphate resistant

Table 13.1 Properties of Reinforcement Bars

Size	# 3	# 4	# 5	# 6	# 7	# 8	# 9	# 10	# 11	# 14	# 18
Diameter (in)	0.375	0.500	0.625	0.750	0.875	1.000	1.128	1.270	1.410	1.693	2.257
Area (in^2)	0.11	0.20	0.31	0.44	0.60	0.79	1.00	1.27	1.56	2.25	4.00
Weight (lb/ft)	0.376	0.668	1.043	1.502	2.044	2.670	3.400	4.303	5.313	7.650	13.60

Reinforcement

Steel reinforcement bars are available in various standard sizes—in the U.S. system of nomenclature, these sizes are designated no. 3, no. 4, etc. up to no. 18 bars.

Nominal diameters, cross-sectional area, and weight of these sizes are given in Table 13.1.

Strength Design Approach

The nominal or theoretical strength (bending, axial, shear, etc.) is denoted as R_n (nominal resistance). The design strength or available strength ϕR_n is less than the nominal strength by a strength reduction factor ϕ, which is less than unity (to reflect material imperfections).

The ACI strength reduction factors are summarized in Table 13.2.

Load Combinations Used in Strength Design Approach (ACI 318)

The design requirement in load and resistance factor design (LRFD) may be stated as *the available strength must be greater than the maximum factored load.* This is shown in Eq. (13.2)

$$P_u = \sum \gamma_i Q_i \leq \phi R_n \tag{13.2}$$

where P_u = factored load

ϕ = strength (resistance) reduction factor

γ_i = individual load factors corresponding to service loads Q_i

R_n = nominal or theoretical strength (resistance)

Table 13.2 Strength Reduction Factor for Reinforced Concrete Design

Stress state	ϕ
Tension-controlled sections ($\varepsilon_t > 0.005$)	0.90
Compression-controlled sections ($\varepsilon_t < 0.002$)	
Members with spiral reinforcement	0.70
Members with tied reinforcement	0.65
Transition sections ($0.002 < \varepsilon_t < 0.005$)	
Members with spiral reinforcement	$0.57 + 67\,\varepsilon_t$
Members with tied reinforcement	$0.48 + 83\,\varepsilon_t$
Shear and torsion	0.75
Bearing on concrete	0.65

The load combinations (adopted from ASCE 7) are given below.

$$P_u = 1.4D \quad (13.3a)$$

$$P_u = 1.2D + 1.6L + 0.5(L_r \text{ or } S \text{ or } R) \quad (13.3b)$$

$$P_u = 1.2D + 1.6(L_r \text{ or } S \text{ or } R) + (L \text{ or } 0.8W) \quad (13.3c)$$

$$P_u = 1.2D + 1.6W + \mathrm{L} + 0.5(L_r \text{ or } S \text{ or } R) \quad (13.3d)$$

$$P_u = 1.2DL + 1.0E + \mathrm{L} + 0.2S \quad (13.3e)$$

$$P_u = 0.9D + (1.6W \text{ or } 1.0E) \quad (13.3f)$$

The load combination that produces the maximum load effect for the member is said to "govern."

Example 13.1

What is the design load for a short reinforced concrete column whose service loads (axial) are as shown below?

$$P_D = 100 \text{ kips}, P_L = 250 \text{ kips}, P_W = 180 \text{ kips}$$

A. 256 kips

B. 520 kips

C. 533 kips

D. 658 kips

Solution

$P_u = 1.4D = 140$ kips

$P_u = 1.2D + 1.6L = 520$ kips

$P_u = 1.2D + 1.6W + 1.0L = 658$ kips

Answer is D.

Analysis and Design of Reinforced Concrete Beams in Flexure

Concrete beams may be reinforced with tension reinforcement only (singly reinforced beams) or both tension and compression steel (doubly reinforced beams). The reinforcement bars are placed with adequate clear cover. For those locations of the beam where the maximum permitted tension reinforcement does not give adequate moment capacity, the deficit may be made up by using additional reinforcement in the compression zone.

Typical progressive behavior of reinforced concrete beams in flexure is summarized in the strain and stress distributions in Fig. 13.1. A rectangular reinforced concrete beam is shown loaded in third-point loading, resulting in a state of pure bending in the middle third

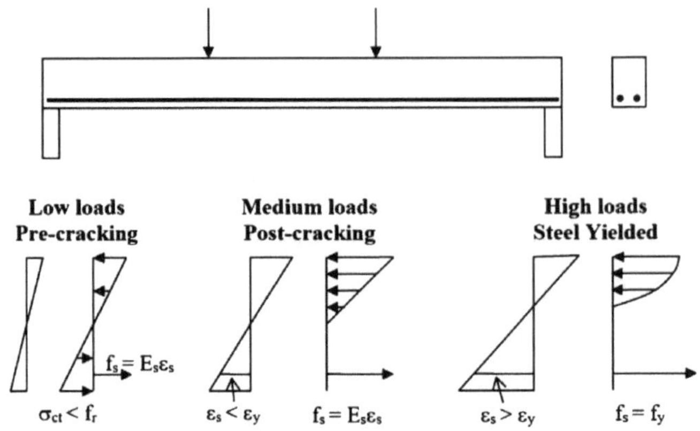

Figure 13.1 Strain and stress progression in a reinforced concrete beam.

of the span. At low loads, when the maximum tensile stress in the concrete σ_{ct} is below the rupture (tensile) strength f_r, the strain and stress distributions are as shown in the first pair.

As the load increases, the tensile stress in the concrete reaches the rupture strength, causing it to form hairline cracks. The concrete in the tension zone becomes ineffective and the compression in the concrete (possibly still linear or beginning to show some nonlinearity) is balanced solely by the tension in the steel reinforcement. This is shown by the second pair. The location of the neutral axis will depend on the amount of reinforcement and the relative strengths of concrete and steel.

As the load increases further, the compressive stress in the concrete becomes nonlinear and the steel yields (strain ε_s exceeds the yield strain ε_y). This is shown by the third pair. The limit of this stress diagram is when the compressive strain in the concrete reaches the limiting value of 0.003, which is defined in ACI as the crushing limit state.

Use of "balanced reinforcement" will cause these limit states—yielding of steel and crushing of concrete—to be reached simultaneously. However, for the failure to be ductile, it is desirable for the beam to be proportioned such that the steel yields well before the crushing strain is reached in the concrete. Such a section is called an under-reinforced section. In the current ACI specifications, this is ensured by imposing a lower limit on the tensile strain in the outermost steel layer ($\varepsilon_t > 0.004$). Prior to the 2002 edition of ACI 318, this criterion as stated as "maximum steel ratio = 75% of the balanced steel ratio ρ_b."

Clear Cover Guidelines

Section 7.7.1 of ACI 318 specifies the minimum concrete cover to reinforcement. For the purposes of the FE exam, it is sufficient to know that minimum clear cover for (a) beams and columns = 1.5 in, (b) slabs and walls (with reinforcement smaller than no. 11) = 0.75 in, and (c) for concrete cast against and permanently exposed to earth (such as footings and retaining walls) = 3.0 in.

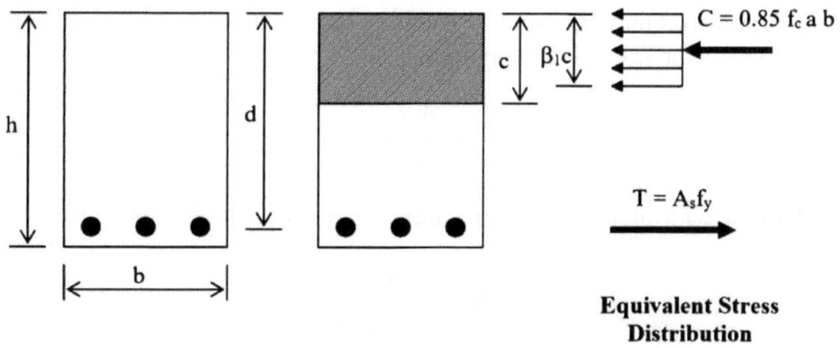

Figure 13.2 Rectangular reinforced concrete beam–stress distribution.

Flexural Capacity of Singly Reinforced Concrete Beams

Figure 13.2 shows a rectangular reinforced concrete beam with tension steel only. Let us assume the beam is simply supported and carrying downward transverse loads, the deflected beam is concave on the top fiber, thus producing compression above the neutral axis and tension below. In such a case, the steel reinforcement is placed near the bottom edge with adequate cover. If the beam is a cantilever beam subjected to downward loads, the curvature is concave on the bottom fiber and, therefore, the tension steel is located near the top fiber.

The bending moment in the beam cross section develops as in Fig. 13.1. The concrete progresses toward the compressive limit stress f_c' while the steel progresses toward yield f_y. For an under-reinforced section, the steel reaches yield first and at ultimate load, the concrete reaches strain $\varepsilon_c = 0.003$. The actual compression stress diagram on the concrete is nonlinear and is replaced by an equivalent stress, as proposed by Whitney and adopted by ACI-318. The ultimate stress condition assumes that the tension steel has previously yielded and the limit state corresponds to concrete reaching the compressive strain limit ($\varepsilon_c = 0.003$). The factor β_1 is proposed to be a function of f_c' as shown in Eq. (13.4) and Fig. 13.3.

$$\beta_1 = 0.85 \geq 0.85 - 0.05\left(\frac{f_c' - 4000}{1000}\right) \geq 0.65 \tag{13.4}$$

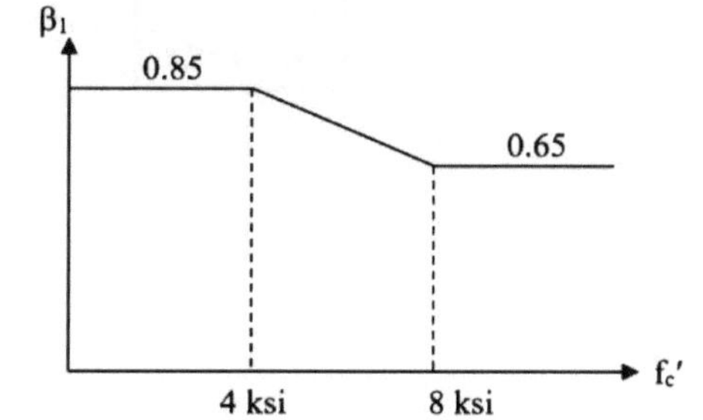

Figure 13.3 Equivalent compressive stress block depth factor (Whitney).

The depth of neutral axis may be found by equating the resultant compression to the resultant tension, which yields the following expression

$$c = \frac{A_s f_y}{0.85\beta_1 f_c' b} \tag{13.5}$$

The depth of the equivalent (Whitney) compression stress block is given by

$$a = \beta_1 c = \frac{A_s f_y}{0.85 f_c' b} \tag{13.6}$$

The tensile strain is calculated from

$$\varepsilon_t = \frac{0.003(d_t - c)}{c} \tag{13.7}$$

The nominal (theoretical) moment capacity M_n is given by the strength of the couple formed by the equal and opposite forces C and T

$$M_n = A_s f_y\left(d - \frac{a}{2}\right) = 0.85 f_c' ab\left(d - \frac{a}{2}\right) \tag{13.8}$$

The design moment capacity is then given by ϕM_n.

Example 13.2

Interior concrete beams for an enclosed frame structure have the rectangular 12 × 24-in section as shown in the figure. Use $f_c' = 5$ ksi and $f_y = 60$ ksi. What is the *nominal* moment capacity of the section?

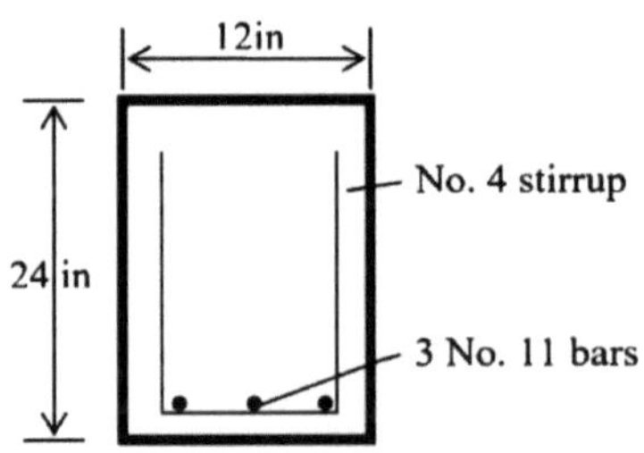

A. 380 kip · ft

B. 400 kip · ft

C. 417 kip · ft

D. 435 kip · ft

Solution According to ACI 318, clear cover should be 1.5 in.
Effective depth $d = 24 - 1.5 - 0.5 - 1.41/2 = 21.3$ in
For $f_c' = 5$ ksi, $\beta_1 = 0.8$
A_s = area of 3 no. 11 bars = 4.68 in^2

Depth of equivalent compression block: $a = \dfrac{A_s f_y}{0.85 f_c' b} = \dfrac{4.68 \times 60}{0.85 \times 5 \times 12} = 5.5$ in

Nominal moment capacity:

$$M_n = A_s f_y \left(d - \frac{a}{2}\right) = 4.68 \times 60 \times \left(21.3 - \frac{5.5}{2}\right) = 5209 \text{ kip} \cdot \text{in} = 434 \text{ kip} \cdot \text{ft}$$

Answer is D.

Example 13.3

A reinforced concrete ($f_c' = 4$ ksi, $f_y = 60$ ksi) beam is simply supported over a span $L = 25$ ft. The width of the rectangular section is 15 in and depth is 24 in. The tensile reinforcement is 4 no. 10 bars. Shear reinforcement is in the form of no. 4 U-shaped stirrups. What is the tensile strain in the steel?

A. 0.0055

B. 0.0061

C. 0.0067

D. 0.0082

Solution If no. 4 stirrups are used and no. 10 bars as longitudinal reinforcement and clear cover is 1.5 in, then effective depth $d = 24 - 1.5 - 0.5 - 1.27/2 = 21.365$ in.

Depth of neutral axis is calculated as: $c = \dfrac{A_s f_y}{0.85 \beta_1 f_c' b} = \dfrac{4 \times 1.27 \times 60}{0.85 \times 0.85 \times 4 \times 15} = 7.03$ in

Tensile strain is calculated as: $\varepsilon_t = \dfrac{0.003(d - c)}{c} = \dfrac{0.003(21.365 - 7.03)}{7.03} = 0.0061$

Answer is B.

Example 13.4

Find the design moment capacity of a reinforced concrete beam section with width $b = 12$ in, depth $h = 20$ in, reinforcement 3 no. 9 bars, $f_c' = 3$ ksi, $f_y = 60$ ksi.

Solution All parameters of the beam section are fully specified in this case. The depth of the neutral axis is given by

$$c = \frac{A_s f_y}{0.85 \beta_1 f_c' b} = \frac{3.0 \times 60}{0.85 \times 0.85 \times 3 \times 12} = 6.92$$

The equivalent depth of compressive stress (Whitney) is given by

$$a = \beta_1 c = 0.85 \times 6.92 = 5.88 \text{ in}$$

Assuming bars in a single layer, the effective depth is

$$d = d_t \approx 20 - 2.5 = 17.5$$

The tensile strain is calculated as

$$\varepsilon_t = \frac{0.003(d_t - c)}{c} = \frac{0.003(17.5 - 6.92)}{6.92} = 0.0046$$

The strength reduction factor is given by $\phi = 0.48 + 83\varepsilon_t = 0.86$.

Nominal moment capacity

$$M_n = A_s f_y \left(d - \frac{a}{2}\right) = 3.0 \times 60 \times \left(17.5 - \frac{5.88}{2}\right) = 2621 \text{ kip} \cdot \text{in} = 218 \text{ kip} \cdot \text{ft}$$

Design moment capacity

$$\phi M_n = 0.86 \times 218 = 187.5 \text{ kip} \cdot \text{ft}$$

NOTE As the tensile stress varies between the minimum permitted value of 0.004 and 0.005, the strength reduction factor varies from 0.81 to 0.90. So, if the answer choices are well separated, one could avoid the computation of the neutral axis depth, tensile strain, and strength reduction factor and still get a fairly good estimate (196 kip · ft) of the design strength by using $\phi = 0.90$, which is appropriate for $\varepsilon_t > 0.005$.

Example 13.5

A rectangular reinforced concrete beam has a simple span L = 18 ft and a uniformly distributed (factored) load w_u = 5 kips/ft. The beam section is rectangular with width = 12 in and depth = 20 in. Use f_c' = 5000 psi and f_y = 60,000 psi. What is the required flexural reinforcement?

A. 2.5 in^2

B. 3.0 in^2

C. 3.5 in^2

D. 4.0 in^2

Solution Assuming effective depth $d \approx h - 2.5 = 17.5$ in

$$\text{Design moment: } M_u = \frac{w_u L^2}{8} = \frac{5 \times 18^2}{8} = 202.5 \text{ kip} \cdot \text{ft}$$

This is a type of problem that is best done using design aids, which are not in the *Handbook*. Thus, the quickest way would be try out answer choices.

- Try $A_s = 4\ \text{in}^2$

$$\text{Depth of compression block: } a = \frac{A_s f_y}{0.85 f_c' b} = \frac{4 \times 60}{0.85 \times 5 \times 12} = 4.7 \text{ in}$$

$$\text{Nominal moment capacity: } M_n = A_s f_y \left(d - \frac{a}{2}\right) = 4 \times 60 \times \left(17.5 - \frac{4.7}{2}\right) = 3635 \text{ kip} \cdot \text{in} = 303 \text{ kip} \cdot \text{ft}$$

$$\text{Design moment capacity: } \phi_b M_n = 0.9 \times 303 = 272.6 \text{ kip} \cdot \text{ft}$$

This is significantly over target.

- Try $A_s = 3\ \text{in}^2$

$$\text{Depth of compression block: } a = \frac{A_s f_y}{0.85 f_c' b} = \frac{3 \times 60}{0.85 \times 5 \times 12} = 3.53 \text{ in}$$

$$\text{Nominal moment capacity: } M_n = A_s f_y \left(d - \frac{a}{2}\right) = 3 \times 60 \times \left(17.5 - \frac{3.53}{2}\right) = 2832 \text{ kip} \cdot \text{in} = 236 \text{ kip} \cdot \text{ft}$$

$$\text{Design moment capacity: } \phi_b M_n = 0.9 \times 236 = 212 \text{ kip} \cdot \text{ft}$$

This is just about right.

Answer is B.

Design of Reinforced Concrete Beams for Shear

Unless the beam qualifies as a deep beam ($L/d < 6$), the effect of shear deformations will be small compared to the flexural deformations. The primary design of the beam is therefore for flexure. Once the beam has been sized so that the flexural capacity is greater than the design moment, the section is checked for shear capacity.

The nominal shear capacity of a rectangular reinforced concrete beam section without shear reinforcement is given by

$$V_c = 2\sqrt{f_c'}bd \tag{13.9}$$

where V_c = nominal shear capacity of the concrete section alone (lb)
f_c' = 28-day compressive strength of the concrete (psi)
b = beam width (in)
d = effective depth (in)

The design shear V_u is then compared to V_c to determine whether the section needs no shear reinforcement, minimum shear reinforcement, or calculated shear reinforcement.

The shear design statement according to ACI's strength design guidelines is

$$\phi V_n \geq V_u \Rightarrow \phi(V_c + V_s) \geq V_u \tag{13.10}$$

The strength reduction factor ϕ for shear is 0.75 (see Table 13.1)

$$\text{Case 1:} \qquad V_u < \frac{1}{2}\phi V_c$$

For those regions of the beam where this relationship is satisfied, no shear reinforcement is required

$$\text{Case 2:} \qquad \frac{1}{2}\phi V_c < V_u < \phi V_c$$

For locations where the shear is within the range shown above, use minimum shear reinforcement. Provide stirrups at spacing no greater than the smaller of

$$s = \frac{A_v f_y}{50 b_w} \qquad \text{or} \qquad s = \frac{A_v f_y}{0.75 b_w \sqrt{f_c'}} \tag{13.11}$$

where A_v is the area of the shear reinforcement normal to the plane of horizontal shear (Fig. 13.4). If single-loop stirrups are used, A_v is twice the area of the stirrup bar.

Maximum permitted spacing of stirrups is not to exceed $s_{max} = d/2$ or 24 in

$$\text{Case 3:} \qquad V_u > \phi V_c$$

Calculate the required nominal shear strength of shear reinforcement as

$$V_s = \frac{V_u}{\phi} - V_c \le 8\sqrt{f_c'}bd \tag{13.12}$$

If the magnitude of V_s exceeds $8\sqrt{f_c'}bd$, the beam must be resized.

The required spacing of the stirrups to provide this capacity is given by

$$s = \frac{A_v f_y d}{V_s} \tag{13.13}$$

If $V_s \le 4\sqrt{f_c'}b_w d$, the maximum permitted spacing is not to exceed $d/2$ nor 24 in.

If $V_s > 4\sqrt{f_c'}b_w d$, the maximum permitted spacing is not to exceed $d/4$ nor 12 in.

Figure 13.4
Area of stirrups resisting longitudinal shear.

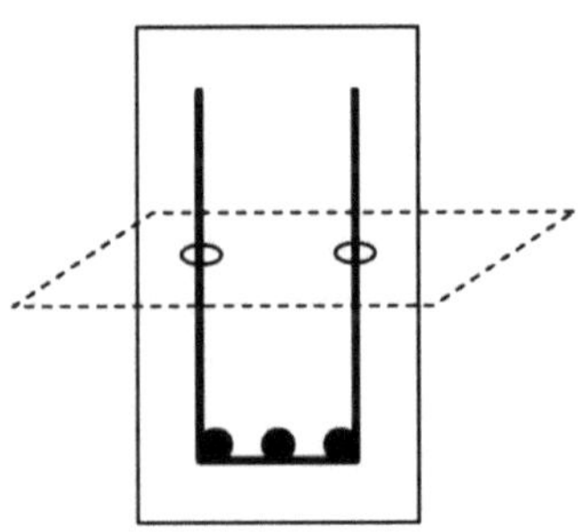

Example 13.6

A simply supported reinforced concrete beam spans $L = 26$ ft. The factored load on the beam is $w_u = 7.2$ kips/ft. The cross section of the beam if rectangular (width = 14 in, depth = 25 in). Use $f_c' = 4500$ psi and $f_y = 60{,}000$ psi. What is the total length of the zone where no shear reinforcement is required?

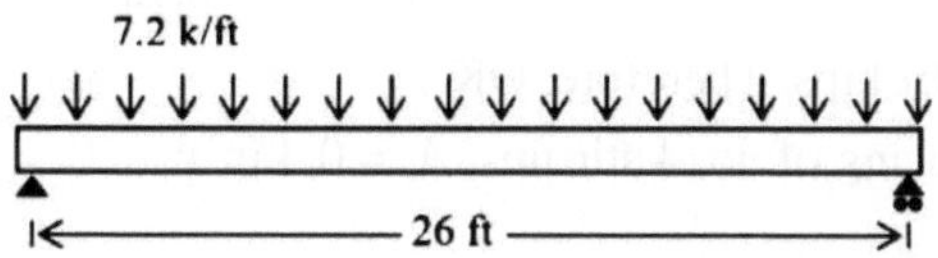

A. 40 in

B. 50 in

C. 60 in

D. 70 in

Solution The effective depth, $d \approx 25 - 2.5 = 22.5$ in

Shear capacity of concrete section, $V_u = 2\sqrt{f_c'}bd = 2 \times \sqrt{4500} \times 14 \times 22.5 = 42261.7$ lb

$$\frac{1}{2}\phi V_c = 0.5 \times 0.75 \times 42.26 = 15.85 \text{ kips}$$

The distance on either side of the midspan, where this shear occurs = 15.85/7.2 = 2.2 ft

The total distance over which no shear reinforcement is required = 4.4 ft = 52.8 in

Answer is B.

Example 13.7

What is the required spacing of no. 4 stirrups in the regions of maximum shear of the beam section shown? The design shear is $V_u = 80$ kips. Assume clear cover = 1.5 in. Use $f_c' = 4500$ psi and $f_y = 60{,}000$ psi.

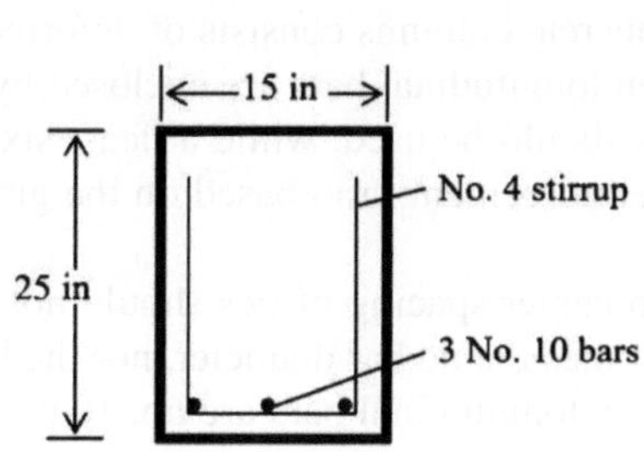

A. 9.0 in

B. 4.5 in

C. 4.25 in

D. 8.5 in

Solution Effective depth: $d = 25 - 1.5 - 0.5 - \dfrac{1.27}{2} = 22.37$ in

Shear capacity of concrete section: $V_c = 2\sqrt{f_c'}bd = 2\sqrt{4500} \times 15 \times 22.37 = 45{,}019 \text{ lb} = 45.02 \text{ kips}$

Required capacity of shear reinforcement: $V_s = \dfrac{V_u}{\phi} - V_c = \dfrac{80}{0.75} - 45 = 61.67$ kips (This does not exceed $8\sqrt{f_c'}bd = 180.1$ kips. Therefore, OK.)

Required (maximum) spacing of no. 4 stirrups ($A_v = 0.4$ in^2):

$$s = \frac{A_v f_y d}{V_s} = \frac{0.4 \times 60 \times 22.37}{61.67} = 8.7 \text{ in}$$

Since $V_s < 4\sqrt{f_c'}bd = 90$ kips, spacing is not to exceed $d/2$ (11.2 in) nor 24 in (neither controls). Therefore, spacing in the maximum shear zone is not to exceed 8.7 in.

Answer is D.

Analysis and Design of Reinforced Concrete Columns

Reinforced concrete columns may be divided into two categories: (1) columns with a purely concentric load (no moment) and (2) columns carrying eccentric loads. A broader name for columns is a compression member—as long as the dominant load effect is axial compression. Columns may be considered "short" or "long," based on the slenderness ratio of the column. For medium and long columns, buckling or the tendency to buckle may diminish the load capacity of the column.

Concrete columns are reinforced with (1) longitudinal reinforcement and lateral ties, (2) longitudinal reinforcement and a continuous spiral, or (3) structural steel shapes, with or without additional longitudinal bars.

Guidelines on Longitudinal Reinforcement

The main reinforcement in concrete columns consists of deformed bars arranged in a rectangular or circular pattern. When longitudinal bars are enclosed by rectangular or circular ties, at least four longitudinal bars should be used, while at least six bars are needed if confined by a continuous spiral. The reinforcement ratio based on the gross area ($\rho_g = A_s/A_g$) must be between 1% and 8%.

For tied columns, center-to-center spacing of ties should not be greater than 16 times the longitudinal bar diameter, 48 times the tie bar diameter, nor the least column dimension. Ties must be at least no. 3 bars when longitudinal bars are no. 10 or smaller. Ties must be at least no. 4 size when longitudinal bars are no. 11 or larger.

Clear distance between longitudinal bars must be at least 1.5 times the bar diameter or 1.5 in. Clear cover to longitudinal reinforcement must be at least 1.5 in.

For columns whose longitudinal reinforcement is confined by a continuous spiral, the clear distance between spirals must be between 1 and 3 in. The minimum diameter of spiral reinforcement is 3/8 in.

Short versus Long Columns

In braced (non-sway) frames, a column is considered "short" (i.e., effects of slenderness may be ignored) if

$$\frac{Kl_u}{r} \leq 34 - 12\left(\frac{M_1}{M_2}\right) \leq 40 \tag{13.14}$$

where M_1/M_2 is the ratio of the smaller end moment to the larger end moment (to be taken no smaller than –0.5). This ratio is taken to be positive if column is in single curvature and negative if the column is in reverse curvature.

In unbraced frames (sway uninhibited), a column is considered "short" if

$$\frac{Kl_u}{r} \leq 22 \tag{13.15}$$

where K is the effective length factor and l_u is the unsupported length, taken as the clear distance between floor slabs, beams, or other members providing lateral support.

Short columns may then be further subdivided into two groups: (1) columns with a purely concentric load (no moment) and (2) columns carrying eccentric loads.

Axial Load Capacity of Short Reinforced Concrete Columns

Even for columns carrying loads with "small" eccentricity, the ACI 318 specifications allow the calculation of the axial load capacity as a concentrically loaded column. In other words, it allows the designer to ignore the resulting moment on the column. The eccentricity of the load may be neglected (i.e., column designed for the "axial" load only) if the eccentricity is less than 0.10 h for tied columns or 0.05 h for spiral columns.

The axial load capacity of a short, reinforced concrete column with small eccentricity is given by

$$\phi P_n = \phi\beta P_o = \phi\beta[0.85 f_c'(A_g - A_s) + f_y A_s] \tag{13.16}$$

which can be also written as

$$\phi P_n = \phi\beta A_g[0.85 f_c'(1 - \rho_g) + f_y \rho_g] \tag{13.17}$$

The factors ϕ and β are given in Table 13.3

Table 13.3 Strength Reduction Factors ϕ and β for Short Columns

	ϕ	β
Tied column	0.65	0.80
Spiral column	0.70	0.85

Example 13.8

A circular concrete column supports a 300-kip dead load and a 350-kip live load. All loads are concentric. The longitudinal reinforcement is laterally confined by a continuous spiral. The compressive strength of the concrete is 4000 psi and the yield stress of the reinforcement steel is 60 ksi. If the maximum (permitted by ACI) reinforcement is used, the required (minimum) cross-sectional area of the column (in^2) is most nearly.

A. 130

B. 150

C. 200

D. 230

Solution Maximum reinforcement = 8%

$$P_u = 1.2 \times 300 + 1.6 \times 350 = 920 \text{ kips}$$

$$\phi P_n = \phi \beta A_g[0.85 f_c'(1-\rho_g) + f_y \rho_g]$$

$$= 0.75 \times 0.85 \times A_g[0.85 \times 4 \times (1-0.08) + 60 \times 0.08] = 5.054 A_g$$

$A_g > 182 \text{ in}^2$

Answer is C.

Example 13.9

What is the smallest factored moment that can be ignored in the design of a square reinforced concrete column that carries a factored compression load P_u = 600 kips? The column has a 15 × 15-in cross section. Use f_c' = 4000 psi and f_y = 60,000 psi.

Solution A rectangular column will necessarily have lateral confinement in the form of ties. Therefore, according to the ACI, the maximum eccentricity that can be ignored is equal to 0.1 h = 0.1 × 15 in = 1.5 in.

Therefore, the maximum factored moment that can be ignored is

$$M_u = P_u e = 600 \times 1.5 = 900 \text{ kip} \cdot \text{in} = 75 \text{ kip} \cdot \text{ft}$$

Column Interaction Diagrams

The load capacity of short, reinforced concrete columns with significant eccentricity can be calculated using column interaction diagrams. The *Handbook* has a single example of a column interaction diagram (rectangular with reinforcement on two parallel faces). For such a column geometry, lateral confinement will always be in the form of ties, whereas for circular columns, it may be in the form of ties or a continuous spiral.

These charts are classified by three parameters:

f_c': 28-day compression strength of concrete
f_y: yield stress in tension of reinforcement
γ: factor related to cover to reinforcement

There are three types of problems that may be solved using these diagrams:

1. Given a column cross section (size and reinforcement), what is the maximum eccentricity of a given load?
2. Given an eccentric load and column size, what is the required reinforcement?
3. Given a column cross section (size and reinforcement), what is the maximum load if the eccentricity is predefined?

Example 13.10

A short reinforced concrete column has an 18 × 18-in cross section. It is reinforced with two lines of reinforcement bars with a total of 12 no. 10 bars, as shown. The service level axial loads are $P_{DL} = 220$ kips, $P_{LL} = 400$ kips. (Assume $f_c' = 4000$ psi, $f_y = 60{,}000$ psi.)

What is the maximum permitted eccentricity of the load?

A. 1.8 in

B. 2.5 in

C. 3.1 in

D. 3.7 in

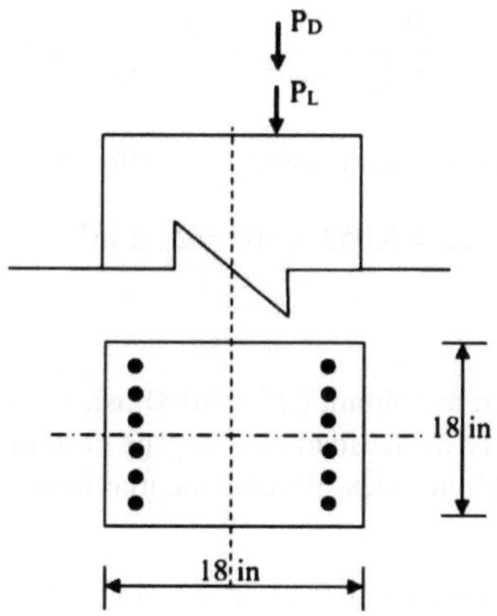

Solution The factored load is $P_u = 1.2 \times 220 + 1.6 \times 400 = 904$ kips
Assuming approximately 2 in of effective cover, the center-to-center distance between parallel lines of reinforcement = $18 - 2 \times 2 = 14$ in.

Parameter $\gamma = 14/18 = 0.78$. Let us use the diagram for $f_c' = 4$ ksi, $f_y = 60$ ksi, $\gamma = 0.80$ (actually, this is the only column interaction diagram we have).

Area of steel $A_s = 12 \times 1.27 = 15.24$ in^2 and gross area $A_g = 18 \times 18 = 324$ in^2

Reinforcement ratio $\rho_g = 15.24/324 = 0.047$

The parameter $K_n = \dfrac{P_u}{\phi f_c' A_g} = \dfrac{900}{0.65 \times 4 \times 324} = 1.07$

Using $K_n = 1.07$ and $\rho_g \approx 0.047$ (visual interpolation between 0.04 and 0.05), we get $e/h \approx 0.17$. Therefore, the tolerable eccentricity $e = 0.17$, $h = 3.1$ in.

Answer is C.

Example 13.11

What is the required reinforcement for a 20 × 20-in reinforced concrete column ($f_c' = 4000$ psi, $f_y = 60{,}000$ psi) subjected to the following loads:

$P_{DL} = 250$ kips, $P_{LL} = 450$ kips. Assume the eccentricity of the load is 4 in.

Solution

$$P_u = 1.2 \times 250 + 1.6 \times 450 = 1020 \text{ kips}$$

Eccentricity is greater than 0.1 h (4 in > 0.1 × 20 in). Therefore, the column must be designed for a combination of P_u and M_u ($e/h = 4/20 = 0.20$).

$$K_n = \frac{P_u}{\phi_c f_c' A_g} = \frac{1020}{0.65 \times 4 \times 400} = 0.98$$

From the column interaction diagram, using $K_n = 0.98$ and $e/h = 0.2$, $\rho_g = 5.3\%$.

$$A_s = 0.053 \times 400 = 21.2 \text{ in}^2$$

Example 13.12

A 20 × 20-in reinforced concrete column ($f_c' = 4000$ psi, $f_y = 60{,}000$ psi) is subjected to an eccentric load. The eccentricity of the load is 6 in. The column is reinforced with 12 no. 11 bars distributed equally on all four faces. What is the maximum factored load that the column can carry?

Solution Eccentricity is greater than 0.1 h (6 in > 0.1 × 20 in). Therefore, the column must be designed for a combination of P_u and M_u ($e/h = 6/20 = 0.30$).

Area of steel $A_s = 12 \times 1.56 = 18.72$ in^2, and gross area $A_g = 20 \times 20 = 400$ in^2.

Reinforcement ratio $\rho_g = 18.72/400 = 0.047$

Assuming approximately 2 in of effective cover, the center-to-center distance between parallel lines of reinforcement = 20 − 2 × 2 = 16 in.

Parameter $\gamma = 16/20 = 0.80$. Let us use the diagram for $f_c' = 4$ ksi, $f_y = 60$ ksi, $\gamma = 0.80$.

From the column interaction diagram, for $e/h = 0.30$ and $\rho_g = 4.7\%$, $K_n = 0.76$.

$$P_u = K_n \phi_c f_c' A_g = 0.76 \times 0.65 \times 4 \times 400 = 790 \text{ kips}$$

Example 13.13

A 16 × 20-in reinforced concrete ($f_c' = 4000$ psi, $f_y = 60{,}000$ psi) column is subjected to the following loads.

$P_D = 300$ kips (concentric)

$P_L = 180$ kips (eccentric; $e = 4$ in)

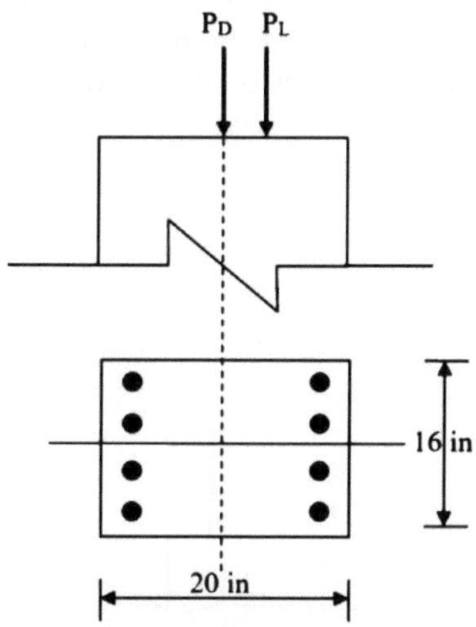

The required area of steel (in²) is most nearly

A. 3.2 in²

B. 4.6 in²

C. 5.8 in²

D. 7.2 in²

Solution Factored axial load: $P_u = 1.2P_D + 1.6P_L = 1.2 \times 300 + 1.6 \times 180 = 648$ kips

Factored moment: $M_u = 1.2P_D e_D + 1.6P_L e_L = 1.2 \times 300 \times 0 + 1.6 \times 180 \times 4 = 1152$ kip·in

The dimensionless parameters to be used in the interaction diagram are

$$K_n = \frac{P_u}{\phi f_c' A_g} = \frac{648}{0.65 \times 4 \times 320} = 0.78$$

$$R_n = \frac{M_u}{\phi f_c' A_g h} = \frac{1152}{0.65 \times 4 \times 320 \times 20} = 0.07$$

Assuming effective cover of about 2.5 in, $\gamma h = 20 - 2 \times 2.5 = 15$ in $\rightarrow \gamma = 0.75$.

For the above data, the column interaction diagram yields $\rho_g < 1\%$. However, minimum of 1% steel must be used.

$$\text{Therefore, } A_s = 0.01 \times 16 \times 20 = 3.2 \text{ in}^2$$

Answer is A.

Example 13.14

The load acting on a rectangular reinforced concrete column has an eccentricity $e = 4$ in. Use $f_c' = 4000$ psi and $f_y = 60{,}000$ psi. The column has dimensions 20 × 14 in and is reinforced with 8 no. ten bars as main reinforcement, as shown. What is the maximum factored load P_u that the column can carry?

A. 650 kips

B. 790 kips

C. 985 kips

D. 2115 kips

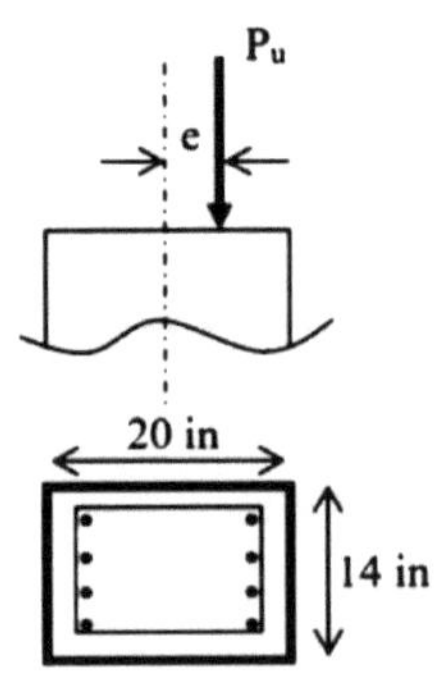

Solution For a tied column, $e/h < 0.1$ would allow us to use the small eccentricity equation. Since $e/h = 4/20 = 0.2$, the eccentricity cannot be ignored.

Reinforcement ratio: $\rho_g = \dfrac{8 \times 1.27}{14 \times 20} = 0.036$

$$\gamma h = 20 - 2 \times (1.5 + 0.5 + 0.635) = 14.73 \Rightarrow \gamma = \frac{14.73}{20} = 0.74; \text{ Use } \gamma = 0.8$$

From the column interaction diagram on page 156 of the *Handbook*, $K_n = 0.9$

$$K_n = \frac{P_u}{\phi_c f_c' A_g} \Rightarrow P_u = K_n \phi_c f_c' A_g = 0.9 \times 0.65 \times 4 \times 280 = 655.2 \text{ kips}$$

Answer is A.

Steel Design

Load and Resistance Factor Design

The AISC Steel Construction Manual outlines procedures for both Allowable Strength Design (ASD) and Load and Resistance Factor Design (LRFD). However, only the LRFD method is outlined in the *Handbook*.

The FE examination tests candidates on the design of the following types of steel components:

Beams—subject to flexure and shear

Columns—subject to concentric compression

Tension members—limit states of yielding, fracture, and block shear

The following supporting pages are given in the *Handbook*. All page numbers are for version 9.4.

Page 159: Table 1-1: W-Shapes Dimensions and Properties

Page 160: AISC Table 3-2 (excerpt): W-Shapes—Selection by Z_x

Page 161: AISC Table 3-10 (excerpt): W-Shapes—Available Moment versus Unbraced Length

Page 162: AISC Table C-A-7.1: Approximate Values of Effective Length Factor, K

Page 162: AISC Figs. C-A-7.1 and C-A-7.2: Alignment Charts for K Values

Page 163: AISC Table 4-22 (excerpt): Available Critical Stress ($\phi_c F_{cr}$) for Compression Members

Page 164: AISC Table 4-1 (excerpt): W-Shapes—Available Strength in Axial Compression ($\phi_c P_n$)

Analysis and Design of Beams (Flexure)

This section is dedicated to the design procedures as suggested by American Institute of Steel Construction (AISC). The design method outlined is the LRFD, which is the one summarized in the *Handbook*.

The LRFD load combinations (adopted from ASCE 7) are given in the first part of this chapter [Eq. (13.3)].

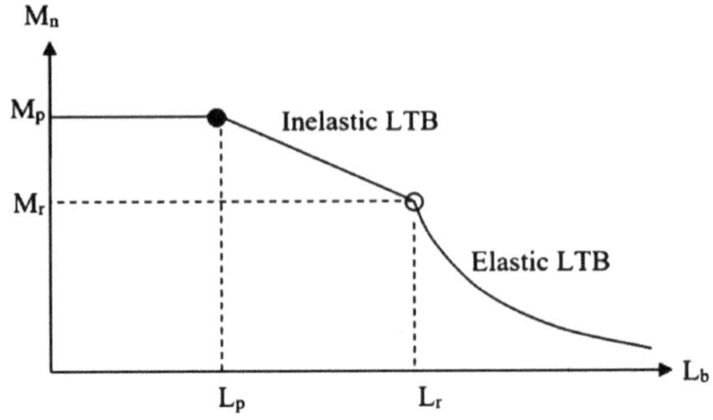

Figure 13.5 Nominal moment capacity of steel beams ($C_b = 1$).

For LRFD, the design requirement may be stated as

$$M_u \le \phi_b M_n = 0.9M_n \tag{13.18}$$

where M_u is the required moment strength = maximum moment calculated from the LRFD load combinations.

Nominal Moment Capacity M_n

Figure 13.5 shows a schematic diagram of the variation of nominal flexural strength M_n with the unbraced length of the compression flange L_b. The nominal moment capacity of steel beams is based on the slenderness of the compression flange and its ability to resist buckling. The full plastic moment capacity M_p can be realized if the unbraced length of the compression flange L_b is within a certain limit L_p.

In other words, for $L_b < L_p$, the nominal moment capacity of the section is equal to the plastic moment capacity M_p. When $L_b > L_p$, the full plastic moment capacity cannot be realized and the section fails due to lateral-torsional buckling (LTB).

When $L_p < L_b < L_r$, the section fails due to inelastic LTB. When $L_b > L_r$, the section fails due to elastic LTB. L_p and L_r are functions of the section geometry and the steel material properties (F_y, E).

Example 13.15

Find the plastic moment capacity of the built-up section shown below. Use $F_y = 36$ ksi,

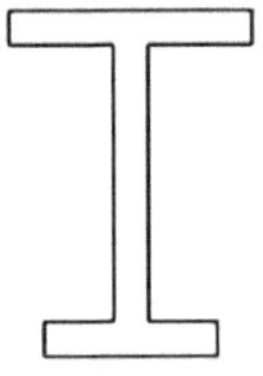

Top Flange Width = 12 in
Bottom Flange Width = 10 in
Top Flange thickness = 1 in
Bottom Flange thickness = 2 in
Web depth = 18 in
Web thickness = ¾ in

A. 975 kip·ft

B. 1050 kip·ft

C. 1125 kip·ft

D. 1200 kip·ft

Solution Total area = $12 \times 1 + 18 \times ¾ + 10 \times 2 = 45.5 \text{ in}^2$
The plastic neutral axis (PNA) is located such that it divides the area into equal halves. Half area = 22.75 in^2.

Since both flange areas are less than half the area, the PNA lies in the web.
If PNA is a distance y above the bottom edge, $20 + 0.75(y - 2) = 22.75 \rightarrow y = 5.667$ in

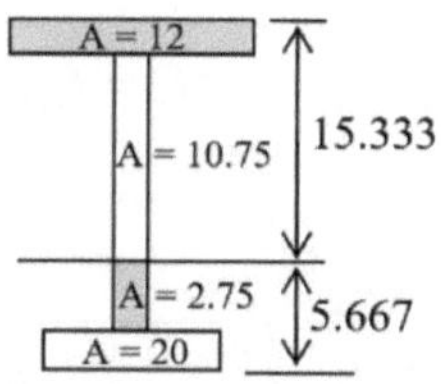

This line divides the section into the four subareas (rectangles) shown above.
Plastic section modulus is calculated as the sum of the first moments of these areas about the PNA:

$$Z = \Sigma A_p y_i = 12 \times 14.833 + 10.75 \times 7.167 + 2.75 \times 1.833 + 20 \times 4.667 = 353.4 \text{ in}^3$$

Plastic moment capacity: $M_P = Z_x F_y = 353.4 \times 36 = 12723.2 \text{ kip} \cdot \text{in} = 1060.3 \text{ kip} \cdot \text{ft}$

Answer is B.

The following is a summary of the nominal flexural strength of compact I- and C-shaped sections.

Failure by Reaching Fully Plastic State

$$\text{For } L_b \le L_p \qquad M_n = M_p = F_y Z_x \tag{13.19}$$

where F_y specified minimum yield stress of the chosen grade of steel and Z_x plastic section modulus about the axis of bending (x axis).

Inelastic Lateral Torsional Buckling

$$\text{For } L_p \le L_b \le L_r \qquad M_n = C_b\left[M_p - (M_p - 0.7F_yS_x)\left(\frac{L_b - L_p}{L_r - L_p}\right)\right] \le M_p \tag{13.20}$$

where C_b (bending coefficient) is the factor to account for the nonuniform bending moment within the unbraced length L_b.

Conservatively, C_b may be taken as 1.0. In fact, the value of C_b depends on the shape of the bending moment diagram and the AISC design charts are based on $C_b = 1.0$, leaving it up to the user to take advantage of a higher value of C_b, if needed.

Elastic Lateral Torsional Buckling

NOTE This case is not given in the *Handbook.*

$$\text{For } L_b > L_r \qquad M_n = F_{cr} S_x \le M_p \tag{13.21}$$

where the critical stress F_{cr} is given by

$$F_{cr} = \frac{C_b \pi^2 E}{(L_b / r_{ts})^2} \sqrt{1 + 0.078 \frac{Jc}{S_x h_o} \left(\frac{L_b}{r_{ts}} \right)^2} \tag{13.22}$$

The design charts in the AISC manual reflect the $C_b = 1$ case, which is conservative. The effect of incorporating the bending coefficient C_b into the design charts is to increase the plotted values of the moment capacity M_n as shown by the broken line in Fig. 13.7. The only limitation to this increase is that the value of $C_b M_n$ cannot be greater than M_p.

Bending Coefficient C_b

If the bending moment within the unbraced length is nonuniform, the bending coefficient C_b is incorporated into the design to account for the fact that spanwise variations in the bending moment mean that the critical bending moment is not simultaneously reached at all locations, thus making the beam more resistant to compression flange buckling than if C_b were to equal 1.0 (for constant bending moment). The coefficient is defined as

$$C_b = \frac{12.5 M_{max}}{2.5 M_{max} + 3M_A + 4M_B + 3M_C} \tag{13.23}$$

where M_A, M_B, and M_C are the absolute values of the bending moments at $\frac{1}{4}L_b$, $\frac{1}{2}L_b$, and $\frac{3}{4}L_b$, respectively, and M_{max} is the absolute value of the maximum bending moment within the unbraced length L_b (see Fig. 13.6)

For some standard load cases involving a simply supported beam, the values of the bending coefficient C_b are given on page 157 of the *Handbook.*

Figure 13.6 Parameters used to calculate C_b.

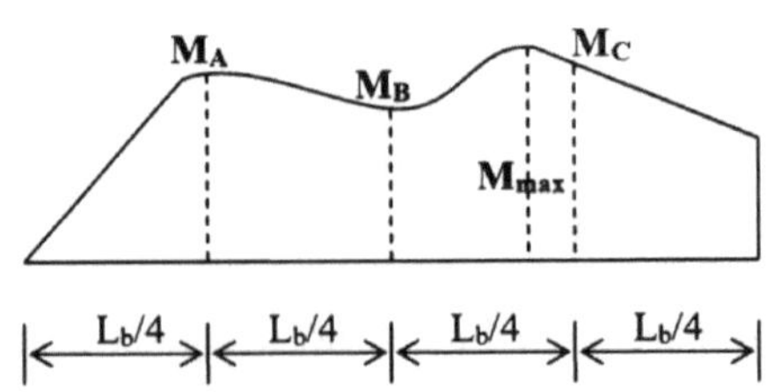

Figure 13.7 Nominal moment capacity of steel beams ($C_b > 1$).

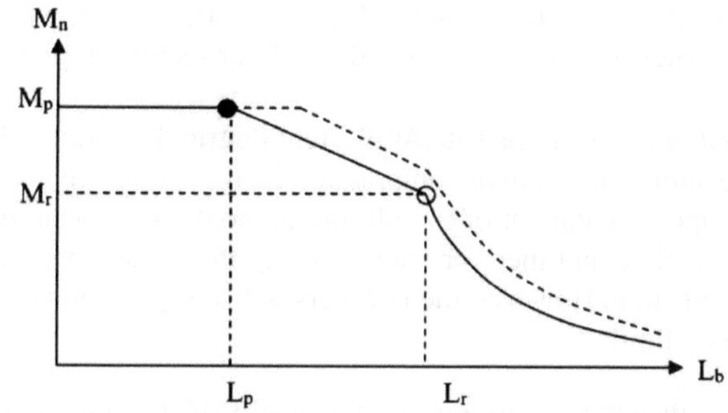

Beam Design Using Z_x Tables

When the unbraced length is very short, such as when the compression flange has "continuous" lateral support from the slab (via shear connectors) or when relatively closely spaced lateral members frame into the compression zone, L_b may be considered to be zero and the full plastic moment capacity of the beam may be realized.

$$\text{Thus, } M_n = M_p = F_y Z_x$$

This translates into the following requirement for minimum plastic section modulus Z_x

$$\text{LRFD:} \quad Z_x \geq \frac{M_u}{\phi F_y} = \frac{M_u}{0.9F_y} \tag{13.24}$$

The Z_x tables may be then used to select the section. The *Handbook* has a one-page extract of the Z_x table on page 160.

Example 13.16

The top flange of a simply supported beam (span = 20 ft) is continuously braced against buckling by the supported slab. If the service loads (uniformly distributed over the entire span) are dead load = 2 kips/ft, live load = 3 kips/ft, and wind load = 2 kips/ft, what is the lightest W-section which is satisfactory? Use $F_y = 50$ ksi.

Solution (LRFD) Load combinations are

(i) LRFD1 $w_u = 1.4\text{DL} = 2.8$ kips/ft

(ii) LRFD2 $w_u = 1.2\text{DL} + 1.6\text{LL} = 1.2 \times 2 + 1.6 \times 3 = 7.2$ kips/ft. This value controls.

(iii) LRFD3 $w_u = 1.2\text{DL} + 0.8\text{WL} = 1.2 \times 2 + 0.8 \times 2 = 4.0$ kips/ft

(iv) LRFD4 $w_u = 1.2\text{DL} + 1.6\text{WL} + 0.5\text{LL} = 1.2 \times 2 + 1.6 \times 2 + 0.5 \times 3 = 7.1$ kips/ft

$$M_u = \frac{w_u L^2}{8} = \frac{7.2 \times 20^2}{8} = 360 \text{ kip} \cdot \text{ft} = 4320 \text{ kip} \cdot \text{in}$$

$$Z_{x,\text{reqd}} = \frac{M_u}{0.9F_y} = \frac{4320}{0.9 \times 50} = 96 \text{ in}^3$$

From page 160 of the *Handbook*, W12 × 65 is the section with the smallest Z_x (96.8 in^3) that meets the criterion but W21 × 48 is the *lightest* section (top of the group—in bold).

Beam Design Using Table 3.10 (AISC Steel Conrtuction Manual)

When the unbraced length of compression flange is not zero, the beam's flexural strength could be based upon attainment of the plastic moment, or inelastic or elastic LTB. Without having chosen a section, and therefore not knowing the values of L_p and L_r, it is difficult to make a member selection. However, the ϕM_n versus L_b design charts (page 161 of the *Handbook*) may be used as follows:

1. Calculate the required moment capacity M_u from the load combinations.
2. The unbraced length is the longest beam length over which the compression flange is unsupported. For a beam having no intermediate bracing, the unbraced length is equal to the total span.
3. The unbraced length versus moment capacity chart plots ϕM_n versus unbraced length L_b. Identify the point corresponding to this required capacity.
4. Move vertically upward (keeping L_b constant) from this point until the *first solid line* is encountered. This is the *lightest* section which has adequate flexural strength. Any line which plots above the point represents a section which is adequate ($\phi M_n > M_u$) for that particular unbraced length.

Example 13.17

The top flange of a simply supported beam (span = 20 ft) is laterally braced at the supports and at midspan only. If the service loads (uniformly distributed over the entire span) are—dead load = 2 kips/ft, live load = 3 kips/ft, and wind load = 2 kips/ft, what is the lightest W-section which is satisfactory (use F_y = 50 ksi and C_b = 1.0)?

Solution (LRFD) Load combinations are

(i) LRFD1 $w_u = 1.4\text{DL} = 2.8$ kips/ft

(ii) LRFD2 $w_u = 1.2\text{DL} + 1.6\text{LL} = 1.2 \times 2 + 1.6 \times 3 = 7.2$ kips/ft. This value controls.

(iii) LRFD3 $w_u = 1.2\text{DL} + 0.8\text{WL} = 1.2 \times 2 + 0.8 \times 2 = 4.0$ kips/ft

(iv) LRFD4 $w_u = 1.2\text{DL} + 1.6\text{WL} + 0.5\text{LL} = 1.2 \times 2 + 1.6 \times 2 + 0.5 \times 3 = 7.1$ kips/ft

$$M_u = \frac{w_u L^2}{8} = \frac{7.2 \times 20^2}{8} = 360 \text{ kip}\cdot\text{ft} = 4320 \text{ kip}\cdot\text{in}$$

From the chart on page 161 of the *Handbook*, if the point L_b = 10 ft, ϕM_n = 360 kip·ft is plotted, W18 × 55 is the first solid line above that point. This is the lightest section that is satisfactory for the given situation. This happens to be in the second zone.

Verification

$$\phi_b M_n = \phi_b M_p - BF \times (L_b - L_p) = 420 - 13.9 \times (10 - 5.9) = 363$$

Therefore, $\phi_b M_n > 360$ (OK)

Example 13.18

The top flange of a simply supported beam (span = 20 ft) has no intermediate lateral support. If the service loads (uniformly distributed over the entire span) are—dead load = 2 kips/ft, live load = 3 kips/ft, wind load = 2 kips/ft, what is the lightest W-section which is satisfactory? For this example, use the C_b value as appropriate for the loading condition. Use $F_y = 50$ ksi.

Solution (LRFD) Load combinations are

(i) LRFD1 $w_u = 1.4\text{DL} = 2.8$ kips/ft

(ii) LRFD2 $w_u = 1.2\text{DL} + 1.6\text{LL} = 1.2 \times 2 + 1.6 \times 3 = 7.2$ kips/ft. This value controls.

(iii) LRFD3 $w_u = 1.2\text{DL} + 0.8\text{WL} = 1.2 \times 2 + 0.8 \times 2 = 4.0$ kips/ft

(iv) LRFD4 $w_u = 1.2\text{DL} + 1.6\text{WL} + 0.5\text{LL} = 1.2 \times 2 + 1.6 \times 2 + 0.5 \times 3 = 7.1$ kips/ft

$$M_u = \frac{w_u L^2}{8} = \frac{7.2 \times 20^2}{8} = 360 \text{ kip} \cdot \text{ft} = 4320 \text{ kip} \cdot \text{in}$$

For a uniformly loaded simply supported beam having no intermediate lateral support, $C_b = 1.14$ (page 157 of the *Handbook*). Since AISC Table 3.10 is for $C_b = 1.0$, we modify the target strength as follows:

$$\phi_b M_n \geq \frac{M_u}{C_b} = \frac{360}{1.14} = 315.8 \text{ kip} \cdot \text{ft}$$

From the chart on page 161 of the *Handbook*, if the point $L_b = 10$ ft, $\phi M_n = 315.8$ kip · ft is plotted, we see that there are several lines (broken) just above this (modified) target. Some of these are W10 × 68, W18 × 50, W21 × 50, and so on. However, tracing these lines back (to the left) indicates that $\phi_b M_p$ values are as follows:

For the W10 × 68, $\phi_b M_p = 320$ kip · ft. In other words, the moment capacity of a W10 × 68 can never be taken as more than 320 kip · ft. This is inadequate, since our actual target for moment capacity is 360 kip · ft.

For the W18 × 50, $\phi_b M_p = 379$ kip · ft. In other words, the moment capacity of a W18 × 50 can never be taken as more than 379 kip · ft. This could work, since our *actual target* for moment capacity is 360 kip · ft.

Verification (W18 × 50)

From page 160 of the *Handbook*, $\phi_b M_p = 379$ kip · ft, $\phi_b \text{BF} = 13.1$ kips, $L_p = 5.83$ ft, $L_r = 17.0$ ft

$$C_b \phi_b M_n = 1.14 \times [379 - 13.1 \times (10 - 5.83)] = 369.8 \, \text{kip} \cdot \text{ft}$$

This "magnified" value is acceptable, since it does not exceed $\phi_b M_p = 379$ kip · ft. If it were to exceed 379 kip · ft, it would be "capped" to a maximum of 379 kip · ft.

Therefore, $C_b \phi_b M_n = 369.8 > 360$ (OK)

Example 13.19

A W16 × 67 beam has a simple span = 30 ft and carries a uniformly distributed load over the entire span. Steel grade is A992 grade 50. The compression flange of the beam has lateral support at the supports and at midspan only. Ignore C_b. The maximum factored load (kips/ft) the beam can carry is most nearly

A. 2.75

B. 3.75

C. 4.50

D. 5.15

Solution Since the beam is braced at midspan, $L_{\text{unbraced}} = 30/2 = 15$ ft.
From the Z_x table extract in the *Handbook* (page 156), we get the following parameters for a W16 × 67

$$\phi_b M_p = 488 \text{ kip}\cdot\text{ft};\ L_p = 8.69 \text{ ft};\ L_r = 26.1 \text{ ft};\ \phi_b \text{BF} = 10.4 \text{ kips}$$

Since the unbraced length ($L_b = 15$ ft) is between L_p and L_r, the moment capacity $\phi_b M_p$ is given by

$$\phi_b M_n = 488 - 10.4 \times (15 - 8.69) = 422.4 \text{ kip}\cdot\text{ft}$$

Max. factored moment capacity = 422.4 kip · ft (approx.)

For a simple span, maximum moment for uniformly distributed load: $M_u = \dfrac{1}{8} wL^2$

$$\frac{1}{8} wL^2 = 422.4 \Rightarrow w = \frac{422.4 \times 8}{30^2} = 3.75 \text{ kip}\cdot\text{ft}$$

Answer is B.

Design for Shear

The exact calculation for shear stress in a transversely loaded member is given by

$$f_v = \frac{VQ}{Ib} \tag{13.25}$$

where f_v is the transverse and longitudinal shear stress at a point in the beam, V is the transverse shear force, Q is the first moment (about the neutral axis) of either "half" of the cross section, I is the moment of inertia of the section about the neutral axis, and b is the width of the beam at the point of interest. For more details, see Chap. 8.

For wide-flange sections, the average shear stress calculated by distributing the shear force over the area of the web alone yields a good estimate of the actual maximum shear stress (at the neutral axis). This simplified equation [Eq. (13.26)] is adopted by the AISC design specifications. The effective area of the web resisting shear is calculated as $A_w = dt_w$, where d = overall depth and t_w = web thickness (Fig. 13.8).

$$\text{Average shear stress } f_v = \frac{V}{A_w} \tag{13.26}$$

Figure 13.8
W-shape steel beam.

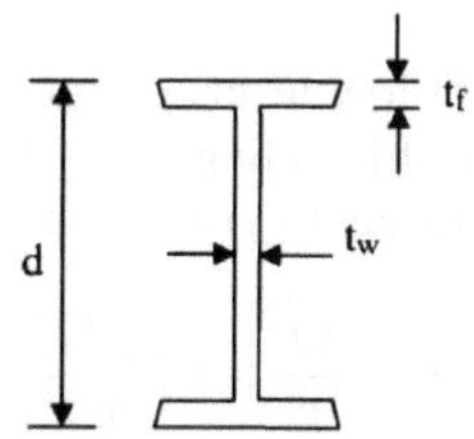

The nominal shear capacity of the section is given by

$$V_n = 0.6F_y A_w = 0.6F_y d t_w \qquad (13.27)$$

For LRFD, the design requirement may be stated as

$$V_u \le \phi_v V_n \qquad (13.28)$$

where V_u is the required shear strength = maximum shear calculated from the LRFD load combinations and ϕ_v is the strength reduction factor for shear. Even though the strength reduction factor for shear can take on the value 0.9 under certain conditions, the *Handbook* simply states $\phi_v = 1.0$.

Analysis and Design of Steel Compression Members

The fundamental theory about buckling under axial compression by Euler is based upon the assumption of elastic material behavior. Euler's elastic buckling stress is given by

$$F_e = \frac{\pi^2 E}{(KL/r)^2} \qquad (13.29)$$

where E = modulus of elasticity of the material (page 84 of the *Handbook* gives E_{steel} = 29,000 ksi = 200 GPa)
K = effective length factor (page 162 of the *Handbook*)
L = physical length of the column
$r = \sqrt{\frac{I}{A}}$ = radius of gyration about the buckling axis

The critical buckling stress of a steel column F_{cr} depends on whether buckling is classified as inelastic (low slenderness ratio, allowing the steel to reach higher stress levels until buckling occurs) or elastic (more slender columns buckle at low load level, thereby ensuring elastic behavior). The equations outlined in AISC are given below.

Elastic Buckling

Slender columns buckle at low levels of stress (significantly less than F_y) and therefore the AISC recommended critical buckling stress F_{cr} is very close to the Euler buckling stress F_e, with an additional factor of safety to account for geometric imperfection of the column.

$$\text{For } \frac{KL}{r} > 4.71\sqrt{\frac{E}{F_y}} \qquad F_{cr} = 0.877F_e \tag{13.30}$$

Inelastic Buckling

As the column becomes less slender, the stress level needed to cause buckling increases and the column behaves inelastically, causing significant departure from Euler's (elastic) theory. The following equation is given in the AISC specifications.

$$\text{For } \frac{KL}{r} \leq 4.71\sqrt{\frac{E}{F_y}} \qquad F_{cr} = \left(0.658^{\frac{F_y}{F_e}}\right)F_y \tag{13.31}$$

Squash Load for Pedestals

When the slenderness ratio KL/r approaches zero, as for a pedestal or stub column, the critical stress F_{cr} approaches the yield stress F_y. The design load corresponding to this limit is termed the squash load, given by $0.9F_y A_g$.

Example 13.20

What is the *nominal compression load capacity* of a W12 × 50 column whose governing slenderness ratio = 90? Use $F_y = 38$ ksi and $F_u = 55$ ksi.

A. 356 kips

B. 376 kips

C. 396 kips

D. 416 kips

Solution W12 × 50 section has cross-sectional area $A_g = 14.7 \text{ in}^2$

$$4.71\sqrt{\frac{E}{F_y}} = 4.71\sqrt{\frac{29000}{38}} = 130.1$$

Euler buckling stress:

$$F_e = \frac{\pi^2 E}{(KL/r)^2} = \frac{\pi^2 \times 29,000}{90^2} = 35.34 \text{ ksi}$$

Since $KL/r < 130.1$, critical buckling stress:

$$F_{cr} = (0.658^{F_y/F_e})F_y = (0.658^{38/35.34})38 = 24.23 \text{ ksi}$$

Nominal load capacity (compression): $P_n = F_{cr}A_g = 24.23 \times 14.7 = 356$ kips.

Answer is A.

Example 13.21

A steel column ($F_y = 50$ ksi) has effective length = 20 ft. Find the smallest satisfactory W12 section that can support an axial load of 200 kips dead load + 215 kips live load.

A. W12 × 53

B. W12 × 58

C. W12 × 65

D. W12 × 72

Solution

$$P_u = 1.2(DL) + 1.6(LL) = 1.2(200) + 1.6(215) = 584 \text{ kips}$$

Required design strength = $\phi_c P_n = 584$ kips

From page 164, for effective length = 20 ft, the design capacity of the W12 × 58 section is 393 kips, which is not adequate. So, it must be one of the heavier sections (W12 × 65 or W12 × 72).

For W12 × 65 (from page 159), $r_y = 3.02$ in; $A_g = 19.1$ in^2; $KL/r = 20 \times 12/3.02 = 79.5$. From page 163, for $KL/r = 79.5$, $\phi_c F_{cr} = 28.35$ ksi. Therefore, $\phi_c P_n = \phi_c F_{cr} A_g = 28.35 \times 19.1 = 541.5$ kips, which is not adequate.

For W12 × 72 (from page 159), $r_y = 3.04$ in; $A_g = 21.1$ in^2; $KL/r = 20 \times 12/3.04 = 78.95$. From page 163, for $KL/r = 78.95$, $\phi_c F_{cr} = 28.5$ ksi. Therefore, $\phi_c P_n = \phi_c F_{cr} A_g = 28.5 \times 21.1 = 601.4$ kips.

Use W12 × 72 (capacity = 601 kips).

Answer is D.

Effective Length for Columns in a Frame

The predetermination of the effective length factor K is not possible for columns which are not "standalone" but rather are part of a structural frame. For example, if the column being considered is designated AB, the rotational rigidity of ends A and B depends on the relative bending stiffness of all members (columns and beams) framing into these ends. The AISC specifications outline the following procedure:

1. At either end of the column, calculate a coefficient G given by

$$G = \frac{\left(\sum I/L\right)_{\text{columns}}}{\left(\sum I/L\right)_{\text{beams}}} \tag{13.32}$$

For column ends supported by, but not rigidly connected to, a footing or foundation, G is theoretically infinite, but the AISC specifications recommend using $G = 10.0$.

For a column end rigidly attached to a properly designed footing, G is theoretically zero, but the AISC specifications recommend using $G = 1.0$.

2. The two G values are then used in one of two alignment charts (page 162 of the *Handbook*) to obtain the effective length factor K—depending on whether the frame is sidesway inhibited (Fig. C-A-7.1) or sidesway uninhibited (Fig. C-A-7.2) of the AISC Steel Construction Manual.
3. The K obtained from the alignment chart is then used to determine the governing slenderness ratio for the column.

Example 13.22

For the unbraced steel building frame shown in the figure, the numbers represent I/L ratios (in^4/ft) of the beams and columns. What is the effective length factor K for an interior column on the second floor (indicated in bold type)? Ignore inelastic behavior.

A. 0.64

B. 1.10

C. 0.90

D. 1.80

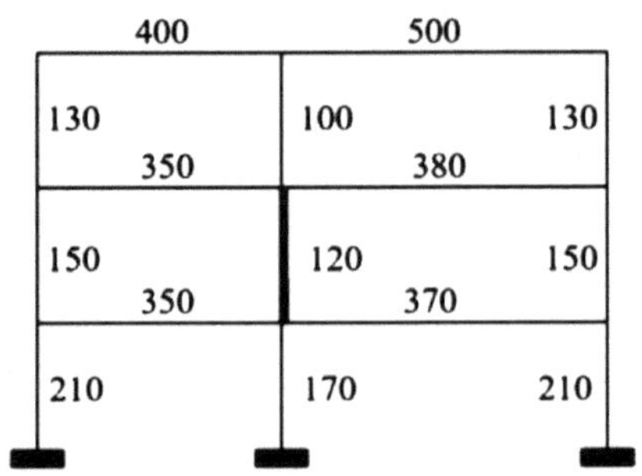

Solution

$$G_{\text{top}} = (100 + 120)/(350 + 380) = 0.3$$

$$G_{\text{bottom}} = (120 + 170)/(350 + 370) = 0.4$$

For a braced (sidesway uninhibited) frame, Fig. C-A7.2 (page 158 of the *Handbook*) yields $K = 1.11$.

Answer is B.

Analysis and Design of Steel Tension Members

This chapter reviews the analysis and design of steel tension members. Steel tension members are designed for two limit states: (1) yielding of the gross section and (2) fracture of the critical net section. In addition, the connection must be checked for the mechanism of block shear.

In the current specification, it is recommended that the slenderness ratio of a steel tension member (L/r_{min}) be less than 300. The designer must exercise caution specifying members with high slenderness ratio, as these may exhibit instability problems during erection.

For LRFD, the design requirement may be stated as

$$P_u \le \phi_t P_n \tag{13.33}$$

where P_u is the required strength in tension = maximum axial tension calculated from the LRFD load combinations.

Nominal Member Strength in Tension

The nominal strength in tension P_n of a steel member is the lesser of two limit states—yielding of the gross section and fracture of the net section.

Nominal strength in yielding of the gross area is given by

$$P_n = F_y A_g \tag{13.34}$$

Nominal strength in fracture of the effective net area is given by

$$P_n = F_u A_e = F_u U A_n \tag{13.35}$$

where A_g = gross area of the member cross section
A_n = net area of the member cross section
U = shear lag factor
F_y = yield stress (tension) of steel
F_u = ultimate stress (tension) of steel

For LRFD, the strength reduction factor for yield is $\phi_t = 0.90$, resulting in

$$P_u \le 0.9 F_y A_g \tag{13.36}$$

and the strength reduction factor for fracture is $\phi_t = 0.75$, resulting in

$$P_u \le 0.75 F_u A_e \tag{13.37}$$

Example 13.23

What is most nearly the axial *design capacity* of a W18 × 71 tension member based on the yield and fracture limit states? Assume $A_e = 0.75A_g$. Use $F_y = 42$ ksi and $F_u = 60$ ksi.

A. 700 kips
B. 784 kips
C. 868 kips
D. 933 kips

Solution From page 155 of the *Handbook*, gross area of a W18 × 71 is $A_g = 20.8$ in^2. The axial design capacity based on yield of the gross area is given by

$$P_u = 0.9F_yA_g = 0.9 \times 42 \times 20.8 = 786.2 \text{ kips}$$

The axial design capacity based on fracture of the net area is given by

$$P_u = 0.75F_uA_e = 0.75F_u(0.75A_g) = 0.75 \times 60 \times 0.75 \times 20.8 = 702.0 \text{ kips. This governs.}$$

Normally, the gross area of the section is first reduced to the governing (minimum) net area and then the net area is reduced to the effective net area by taking into account the shear lag factor U. The first step requires knowledge of the number, size, and arrangement of bolt holes, and the second step requires knowledge of the connectivity of various parts of the section. Since these specifics are not given the question instructs to lump these two reductions into one ($A_e = 0.75A_g$). This assumption is consistent with the tables in the AISC manual.

Answer is A.

Net Area in Tension

The governing net area is calculated as the net area at the weakest section. For welded members, the net area is equal to the gross area A_g. When connections in a tension member consist of bolts (fasteners), the net area is calculated by subtracting areas removed by drilling for bolts. The usual practice is to drill or punch standard sized holes with diameter 1/16 in larger than the nominal bolt diameter. To account for roughness around the edges of the hole, AISC specifications require an additional 1/16 in deduction for strength computations. Thus, the diameter of the hole is calculated as the nominal diameter of the bolt plus 1/8 in

$$d_{\text{hole}} = d_{\text{bolt}} + \frac{1}{8}'' \tag{13.38}$$

For bolted connections, the governing net area is calculated from the fracture pattern which yields the smallest reduced area after subtracting areas of the holes. This reduction is calculated as the product of the hole diameter d and the thickness of the part with the hole t. For example, for the bolt pattern (line 1) in the plate shown in Fig. 13.9, the net width of the plate is equal to the total width minus two hole diameters. This net width can then be multiplied by the plate thickness to get the net area. Calculating the net area along a line which is not normal to the direction of the load is not as straightforward. The AISC specifications allow the use of a mathematically simple method that closely approximates the strength of such "staggered" bolt lines. This approach is presented in the next section.

Net Area for Staggered Bolt Lines

Figure 13.9 shows a plate used as a member in tension. The connection is composed of five bolts as shown. The three sections marked 1, 2, and 3 all experience the full member load P_u. In section 1, the net area has a single bolt, in section 2 the net area has two bolts and one staggered segment (dimensions s_1 and g_1), and in section 3, the net area has three bolts with two staggered segments (dimensions s_1, g_1, and s_2, g_2, respectively).

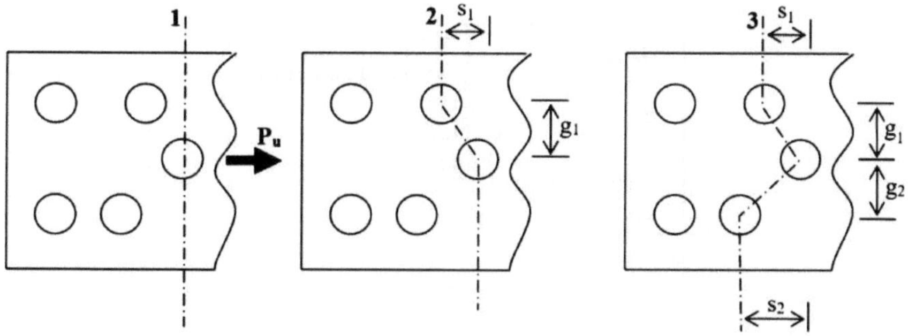

Figure 13.9 Net area for tension members.

The connection shown in Fig. 13.9 must be checked for three fracture patterns: lines 1, 2, and 3. For staggered lines such as 2 and 3, the AISC specifications suggest the following simplified method.

For each staggered line segment, a correction term, given by $\frac{s^2}{4g}t$, is added to the net area.

Thus, for a pattern that consists of several staggered segments (line 2 above has two), the correction is calculated as the summation of all such terms, calculated individually for each segment. In this calculation, s is the longitudinal spacing between the bolts and g is the transverse or gage distance between the bolts.

$$A_{\text{net}} = A_g - A_{\text{holes}} + \sum \frac{s^2}{4g}t \tag{13.39}$$

Thus, for the plate shown in Fig. 13.9, the net area along sections 1, 2, and 3 are

$$A_{\text{net},1} = A_g - A_{\text{hole}} = (b - d_H)t$$

$$A_{\text{net},2} = \left(b - 2d_H + \frac{s_1^2}{4g_1}\right)t$$

$$A_{\text{net},3} = \left(b - 3d_H + \frac{s_1^2}{4g_1} + \frac{s_2^2}{4g_2}\right)t$$

Example 13.24

Find the governing net area for the tension member (7/8-in-thick steel plate) shown below.

A. 6.21 in^2

B. 6.74 in^2

C. 6.85 in^2

D. 7.04 in^2

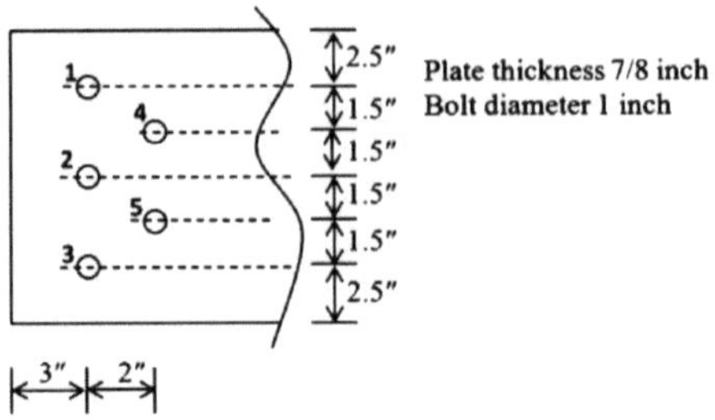

Solution

Line 4-5: $A_{net,2} = 11 \times 0.875 - 2 \times 7/8 \times 9/8 = 7.656$

Line 1-4-5: $A_{net,3} = 11 \times 0.875 - 3 \times 7/8 \times 9/8 + 1 \times 0.875 \times (2^2/4 \times 1.5) = 7.255$

Line 1-4-2-5: $A_{net,4} = 11 \times 0.875 - 4 \times 7/8 \times 9/8 + 3 \times 0.875 \times (2^2/4 \times 1.5) = 6.854$

Line 1-4-2-5-3: $A_{net,5} = 11 \times 0.875 - 5 \times 7/8 \times 9/8 + 4 \times 0.875 \times (2^2/4 \times 1.5) = 7.036$

Governing net area, $A_{net} = 6.85$ in^2.

Answer is C.

Effective Net Area

The effective net area A_e is given by the product of the governing net area A_{net} and the shear lag coefficient U. The shear lag coefficient represents the inefficiency of a connection in transmitting tension when only certain parts of the section are connected. This inefficiency may exist for both bolted and welded connections.

Shear lag coefficient: The area used in the calculation of the fracture limit state capacity is called the effective net area A_e. This area is calculated as a fraction of the minimum net area A_{net}.

$$A_e = UA_{net} \tag{13.40}$$

where U is the shear lag coefficient.

Bolted Members For plates connected by bolts, where the connection is in one plane, $U = 1.0$. For other sections, if all parts are connected (such as for a single angle which is connected in both legs), $U = 1.0$.

For all other sections, when all parts of the tension member are not directly connected, the shear lag factor U is less than 1.0.

The AISC specifications outline several categories of connections for calculating or assigning the shear lag coefficient.

Welded Members

1. *For any type of tension member except plates and round HSS with $l \geq 1.3D$*

The coefficient is calculated using Eq. (13.41)

$$U = 1 - \frac{\bar{x}}{L} \tag{13.41}$$

where L is the length of the connection (distance between outermost bolts) and $\bar{x}$ is the distance from the plane of the connection to the centroid of the section.

If a section has two symmetrically placed planes of connection, then $\bar{x}$ is measured from the centroid of the nearest one-half of the area.

2. *Plates*

For plates where the tension load is transmitted by longitudinal welds (length l and width of plate w) only, as shown in Fig. 13.10

for $l \geq 2w$	$U = 1.0$
for $2w > l \geq 1.5w$	$U = 0.87$
for $1.5w > l \geq w$	$U = 0.75$

3. *Round HSS with $l \geq 1.3D$*

$U = 1.0$

4. *Single angles*

In lieu of the equation-based approach, the following values are permitted for single angles.

For four or more fasteners in the direction of the load, $U = 0.80$.

For two or three fasteners in the direction of the load, $U = 0.60$.

5. *W-, M-, S-, HP-, and T-sections cut from them*

In lieu of the equation-based approach, the following values are permitted for the sections named above.

- Connected through the flange with three or more fasteners in the direction of the load, with a width at least equal to 2/3 of the depth: $U = 0.90$.
- Connected through the flange with three or more fasteners in the direction of the load, with a width less than 2/3 of the depth: $U = 0.85$.
- Connected through the web with four or more fasteners in the direction of the load, with a width less than 2/3 of the depth: $U = 0.70$.

Figure 13.10 Plate tension member with longitudinal welds only.

Example 13.25

A C10 × 30 is used as a tension member as shown. Bolts are ¾ in diameter A325 high strength bolts. $F_y = 36$ ksi, $F_u = 58$ ksi. Geometric properties of the channel are

$d = 10.0$ in

$A = 8.82\ \text{in}^2$

$b_f = 3.03$ in

$t_f = 0.436$ in

$t_w = 0.673$ in

The design strength of the member is most nearly

A. 195 kips

B. 275 kips

C. 315 kips

D. 355 kips

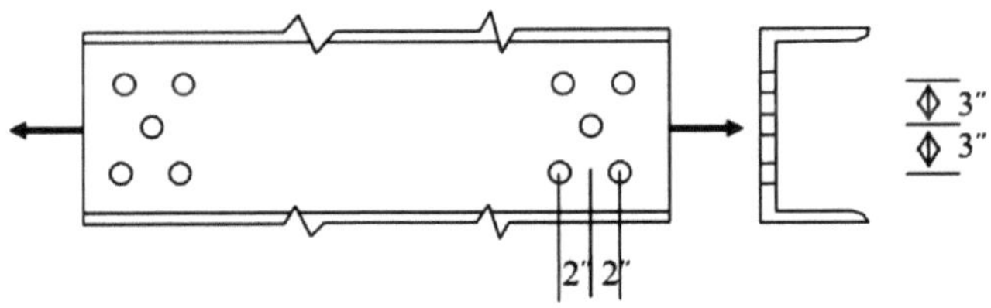

Solution For a C10 × 30, $A_g = 8.81\ \text{in}^2$; $t_w = 0.673$ in
Hole diameter = bolt diameter + 1/8 in = 7/8 in
Design strength (based on yield capacity): $\phi_t P_n = 0.9F_y A_g = 0.9 \times 36 \times 8.81 = 285.4$ kips

Net area no. 1: $A_{\text{net},1} = 8.81 - 2 \times \frac{7}{8} \times 0.673 = 7.63\ \text{in}^2$

Net area no. 2: $A_{\text{net},2} = 8.81 - 3 \times \frac{7}{8} \times 0.673 + 2 \times \frac{2^2}{4 \times 3} \times 0.673 = 7.49\ \text{in}^2$

Shear lag factor: $U = 1 - \frac{0.649}{4} = 0.84$

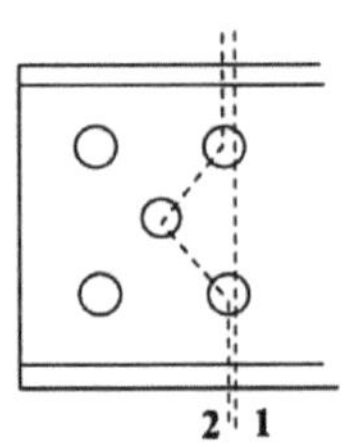

Design strength (fracture through effective net area): $\phi_t P_n = 0.75 F_u A_e = 0.75 \times 58 \times 0.84 \times 7.49 = 273.7$ kips. This governs.

Answer is B.

Block Shear

The adequacy of a structural member in tension must be established in three aspects:

1. The adequacy of the member itself (yield of the gross section and fracture of the net section)
2. The adequacy of the connection elements (welds or bolts) to withstand the member force
3. The capacity of the configuration of the connection that may create weak "zones" due to inadequate separation and edge distance. This failure, called block shear, involves the "tearing out" of a "block" or zone under the combined effect of shear and tension. In Fig. 13.11, the net area subjected to shear fracture is shown with a cross-hatched pattern while the net area subjected to tensile fracture is shown with a dotted pattern.

The nominal capacity in block shear is given as the smaller of

$$R_n = 0.60 F_u A_{nv} + U_{bs} F_u A_{nt}$$
$$R_n = 0.60 F_y A_{gv} + U_{bs} F_u A_{nt}$$

U_{bs} is a factor to reflect unequal distribution of tensile stress, but is taken to be 1.0 for flat bars and angles.

A_{gv} = gross area subject to shear

A_{nv} = net area subject to shear

A_{nt} = net area subject to tension

The design strength is given by ϕR_n ($\phi = 0.75$).

Figure 13.11
Block shear.

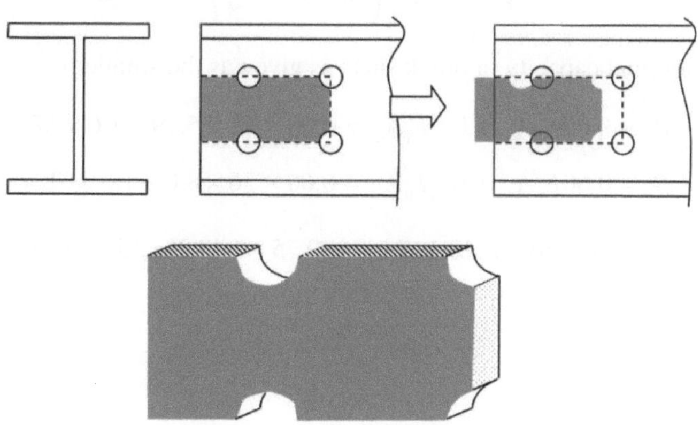

Example 13.26

A L8X8X1 truss member ($F_y = 36$ ksi; $F_u = 58$ ksi) is loaded in tension as shown. Fasteners are 1-in-diameter high strength bolts.

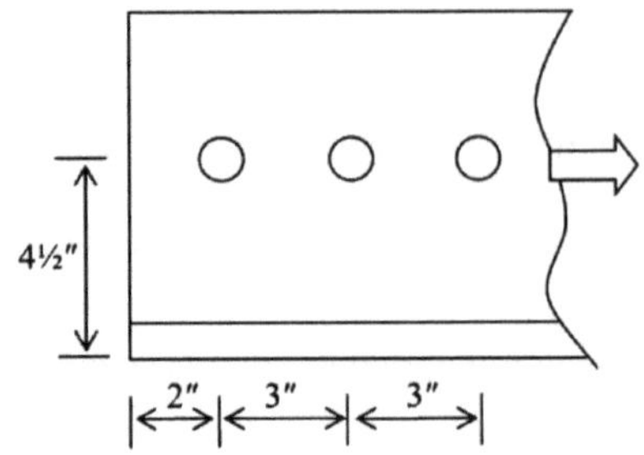

The available strength in *block shear* (kips) is most nearly

A. 250

B. 300

C. 350

D. 400

Solution For the "block" shown below:

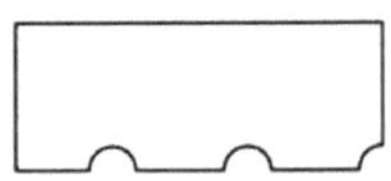

$$A_{gv} = 8 \times 1.0 = 8.0 \text{ in}^2$$

$$A_{nv} = \left(8 - 2.5 \times \frac{9}{8}\right) \times 1.0 = 5.19 \text{ in}^2$$

$$A_{nt} = \left(3.5 - 0.5 \times \frac{9}{8}\right) \times 1.0 = 2.94 \text{ in}^2$$

The nominal capacity in block shear is given as the smaller of

$$R_n = 0.60 F_u A_{nv} + U_{bs} F_u A_{nt} = 0.60 \times 58 \times 5.19 + 1.0 \times 58 \times 2.94 = 351.1 \text{ kips}$$

$$R_n = 0.60 F_y A_{gv} + U_{bs} F_u A_{nt} = 0.60 \times 36 \times 8.0 + 1.0 \times 58 \times 2.94 = 343.3 \text{ kips}$$

The design capacity in block shear is $0.75 \times 343.3 = 257.5$ kips.

Answer is A.

CHAPTER 14

Geotechnical Engineering

Mineralogy

Soil is the uncemented aggregate of mineral grains (solid particles) with liquid and gas in the void spaces between the solid particles. Mineral grains are the product of weathering, which is the breakdown of rock masses due to physical and chemical processes. The three major classifications of rocks—based on the processes leading to their formation—are igneous, sedimentary, and metamorphic.

Igneous Rocks

Igneous rocks are formed by the solidification of molten magma ejected from the earth's mantle. The rate at which magma cools determines the sequence in which different minerals are formed. Extrusive igneous rocks form when magma makes its way to the earth's surface as lava and then cools forming rock. Typical time for lava to cool ranges from a few days to weeks and minerals have enough time to form but not enough to grow into large crystals. The most abundant type of extrusive igneous rock on earth is basalt. Intrusive igneous rocks are formed from magma that cools and solidifies within the earth's crust. Such magma cools very slowly and as a result, these rocks are coarse-grained. The most abundant type of intrusive igneous rock on earth is granite. Table 14.1 lists various igneous rocks, their mode of occurrence and mineral content.

Weathering

Weathering is the breaking down of rocks into smaller particles due to *mechanical and chemical processes*. Physical weathering processes may be due to expansion and contraction from continuous gain and loss of heat. Other physical agents are glacier, wind, ice, rivers, and ocean waves. The only change that occurs is in physical size and not in chemical composition.

In chemical weathering, original rock minerals are transformed by chemical reaction. Water and carbon dioxide from the atmosphere form carbonic acid, which in turn reacts

Table 14.1 Classification and Mineral Content of Igneous Rocks

Name	Mode of occurrence	Texture	Minerals
Granite	Intrusive	Coarse	Quartz, sodium feldspar, potassium feldspar
Rhyolite	Extrusive	Fine	
Gabbro	Intrusive	Coarse	Plagioclase, pyroxenes, olivine
Basalt	Extrusive	Fine	
Diorite	Intrusive	Coarse	Plagioclase, hornblende
Andesite	Extrusive	Fine	
Syenite	Intrusive	Coarse	Potassium feldspar
Trachyte	Extrusive	Fine	
Peridotite	Intrusive	Coarse	Olivine, pyroxenes

with minerals. For example, orthoclase undergoes chemical weathering to turn into kaolinite, which is a clay mineral.

Transportation of weathering products occurs due to glaciers (loess and till), running water in streams (alluvial), deposition in quiet lakes (lacustrine), deposition in seas (marine), wind (aeolian), and soil movement due to gravity (colluvial).

Sedimentary Rocks

Deposited soil particles become compacted by overburden pressure and cemented by agents such as iron oxide. The cementing agents are usually in solution. Soils formed in this way are called *detrital sedimentary rocks*. Examples of sedimentary rocks are calcite, dolomite and quartz.

Sedimentary rocks can also be formed by chemical processes. Rocks such as limestone, chalk, dolomite, gypsum, anhydrite are called *chemical sedimentary rocks*.

Diagenesis is also a change in form that occurs in sedimentary rocks. In geology, however, we restrict diagenetic processes to those which occur at temperatures below 200°C and pressures below about 300 MPa (this is equivalent to about 3 kilobars of pressure).

Metamorphic Rocks

Metamorphic rocks are formed by a change of chemical composition and texture of soils due to pressure and heat. New minerals are formed and mineral grains are sheared to give a foliated texture. Granite, diorite, and gabbro transform to gneiss by high-grade metamorphism. Shales and mudstones are transformed into slates and phyllites by low-grade metamorphism.

Metamorphism occurs at temperatures and pressures higher than 200°C and 300 MPa. These conditions are present for rocks buried deep in the earth, usually as a result of tectonic processes such as continental collisions or subduction.

Low-grade metamorphism takes place at temperatures between about 200 to 320°C, and relatively low pressure. Low-grade metamorphic rocks are generally characterized by an abundance of hydrous minerals. High-grade metamorphism takes place at temperatures greater than 320°C and relatively high pressure. As grade of metamorphism increases, hydrous minerals lose water and nonhydrous minerals become more common. The upper limit of metamorphism occurs at the pressure and temperature where melting of the rock in question begins, at which point, the process changes to an igneous process rather than a metamorphic process.

The different types of metamorphism are (1) contact with igneous intrusion creating fine-grained rock with little foliation, (2) regional metamorphism over large area showing extensive foliation, (3) cataclastic metamorphism due to large mechanical deformation, such as in the fault zone, (4) hydrothermal, and (5) shock or impact metamorphism.

Most metamorphic textures involve foliation. Foliation is generally caused by a preferred orientation of sheet silicates. If a rock has a slatey cleavage as its foliation, it is termed a "slate"; if it has a phyllitic foliation, it is termed a "phyllite"; if it has a schistose foliation, it is termed a "schist." A rock that shows a banded texture without a distinct foliation is termed a "gneiss."

Schists are a type of metamorphic rock with well-foliated texture. Dolomite and calcite transform into marble by recrystallization. Quartzite is a metamorphic rock formed from quartz-rich sandstones. Silica enters into the void spaces between quartz and sand grains and acts as a cementing agent.

Clay Minerals

Clay minerals are complex aluminum silicates composed of two basic units (1) silica tetrahedron (4 oxygen atoms surrounding a silicon atom) and (2) alumina octahedron (6 hydroxyl ions surrounding an aluminum atom). Clay particles carry a net negative charge on their surfaces. There are three important clay minerals: kaolinite, illite, and montmorillonite.

Kaolinite (repeating sheets of elemental silica-gibbsite sheets in a 1:1 lattice, each layer about 7.2 Å thick). Specific surface (surface area per unit mass) is about 15 m^2/g. The layers are held together by hydrogen bonding.

Illite (sometimes called clay mica) consists of a gibbsite sheet bonded to two silica sheets bonded by potassium ions. Illite particles are typically between 50 and 500 Å thick and have specific area of about 80 m^2/g.

Montmorillonite is similar to illite in structure. Montmorillonite particles are typically between 10 and 50 Å thick and have specific area of about 800 m^2/g.

Soil as a Three-Phase System

Soil may be described as a three-phase system consisting of soil solids, voids containing pore water and empty voids filled with air. A saturated soil may be assumed to have only two of these phases, as voids are completely saturated with water.

The default value of specific weight (weight per unit volume) of water is taken as $\gamma_w = 62.4\ \text{lb/ft}^3$ (9810 N/m^3). The weight of water is converted to an equivalent volume by dividing by γ_w.

The specific gravity of soil solids G_s is typically around 2.60. The weight of solids is converted to an equivalent volume by dividing by $\gamma_s = G_s\,\gamma_w$.

Figure 14.1 shows a schematic view of the phase distribution in a soil mass. The three-phase system therefore has five unknown quantities (three volumes and two weights).

Figure 14.1 Component phases in a soil.

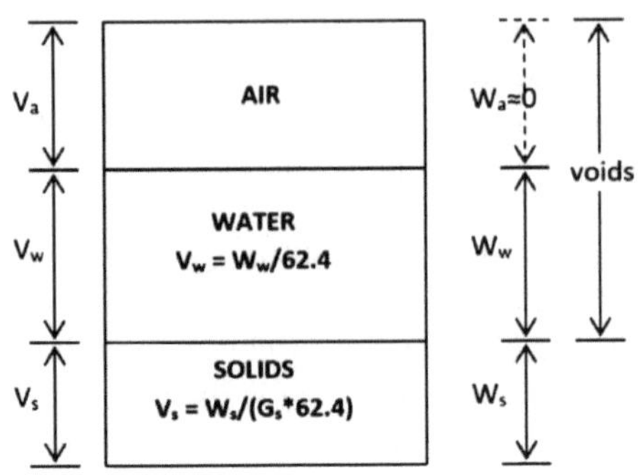

There are two implicit equations: one relating the volume of water to the weight of water and the other relating the volume of solids to the weight of solids.

$$V_w = \frac{W_w}{\gamma_w} \tag{14.1}$$

$$V_s = \frac{W_s}{G_s \gamma_w} \tag{14.2}$$

Thus, if three parameters are specified, we have adequate information to solve for the five system parameters. Once all five-phase parameters have been determined, any of the derived parameters—such as void ratio, porosity, degree of saturation—can be calculated.

Fundamental Definitions

The fundamental definition of the phase parameters is given below. In addition, cross-relationships between these parameters are also given.

The water content w is the dimensionless ratio of the weight of water to the weight of the solids

$$w = \frac{W_w}{W_s} \tag{14.3}$$

The void ratio e is the dimensionless ratio of the volume of the voids to the volume occupied by the soil solids

$$e = \frac{V_{\text{voids}}}{V_s} = \frac{V_a + V_w}{V_s} = \frac{n}{1-n} = \frac{wG_s}{S} = \frac{G_s \gamma_w}{\gamma_d} - 1 \tag{14.4}$$

The porosity n is the dimensionless ratio of the volume of the voids to the total volume

$$n = \frac{V_{\text{voids}}}{V_{\text{total}}} = \frac{V_a + V_w}{V_{\text{total}}} = \frac{e}{e+1} \tag{14.5}$$

The degree of saturation S is the fraction of the voids that are saturated with water

$$S = \frac{V_w}{V_{\text{voids}}} \tag{14.6}$$

Volume occupied by soil solids is

$$V_s = \frac{W_s}{G_s \gamma_w} = V_{\text{total}}(1-n) = \frac{V_{\text{total}}}{1+e} = \frac{V_{\text{voids}}}{e} \tag{14.7}$$

Specific gravity of solids: $G_s = \dfrac{\gamma_s}{\gamma_w} = \dfrac{Se}{w}$ (14.8)

Saturated unit weight is given by $\gamma_{\text{sat}} = \dfrac{G_s + e}{1+e}\gamma_w$ (14.9)

And the buoyant unit weight is $\gamma' = \gamma_{\text{sat}} - \gamma_w = \dfrac{G_s - 1}{1+e}\gamma_w$ (14.10)

To solve all five nonzero components of the three-phase system, it is sometimes convenient to take either a unit volume or unit weight approach. For example, if the water content of the soil w is given and we see that water content is defined as the ratio W_w/W_s, it may be convenient to assign $W_s = 1$. This makes the weight of water $W_s = w$. On the other hand, if the porosity n is given and we see that porosity is defined as the ratio $V_{\text{voids}}/V_{\text{total}}$, it may be convenient to assign $V_{\text{total}} = 1$. This makes the volume of voids $V_{\text{voids}} = n$. If both options are available, it is often more convenient to use the unit weight approach rather than unit volume, as there are only two unknown weight components and three volume components.

Example 14.1

What is the dry unit weight of a soil with a porosity $n = 0.23$, specific gravity of solids = 2.70 and water content = 20%?

Solution If the step-by-step approach is followed, since given data has porosity (denominator total volume), we could start with

Total volume = 1.0 ft^3

Volume of voids = 0.23 ft^3

Volume of solids = 0.77 ft^3

Weight of solids = 0.77 × 2.70 × 62.4 = 129.7 lb

Dry density = 129.7 lb/ft^3

Alternative: Using the formula $\gamma_d = (1-n)G_s\gamma_w = 0.77 \times 2.70 \times 62.4 = 129.7$ lb/ft^3

Example 14.2

A soil has a unit weight of 128 lb/ft^3, void ratio = 0.45 and specific gravity of solids = 2.65. What is the degree of saturation of the soil?

Solution If the step-by-step approach is followed, since given data has void ratio (denominator volume of solids), we could start with

Volume of solids = 1.0 ft^3

Volume of voids = 0.45 ft^3

Weight of solids = $1.0 \times 2.65 \times 62.4 = 165.4$ lb

Total weight = $128 \times 1.45 = 185.6$ lb

Weight of water = $185.6 - 165.4 = 20.2$ lb

Volume of water = $20.2 \div 62.4 = 0.324$ ft^3

Saturation = $0.324 \div 0.45 = 0.72$

Alternative: Using the formula $\gamma = \left(\frac{G_s + Se}{1+e}\right)\gamma_w \Rightarrow 128 = \left(\frac{2.65 + 0.45S}{1.45}\right)62.4 \Rightarrow S = 0.72$

Example 14.3

Results from a sand cone test are listed below:

Weight of soil obtained from the test hole:	3.65 lb
Moisture content of soil from test hole:	17.3%
Maximum dry density (standard Proctor):	122.3 lb/ft^3
Unit weight of dry test sand:	84.7 lb/ft^3
Initial weight of sand cone apparatus filled with test sand:	13.36 lb
Final weight of sand cone apparatus after sand fills test hole:	11.13 lb

The in-place percent compaction is most nearly

A. 97

B. 104

C. 114

D. 133

Solution

Weight of test sand in test hole = $13.36 - 11.13 = 2.23$ lb

Volume of test sand in test hole = $2.23 \div 84.7 = 0.02633$ ft^3

Weight of soil obtained from the test hole = 3.65 lb

Unit weight of soil obtained from the test hole = $3.65 \div 0.02633 = 138.6$ lb/ft^3

Dry unit weight of soil = $138.6 \div 1.173 = 118.2$ lb/ft^3

Percent compaction = $118.2 \div 122.3 = 97\%$

Answer is A.

Shrinkage and Swell

When earth is excavated, its volume is initially greater than its in situ volume (also called bank volume). However, after being placed in an embankment and compacted, the earth shrinks so that its volume is less than the volume in the borrow pit. The ratio of the shrunk volume to the in situ volume is termed the shrinkage factor. The approximate shrinkage factors for the various types of soils are sand and gravel 0.92, clay 0.90, loam 0.88, and organic soil 0.85.

When an embankment is made of broken-tip hard rock, it has a volume from 40% to 80% larger than its original volume in the cut, and there is practically no subsequent settling of the embankment. This increase in volume is called the swell of rock. It varies a great deal with conditions, but 50% may be taken as the average swell.

Thus, a unit volume (bank measure) of soil, upon being excavated, swells to a volume = 1 + bulking fraction. This soil, upon being placed in its destined location and compacted, shrinks to a final volume = 1 − shrinkage fraction.

Example 14.4

A highway project requires 38,000 yd^3 of compacted fill material for construction of an embankment. The contractor has access to a borrow pit of capacity 26,000 yd^3. The soil in this location has 12% shrinkage. The rest of the soil has to come from a borrow pit which has a clay loam with an in-place density of 124 lb/ft^3, shrinkage of 15% and a bulking of 16%. Trucks used to transport the borrow soil have a net capacity of 24 tons and a volumetric capacity of 15 yd^3. How many truckloads are needed?

Solution

Fill needed	38,000 yd^3
Soil in local borrow pit (in situ volume)	26,000 yd^3
Yield volume of local soil (after shrinkage) = $0.88 \times 26{,}000$ =	22,880 yd^3
Clay loam yield needed = 38,000 − 22,880 =	15,120 yd^3
In place volume of clay loam = 15,120/0.85 =	17,788 yd^3

In place weight of clay loam = $17{,}788 \times 27 \times 124/2000 = 29{,}777$ tons

Therefore, by weight capacity of trucks, we need 1240.7 truckloads.

Transported volume of borrow soil = $17{,}788 \times 1.16$ = 20,634 yd^3

Therefore, by volume capacity of trucks, we need 1375.6 truckloads.

Therefore, 1376 truckloads (based on volume criterion) will be required.

Effective Stress

The normal stress of contact between adjacent soil particles is called the effective stress. Since the water table creates a hydrostatic pressure condition that grows linearly with depth below the water table elevation, this relieves some of the interparticle stress, causing them to "float apart." Therefore, the effective contact stress can be used using the submerged unit weight of the soil, wherever appropriate.

The vertical effective stress at a certain point within the soil is calculated as the weight per unit area of the soil prism above it and minus the hydrostatic pressure created by the static groundwater. It can be calculated as the summation $\gamma'\Delta z$, using the submerged unit weight γ' wherever appropriate.

$$\sigma' = \sum \gamma' \Delta z = \sum (\gamma - \gamma_w)\,\Delta z \tag{14.11}$$

Example 14.5

A 12-ft-thick clay layer is overlain by 13-ft-thick layer of sand as shown in the figure below. The water table is originally at a depth of 8 ft below the surface. Calculate the following at the bottom of the clay layer: (a) total vertical stress, (b) effective vertical stress, (c) effective horizontal stress, and (d) total horizontal stress.

Assume lateral earth pressure coefficient = 0.4.

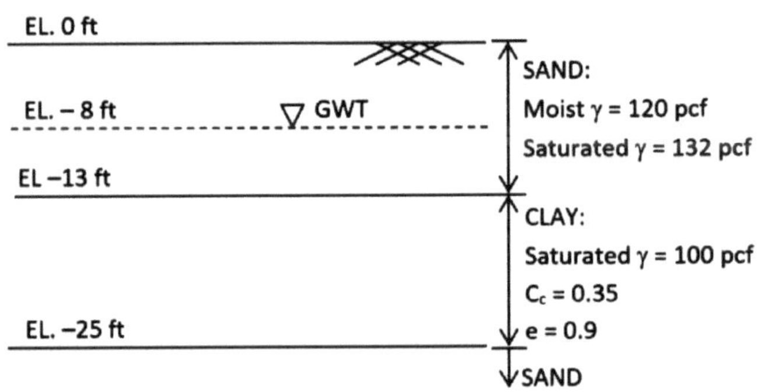

Solution

A. At a depth of 25 ft, the total vertical stress is calculated as the total weight of the soil prism above $z = 25$ ft.

$$\sigma_v = 120 \times 8 + 132 \times 5 + 100 \times 12 = 2820 \text{ psf}$$

B. The effective vertical stress is calculated as the submerged weight of the soil prism above $z = 25$ ft.

$$\sigma_v' = 120 \times 8 + (132 - 62.4) \times 5 + (100 - 62.4) \times 12 = 1759.2 \text{ psf}$$

C. The effective horizontal stress is related to the effective vertical stress

$$\sigma_h' = K\sigma_v' = 0.4 \times 1759.2 = 703.7 \text{ psf}$$

D. The total horizontal stress is calculated as the effective horizontal stress plus the hydrostatic pressure at that depth

$$\sigma_h = \sigma_h' + \sigma_w = 703.7 + 17 \times 62.4 = 1764.5 \text{ psf}$$

Note that this last value is *not* equal to $0.4 \times 2820 = 1128$ lb/ft²

Shear Strength (Mohr-Coulomb)

The shear strength of a soil determines its suitability as a foundation material and its stability characteristics. The soil parameters that are most significant for shear strength are internal friction ϕ and cohesion c. The most commonly adopted model for shear strength of a soil is given by the Mohr-Coulomb theory, which predicts the shear strength τ to be function of the cohesion c and the angle of internal friction ϕ as follows:

$$\tau = c + \sigma \tan\phi \qquad (14.12)$$

The parameters c and ϕ can be obtained through various standard tests. For example, the direct shear test yields (σ, τ) pairs to yield a line of best fit (See example 14.6). On the other hand, the triaxial test yields (σ_1, σ_3), which can be used to draw Mohr's circles. A common tangent to several Mohr's circles is then considered the Mohr-Coulomb shear strength envelope (Fig. 14.2).

Example 14.6

A direct shear test is performed on a silty sand sample. Sample *A* tested at a vertical stress equal to 1000 lb/ft² failed at a shear stress of 675 lb/ft² while sample *B* tested at a vertical stress of 3000 lb/ft² failed at a shear stress of 2025 lb/ft². Based upon the results of this direct shear test the shear strength parameters for this silty sand are most nearly:

A. $c = 0$ lb/ft² and $\phi = 34°$

B. $c = 200$ lb/ft² and $\phi = 34°$

C. $c = 0$ lb/ft² and $\phi = 42.5°$

D. $c = 200$ lb/ft² and $\phi = 42.5°$

Figure 14.2 Mohr-Coulomb shear strength envelope.

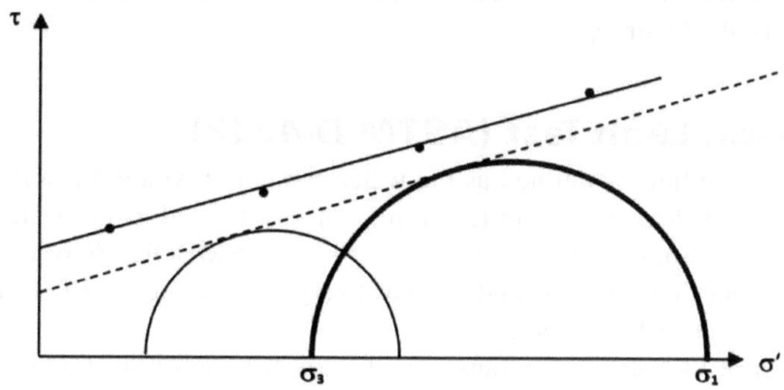

Solution

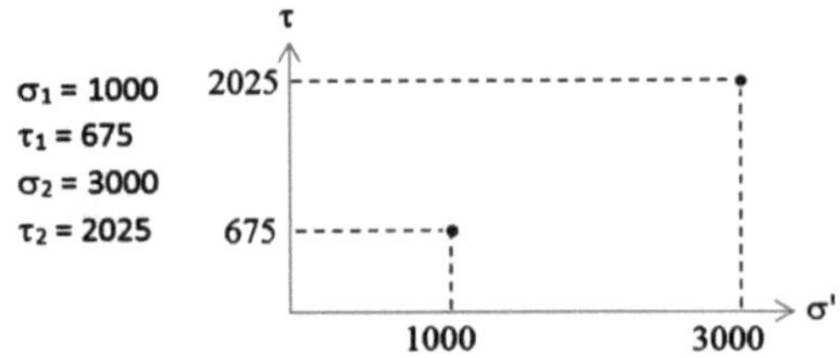

Friction angle: $\tan\phi = \dfrac{\tau_2 - \tau_1}{\sigma_2 - \sigma_1} = \dfrac{2025 - 675}{3000 - 1000} = 0.675 \Rightarrow \phi = 34°$

Using coordinates of point 1: Cohesion: $c = 675 - 1000 \times \tan\phi = 0$

Answer is A.

Standard Soil Tests

Liquid Limit Test (ASTM D-4318)

The Atterberg limits w_L (liquid limit) and w_P (plastic limit) are important for classification of cohesive soils. The apparatus for the liquid limit test consists of a brass cup that is dropped onto a hard rubber base by a cam operated by a crank. An 11-mm-wide groove is cut at the center of the soil paste in the brass cup. The liquid limit is defined as the moisture content at which it takes exactly 25 blows to close the groove over a distance of ½ in (12.7 mm). The test is performed for various values of moisture content. The plot of water content w versus log N is typically linear. The slope of this plot is called the flow index I_F defined as

$$I_F = \frac{w_1 - w_2}{\log(N_1/N_2)} \tag{14.13}$$

The water content corresponding to $N = 25$ is interpolated from this plot (Fig. 14.3). This is defined as the water content at which the soil transitions from a plastic state to liquid state. SYMBOL: *LL* or w_L.

Plastic Limit Test (ASTM D-4318)

The plastic limit is defined as the water content at which the soil crumbles when rolled into a thread of diameter 1/8 in (3.2 mm). This is defined as the water content at which the soil transitions from a semisolid to a plastic state. SYMBOL: *PL* or w_P.

Nonplastic soils (sand and silt) have very small values of *PL*. Clays and plastic silt can have *PL* from 0 to 100 or more.

Figure 14.4 shows the transition of a soil from a brittle solid to a liquid state.

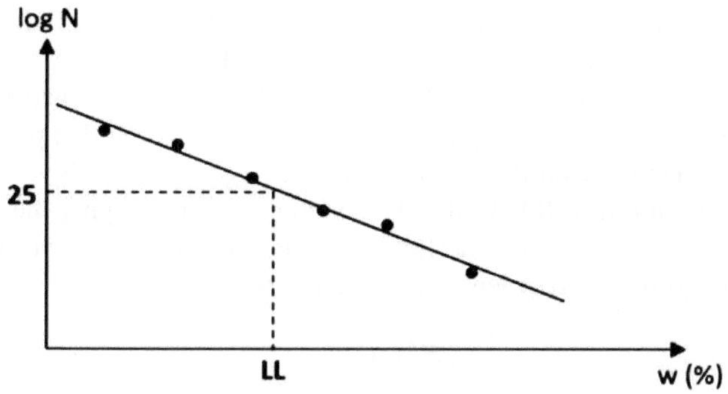

Figure 14.3 Determination of liquid limit.

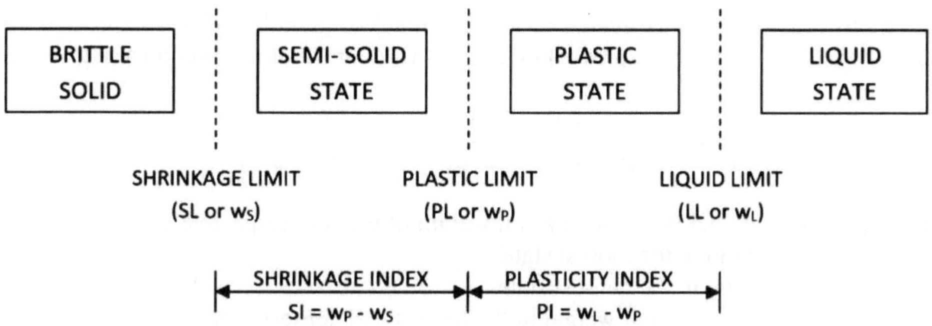

Figure 14.4 Transitions in consistency of a fine-grained soil.

The shrinkage limit is that water content below which, removal of water does not result in a reduction of volume. When the moisture content increases beyond the shrinkage limit, the volume of the soil increases. SYMBOL: *SL* or w_S.

The plasticity index *PI* of a cohesive soil is the range of moisture content over which the soil behaves as a plastic material.

$$PI = w_L - w_P \tag{14.14}$$

Standard Penetration Test

The standard penetration test (SPT) is one of various standardized tests to measure in situ compactness of soils. It can be used for all types of soil, but is preferred for sand deposits. In the standard penetration test, a split-spoon sampler (outer diameter = 2 in, inner diameter = 1.5 in) is driven by a 140-lb hammer falling 30 in through three successive 6-in penetrations. The field recorded result for the SPT consists of three numbers: the number of blows required to drive the sampler through 0 to 6 in, 6 to 12 in, and 12 to 18 in penetration or until normal maximum resistance (refusal) is reached, using the standard hammer and drop. Refusal is defined as a penetration of less than 6 in for 100 hammer blows. The standard penetration resistance (also known as the SPT N value or blow count) from the field is the sum of the counts (blows per foot) from the second and third intervals.

The field measured N value needs to be corrected for equipment factors such as hammer efficiency, borehole diameter, rod length as well as existing overburden pressure.

The adjustment due to overburden is applied so that the corrected blow count is normalized to an overburden of 1 ton/ft^2 (2000 lb/ft^2). This is done by multiplying the field N value by a correction factor C_n, which corrects the field N value upward (if the overburden is less than 1 tsf) or downward (if the overburden is more than 1 tsf)

$$N' = C_n N \tag{14.15}$$

Relative Density

The in situ density (often expressed as a percentage) of granular soils is expressed by the relative density R_D. It can be expressed in terms of void ratio e or unit weight γ as given below

$$R_D = \frac{e_{\max} - e}{e_{\max} - e_{\min}} = \left(\frac{\gamma_d - \gamma_{d,\min}}{\gamma_{d,\max} - \gamma_{d,\min}}\right)\left(\frac{\gamma_{d,\max}}{\gamma_d}\right) \tag{14.16}$$

where e, γ_d = in situ void ratio and dry unit weight of the soil, respectively
$e_{\max}$ = void ratio in the loosest state
$e_{\min}$ = void ratio in the densest state
$\gamma_{d,\min}$ and $\gamma_{d,\max}$ = dry unit weight in the loosest and densest states

The dry unit weight of granular soils is measured according to ASTM test D-2049. For sand, this test involves a 0.1 ft^3 mold. Sand is poured loosely into the mold until the mold is filled without using any compacting effort. The minimum dry unit weight $\gamma_{d,\min}$ is then determined by dividing the net weight of the sand by the mold volume.

The maximum dry unit weight $\gamma_{d,\max}$ is determined by vibrating sand in the mold for 8 min (at 60 Hz, amplitude 0.025 in). A surcharge of 2 psi is added to the top of the sand. Soil is added to the mold until full. The value of $\gamma_{d,\max}$ is determined at the end of the vibrating period.

The relative density of a soil can have a significant impact on its shear strength, and therefore, on its bearing capacity.

Example 14.7

A sand sample has an in place dry unit weight = 121 lb/ft^3. When this sample is excavated and loosely placed in a laboratory apparatus, the dry unit weight is measured as 105 lb/ft^3. When the sample is then placed in a vibrating mold and packed into a dense state, the dry unit weight is observed to be 125 lb/ft^3. What is the relative density of the sand in situ?

A. 80%

B. 83%

C. 91%

D. 97%

Solution The formula for relative density is

$$D_r = \left(\frac{\gamma_{D,\text{field}} - \gamma_{D,\text{min}}}{\gamma_{D,\text{max}} - \gamma_{D,\text{min}}}\right)\left(\frac{\gamma_{D,\text{max}}}{\gamma_{D,\text{field}}}\right) = \left(\frac{121-105}{125-105}\right)\left(\frac{125}{121}\right) = 82.6\%$$

Relative compaction on the other hand is

$$RC = \frac{\gamma_{D,\text{field}}}{\gamma_{D,\text{max}}} = \frac{121}{125} = 96.8\%$$

Answer is B.

Cone Penetrometer Test

This test involves forcing a cone into the ground and measuring the driving pressure needed for each increment of penetration. Resistance to penetration is the sum of point resistance and frictional resistance on the sides of the shaft. The more sophisticated systems can differentiate between the point and frictional components of the resistance, and the ratio between frictional and point resistance (friction ratio) is one aid in differentiating between various soil types. Clean sands generally exhibit very low ratios (low-friction component in comparison to point resistance), while an increase in clay content will usually result in a higher ratio, which is usually due to a reduction in point resistance rather than an increase in frictional component.

The cone is first pushed into the soil to the desired depth and then a force is applied to the rod which moves the cone downward. The force required to move the cone into the extended position divided by the projected area of the cone (on a horizontal plane) is termed the cone penetration resistance q_c.

The static cone test can be used as a partial replacement for conventional borings. The speed of operation allows considerable data to be obtained in a short period of time. The major drawbacks of static cone tests are the non-recoverability of samples for identification, difficulty in advancing the cone in dense or hard deposits, and need for stable and fairly strong working surface to jack the rig against.

Direct Shear Test

In the direct shear test, a soil specimen (typical diameter = 2.5 in, height = 1 in) is placed between porous plates in the center of the direct shear apparatus. See Fig. 14.5. Vertical load is applied to the specimen and allowed to consolidate fully. A dial gauge is used to measure horizontal deflection. A horizontal force is applied to the upper half of the apparatus to cause shear failure on the middle plane of the sample.

Drained conditions maintained by the porous plates (escape of pore water allowed under slow loading) ensure that total stresses and effective stresses are the same. At various values of the normal force N, the soil is loaded with increasing horizontal shear forces P until the soil fails in shear. The normal stress σ is calculated as the vertical force divided by the cross sectional area, while the shear force at failure τ is calculated from the horizontal force at failure. The failure envelope may be obtained from plotting several σ-τ pairs.

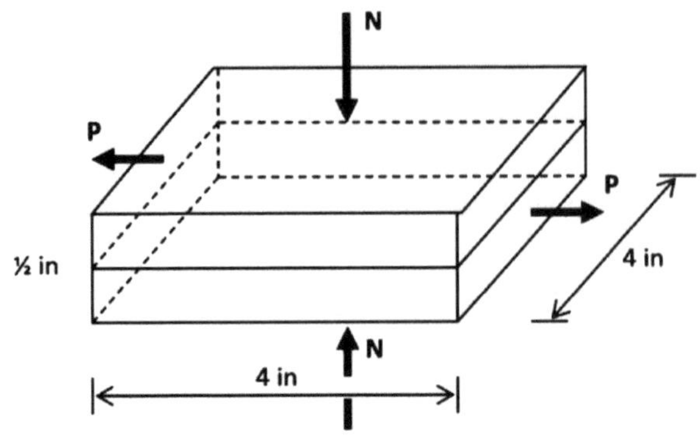

Figure 14.5 Direct shear test.

This is illustrated in Fig. 14.5. A 4 × 4-in soil sample is subjected to shear in a direct shear box. Upper and lower surfaces are capped by porous plates, thereby maintaining drained conditions. A normal force N is maintained on the sample, while shear forces P are applied until failure occurs. Results from tests conducted on several samples are tabulated and plotted. Cohesion and angle of internal friction can be estimated from this plot. See example 14.8.

Example 14.8

A direct shear test is conducted on soil samples placed in a split box with a 2 × 2-in cross section. Porous plates are used to maintain drained conditions as a normal force N is applied evenly to the sample. With the soil sample under the normal force, a horizontal force pair F is applied to the split box until the sample fails in shear.

The table below summarizes the data for three samples.

Sample	N (lb)	F (lb)
1	34.0	63.0
2	56.0	75.0
3	90.0	93.0

Calculate the (a) angle of internal friction and (b) the cohesion of the soil.

Solution

N (lb)	P_f (lb)	Normal stress σ (lb/ft²)	Shear stress τ (lb/ft²)
34.0	63.0	$\frac{34}{(4/12)^2} = 306$	$\frac{63}{(4/12)^2} = 567$
56.0	75.0	$\frac{56}{(4/12)^2} = 504$	$\frac{75}{(4/12)^2} = 675$
90.0	93.0	$\frac{90}{(4/12)^2} = 810$	$\frac{93}{(4/12)^2} = 837$

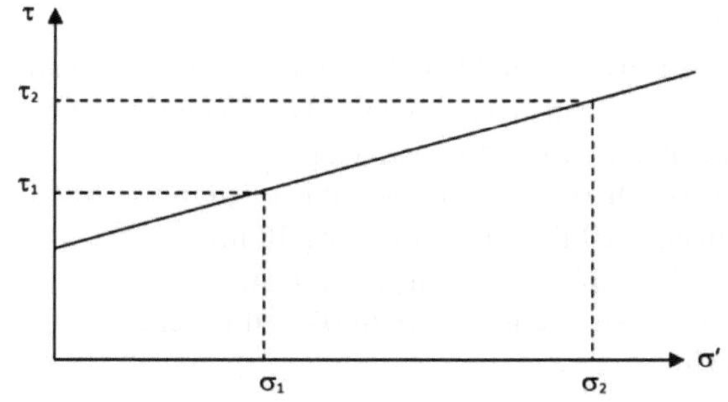

From points 1 and 2 that are on the failure envelope, the angle of internal friction ϕ and the cohesion c can be calculated from

$$\tan\phi = \frac{\tau_2 - \tau_1}{\sigma_2 - \sigma_1} = \frac{675 - 567}{504 - 306} = 0.545 \Rightarrow \phi = 28.6°$$

$$c = \frac{\sigma_2\tau_1 - \sigma_1\tau_2}{\sigma_2 - \sigma_1} = \frac{504 \times 567 - 306 \times 675}{504 - 306} = 400 \text{ lb/ft}^2$$

NOTE Obviously, in a real situation, one would test several samples and draw a line of best fit through these data points. The best fit line may or may not pass through the data points, and the parameters (slope and intercept) of the line should be used rather than coordinates of the test points themselves.

Unconfined Compression Test

In the unconfined compression test, a cylindrical sample is loaded in axial compression without any radial confining pressure. Generally failure occurs along diagonal planes where the greatest ratio of shear stress to shear strength occurs. Very soft material may not show well-defined diagonal planes of failure. In such cases, failure is generally assumed to have occurred when the axial strain has reached a value of 20%. The unconfined compression test is performed only on cohesive soil samples. For soils with negligible internal friction, the cohesion is taken as one-half the unconfined compressive strength.

Compaction

Compaction of soils serves to increase soil shear strength and improve stability, decrease erosion potential as well as decrease permeability and compressibility.

In the field, soil is placed in lifts (layers) and then mechanically compacted using one or more of the following: static loading, impact, vibration, kneading.

The in situ dry density is often expressed as a percentage of the maximum dry density as obtained in the laboratory. Thus, if the dry density in the field is 95 lb/ft^3 and the maximum dry density is 100 lb/ft^3, then the soil is said to exist at 95% relative compaction.

$$\text{Relative compaction} = \frac{\gamma_d}{\gamma_{d,\max}} \times 100\% \qquad (14.17)$$

where γ_d is the dry unit weight achieved on site and $\gamma_{d,\max}$ is the maximum dry unit weight from the laboratory.

Standard Proctor Test

The standard Proctor test uses a mold with a diameter of 4 in and a volume of 1/30 ft^3 (943.3 cm^3)—filled in 3 lifts, each layer compacted by 25 blows of a hammer (a 5.5-lb. weight dropping 12 in). In the modified Proctor test, adapted to better represent current field conditions, the mold volume is still 1/30 ft^3; however, it is filled in five lifts—each layer compacted by 25 blows of a hammer (a 10-lb weight dropping 18 in).

In either test, several samples with varying moisture content are tested. The total unit weight (or density) and moisture content are measured for each sample. The dry unit weight (or density) is calculated as

$$\text{Dry unit weight } \gamma_d = \frac{\gamma}{1+w} \tag{14.18}$$

A typical plot (silty-clayey soil) of dry unit weight versus moisture content is shown in Fig. 14.6.

Figure 14.6 shows a peak for the dry density at optimum moisture content (w_{opt}). The dry density can also be expressed as the following function of the degree of saturation S

$$\gamma_d = \frac{G_s \gamma_w}{1 + \dfrac{G_s w}{S}} \tag{14.19}$$

G_s = specific gravity of soil solids
gamma$_w$ = unit weight of water = 62.4 lb/ft^3
w = water content
S = degree of saturation

At any moisture content, the theoretical maximum dry unit weight is obtained when there is no air in the void spaces, i.e. when the degree of saturation S = 1. This yields the following formula for the zero air-voids unit weight γ_z (fully compacted)

$$\gamma_z = \frac{G_s \gamma_w}{1 + wG_s} = \frac{\gamma_w}{w + \dfrac{1}{G_s}} \tag{14.20}$$

Figure 14.6 Variation of dry density with water content.

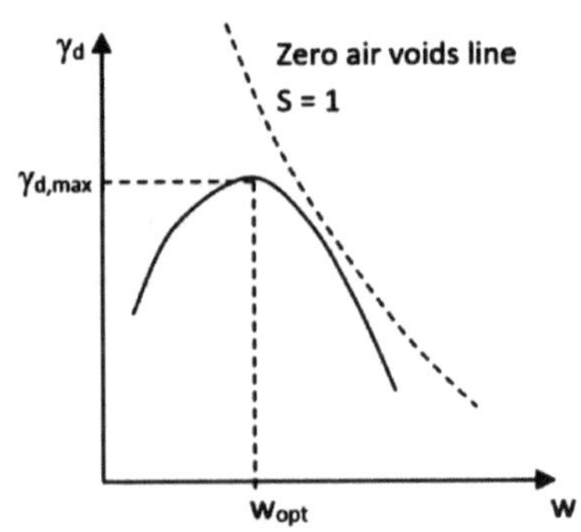

Sandy soil typically exhibits greatest dry density at a moisture content of about 8% while clay compacts best at moisture content of approximately 20%.

Example 14.9

Results from a standard Proctor test on six soil samples are tabulated below. Determine the optimum moisture content, the maximum dry unit weight, the void ratio, and the degree of saturation at the optimum moisture content. Also calculate the range of water content for which 95% compaction can be achieved. Assume specific gravity of solids = 2.65

Sample	Net weight of wet soil (lb)	Moisture content (%)
1	3.45	12
2	3.76	14
3	3.96	16
4	3.98	18
5	3.94	20
6	3.85	22

Solution

Sample	Net weight of wet soil (lb)	Moisture content (%)	Moist unit weight (lb/ft^3)	Dry unit weight (lb/ft^3)
1	3.45	12	103.5	92.4
2	3.76	14	112.8	98.9
3	3.96	16	118.8	102.4
4	3.98	18	119.4	101.2
5	3.94	20	118.2	98.5
6	3.85	22	115.5	94.7

In the table above, the shaded cells represent given data. The white cells are calculated. Standard volume of Proctor mold is 1/30 ft.^{3}

Plotting the above data—moisture content versus dry unit weight—we obtain the following graph. Note that once the maximum dry unit weight has been calculated as 102.4 (approx.) and the 95% level as 97.3, it may be seen from the last column in the table that this would intersect the plot just before the 2nd data point, i.e. at w = 13.5 (approx.) and just after the 5th data point, i.e. at w = 20.5 (approx.). Thus, unless a very precise read of these values is required, the conclusions can be drawn from the tabulated data itself, without resorting to the much greater effort of plotting.

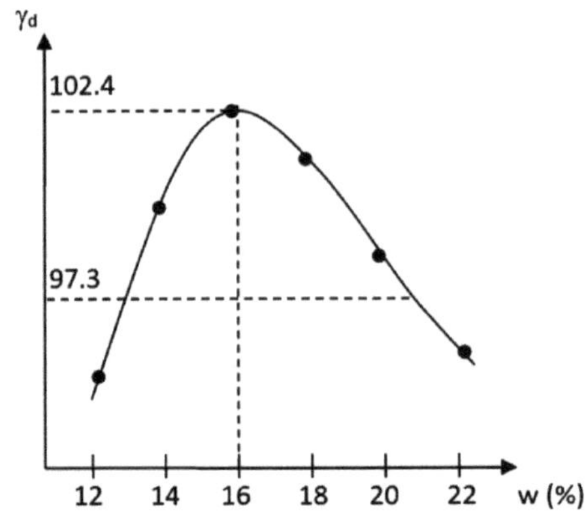

The following results can be obtained:

Optimum moisture content = 16%

Maximum dry unit weight = 102.5 lb/ft^3 (approx.)

95% maximum dry unit weight = 97.3 lb/ft^3

The range of water content values over which greater than 95% compaction can be achieved is w_1 = 13.5% to w_2 = 20.5%.

Triaxial Test Fundamentals

In the triaxial test, a soil sample is forced to undergo shear failure on an inclined plane by independently controlling a radial confining pressure σ_R and an axial stress σ_A. See Fig. 14.7. The sample fails in shear due to the deviator stress $(\sigma_A - \sigma_R)$. The soil specimen is cylindrical, with a diameter of about 1.4 in and length of 3 in, and is encased by a thin rubber membrane. The specimen is placed in a plastic chamber that is equipped with connections that either (a) allow drainage from the soil specimen or (b) measurement of pore pressure when undrained conditions are maintained.

Three types of triaxial test are commonly conducted:

1. *Consolidated drained (CD) or S test:* This test usually takes a very long time because the deviator stress must be applied very slowly to ensure proper drainage. The rate of strain is controlled to prevent the build-up of pore pressure in the specimen. CD tests are generally performed on well-draining soils. For slow-draining soils, several weeks may be required to perform a CD test.
2. *Consolidated undrained (CU) or R test:* This is the most common type of triaxial test. The soil sample is first consolidated by radial pressure. After the pore pressure generated by this pressure is dissipated, the drainage line is closed and the soil is tested to failure. Pore water pressure and deviator stress are measured simultaneously. Specimens must be completely saturated before application of the deviator stress.
   In the absence of pore pressure measurements CU tests can provide only total stress values cohesion c and friction angle ϕ.

Figure 14.7
Soil sample subject to triaxial test.

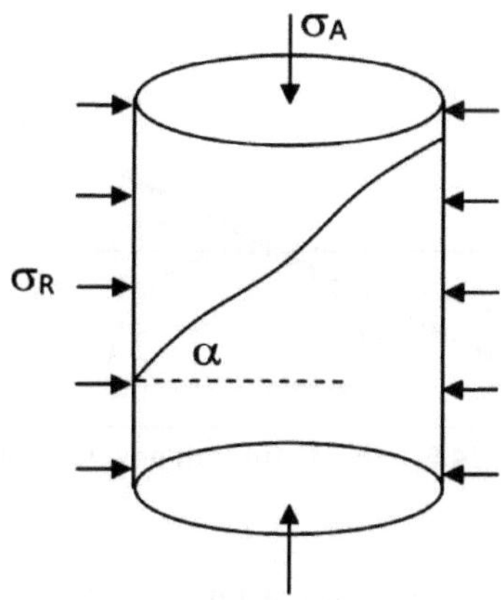

3. *Unconsolidated undrained (UU) or quick (Q) test:* In this test, drainage is not permitted during the application of the radial pressure. Because drainage is not allowed during any stage of the test, it can be completed quickly. The shear strength of soil as determined in UU tests corresponds to total stress, and is applicable only to situations where little consolidation or drainage can occur during shearing. It is applicable primarily to soils having permeability less than 10^{-3} cm/s.

Triaxial test results can be affected by added confinement due to membrane stiffness, friction in the loading piston and other factors. The shear strength of soft sensitive soils is greatly affected by sample disturbance. For UU tests, the laboratory-measured shear strength of disturbed samples will be lower than the in-place strength. For CU or CD tests, the strength may be higher because of the consolidation permitted.

For a cohesionless soil (clean sand) sample, the failure plane is inclined at an angle $\alpha = 45° + \phi/2$ with the radial plane (see Fig. 14.7) and the axial stress and the radial stress at failure are related by

$$\frac{\sigma_A}{\sigma_R} = \frac{1+\sin\phi}{1-\sin\phi} \tag{14.21}$$

A typical stress-strain curve obtained from the triaxial test is shown in Fig. 14.8. The deviator stress $\sigma_D = \sigma_A - \sigma_R$ is plotted on the *y* axis. The deviator stress σ_D is responsible for developing shear stress in the soil.

Figure 14.8
Typical result from soil tested to failure in triaxial test.

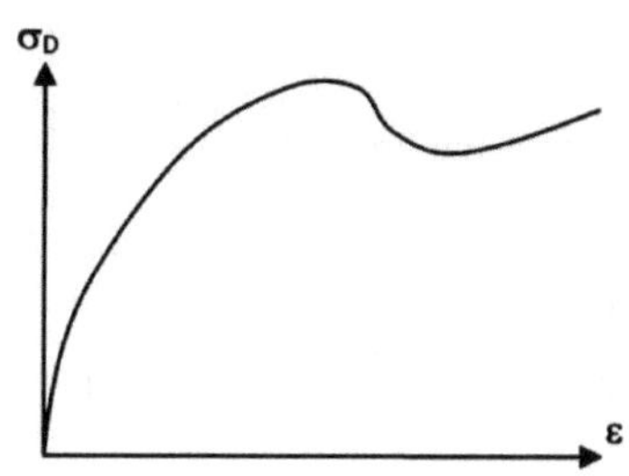

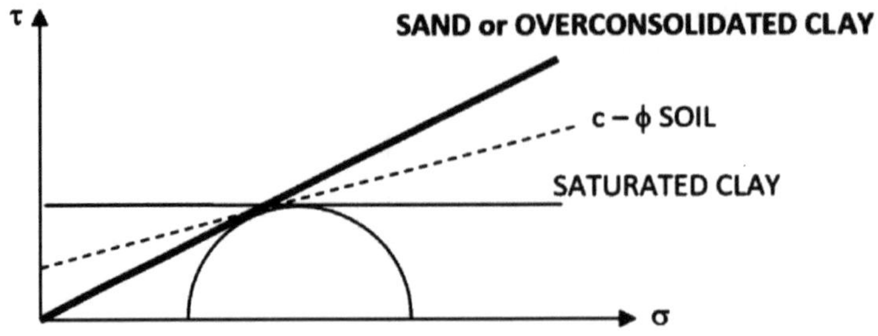

Figure 14.9 Shear strength envelope and Mohr's circles.

The radial and axial stresses at failure for a soil sample represent principal stresses. It is conventional to represent them as σ_1 (axial) and σ_3 (radial). Using these normal stresses as the ends of the diameter, Mohr's circle may be drawn for each stress pair. If several Mohr's circles are drawn based on several samples, a shear failure envelope may be drawn tangent to these circles. Figure 14.9 shows typical (tangent) envelopes for various type of soil.

The geometric relationship between the parameters of the Mohr's circle (σ_1 and σ_3) and the parameters of the shear envelope (c and ϕ) is shown below for (1) a cohesionless soil and (2) a c-ϕ soil.

Example 14.10

A cylindrical sample of sand is tested in a triaxial apparatus. The following data is given for the undrained test.

Sample diameter = 2 in

Sample length = 4 in

Confining pressure = 50 psi

Axial load at failure = 240 lb

Pore water pressure at failure = 16 psi

The angle of internal friction of the sand is most nearly

A. 30°

B. 32°

C. 34°

D. 36°

Solution Cross sectional area: $A = \dfrac{\pi D^2}{4} = \dfrac{\pi \times 2^2}{4} = 3.14 \text{ in}^2$

The 240-lb load represents an added axial load *after* the confining pressure is exerted hydrostatically on the sample.

Axial stress at failure: $\sigma_1 = \dfrac{240}{3.14} + 50 = 126.4 \text{ psi}$

Confining stress: $\sigma_3 = 50$ psi

Effective stresses: $\sigma_1' = 126.4 - 16 = 110.6$ psi; $\sigma_3' = 50 - 16 = 34$ psi

The distance from the origin to the center if the Mohr's circle is ½(σ_1' + σ_3') and radius of the Mohr circle is ½(σ_1' − σ_3').

Therefore, angle of internal friction: $\sin\phi = \dfrac{\sigma_1' - \sigma_3'}{\sigma_1' + \sigma_3'} \Rightarrow \phi = \sin^{-1}\left(\dfrac{110.6 - 34}{110.6 + 34}\right) = 32°$

Answer is B.

Consolidation

The compression of a soil due to stress increases is caused by deformation of soil particles, expulsion of air and water from the voids and the restructuring of the soil "skeleton." Whereas immediate settlements occur primarily due to the elastic deformation without any accompanying moisture changes, consolidation settlements occur more slowly due to the volume change resulting from expulsion of water from the voids of saturated cohesive soils.

Consolidation Test

The consolidation characteristics of a cohesive coil are determined using a consolidometer (or oedometer), where a soil sample is placed inside a metal ring and loaded with two porous loading plates, one at each end. Specimens are usually 2.5 in. in diameter and 1 in thick. The specimen is kept submerged during the test. Each load is usually maintained for 24 h. Successive loads are usually double the previous load.

Figure 14.10 shows a typical plot of void ratio versus the effective vertical stress (log scale). The plot shows two tangents that have been fitted to the actual curve. The two tangents appear to intersect a point which corresponds to the preconsolidation pressure p_o', which is the maximum pressure that the soil has experienced in its loading history. If the preconsolidation pressure for a soil layer is not specified, it is customary to take it as the existing effective stress due to the overburden and consider all subsequent consolidation to be normal consolidation.

Figure 14.10 Primary consolidation curve.

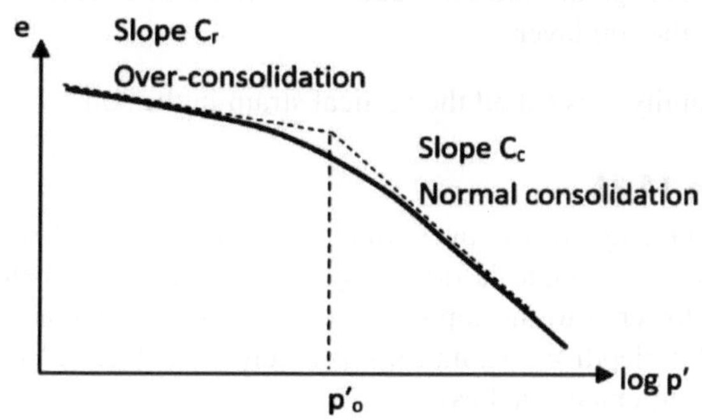

For pressures less than the preconsolidation pressure, the soil experiences overconsolidation (slope C_r). The change in void ratio during overconsolidation from effective stress p_1' to p_2' is given by

$$\Delta e = e_1 - e_2 = C_r \log_{10}\left(\frac{p_2'}{p_1'}\right) \tag{14.22}$$

For pressures exceeding the preconsolidation pressure, the soil experiences normal (virgin) consolidation (slope C_c). The change in void ratio during normal consolidation from effective stress p_1' to p_2' is given by

$$\Delta e = e_1 - e_2 = C_c \log_{10}\left(\frac{p_2'}{p_1'}\right) \tag{14.23}$$

For normally consolidated clay, Terzaghi and Peck suggested the following correlation between the compression index C_c and the liquid limit.

$$C_c = 0.009(LL - 10) \tag{14.24}$$

A typical estimate of the recompression index C_r is

$$C_r = \frac{C_c}{6} \tag{14.25}$$

Settlement due to Primary Consolidation

The settlement of a soil layer of thickness H due to primary consolidation is given by

$$s = H\varepsilon_v = H\left(\frac{\Delta e}{1+e_o}\right) = \frac{H\Delta e}{1+e_o} \tag{14.26}$$

where H = layer thickness
e_o = initial void ratio
Δe = change of void ratio due the increase of effective vertical stress at the midpoint of the soil layer

The quantity ε_v is called the vertical strain in the soil.

Example 14.11

A 12-ft-thick clay layer is overlain by 13-ft-thick layer of sand as shown in the figure below. The ground water table is originally at a depth of 8 ft below the surface. The groundwater is first lowered to the top of the clay layer and then a mat foundation (plan dimensions 200 × 150 ft, depth = 3 ft; uniform pressure = 1000 lb/ft^2) is constructed. Calculate the consolidation settlement (inches).

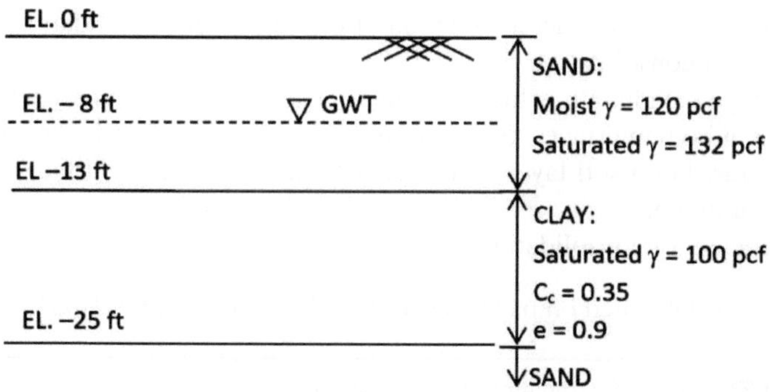

Solution Assuming the behavior of the clay to be normally consolidated (preexisting overburden pressure is the preconsolidation pressure), the effective vertical stress at the center of the clay layer under original conditions is given by

$$p_1' = \sigma_v' = 120 \times 8 + (132 - 62.4) \times 5 + (100 - 62.4) \times 6 = 1534 \text{ lb/ft}^2$$

Once the water table has been lowered and the surcharge load is applied, the effective vertical stress at the same location is calculated as

$$p_2' = \sigma_v' = 1000 + 120 \times 13 + (100 - 62.4) \times 6 = 2786 \text{ psf}$$

Note that because the horizontal extent of the mat foundation load is large compared to the depth (center of clay layer is 16 ft below application of this uniform pressure), the 1000 lb/ft^2 is carried down at full strength (without dissipation)

The change in void ratio is given by

$$\Delta e = C_c \log \frac{p_2'}{p_1'} = 0.35 \log \frac{2786}{1534} = 0.091$$

The consolidation settlement is therefore given by

$$s = \frac{H \Delta e}{1 + e_0} = \frac{12 \times 0.091}{1.9} = 0.575 \text{ ft} = 6.9 \text{ in}$$

Consolidation Rate

The time required for primary consolidation to occur can be expressed as

$$t = \frac{T_v H_d^2}{c_v} \tag{14.27}$$

where T_v = consolidation time factor (function of the degree of consolidation U_z)
U_z = degree of consolidation
H_d = drainage path length, which is equal to half the layer thickness when the layer can drain at both interfaces, or the entire layer thickness if drainage can occur on only one side. For a soil layer at the ground surface, the free surface is assumed to be free-draining.
c_v = coefficient of consolidation

The time factor T_v is tabulated on page 148 of the *FE Reference Handbook*, version 9.4.

NOTE Units of parameters C_c, C_r, c_v, a_v, U_z, T_v.

C_c and C_r are dimensionless. U_z (degree of consolidation) and T_v (time factor for consolidation) are dimensionless.

c_v has units of L^2/T (ft²/day typical)

Example 14.12

What are expected values of compression and recompression index for clay with $LL = 45\%$ and $PL = 15\%$?

A. $C_c = 0.32$; $C_r = 0.05$

B. $C_c = 0.32$; $C_r = 0.10$

C. $C_c = 0.42$; $C_r = 0.08$

D. $C_c = 0.42$; $C_r = 0.10$

Solution According to the correlation between liquid limit and the compression index,

$$C_c = 0.009(LL - 10) = 0.009 \times (45 - 10) = 0.315$$

According to the correlation between compression index and the recompression index,

$$C_r = C_c/6 = 0.053$$

Answer is A.

Soil Classification

Sieve Sizes

In the United States, soil classification methods employ the standard sieve sizes listed in Table 14.2. Also in use, but somewhat less common, are metric sieve sizes. For fine-grained

Table 14.2 Standard U.S. Sieve Sizes

Sieve size	Sieve opening (mm)
2 in	50
1 in	25
½ in	12.5
No. 4	4.75
No. 10	2.00
No. 20	0.85
No. 40	0.425
No. 60	0.25
No. 100	0.15
No. 140	0.106
No. 200	0.075

soils, plasticity characteristics (Atterberg limits) play a more significant role than the particle size distribution, which is more significant for coarse grained soils.

Particle Size Distribution Curves

Figure 14.11 shows a typical particle size distribution curve for a soil aggregate.

The shape of the particle size distribution curve is expressed by two parameters—the uniformity coefficient C_u and the coefficient of curvature or coefficient of concavity C_c. These coefficients are defined as

$$\text{Coefficient of uniformity } C_u = \frac{D_{60}}{D_{10}} \tag{14.28}$$

$$\text{Coefficient of curvature } C_c = \frac{D_{30}^2}{D_{10}D_{60}} \tag{14.29}$$

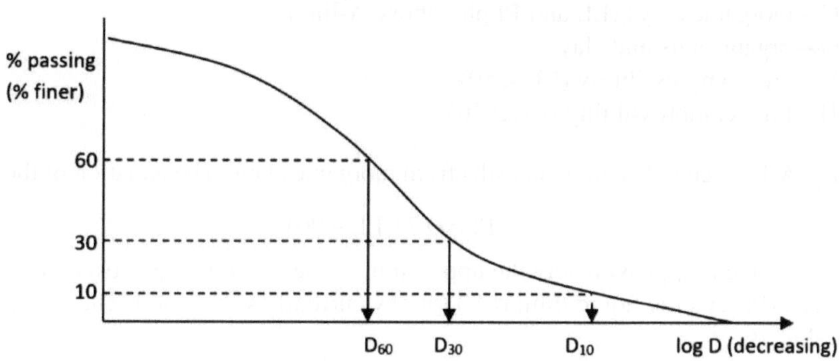

Figure 14.11 Typical particle size distribution for soils.

where D_{10}, D_{30}, and D_{60} are the particle sizes for which 10%, 30%, and 60% (by weight) are finer. These sizes are indicated in Fig. 14.11. D_{10} is called the effective size of the soil (Hazen). The effective size of clean ends and gravels can be correlated to their permeability.

Hydrometer Analysis

Particle size distribution for finer soils (silts and clays with particles finer than a No. 200 sieve) can be established by a hydrometer analysis which uses sedimentation of dispersed soil particles (about 50 g of soil in a distilled water suspension filling a 1-L glass cylinder). The hydrometer test is based on type I settling of particles governed by Stokes' law. The continuous record of settling rates or velocities can be correlated to the particle size distribution.

Unified Soil Classification System

The unified soil classification system (USCS) uses letter symbols to describe soils. The parameters used for classification are

1. F_{200} = percent passing the No. 200 sieve (opening size = 0.075 mm)
2. Uniformity coefficient C_u
3. Coefficient of curvature C_c
4. R_4 = percent retained on the No. 4 sieve (opening size = 4.75 mm)

In general, particles retained on a No. 4 sieve ($d > 4.75$ mm) are *gravel* and particles retained on a No. 200 sieve ($d > 0.075$ mm) but passing the No. 4 sieve are classified as *sand*.

The USCS classification procedure may be outlined by the following steps:

Step 1 If the soil can be visually identified as peat, classify as Pt.

Step 2 If $F_{200} > 50\%$, the soil is predominantly fine-grained. The Atterberg limits liquid limit (LL) and plastic limit (PL) are then used in the Casagrande plasticity chart (page 151 of the *Handbook*) to obtain the appropriate classification symbol. The following symbols are used on the plasticity chart:

M—inorganic silt (LL and PI plot below A-line)
C—inorganic clays (LL and PI plot above A-line)
O—organic silts and clays
L—low compressibility (LL < 50)
H—high compressibility (LL ≥ 50)

The A-line separates inorganic silts from inorganic clays. The equation of the A-line is

$$\text{PI} = 0.73(\text{LL} - 20) \tag{14.30}$$

The U-line is approximately the upper limit of the relationship between the plasticity index (PI) and the liquid limit (LL) for any currently known soil. The equation of the U-line is

$$\text{PI} = 0.90(\text{LL} - 8) \tag{14.31}$$

Step 3 If $F_{200} < 50$, the soil is predominantly coarse grained. The first letter of the USCS classification is *G* or *S*. Identify cumulative % retained on No. 4 (R_4).

Step 4 If $R_4 > ½(100 - F_{200})$, first letter is *G*, otherwise *S*.

The following symbols are used for classification of coarse grained soils:

First letter:

G—gravel (>50% of coarse fraction retained on No. 4 sieve)
S—sand (<50% of coarse fraction retained on No. 4 sieve)

Second letter:

W—well graded, clean
P—poorly graded, clean
C—significant amounts of clay
M—significant amount of silt

Step 5 If first letter is *G*, use the following guidelines to identify second letter

If $F_{200} < 5\%$, the soil is classified as GW if $C_u \geq 4$ and $1 < C_c < 3$, GP otherwise.

If $5\% < F_{200} < 12\%$, the soil has a dual classification reflecting both gradation and plasticity. Examples are GP-GM, GP-GC, etc.

If $12\% < F_{200} < 50\%$, clayey gravel (GC) or silty gravel (GM).

Step 6 If first letter is *S*, use the following guidelines to identify second letter.

If $F_{200} < 5\%$, the soil is classified as SW if $C_u \geq 6$ and $1 < C_c < 3$, SP otherwise.

If $5\% < F_{200} < 12\%$, the soil has a dual classification reflecting both gradation and plasticity. Examples are SP-SM, SP-SC, etc.

If $12\% < F_{200} < 50\%$, clayey sand (SC) or silty sand (SM).

AASHTO Soil Classification

The AASHTO (American Association of State Highway and Transportation Officials) soil classification criteria are summarized in a table on page 150 of the *Handbook*. The soil classification protocol defined by AASHTO utilizes the following criteria:

Percent passing No. 10 sieve (F_{10})

Percent passing No. 40 sieve (F_{40})

Percent passing No. 200 sieve (F_{200})

Liquid limit (LL)

Plasticity index (PI) = liquid limit (LL) – plastic limit (PL)

The broad distinction between granular materials (groups A-1, A-2, and A-3) and fine-grained materials (groups A-4, A-5, A-6, and A-7) is made on the basis of the percent (by weight) passing the No. 200 sieve (F_{200}).

Less than or Equal to 35% Passing No. 200 Sieve—Predominantly Granular

Of the three groups meeting this criterion, the A-2 group represents a transition between coarse- and fine-grained soils. Plasticity characteristics such as LL and PI are necessary parameters to determine the subgroup under group A-2 (choices are A2-4, A2-5, A2-6, and A2-7).

Is $F_{40} > 50$? If the answer is *yes*, the soil group is A-3. This group is assigned to fine sands which are nonplastic (NP). If the answer is *no*, the soil group is A-1 or A-2.

Group A-1 is subdivided into A-1-a and A-1-b. Values of F_{10}, F_{40}, and F_{200} are used to differentiate between A1-a and A1-b.

More than 35% Passing No. 200 Sieve—Predominantly Fine-Grained

If the fraction passing No. 200 sieve (F_{200}) is greater than 35, the soil is A-4, A-5, A-6, or A-7 and only LL and PI are used for the classification.

Group A-7 is subdivided into A-7-5 and A-7-6, depending on the PL. For PL < 30, classification is A-7-6. For PL ≥ 30, the classification is A-7-5.

Group Index

In the AASHTO soil classification system, the group index (GI) is used as an indicator of the suitability of soils as a highway subgrade material. A low value of GI indicates good subgrade properties. For A4 to A7 soils, the group index is calculated using Eq. (14.32), where F_{200} is the percent weight passing the No. 200 sieve. The value is then reported as the nearest integer. Negative values are reported as zero.

$$\mathrm{GI} = (F_{200} - 35)[0.2 + 0.005(\mathrm{LL} - 40)] + 0.01(F_{200} - 15)(\mathrm{PI} - 10) \tag{14.32}$$

For groups A-2-6 and A-2-7, only the second term is retained for the calculation. In other words, for these groups, the partial group index GI is defined as

$$\mathrm{GI} = 0.01(F_{200} - 15)(\mathrm{PI} - 10) \tag{14.33}$$

For groups A-1-a, A-1-b, A-2-4, A-2-5, and A-3, the group index is zero.

Example 14.13

The table on the left shows sieve analysis data for a soil sample (total mass 200 g). The table on the right shows data derived from the raw data.

Sieve size	Opening (mm)	Weight retained (g)
2 in	50	10
1 in	25	20
½ in	12.5	49
No. 4	4.75	38
No. 10	2.00	22
No. 40	0.425	18
No. 200	0.075	24
Pan		19

Cumulative weight ret. (g)	Cumulative % retained (R)	% Passing (F)
10	05.0	95.0
30	15.0	85.0
79	39.5	60.5
117	58.5	41.5
139	69.5	30.5
157	78.5	21.5
181	90.5	09.5
200	100	00.0

What is the soil classification using (a) USCS and (b) AASHTO guidelines?

Solution Looking for 10%, 30%, and 60% passing in last column (F) and corresponding sizes, we draw the following conclusions:

$$
\begin{array}{ll}
0.075 < D_{10} < 0.425 & \text{say } D_{10} = 0.08 \text{ mm} \\
0.425 < D_{30} < 2.000 & \text{say } D_{30} = 1.9 \text{ mm} \\
4.750 < D_{60} < 12.50 & \text{say } D_{60} = 12.0 \text{ mm}
\end{array}
$$

Uniformity coefficient $C_u = \dfrac{D_{60}}{D_{10}} = \dfrac{12.0}{0.08} = 150$

Coefficient of curvature $C_c = \dfrac{D_{30}^2}{D_{10}D_{60}} = \dfrac{1.9^2}{0.08 \times 12.0} = 3.76$

USCS Classification

F_{200} = fraction passing No. 200 sieve = 9.5% (since this percentage is between 5% and 12%, the soil should have dual classification)

Coarse fraction = 100 − 9.5 = 90.5%, which is greater than 50%. Therefore, *soil is either G or S (gravel or sand).*

% retained on No. 4 sieve = 58.5% which is more than half of coarse fraction (0.5 × 90.5 = 45.3%). Therefore, *first letter is G.*

Since both conditions $C_u > 4$ and $1 < C_c < 3$ are not satisfied, *the soil is GP.*

Narrowing down the second designation of the soil, that is, whether it is GP-GM or GP-GC needs further information about the plasticity characteristics of the soil.

AASHTO Classification

Percent passing No. 10 = 30.5

Percent passing No. 40 = 21.5

Percent passing No. 200 = 9.5

This soil sample has less than 35% passing No. 200 sieve. Therefore, it is predominantly granular. From the table, this soil meets the criteria for A1-a. Group Index for group A1-a is zero (0).

Example 14.14

The sieve analysis result for a soil sample is shown in the table below. Atterberg tests resulted in LL = 54, PL = 23. What is the USCS classification for the soil?

A. GW-GC

B. GW-GM

C. GP-GM

D. GP-GC

Sieve size	% Passing
2 in	95
1 in	85
½ in	60
No. 4	41
No. 10	31
No. 40	22
No. 200	10

Solution

$F_{200} = 10$. This is less than 50, so it is predominantly a coarse grained soil (first letter *G* or *S*).

Coarse fraction = 90%

$R_4 = 100 - 41 = 59$. This is more than half of the coarse fraction. Therefore, first letter is *G*.

D_{10} = No. 200 size = 0.075 mm

D_{30} = No. 10 size = 2.0 mm (slightly less)

D_{60} = 0.5 in = 12.7 mm

$$C_u = \frac{D_{60}}{D_{10}} = \frac{12.5}{0.075} = 166.7$$

$$C_c = \frac{D_{30}^2}{D_{10}D_{60}} = \frac{2.0^2}{0.075 \times 12.5} = 4.3$$

Since F_{200} is between 5% and 12%, the soil has a dual classification. The first part of the classification is based on gradation. Since BOTH criteria ($C_u > 4$ and $1 < C_c < 3$) for GW are not met, the soil must be classified GP.

Plasticity index: PI = 54 − 23 = 31

PI on the A-line: PI = 0.73(LL − 20) = 24.8

GP-GM if PI < 0.73(LL − 20) *or* PI < 4

GP-GC if PI > 0.73(LL − 20) *and* PI > 7

Thus, the soil meets *both* criteria for GP-GC.

Answer is D.

Bearing Capacity

The failure of a substructure can occur in three different ways: (1) shear failure of the soil due to excessive soil pressure, (2) excessive settlement of the foundation due to excessive soil pressure, or (3) structural failure of elements of the footing.

Foundation Types

A spread footing simply spreads the load from a wall or column over an area such that the soil pressure is within some prescribed limit. For soils with low bearing capacity, it is more cost effective to provide a mat foundation to support all walls and columns. If the soil is still too weak to support the structural loads, it may be necessary to transfer the loads to a stronger stratum at greater depth. This may be done using piles and drilled shaft foundations.

Pad Foundations

Pad foundations are used to support an individual point load such as that due to a structural column. They usually consist of a block or slab of uniform thickness, but they may be stepped if they support the load from a heavily loaded column. Pad foundations are usually shallow, but deep pad foundations can also be used.

Strip Foundations

Strip foundations are used to support a line of loads, either due to a load-bearing wall, or if a line of columns need supporting where columns are so close that individual pad foundations would be impractical.

Raft Foundations

Raft foundations are used to spread the load from a structure over a large area, typically the entire area of the structure. They are used when column loads or other structural loads are close together and individual pad foundations would interact. A raft foundation normally consists of a concrete slab which extends over the entire loaded area. It may be stiffened by ribs or beams incorporated into the foundation.

Ultimate Bearing Capacity

The ultimate bearing capacity q_{ult} is the value of bearing stress which causes a sudden catastrophic settlement of the foundation (general shear failure).

Terzaghi's Bearing Capacity Theory

Terzaghi's bearing capacity equation predicts the ultimate bearing capacity of a soil medium experiencing a general bearing capacity failure under a strip footing ($L/B \to \infty$). The ultimate bearing capacity of a strip footing is given by

$$q_{\text{ult}} = cN_c + \gamma D_f N_q + 0.5\gamma B N_\gamma \tag{14.34}$$

Terzaghi's bearing capacity factors N_c, N_q, and N_γ depend only on the soil friction angle ϕ. Also, the general bearing capacity theory includes corrections to each of these terms for depth, footing shape, and load inclination. The *Handbook* does not provide bearing capacity factors, so any questions related to this must include these values.

Example 14.15

What is the smallest isolated square footing to carry a column load of 90 kips? The depth of the footing is 3 ft and the unit weight of the sandy soil ($\phi = 29°$) is 120 lb/ft^3. The corrected bearing capacity factors for $\phi = 29°$ are $N_c = 45$, $N_q = 20$, $N_\gamma = 13$. Minimum factor of safety based on ultimate bearing capacity = 3.0.

Solution The given bearing capacity factors have been corrected (shape, depth, etc.). The ultimate bearing capacity under the footing is given by

$$q_{\text{ult}} = cN_c + \gamma D_f N_q + 0.5\gamma B N_\gamma = 0 + 120 \times 3 \times 20 + 0.5 \times 120 \times B \times 13 = 7200 + 780B$$

The allowable soil pressure is $q_{\text{all}} = \dfrac{q_{\text{ult}}}{3} = 2400 + 260B$

The design equation (total soil pressure due to superstructure load + overburden < allowable soil pressure) is

$$\frac{90000}{B^2} + 120 \times 3 \le 2400 + 260B \Rightarrow 2040 + 260B \ge \frac{90000}{B^2}$$

By trial and error, $B > 4.942$ ft. Using a 5 × 5-ft footing is adequate

---

NOTE Given the multiple choice format of the FE exam, substituting values of footing size B from the four answer choices is probably more time efficient that trying to solve this cubic equation.

Example 14.16

A concrete footing is subject to the following column loads:

Concentric load, P	80 kips
Moment due to wind, M	35 kip·ft

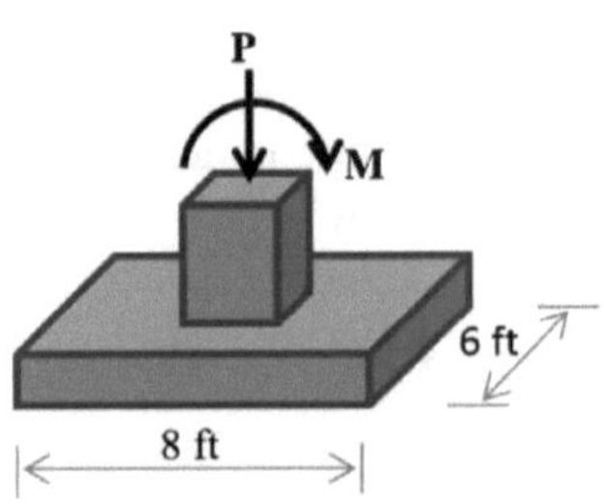

What is the maximum soil pressure under the footing?

A. 1734 lb/ft²

B. 2215 lb/ft²

C. 2564 lb/ft²

D. 2914 lb/ft²

Solution Concentric load = 80 kips. Moment = 35 kip·ft. As a result, eccentricity $e = M/P = 0.44$ ft. This is less than $B/6$, so the entire footing is effective in compression.

$$\text{Maximum soil pressure: } \sigma = \frac{P}{A} + \frac{Mc}{I} = \frac{80}{6\times 8} + \frac{35\times 4}{\frac{1}{12}\times 6\times 8^3} = 2.214 \text{ kips/ft}^2$$

Answer is B.

Effect of Water Table on Bearing Capacity

The preceding formulations for bearing capacity were based on the assumption that the water table is at a significant depth below the bottom of the footing. However, in locations where the water table is shallow, one or more terms in the equation need to be modified. Two cases are discussed below (1) when the water table is at the ground surface and (2) when the water table is at the bottom of the footing.

Case 1—Water Table at Ground Surface

When the water table is at the ground surface, the unit weight γ should be replaced with the buoyant unit weight γ' in the second and third terms of the equation (Fig. 14.12).

$$q_{\text{ult}} = cN_c + (\gamma_{\text{sat}} - \gamma_w)D_f N_q + \frac{1}{2}(\gamma_{\text{sat}} - \gamma_w)BN_\gamma \tag{14.35}$$

Case 2—Water Table at Bottom of Footing

When the water table is at the bottom of the footing, only the last term of the bearing capacity equation is modified by replacing the unit weight of the soil with the buoyant unit weight (Fig. 14.13)

$$q_{\text{ult}} = cN_c + \gamma D_f N_q + \frac{1}{2}(\gamma - \gamma_w)BN_\gamma \tag{14.36}$$

Figure 14.12 Groundwater table at ground surface.

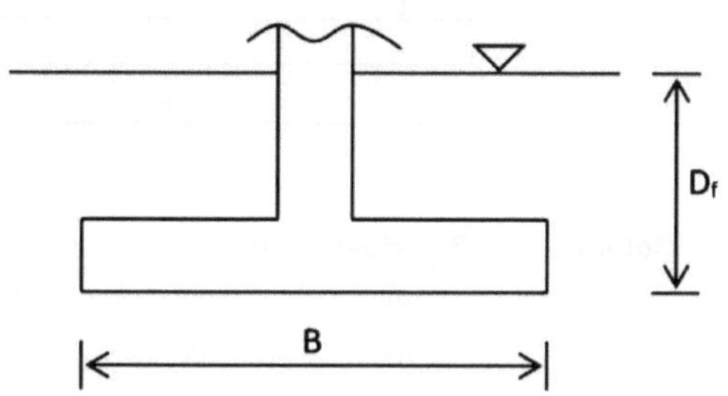

Figure 14.13 Groundwater table at bottom of footing.

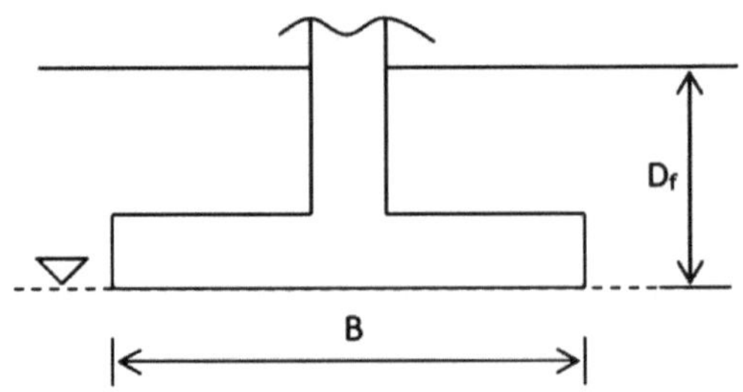

NOTE If the depth of the water table below the bottom of the footing is greater than the footing width *B*, no modification to the original bearing capacity equation is needed.

Combined Footing

When loads on individual columns are large enough, providing individual footings for them may cause them to encroach upon each other or get so close that providing a combined footing may be more economical. Such a combined footing for a pair of columns is shown in the figure below. The combined footing can be sized to satisfy bearing capacity and settlement criteria in the same fashion as an isolated footing.

Example 14.17

A combined footing is to be designed for columns 1 and 2 as shown. Column loads are shown in the table below.

(i) What must be the minimum length of the footing to avoid uplift under the footing?

(ii) What must be the length of the footing to produce uniform pressure under it?

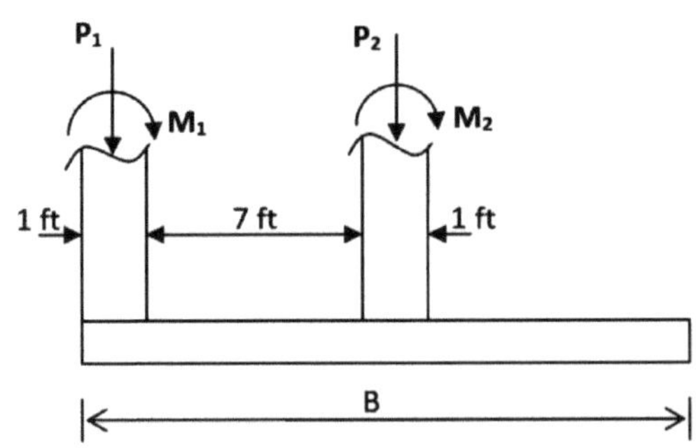

Column	P_{DL} (kips)	P_{LL} (kips)	M_{DL} (kip·ft)	M_{LL} (kip·ft)
1	50	105	80	90
2	70	85	50	100

Solution Taking moments about the centerline of column 1, the effective location of the resultant is calculated as

$$x_R = \frac{155 \times 0 + 170 + 155 \times 8 + 150}{155 + 155} = 5.03 \text{ ft}$$

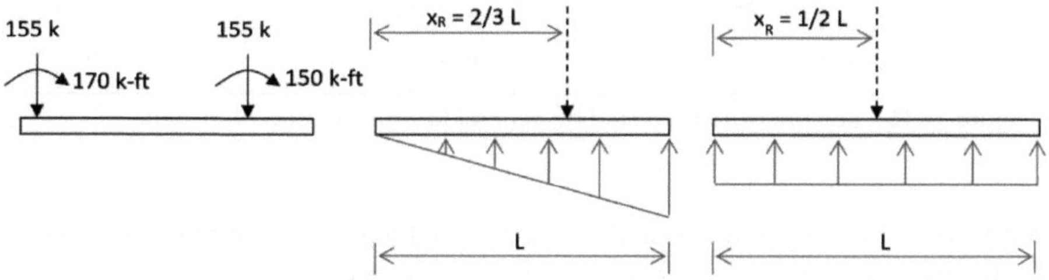

Thus, the resultant of the column loads acts at 5.03 ft from the centerline of column 1, that is, at 5.53 ft from the left edge of the column.

(i) In order to avoid uplift, resultant must be within the middle third of the footing base. Thus $2L/3 = 5.53$ ft. Therefore, $L = 8.3$ ft. This is illustrated in the second figure above.

(ii) In order to have uniform pressure under footing, resultant should be at footing center, that is, $L/2 = 5.53$ ft. Therefore, $L = 11.06$ ft. This is illustrated in the third figure above.

Eccentric Load on a Shallow Footing

Consider a strip footing of width *B* as shown in Fig. 14.14. Superstructure loads consist of a vertical load *P* and a moment (about the footing centerline) *M*. The loads *P* and *M* are for unit depth (perpendicular to plane of figure) of footing. Thus, *P* would have units of lb/ft and *M* would be lb·ft/ft. This represents an eccentricity of *e*, as given by Eq. (14.37)

$$e = \frac{M}{P} \tag{14.37}$$

Figure 14.14 Shallow footing with eccentric load.

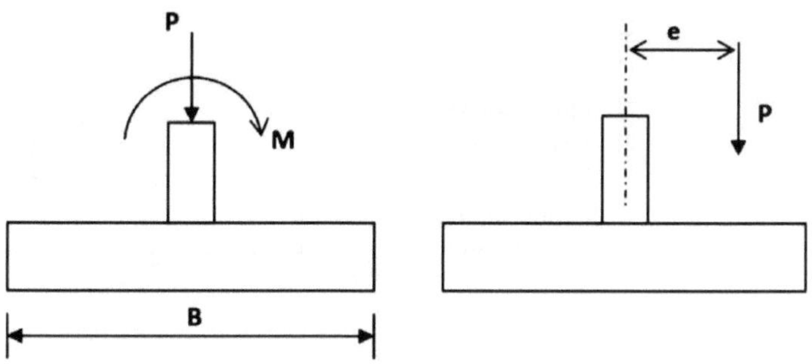

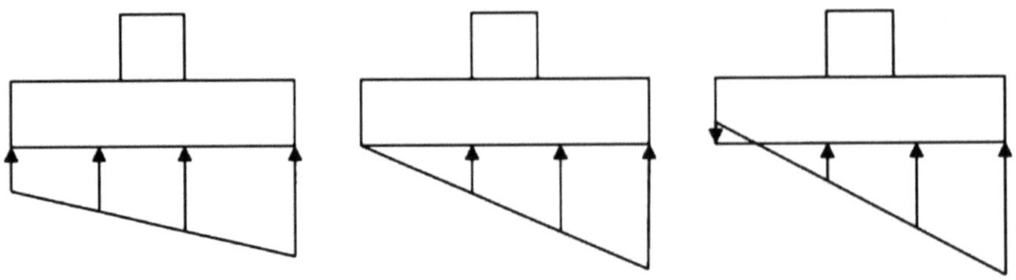

Figure 14.15 Soil pressure predicted by general formula for eccentric load.

When the load is located within the middle third of the footing, the entire width of the footing is effective (soil in compression). This occurs when the eccentricity of the load is less than one-sixth of the footing width. The resulting soil pressure profile is trapezoidal (ranging from triangular, when the eccentricity is *B*/6 to rectangular, when the eccentricity is zero). This result is specific to the rectangular contact area between footing and soil (Fig. 14.15).

The extreme stresses under the footing are given by

$$\sigma_{\text{max,min}} = \frac{P}{B} \pm \frac{6M}{B^2} = \frac{P}{B} \pm \frac{6Pe}{B^2} \tag{14.38}$$

Cases 1 and 2 lead to acceptable diagrams, since the entire footing is in contact with the soil and the soil experiences compressive stresses. However, Case 3 leads to an uplift condition on the left side of the footing, which cannot be sustained by the soil. Thus, the total vertical force P at eccentricity e must be resisted by a triangular stress distribution on a partial width of the footing (Fig. 14.16).

Solving the above equations gives the following results for maximum soil pressure

$$q_{\text{max}} = \frac{4P}{3(B-2e)} \tag{14.39}$$

and effective width of footing (contact width between soil and footing) is given by

$$x = 1.5B - 3e \tag{14.40}$$

Figure 14.16 Soil pressure under footing experiencing partial uplift.

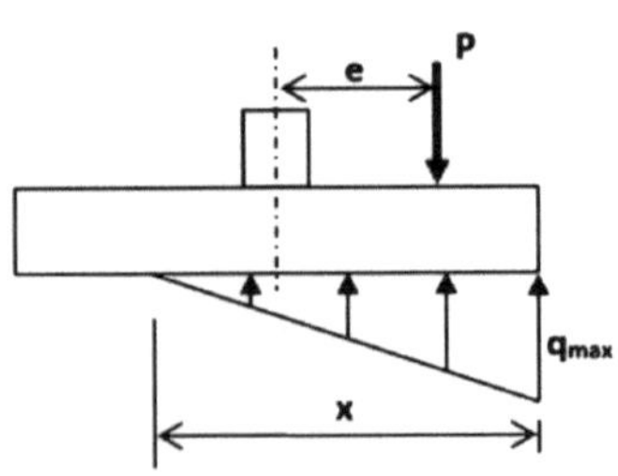

Mat Foundations

For a structure with a large number of columns, a foundation in the form of a slab or a mat can prove to be the most economical solution. The large area of the mat creates a low to moderate soil pressure, resulting in smaller settlement. A mat foundation is particularly useful in reducing differential settlement. An excavated mat foundation that is founded at depth also has the advantage of compensating for the soil that is removed. For example, if the average unit weight of the native soil is 120 lb/ft^3, then placing the mat at a depth of 5 ft below surface requires removal of that soil, which would have produced a prism load of $120 \times 5 = 600$ lb/ft^2 at that location. If the superstructure loads result in a soil pressure of 900 lb/ft^2 (say), then the net pressure increase transmitted to the soil is only $900 - 600 = 300$ lb/ft^2. This is called a partially compensated mat foundation. If the superstructure loads created an average pressure of 600 lb/ft^2 and the soil removal is equivalent to a pressure reduction of 600 lb/ft^2, then the mat would be called a fully compensated mat foundation.

Of course, excavation of a large area to a significant depth is expensive, but it can yield the secondary benefit of locating a subsurface building element such as a basement or a parking garage.

Deep Foundations

Deep foundations are those founded so deep below the ground surface that their bearing capacity is not affected by surface conditions, typically at depths greater than 10 ft below finished ground level. They include piles, piers, and caissons or compensated foundations using deep basements and also deep pad or strip foundations. Deep foundations can be used to transfer the loading to a deeper, more competent stratum at depth if unsuitable soils are present near the surface.

In addition to supporting structures, piles are also used to anchor structures against uplift forces and to assist structures in resisting lateral and overturning forces. Piles get support from both end bearing and skin friction.

End bearing piles: These are those which terminate in hard, relatively impenetrable material such as rock or very dense sand and gravel. They derive most of their carrying capacity from the resistance of the stratum at the toe of the pile.

Friction piles: These obtain a greater part of their carrying capacity by skin friction or adhesion. This tends to occur when piles do not reach an impenetrable stratum but are driven for some distance into a penetrable soil. Their carrying capacity is derived partly from end bearing and partly from skin friction between the embedded surface of the soil and the surrounding soil.

Settlement reducing piles: These are usually incorporated beneath the central part of a raft foundation in order to reduce differential settlement to an acceptable level. Such piles act to reinforce the soil beneath the raft and help to prevent dishing of the raft in the center.

Tension piles: Structures such as tall chimneys and transmission towers can be subjected to large overturning moments and so piles are often used to resist the resulting uplift forces at the foundations. In such cases the resulting forces are transmitted to the soil along the embedded length of the pile.

Laterally loaded piles: Almost all piled foundations are subjected to at least some degree of horizontal loading. These lateral loads are generally small compared to the applied vertical axial loading. However, if the horizontal force is relatively large (such as for wharves and jetties subject to impact from ships), then they may prove critical in design.

Batter piles: A vertical pile has some capacity to resist loads applied normally to the axis, although this lateral capacity is much smaller than the axial capacity. If this lateral capacity is sufficient, it is unnecessary to use "raked" or "battered" piles which are more expensive to install. If needed, battered piles are installed at an angle to the vertical, providing sufficient horizontal resistance by virtue of the horizontal component of axial capacity of the pile.

Displacement piles: These cause the soil to be displaced radially as well as vertically as the pile shaft is driven or jacked into the ground. With non-displacement piles (or replacement piles), soil is removed and the resulting hole filled with concrete or a precast concrete pile is dropped into the hole and grouted in.

Displacement piles are either driven or jacked into the ground. A number of different methods can be used. The dropping weight or drop hammer is the most commonly used method of insertion of displacement piles. Vibratory methods can prove to be very effective in driving piles through noncohesive granular soils. The vibration of the pile excites the soil grains adjacent to the pile making the soil almost free flowing thus significantly reducing friction along the pile shaft. However, the large energy resulting from the vibrations can damage equipment, noise and vibration propagation can also result in the settlement of nearby buildings.

Jacked piles are most commonly used in underpinning existing structures. By excavating underneath a structure short lengths of pile can be inserted and jacked into the ground using the underside of the existing structure as a reaction.

Example 14.18

A concrete pile (diameter $D = 18$ in, length = 40 ft) is embedded in a deep clay (cohesion = 2300 lb/ft^2) deposit. The bearing capacity factors are $N_c = 9$, $N_q = 1$, $N_g = 0$. The unit friction along the sides of the pile is 1400 lb/ft^2. What is the ultimate bearing capacity of the pile?

A. 36,600 lb

B. 160,000 lb

C. 264,000 lb

D. 300,000 lb

Solution The ultimate bearing capacity is composed of two components—side friction over the embedded surface area A_s, and point bearing on the tip area A_p

$$A_s = 2\pi RH = 2 \times \pi \times \frac{9}{12} \times 40 = 188.5 \text{ ft}^2$$

$$A_p = \pi R^2 = \pi \times \left(\frac{9}{12}\right)^2 = 1.77 \text{ ft}^2$$

Ul $Q_p = cN_cA_p = 2300 \times 9 \times 1.77 = 36{,}639$ lb

Ultimate side friction capacity: $Q_s = f_sA_s = 1400 \times 188.5 = 263{,}900$ lb

Total ultimate bearing capacity = 300,539 lb

Answer is D.

Lateral Earth Pressure

Embankments can either be set back on a stable slope or retained by structural elements such as retaining walls. Such elements must be designed to resist the horizontal forces produced as a by-product of the vertical stress existing within the soil. For locations where there is not adequate horizontal space for the provision of a stable slope, earth embankments may be retained using either gravity structures such as gabion walls, flexural structures such as cantilever retaining walls or reinforced earth structures such as mechanically stabilized earth walls.

The effective horizontal stress is related to the effective vertical stress according to

$$\sigma'_h = K\sigma'_v \tag{14.41}$$

The earth pressure coefficient K will be either the at-rest K_o, active K_a, or passive K_p earth pressure coefficient. For surfaces that are not expected to move laterally (such as massive bridge abutments and walls which are adequately restrained), the at-rest earth pressure must be used.

Stability and Strength Checks

The retaining wall must be checked for stability. This includes checks for sliding, overturning, and bearing capacity failure. Once the chosen wall configuration is shown to have adequate factors of safety for these criteria, all components of the wall must also be checked for adequate strength. The typical bending pattern for the various parts of a cantilever retaining wall with backfill on the right is shown in Fig. 14.17. As a result, the typical reinforcement pattern for the various parts of the wall is shown in the figure on the right.

Figure 14.17 Flexure of a cantilever retaining wall.

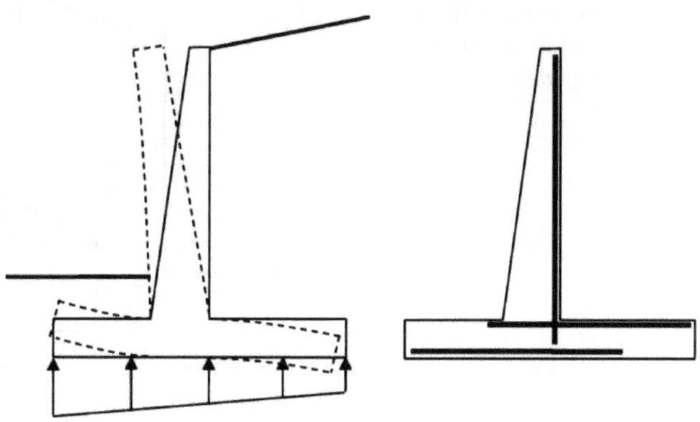

It is to be noted that the stability and bearing capacity checks are performed at service load levels, that is, without applying any load factors. However, the structural design of wall components must be performed using the appropriate design method. For example, for a reinforced concrete retaining wall, structural design of wall components must be according to provisions in ACI 318, which specifies the strength design approach, using factored loads.

Active Earth Pressure

Figure 14.18 shows a cantilever retaining wall with a sloping backfill (angle with horizontal = β) behind it. The back of the wall is considered rough (angle of friction between the wall and the backfill = δ). The back of the wall makes an angle θ with the horizontal.

Under the effect of the lateral earth pressure, the stem of the wall rotates and moves away from the backfill. With adequate movement of the wall, the lateral earth pressure developed behind the wall reaches a minimum value called active earth pressure.

Rankine's Theory for Earth Pressure

Rankine's theory for earth pressure ignores the friction between the wall and the backfill ($\delta = 0$). This simplifies the expressions for active and passive earth pressure coefficient. For a simplified situation—horizontal backfill ($\beta = 0$), negligible friction between wall and backfill ($\delta = 0$), and vertical wall stem ($\theta = 90°$), these expressions become

$$K_a = \frac{1-\sin\phi}{1+\sin\phi} = \tan^2\left(45 - \frac{\phi}{2}\right) \tag{14.42}$$

$$K_p = \frac{1+\sin\phi}{1-\sin\phi} = \tan^2\left(45 + \frac{\phi}{2}\right) \tag{14.43}$$

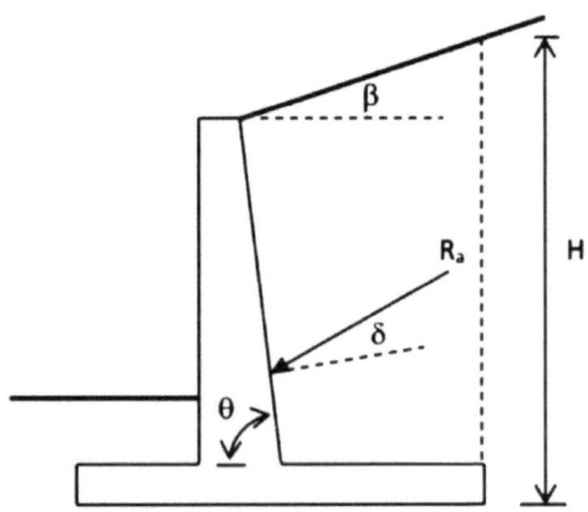

Figure 14.18 Active earth pressure behind cantilever retaining wall.

Steps for Evaluating Stability of a Retaining Wall

The following steps must be followed in order to evaluate the overall stability of a retaining wall.

1. Calculate K_a according to either the Rankine or Coulomb theory.
2. The lateral earth pressure may have two components: the triangular (hydrostatic) earth pressure due to overburden and the rectangular (constant) earth pressure due to surcharge load on the backfill, if any.

   The combined effective vertical stress at a depth z below the surface is then given by

   $$\sigma_h' = K_a(\gamma z + q_s) \tag{14.44}$$

   Note that the right-hand side has been calculated as a total stress, presuming that the water table is at significant depth.
3. If there is groundwater in the backfill zone, add the hydrostatic pore water pressure profile (triangular) to the effective horizontal stress to obtain the total horizontal stress on the wall.
4. The resultant force (per unit length of the wall) on a wall of height H, is then given by (these are shown in the figure below)

   $$R_a = \frac{1}{2}K_a\gamma H^2 + K_a q_s H \tag{14.45}$$

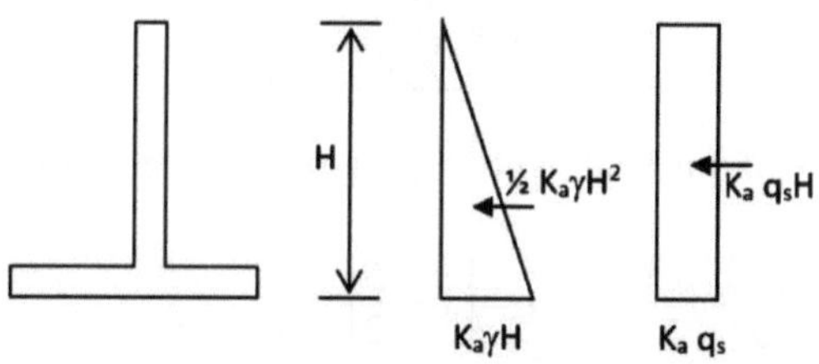

5. The resulting overturning moment (about the toe) is given by

   $$M_{OT} = \frac{1}{2}K_a\gamma H^2\frac{H}{3} + K_a qH\frac{H}{2} = \frac{1}{6}K_a\gamma H^3 + \frac{1}{2}K_a q H^2 \tag{14.46}$$

6. The stabilizing moment is due to the gravity forces (vertical)

   $$M_s = \sum W_i x_i \tag{14.47}$$

7. Factor of safety against sliding is obtained by comparing the friction mobilized under the bottom of the footing to the resultant active force.

$$FS = \frac{\sum W \tan\delta}{R_a} \tag{14.48}$$

where δ is the friction angle between the base of the wall footing and the supporting soil.

NOTE Designers typically tend to make the conservative decision to ignore the lateral resistance from the passive soil in computing factors of safety.

8. Factor of safety against overturning is obtained by comparing the stabilizing moment of the weight components to the overturning moments produced by the lateral forces.

$$FS = \frac{M_s}{M_{\text{OT}}} \tag{14.49}$$

9. The net moment is given by

$$M_{\text{net}} = M_s - M_{OT} \tag{14.50}$$

10. The effective location of the resultant force (measured from the toe) is given by

$$\overline{x} = \frac{M_{\text{net}}}{\sum W} \tag{14.51}$$

and the eccentricity of this load (with respect to the center of the wall footing) is given by

$$e = \left| \overline{x} - \frac{B}{2} \right| \tag{14.52}$$

11. The maximum soil pressure (as long as the eccentricity is less than $B/6$) is

$$q_{\max} = \frac{\sum W}{B} + \frac{6 \times \sum W \times e}{B^2} \tag{14.53}$$

Example 14.19

In the example shown here, assume the unit weight of the backfill soil is $\gamma = 120$ lb/ft^3, unit weight of concrete is 150 lb/ft^3, friction angle between base of wall footing and soil is 26° and the angle of internal friction for the backfill is 32°. Assume the surcharge to be 400 lb/ft^2.

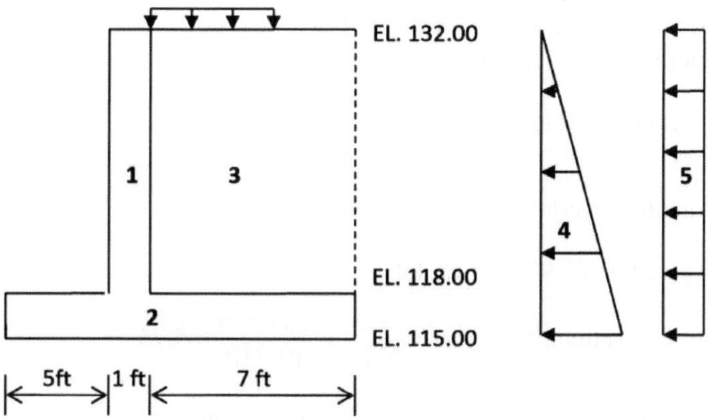

The active earth pressure coefficient is

$$K_a = \tan^2\left(45 - \frac{32}{2}\right) = 0.307$$

The resultant earth pressure has two components: due to overburden and due to surcharge. These are shown as components 4 and 5 in the figure above.

$$R_4 = \frac{1}{2}K_a\gamma H^2 = \frac{1}{2}\times 0.307\times 120\times 17^2 = 5323.4 \text{ lb/ft}$$
$$R_5 = K_a q_s H = 0.307\times 400\times 17 = 2087.6 \text{ lb/ft}$$

The weight components are

$$W_1 = 14\times 1\times 150 = 2100 \text{ lb/ft (concrete)}$$
$$W_2 = 13\times 3\times 150 = 5850 \text{ lb/ft (concrete)}$$
$$W_3 = 7\times 14\times 120 = 11760 \text{ lb/ft (soil)}$$

These values are summarized in the table below. Note that the stabilizing moments are designated as positive and the destabilizing (overturning) moments as negative. This is simply for the purpose of grouping them properly when calculating the factor of safety against overturning and when calculating the net moment.

	Description	F_x (lb)	F_y (lb)	Distance from toe (ft)	Moment about toe (lb·ft)	ΣM
1	Concrete stem	—	2100	5.50	+11,550	+161,295
2	Concrete footing	—	5850	6.50	+38,025	
3	Backfill soil	—	11,760	9.50	+111,720	
4	Lateral E.P. (overburden)	5323.4	—	5.67	–30,184	
5	Lateral E.P. (surcharge)	2087.6	—	8.50	–17,745	–47,929

Total vertical load (excluding surcharge) = 19,710
Coefficient of friction (below footing) = tan 26° = 0.49

$$\text{Factor of safety against sliding} = \frac{19,710 \times 0.49}{7411} = 1.30$$

$$\text{Factor of safety against overturning} = \frac{161,295}{47,929} = 3.36$$

Net moment = 161,295 – 47,929 = 113,366

$$\text{Effective distance of resultant from toe} = \frac{M_{\text{net}}}{\sum F_y} = \frac{113,366}{19,710} = 5.75$$

Eccentricity of resultant = 6.50 – 5.75 = 0.75 ft (*left* of center)

$$\text{Stress at toe} = \frac{19,710}{13} + \frac{6 \times 19,710 \times 0.75}{13^2} = 1516.2 + 524.8 = 2041\ \text{lb/ft}^2$$

$$\text{Stress at heel} = \frac{19,710}{13} - \frac{6 \times 19,710 \times 0.75}{13^2} = 1516.2 - 524.8 = 991.4\ \text{lb/ft}^2$$

Example 14.20

The concrete cantilever retaining wall pictured below has been backfilled with clean granular soil, is free to rotate and has good drainage so that hydrostatic pressure will not build up behind the wall. The backfill, foundation, and concrete have the following properties:

Concrete: $\gamma_c = 150$ lb/ft³

Granular backfill $\gamma = 118$ lb/ft³

$K_A = 0.3$

$K_P = 3.3$

μ (base of wall) = 0.5

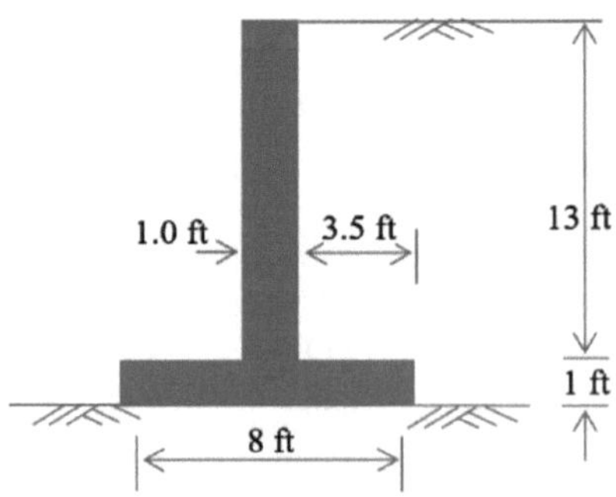

Assuming negligible friction between the wall and the backfill, the factor of safety against sliding is most nearly

A. 2.5

B. 4.25

C. 0.25

D. 1.25

Solution

Weight of wall = $(13 \times 1 + 8 \times 1) \times 150 = 3150$ lb/ft

Weight of soil = $13 \times 3.5 \times 118 = 5369$ lb/ft

Total weight = 8519 lb/ft

Friction force = $0.5 \times 8519 = 4260$ lb/ft

$K_A = 0.3$

Active earth pressure resultant $R_a = ½\, K_a\, \gamma H^2 = 3469$ lb/ft

$FS = 4260 \div 3469 = 1.23$

Answer is D.

Example 14.21

A mass concrete gravity retaining wall (concrete unit weight = 140 lb/ft³) is shown. Ignoring the contribution of the passive soil, the factor of safety for overturning is most nearly

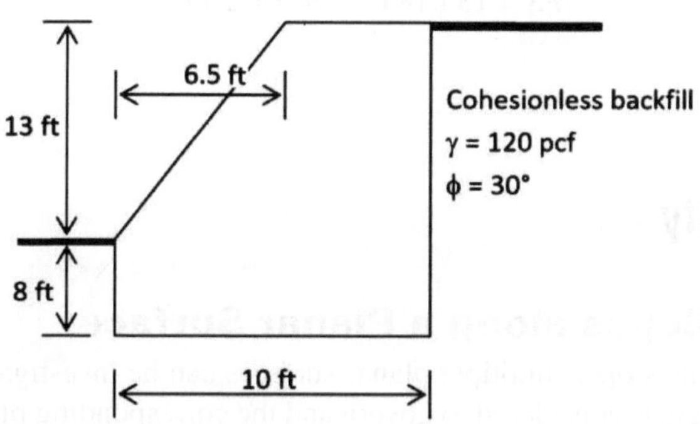

A. 1.6

B. 2.2

C. 2.8

D. 3.2

Solution

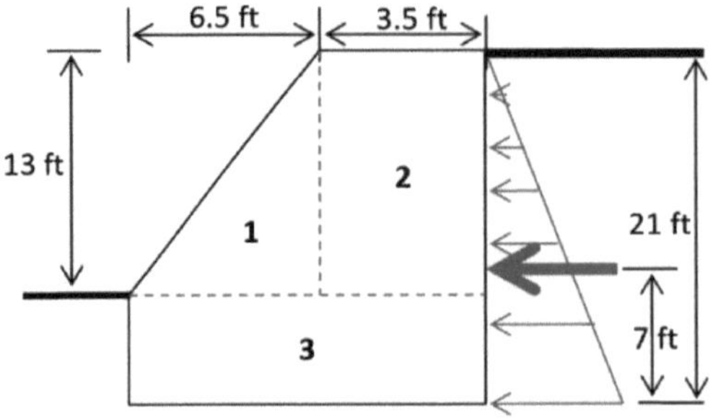

Weights of segments 1, 2, and 3 of the wall are $140 \times \left(\frac{1}{2} \times 6.5 \times 13\right) = 5915$ lb/ft, $140 \times (3.5 \times 13) = 6370$ lb/ft, and $140 \times (8 \times 10) = 11200$ lb/ft, respectively.

The stabilizing moment about the toe (tipping point) is given by

$$5915 \times \frac{2}{3} \times 6.5 + 6370 \times 8.25 + 11{,}200 \times 5 = 134{,}184 \text{ lb} \cdot \text{ft/ft}$$

For $\phi = 30°$, $K_a = 0.333$

$$R_a = \frac{1}{2} K_a \gamma H^2 = 0.5 \times 0.333 \times 120 \times 21^2 = 8820 \text{ lb/ft}$$

Overturning moment $= 8820 \times 7 = 61740$ lb · ft/ft

$$FS = 134{,}184/61{,}740 = 2.17$$

Answer is B.

Slope Stability

Stability of Slopes along a Planar Surface

In the stability of finite slopes, multiple planar surfaces can be investigated for stability. The smallest factor of safety is considered to govern and the corresponding plane is considered the critical plane. For example, in Fig. 14.19, a mass of soil forms a wedge ABC and the weight

Figure 14.19 Stability of a soil wedge.

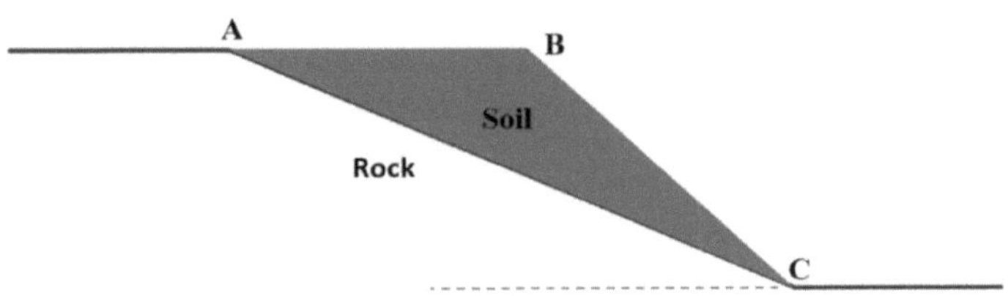

of this soil mass would tend to cause a sliding downslope (along the plane AC). The critical parameters would be: weight of soil wedge ABC (W_M), length of interface AC (L_s), slope of the interface AC (α_s), and the strength characteristics of the soil (angle of internal friction ϕ and cohesion c).

The weight of the wedge of soil (W_M) has a component ($W_M \sin \alpha_s$) parallel to the slope that acts as the destabilizing force (T_{MOB}) while the shearing resistance (T_{FF}) along the slope acts as the resisting force.

$$T_{\text{MOB}} = W_M \sin \alpha_s \tag{14.54}$$

$$T_{FF} = cL_s + W_M \cos \alpha_s \tan \phi \tag{14.55}$$

The factor of safety against slope stability is then given by

$$FS = \frac{T_{FF}}{T_{\text{MOB}}} = \frac{cL_s + W_M \cos \alpha_s \tan \phi}{W_M \sin \alpha_s} \tag{14.56}$$

Example 14.22

A 30-ft-high embankment exists as shown. ABC is a soil mass with the following properties: unit weight $\gamma = 108$ lb/ft^3.

Angle of internal friction $\phi = 20°$.

Cohesion $c = 400$ lb/ft^2.

AC is an interface between the soil and underlying stable rock. For the wedge ABC, determine the factor of safety against sliding along the rock surface.

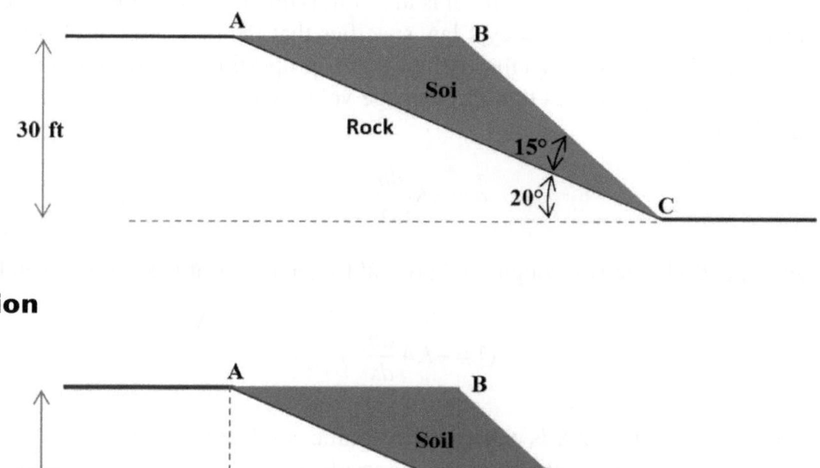

Solution

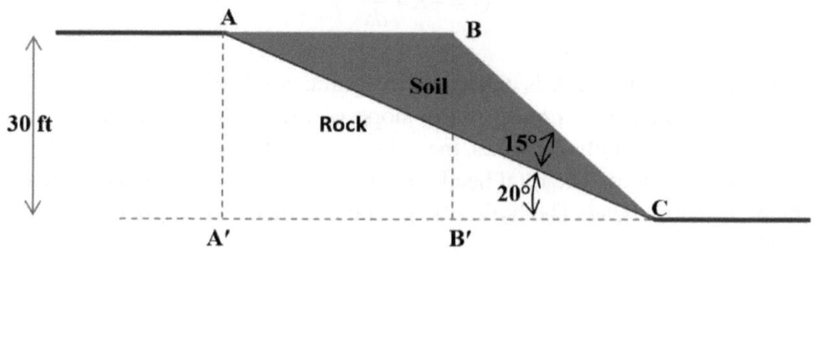

From triangle AA′C, the distance A′C = 30/tan 20 = 82.42 ft
From triangle BB′C, the distance B′C = 30/tan 35 = 42.84 ft
Therefore AB = A′B′ = 39.58 ft
Length of slope AC = 30/sin 20 = 87.71 ft
Area of soil wedge ABC = ½ × 39.58 × 30 = 593.7 ft^2

Weight of soil wedge ABC (unit depth) $= 593.7 \times 108 = 64{,}120 \frac{\text{lb}}{\text{ft}}$

Component of soil weight parallel to slope: $W_T = 64{,}120 \sin 20 = 21{,}930 \frac{\text{lb}}{\text{ft}}$

Shear stress on slope AB: $\tau = \frac{21{,}930}{87.71 \times 1} = 250\ \text{lb/ft}^2$

Component of soil weight perpendicular to slope: $W_N = 64{,}120 \cos 20 = 60{,}253\ \frac{\text{lb}}{\text{ft}}$

Normal stress on slope AB: $\sigma = \frac{60{,}253}{87.71 \times 1} = 687\ \text{lb/ft}^2$

Shear strength along interface: $\tau_{\text{ult}} = c + \sigma \tan\phi = 400 + 687 \tan 20 = 650\ \text{lb/ft}^2$

$FS = \tau_{\text{ult}}/\tau = 650/250 = 2.6$

Permeability of Soils

Darcy's Law for Seepage

Seepage, the migration of water through voids in a soil, is due to a soil property called permeability or hydraulic conductivity. Darcy's law specifies that the (average) discharge velocity (also called the specific discharge q) through the soil is proportional to the hydraulic gradient (dh/dx). The negative sign ensures that the seepage velocity is in the direction of decreasing hydraulic head (h)

$$q = -K\frac{dh}{dx} \tag{14.57}$$

Thus, the seepage discharge Q through (and normal to) an area A of the soil is given by

$$Q = -KA\frac{dh}{dx} \tag{14.58}$$

The constant of proportionality K is called the hydraulic conductivity.

The hydraulic gradient ($i = dh/dx$) is the slope of the energy grade line across the soil boundary. For seepage through soil voids, the velocity is small and therefore the kinetic energy head ($V^2/2g$) can be ignored. The total head h is therefore approximately equal to $h = p/\gamma + z$, which is the piezometric head. The hydraulic gradient is therefore approximated by the slope of the hydraulic grade line.

The (actual) seepage velocity through the soil voids is much higher than the average velocity. The seepage velocity may be expressed as

$$V_s = \frac{V}{n} = -\frac{K}{n}\frac{dh}{dx} \tag{14.59}$$

where n is porosity of the soil.

NOTE If the seepage loss (flow rate) is to be estimated, it would be appropriate to use the average discharge velocity. But, if the objective is to determine the potential for erosion or scour, the seepage velocity should be used, since this is the actual velocity at which water is migrating through the soil.

Liquefaction

The phenomenon of liquefaction occurs when the effective stress at a particular location gets drastically reduced due to seepage forces acting against gravity forces. Since the effective stress with seepage is given by

$$\sigma' = \gamma' z - i\gamma_w z \tag{14.60}$$

The critical condition ($\sigma' \to 0$) results in the critical hydraulic gradient

$$i_c = \frac{\gamma'}{\gamma_w} \tag{14.61}$$

Thus, in a location where the seepage exit gradient is i_e, the factor of safety against liquefaction may be expressed as

$$FS = \frac{i_c}{i_e} \tag{14.62}$$

Laboratory Measurement of Hydraulic Conductivity

The hydraulic conductivity K of a soil can be measured using several laboratory methods. Two of the more common of these methods are

- Constant head test
- Falling head test

Constant Head Test

In the constant head test, the inlet head is controlled to be constant at all times, thereby making the head difference H between inlet and outlet constant. Water is allowed to seep through the soil column and collected at the outlet. The volume of water V flowing through the soil in a measured time interval t is monitored. A schematic of the constant head test is shown in Fig. 14.20.

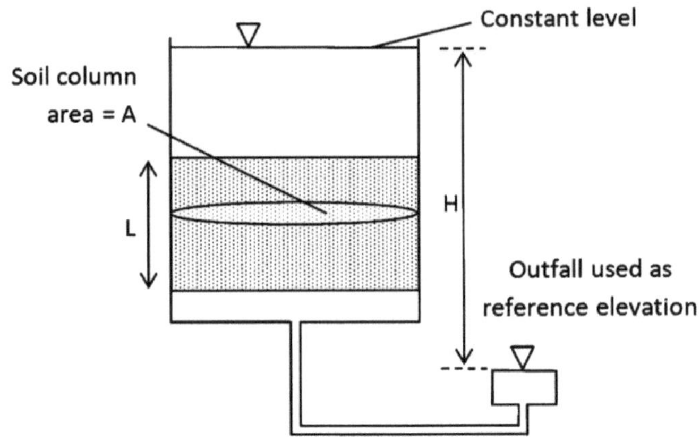

Figure 14.20 Schematic of constant head permeability test.

The length of the soil column (measured parallel to the direction of flow) is L and the cross section of the soil column (normal to the longitudinal direction) is A.

Seepage occurs under constant head H. Therefore, the constant hydraulic gradient is H/L. According to Darcy's law, the average flow rate Q may be written as

$$Q = \frac{V}{t} = K\frac{H}{L}A \tag{14.63}$$

$$K = \frac{VL}{AHt_e} \tag{14.64}$$

where A = cross section of the soil column
L = length of flow path through soil
V = volume of water collected at outlet
t_e = elapsed time (of water collection)
H = constant head

Falling Head Test

In the falling head test, the liquid is allowed to flow through and out of the system without maintaining a constant inlet reservoir elevation. A schematic view of this type of test is shown in Fig. 14.21. The inlet head is decreasing continuously, as liquid flows through the soil and out at the outlet, making the head difference h (between inlet and outlet) variable with time t. This test requires two-point data, that is, the head is to be measured at two different instants t_1 and t_2. The head is measured with respect to the elevation at the free outfall.

The hydraulic conductivity K is then given by

$$K = \frac{2.303aL\log_{10}(h_1/h_2)}{At_e} \tag{14.65}$$

Figure 14.21 Schematic of falling head permeability test.

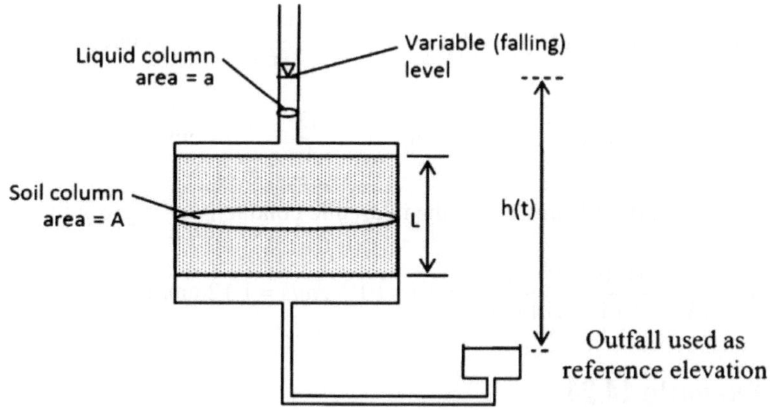

where a = cross section of the falling liquid column
A = cross section of the soil column
L = length of flow path through soil
t_e = elapsed time = $t_2 - t_1$
h_1 and h_2 = head at times t_1 and t_2 ($h_1 > h_2$)

Example 14.23

For the apparatus shown in the figure, the volume of water collected at the outlet during a time interval of 40 min is 5 cm^3. The locations A, B, C, and D are marked with their elevations in cm (with respect to a common datum). Assume that reservoirs A and D have constant surface elevation. What is the permeability of the soil (cm/h)?

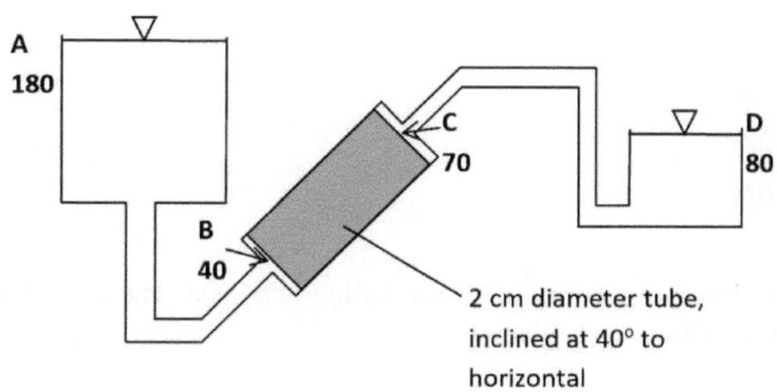

Solution Since there is no energy loss mechanism between A and B, the total head at B is equal to the total head at A = 180 cm

Similarly, the total head at C is equal to the total head at D = 80 cm

The length of the soil column is given by $L = \dfrac{\Delta h}{\sin\theta} = \dfrac{70-40}{\sin 40} = 46.67$ cm

Hydraulic gradient is given by $i = \dfrac{\Delta H}{\Delta L} = \dfrac{180-80}{46.67} = 2.14$ cm/cm

Cross-sectional area of soil column $A = \dfrac{\pi D^2}{4} = 3.14$ cm^2

Therefore, from Darcy's law, the hydraulic conductivity is given by

$$K = \frac{Q}{iA} = \frac{5/(40\times 60)\ \text{cm}^3/\text{s}}{2.14\times 3.14\ \text{cm}^2} = 3.1\times 10^{-4}\ \text{cm/s} = 1.12\ \text{cm/h}$$

Example 14.24

Water seeps through a sandy soil layer (20 in thick, 30 in diameter) and overflows through an outlet maintaining the outfall elevation of 50 in, as shown below. As the water seeps out, the level of water drops in the fine bore tube from an elevation of 85 to 75 in over a 35-min interval. The hydraulic conductivity of the soil is most nearly

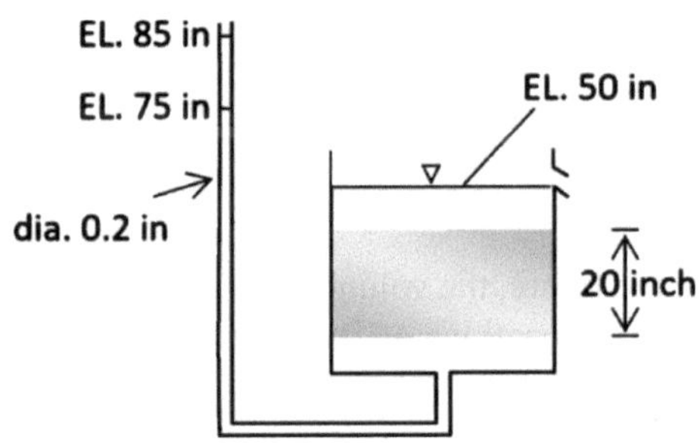

A. 1.5×10^{-5} ft/h

B. 5.2×10^{-4} ft/h

C. 4.3×10^{-5} ft/h

D. 2.5×10^{-4} ft/h

Solution The governing theory is for the falling head test. The area of the tube in which the liquid column is falling (a) is given by

$$a = \frac{\pi \times 0.2^2}{4} = 0.0314\ \text{in}^2$$

Cross-sectional area of the soil column:

$$A = \frac{\pi \times 30^2}{4} = 706.9\ \text{in}^2$$

h_1 and h_2 must be measured from the outfall elevation (50 in). $h_1 = 85 - 50 = 35$. $h_2 = 75 - 50 = 25$. Elapsed time = 35 min = 0.583 h

$$K = 2.303\frac{aL}{A(t_2 - t_1)}\log_{10}(h_1/h_2) = \frac{2.303\times 0.0314\times 20\times \log_{10}(35/25)}{706.9\times 0.583} = 0.000513\frac{\text{in}}{\text{h}}$$
$$= 4.3\times 10^{-5}\ \text{ft/h}$$

Answer is C.

Flow Nets

The governing equation for seepage transport through a porous medium (soil) is the homogeneous Laplace equation in terms of the potential function $h(x, y)$

$$K_x\frac{\partial^2 h}{\partial x^2} + K_y\frac{\partial^2 h}{\partial y^2} = 0 \tag{14.66}$$

As the groundwater seeps through the soil, driven by a potential difference between the upstream and the downstream boundaries, the potential (head) decreases according to the solution of Eq. (14.66). For isotropic soils, the hydraulic conductivity is identical in all directions. This leads to

$$\frac{\partial^2 h}{\partial x^2} + \frac{\partial^2 h}{\partial y^2} = 0 \tag{14.67}$$

The graphical solution to the above equation is a technique known as *flow net*. The flow net is constructed by discretizing the soil domain using two kinds of lines intersecting at right angles. These lines are

1. Equipotential lines—lines of constant total head
2. Flow lines—representing the direction of flow for water

The upstream boundary in the soil domain serves as the first ($i = 0$) equipotential line. The total head at all points along this line is equal to the total head at the free surface of the upstream reservoir. Similarly, the downstream boundary of the soil domain serves as the last ($i = 6$ in this case) equipotential line. See Fig. 14.22.

Rules for constructing a "good" flow net are

1. Equipotential lines and flow lines should intersect at nearly 90°.
2. Cells should have uniform length to width ratio.

The total seepage flow through all flow channels (if flow net elements are approximately square) is given by

$$Q = KH\frac{N_f}{N_d} \tag{14.68}$$

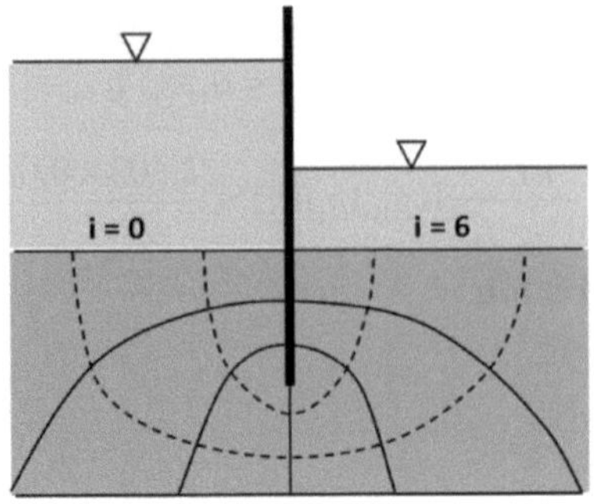

Figure 14.22 Seepage occurring under a sheet pile wall.

where Q = flow rate per unit length (measured perpendicular to flow net)
N_f = number of flow channels
N_d = number of equipotential drops

NOTE Q represents flow rate per unit width (perpendicular to plane of the flow net). It therefore has dimensions L^2/T. To calculate the total seepage flow rate (L^3/T), one must multiply Q by the length normal to the plane in which the flow net is drawn.

Example 14.25

Seepage flow occurs around a sheet pile wall driven to point B (elevation 12 ft) in a soil deposit as shown below. A flow net has been drawn. Points A and C are at elevation 16 ft and 5 ft above the datum shown. Calculate the piezometric pressure at points A, B, and C.

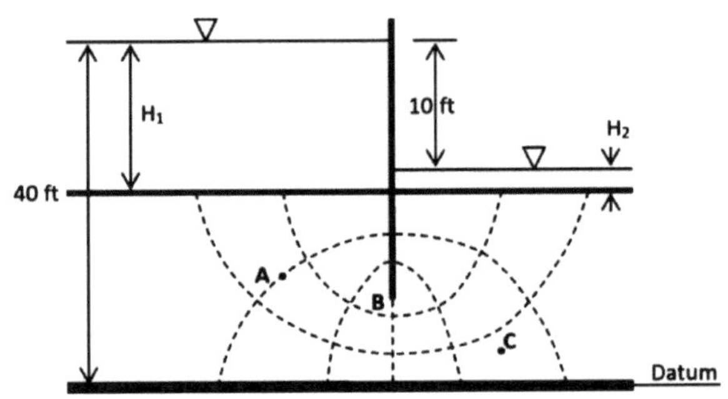

Solution Expressing all pressures as gage pressure (rather than absolute), the pressure head is designated as PH, the velocity head as VH, and the elevation head as EH. The total head (TH) is the sum of these three. These quantities are summarized below for the key locations within the soil domain shown above.

At upstream surface:	PH = 0	VH = 0	EH = 40	TH = 40
At ground upstream:	PH = H_1	VH = 0	EH = 40 − H_1	TH_{US} = 40

At downstream surface: PH = 0 VH = 0 EH = 30 TH = 30

At ground downstream: $PH = H_2$ VH = 0 $EH = 30 - H_2$ $TH_{DS} = 30$

The difference between the total upstream head (TH_{US}) and total downstream head (TH_{DS}) is

$$TH_{US} - TH_{DS} = 40 - 30 = 10$$

Since successive equipotential lines represent equal increments of head loss, the total head at points A, B, and C are given by

$$TH_A = TH_{US} - 1/6 \times (TH_{US} - TH_{DS}) = 40 - 1/6 \times (40 - 30) = 38.33$$

$$TH_B = TH_{US} - 3/6 \times (TH_{US} - TH_{DS}) = 40 - 3/6 \times (40 - 30) = 35.00$$

$$TH_C = TH_{US} - 4.5/6 \times (TH_{US} - TH_{DS}) = 40 - 4.5/6 \times (40 - 30) = 32.50$$

Since the total head at A is 38.33 ft and the elevation at A is 16 ft, assuming velocity head to be negligible, the pressure head at A is 38.33 – 16.00 = 22.33 ft.

This pressure head can be then converted to a pressure using

$$p_A = \gamma H = 62.4 \times 22.33 = 1393.4 \text{ psf} = 9.68 \text{ psi}$$

Similarly, the pressure head at B is 23.00 ft and the pressure is 1435.2 lb/ft^2. The pressure head at C is 27.50 ft and the pressure is 1716.0 lb/ft^2.

The significance of the pressure head at A, B, and C being 22.33, 23.00, and 27.50 ft is that these would be the height of a water column in a fine bore piezometer driven into the soil to these points. In other words, at A, the water column would rise to an elevation 22.33 + 16.00 = 38.33 ft.

Aquifers

Groundwater is extracted principally through wells. Wells may be drilled and installed with a casing or without—the latter situation appropriate for stable soils where a casing or screen may not be required to stop unwanted inflow. A well typically consists of a drilled hole, a casing that prevents caving in of the sides or unwanted inflow, a pump and a section of perforated casing (the screen) to ensure that water is extracted from a specific elevation within the aquifer, a grout seal at the surface (to prevent contaminated surface runoff from percolating along the sides of the casing and reaching the screen) and a gravel pack around the screen to improve the flow to the well.

In a confined aquifer, the hydraulic head is given by the elevation of the piezometric surface, which is above the upper surface of the aquifer. Adding or removing water to the aquifer does not change the saturation of the aquifer, as long as the piezometric surface remains above the upper surface of the aquifer. The change of the hydraulic head is accomplished by a change in pressure. This pressure tries to force individual soil particles apart. Thus, in a confined aquifer, the total storage increases as pressure increases because the granular skeleton expands to create more void space.

If water is added to a groundwater reservoir, hydraulic head increases (water level in wells rises) and fluid pressure increases in the aquifer. Porosity increases as the granular skeleton expands.

In an unconfined aquifer, the hydraulic head is the height of the water table. In an unconfined aquifer, a change in hydraulic head results in both a change in pressure in the saturated portion, as well as a change in thickness of the saturated zone.

The transmissivity (T) is a measure of how much water can be transmitted horizontally through a unit width of a fully saturated aquifer under a hydraulic gradient of 1.0. The transmissivity is calculated as the product of the hydraulic conductivity K and the aquifer thickness Y

$$T = KY \tag{14.69}$$

A steady rate of pumping from an aquifer creates a drawdown of the water table. In Fig. 14.23, Y is the thickness of the unconfined aquifer; y is the water table elevation at a radial distance r, s is the drawdown, steady pumping rate is Q and permeability of the soil in the aquifer is K.

$$Y = s + y \tag{14.70}$$

Dupuit's Formula

The steady-state solution, which represents the equilibrium between the rate of pumped flow out of the well and the rate of radial inflow from the cone of depression, is given by Eq. (14.71)

$$Q = \frac{\pi K\left(y_1^2 - y_2^2\right)}{\ln(r_1/r_2)} \tag{14.71}$$

where y_1 = depth of aquifer at radial distance r_1
y_2 = depth of aquifer at radial distance r_2
Q = pumping rate
K = hydraulic conductivity

Figure 14.23 Radial drawdown in an unconfined aquifer.

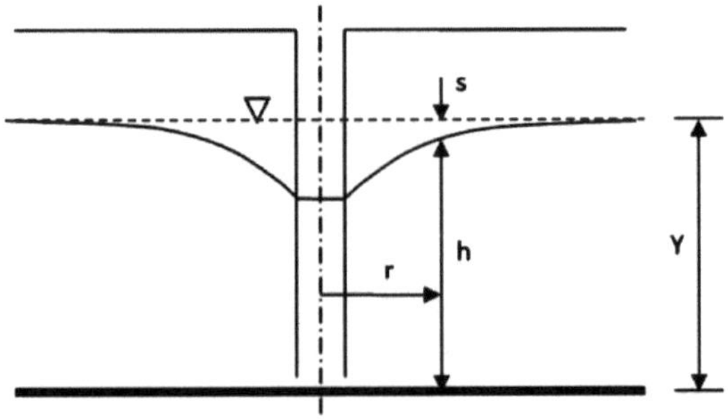

For a confined aquifer, the same expression may be used, except that the variable y represents the elevation of the piezometric surface instead of the free surface as in the unconfined aquifer.

Thiem Equation

For a very thick unconfined aquifer, the following approximation may be used

$$Q \approx \frac{2\pi KY(s_2 - s_1)}{\ln(r_1/r_2)} = \frac{2\pi T(s_2 - s_1)}{\ln(r_1/r_2)} \qquad (14.72)$$

Example 14.26

A pumping well is drilled into an unconfined aquifer at location A and a steady pumping rate of 800 gal/min is established. Observation wells are dug at B and C as shown on the map of the region. The original elevation of the GWT in the underlying soil layer is 295 ft above sea level. The table shows data for the wells. Find the elevation of GWT in the observation well at C.

Well	Type	Diameter (in)	GWT (ft)
A	Pumped	10	288
B	Obs	5	291
C	Obs	6	–

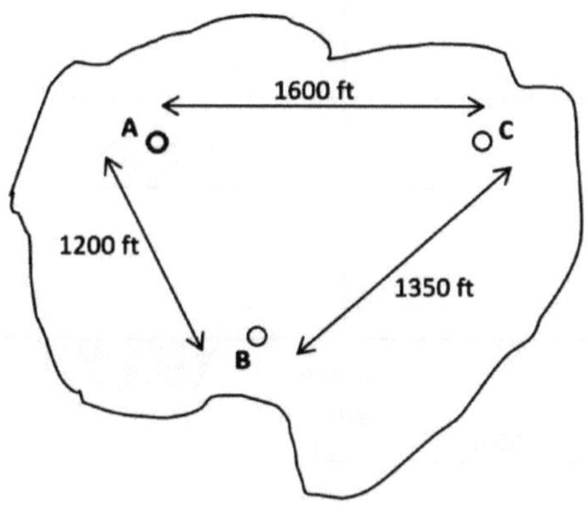

Solution

$Q = 800$ gal/min $= 800 \times 0.002228 = 1.782$ ft³/s

Assuming aquifer depth Y is large, we can approximate the drawdown equation in terms of the drawdown s instead of the water table elevations y

$$Q = \frac{\pi K(y_1^2 - y_2^2)}{\ln(r_1/r_2)} \approx \frac{2\pi KY(s_2 - s_1)}{\ln(r_1/r_2)}$$

For well A, $r_1 = 5$ in $= 0.4167$ ft and the drawdown $s_1 = 295 - 288 = 7$ ft
For well B, $r_2 = 1200$ ft and the drawdown $s_2 = 295 - 291 = 4$ ft
Using these values, we get
$KY = 0.753$ ft²/s
Now using this calculated value of KY for the pair A and C, we get
For well A, $r_1 = 5$ in $= 0.4167$ ft and the drawdown $s_1 = 295 - 288 = 7$ ft
For well C, $r_2 = 1600$ ft and the drawdown s_2
Therefore, $s_2 - s_1 = -3.11$ ft
$s_2 = s_1 - 3.11 = 3.89$
Therefore, elevation of GWT in observation well C is $295 - 3.89 = 291.11$ ft

Example 14.27

A 9-in-diameter well is drilled into a confined aquifer as shown below. The thickness of the aquifer is 32 ft. The impermeable layer above the aquifer is 46 ft thick. The original piezometric surface is at a depth of 30 ft below the ground surface. A steady rate of pumping at 125 gal/min is established from the well until water levels in the pumping well and the observation well stabilizes. Given the recorded drawdown in the wells:

Pumped well	15 ft
Observation well	8 ft

Find the permeability of the soil in the aquifer.

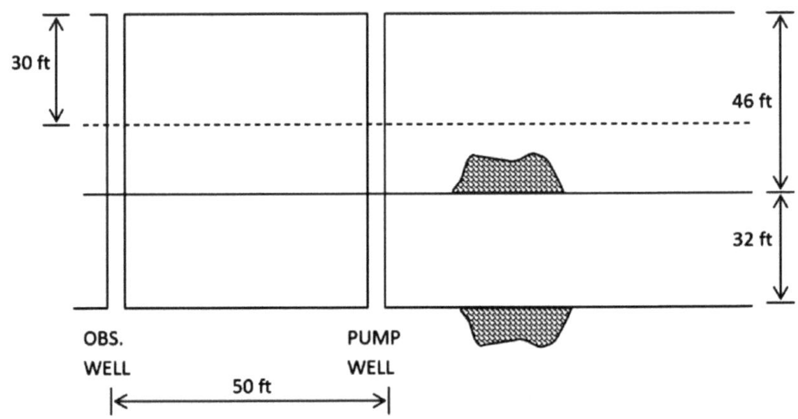

Solution

Initial elevation of piezometric surface $= h_0 = 78$ ft $- 30$ ft $= 48$ ft
Drawn down elevation in pumped well $= 48 - 15$ ft $= 33$ ft
Drawn down elevation in observation well $= 48 - 8$ ft $= 40$ ft
$Q = 125$ gal/min $= 125 \times 0.002228 = 0.2785$ ft³/s

$$K = \frac{Q}{2\pi Y(h_2 - h_1)} \ln\left(\frac{r_2}{r_1}\right) = \frac{0.2785}{2\pi \times 32 \times (40 - 33)} \ln\left(\frac{50}{4.5/12}\right) = 9.68 \times 10^{-4} \text{ ft}^3/\text{s}$$

Soil Stabilization

Cement and lime are the most common stabilizing agents for weak or wet soils. The choice is primarily determined by the plasticity index (PI) of the soil. Cement can be applied to stabilize any type of soil, except those with significant (greater than 2%) organic content or having pH lower than 5.3 Stabilization with lime or cement increases plastic limit but reduces the liquid limit, thereby reducing the plasticity index. Stabilization with cement or lime typically decreases the maximum dry density and increases the optimum water content of soils.

A stabilizing additive is a mechanical, chemical or bituminous additive used to maintain or increase the strength and durability, decrease the moisture sensitivity or otherwise improve the engineering properties of a soil.

Lime (CaO) is a white or grayish white, odorless, lumpy, slightly water soluble solid, which when mixed with water, forms calcium hydroxide (slaked lime), which is used in mortar, plaster, and cements.

Foamed asphalt is expanded in volume by the controlled addition of air and water in an expansion chamber. The foam is then spray-injected into a mixer where the asphalt attaches itself to fines in the mix.

Cement treated base (CTB) is a mixture of aggregate material and/or granular soils combined with measured quantities of Portland cement and water that hardens after compaction and curing to form a durable paving material.

Bitumen is a class of black cementitious substances composed principally of high molecular weight hydrocarbons.

Asphalt emulsion is a mixture of asphalt binder and water that contains a small amount of emulsifying agent to cause the asphalt to become suspended in the water.

Mechanical Stabilization

This involves the pulverization and homogenization of existing materials in place without the addition of an additive. This technique is usually performed when the material is suitable and full depth reclamation results in anew stabilized base of sufficient strength.

Chemical Stabilization

Addition of a single additive such as lime, cement, or bitumen in the form of a slurry is the most common form of chemical stabilization. CTB (cement-treated base) has provided economical and durable pavement foundations. Cement stabilization increased base material strength, reduces deflections and increases pavement life by delaying surface cracking. Use of cement stabilization also increases moisture resistance and reduces probability of subgrade failure.

Cement modified soil (CMS) is a soil material that is treated with a relatively low proportion of Portland cement. The objective of CMS is to improve soil properties such as plasticity index (decrease), California Bearing Ratio (increase), shear strength (increase), shrinkage and swelling potential (decrease).

Soil stabilization with bitumen can be done with either an asphalt emulsion or foamed bitumen. The latter is more appropriate for extremely wet soil conditions.

Erosion Control Methods

Erosion control methods aim to protect the soil surface and prevent soil particles from being dislodged by wind or water. Often featuring biodegradable or natural materials, these methods will help facilitate a natural environment that will allow for plants to take root. Some common strategies for erosion control are (a) natural—vegetative cover, composting, surface roughening, sodding and mulching and (b) artificial—blankets and mats, geotextiles, soil bioengineering.

Geosynthetics

Geosynthetics are usually polymer products used to stabilize ground. They are available as geomembranes, geotextiles, geogrids, geonets, geosynthetic clay liners, geofoam, geocells, and geocomposites.

Geotextiles, which may be woven or knitted, are porous, resistant to biodegradation and perform the functions of separation, reinforcement, filtration and drainage, trapping fine sediment while allowing water to pass through.

Geomembranes are relatively thin, impervious sheets of polymer material used primarily for linings of liquid or solid storage and/or containment facilities such as canals and landfills.

Geogrids are polymers formed into a very open grid and perform almost exclusively as reinforcement materials.

Geosynthetic clay liners (GCL) are created by a thin layer of bentonite clay sandwiched between two geotextiles or bonded to a geomembrane. The structure is made composite by needle punching, stitching or adhesive bonding.

CHAPTER 15

Transportation

Transportation Planning

A major part of transportation planning is the use of empirical models developed from economic data. The term "model" is used to refer to a series of mathematical equations that are used to represent how choices are made when people travel. The coefficients and parameters in the model are set (calibrated) to match existing data. Usually, these relationships are assumed to be valid and to remain constant in the future.

Models are created based on population forecasts, economic forecasts, and land use patterns. They are then used to estimate the number of trips that will be made on a network (trip generation), how these trips will originate and terminate (trip distribution), how these trips will distribute across various modes of travel (modal split), and what routes will be available or need to be made available for these forecast trips (route assignment). These estimates are the basis for transportation plans and are used in major investment analysis, environmental impact statements, and in setting priorities for investments.

For purposes of simulations, an urban area is represented as a series of small geographic areas (zones). Zones are characterized by their population, employment, and other factors and are the places where trips begin (trip producers) or end (trip attractors). Trip making is first estimated at the household level and then aggregated to the zone level. Zones can be as small as a single block but typically are ¼ to 1 mi^2 in area.

Trip Generation

The first step in the traffic forecasting process—trip generation—predicts the number of trips originating from or destined for a particular traffic analysis zone. Typically, trip generation analysis focuses on residential communities, based on an economic model which is a function of the social and economic attributes of households. Residential trip generation analysis is often undertaken using statistical regression using variables such as household size, number of workers in the household, persons in an age group, type of residence (single family, apartment, etc.), auto ownership per household, and so on. Usually, measures on five to seven independent variables are available.

There are two kinds of trip generation models: production models and attraction models. Trip production models estimate the number of home-based trips created from zones where trip makers reside. Information from land use, population, and economic forecasts are used to estimate how many person trips will be made to and from each zone. Traffic analysis zones are also destinations of trips or trip attractors. The analysis of attractors focuses on nonresidential land uses. Trip attractions are typically based on the level of employment in a zone. Trip attraction models estimate the number of home-based trips to each zone at the nonhome end of the trip. Different production and attraction models are used for each trip purpose.

Trip Distribution: Gravity Model

Once the trip generation model(s) predict production P_i and attraction A_j values for each zone within the analysis area, these could be out of balance. In other words, the total number of trips produced within the area may not be equal to the total number of trips produced. An iterative method must then be used to balance the system.

The most commonly used procedure for trip distribution is the "gravity model." The gravity model takes the trips produced at one zone and distributes to other zones based on the size of the other zones (as measured by their trip attractions) and on the basis of the distance to other zones. According to the gravity model, the number of trips between two zones is inversely proportional to the square of the time that it takes to travel between these zones.

$$T_{ij} = P_i \left[\frac{A_j F_{ij} K_{ij}}{\sum_j A_j F_{ij} K_{ij}} \right] \tag{15.1}$$

where T_{ij} = number of trips produced in zone i and attracted to zone j

P_i = total number of trips produced in zone i (produced by trip generation model)

A_j = total number of trips attracted to zone j (produced by trip generation model)

F_{ij} = value representing impedance (usually the inverse of travel time) between zones i and j

K_{ij} = socioeconomic adjustment factor for route between i and j (these factors are used to adjust model predictions when there is a significant difference between predicted and observed values due to unique characteristics of certain zones and/or routes)

The sum of all the trip productions P_i must equal the sum of all the trip attractions A_j.

Example 15.1

A study area consists of three zones: A, B, and C. The trip generation model predicts the following trip productions and attractions.

Zone	A	B	C	Total
Trip production	170	400	330	900
Trip attraction	350	325	225	900

Travel time (minutes) between zones is given in the matrix below. For this example, the F factor is taken as the inverse of the corresponding travel time. Also, for this example, all K_{ij} values are taken as 1.0.

Zone	A	B	C
A	7	4	5
B	4	5	8
C	5	8	3

Solution The first iteration of trip distribution yields the following results:

Zone	A	B	C	Production (P)
A	48	78	43	170
B	194	144	62	400
C	124	72	133	330
Computed A	366	295	239	
Given A	350	325	225	

For instance, the distribution of number of trips (170) produced in zone A is calculated as

$$T_{\mathrm{AA}} = P_A \frac{A_A/t_{\mathrm{AA}}}{\sum_j A_j/t_{Aj}} = 170\left(\frac{350/7}{350/7+325/4+225/5}\right) = 48$$

$$T_{\mathrm{AB}} = P_A \frac{A_B/t_{\mathrm{AB}}}{\sum_j A_j/t_{Aj}} = 170\left(\frac{325/4}{350/7+325/4+225/5}\right) = 78$$

$$T_{\mathrm{AC}} = P_A \frac{A_C/t_{\mathrm{AC}}}{\sum_j A_j/t_{Aj}} = 170\left(\frac{225/5}{350/7+325/4+225/5}\right) = 43$$

As may be seen, this produces an imbalance in the attractions for each zone. The computed attractions to zones A, B, and C are 366, 295, and 239, which are different from the initial values of 350, 325, and 225. The next step will be to apply a correction to the attraction values and reiterate through the distribution process. The process is continued until the values converge.

Modal Split by the Logit Model

Mode choice is one of the most critical parts of the travel demand modeling process. It is the step where trips between a given origin and destination are split into trips using transit, trips by car pool or as automobile passengers, and trips by automobile drivers. The most commonly used process for mode split is to use the "logit" model. This involves a comparison of the "disutility" of travel between two points for the different modes that are available.

Disutility is a term used to represent a combination of the travel time, cost, and convenience of a mode between an origin and a destination. Once disutilities are known for the various mode choices between an origin and a destination, the trips are split among various modes based on the relative differences between disutilities. The logit equation is used in this step. A large advantage in disutility will mean a high percentage for that mode. Traffic assignment is typically done for peak hour travel, while forecasts of trips are done on a daily basis. A ratio of peak hour travel to daily travel is needed to convert daily trips to peak hour travel.

The probability of choosing a particular mode p out of all available modes n is then given by

$$P(p)=\frac{e^{U_p}}{\sum_1^n e^{U_i}} \tag{15.2}$$

where $U(i)$ is the value of the utility function for mode i.

Automobile trips must be converted from person trips to vehicle trips with an auto-occupancy model. Mode split and auto-occupancy analysis can be two separate steps or can be combined into a single step, depending on how a forecasting process is set up.

Once trips have been split into highway and transit trips, the specific path that they use to travel from their origin to their destination must be found. These trips are then assigned to that path in the step called traffic assignment.

Example 15.2

Citizens of a small community have three choices for commuting to the employment zone defined as "downtown," which is a distance of 17 mi away. The choices are (1) driving, (2) commuter rail, and (3) a suburban bus line. The utility values for the three modes, based on a factored sum of attributes such as time, cost, etc., are −1.23 for driving, −2.11 for commuter rail, and −1.89 for the bus line. If the total number of commuters originating from the community is 560, how many are expected to take the bus?

Solution The probability of selecting mode C (bus line) can be written

$$P(C)=\frac{e^{U_C}}{e^{U_A}+e^{U_B}+e^{U_C}}=\frac{e^{-1.89}}{e^{-1.23}+e^{-2.11}+e^{-1.89}}=\frac{0.151}{0.292+0.121+0.151}=0.268$$

Number of commuters expected to use the bus line = 0.268 × 560 = 149.9.

Design Traffic Volume

Average daily traffic (ADT) is an average 24-h volume at a given location for some period of time less than 1 year, but more than 1 day.

Average weekday traffic (AWT) is an average 24-h traffic volume occurring on weekdays for some period less than 1 year.

Average annual weekday traffic (AAWT) is an average 24-h traffic volume occurring on weekdays over a full 365-day year.

Average annual daily traffic (AADT) is an average 24-h traffic volume at a given location over a full 365-day year.

Design hourly volume (DHV) is the 30th highest hourly volume of the year, based on hourly vehicular counts taken throughout a full year.

K is the proportion of AADT on a roadway segment or link that occurs during the design hour.

$$\text{DHV} = K \times \text{AADT} \tag{15.3}$$

D is the proportion of DHV occurring in the heavier direction, and is called the directional split.

The directional design hourly volume, denoted by DDHV, is given by Eq. (15.4)

$$\text{DDHV} = D \times \text{DHV} = D \times K \times \text{AADT} \tag{15.4}$$

Some typical range of values of K and D factors are listed below.

Facility type	*K* factor	*D* factor
Rural	0.15–0.25	0.65–0.80
Suburban	0.12–0.15	0.55–0.65
Urban:		
Radial route	0.07–0.12	0.55–0.60
Circumferential route	0.07–0.12	0.50–0.55

Expansion Factors

To make reasonable estimates of annual traffic volume characteristics on an area-wide basis, different types of periodic counts, with count durations ranging from 15 min to continuous, are conducted.

Periodic volume counts are used to calculate expansion factors needed to estimate the annual traffic volume.

For example, the hourly expansion factor (HEF) is defined as

$$\text{HEF} = \frac{\text{total volume for 24-h period}}{\text{volume for particular h}} \tag{15.5}$$

So, if hourly volume counts are available for a continuous 24-h day, the HEF can be computed for each hour. This factor can be used to extrapolate the hourly count to a daily count.

Similarly, a daily expansion factor (DEF) is defined as

$$\text{DEF} = \frac{\text{average total volume for a week}}{\text{volume for particular day}} \tag{15.6}$$

And, a monthly expansion factor (MEF) is defined as

$$\text{MEF} = \frac{\text{AADT}}{\text{volume for particular month}} \tag{15.7}$$

For example, if the following hourly data are given for a rural road, the 24-h volume = 12,832.

Hour	Volume	HEF	Hour	Volume	HEF
6–7 am	312	41.13	6–7 pm	782	16.41
7–8 am	456	28.14	7–8 pm	702	18.28
8–9 am	554	23.16	8–9 pm	603	21.28
9–10 am	678	18.93	9–10 pm	511	25.11
10–11 am	723	17.75	10–11 pm	432	29.70
11–12 pm	682	18.82	11–12 pm	421	30.48
12–1 pm	654	19.62	12–1 am	249	51.53
1–2 pm	765	16.77	1–2 am	189	67.89
2–3 pm	843	15.22	2–3 am	156	82.26
3–4 pm	892	14.39	3–4 am	102	125.80
4–5 pm	932	13.77	4–5 am	98	130.94
5–6 pm	908	14.13	5–6 am	188	68.26

The HEF values are generated as follows: for 6 to 7 am, HEF = 12,832 ÷ 312 = 41.13. This means that the count during the 6 to 7 am hour should be multiplied by 41.13 to extrapolate to a 24-h period.

Example 15.3

The following traffic counts were observed on a Thursday in April on a rural primary road. The second column in the table shows HEF values corresponding to each hour. What is the AADT of the road? DEF for Thursday = 5.99 and MEF for April = 1.48.

Hour	HEF	Volume
8–9 am	23.16	294
9–10 am	18.93	426
10–11 am	17.75	560
11–12 pm	18.82	657
10–11 am	19.62	722

Solution Using the HEF factors given in the table, the 24-h volume (averaged from the 5 hourly numbers) for Thursday is given by

$$\frac{294 \times 23.16 + 426 \times 18.93 + 560 \times 17.75 + 657 \times 18.82 + 722 \times 19.62}{5}$$

$$= 10{,}269 \text{ vehicles per hour}$$

Since the data were collected on a Thursday (DEF = 5.99), the average (daily) volume for the week is

$$\frac{10{,}269 \times 5.99}{7} = 8787 \text{ vehicles per day}$$

Since the data were collected in the month of April (MEF = 1.48), the AADT is given by

$$\text{AADT} = 8787 \times 1.48 = 13{,}005 \text{ vehicles per day}$$

Travel Speed

Average running speed is based on the observation of vehicle travel times traversing a section of highway of known length. It is the length of the segment divided by the average running time of vehicles to traverse the segment. Running time includes only time that vehicles are in motion.

Average travel speed is based on travel time observed on a known length of highway. It is the length of the segment divided by the average travel time of vehicles traversing the segment, including all stopped delay times. It is also called space mean speed.

Time mean speed is the arithmetic average of speeds of vehicles observed passing a point on a highway; also referred to as the average spot speed.

Free-flow speed (FFS) is the average speed of vehicles on a given facility, measured under low-volume conditions, when drivers tend to drive at their desired speed and are not constrained by control delay.

Operating speed is the speed at which drivers are observed operating their vehicles under free-flow conditions. The 85th percentile of the distribution of observed speeds is most frequently used as the operating speed.

Design speed is defined as "the maximum safe speed that can be maintained over a specified section of highway when conditions are favorable such that the design features of the highway govern" (*AASHTO Green Book*). The design speed assumed for a highway should be based on functional classification, topography, anticipated operating speeds, and adjacent land use. Design speeds range from 20 to 70 mi/h, usually in 10 mi/h increments. Usually, freeways should be designed for 60 to 70 mi/h whereas design speeds for other arterial highways range from 30 to 60 mi/h. A common rule of thumb for the design speed is to add 10 to 15 mi/h (15 to 25 km/h) to the posted speed.

Time-Mean Speed versus Space-Mean Speed

There are two types of mean speed—time-mean speed and space-mean speed. The time-mean speed is the arithmetic mean of the speeds of vehicles passing a specific point on a highway during an interval of time. Thus, the time-mean speed of *n* vehicles is given by

$$\overline{u_t} = \frac{1}{n}\sum_{i=1}^{n} u_i \qquad (15.8)$$

where u_i is the instantaneous speed of the *i*th vehicle.

On the other hand, the space-mean speed is the harmonic mean of the speeds of vehicles passing a point on the highway during an interval of time.

$$\overline{u_s} = \frac{n}{\sum_{i=1}^{n} \frac{1}{u_i}} \tag{15.9}$$

Example 15.4

The speeds of five cars are recorded using a radar gun as follows: 48, 34, 42, 39, and 45 mi/h. What is the (a) time-mean speed and (b) space-mean speed?

Solution The time-mean speed is given by Eq. (15.8)

$$\overline{u_t} = \frac{1}{n}\sum_{i=1}^{n} u_i = \frac{48+34+42+39+45}{5} = 41.6\,\text{mi/h}$$

The space-mean speed is given by Eq. (15.9).

$$\overline{u_s} = \frac{n}{\sum_{i=1}^{n} \frac{1}{u_i}} = \frac{5}{\frac{1}{48}+\frac{1}{34}+\frac{1}{42}+\frac{1}{39}+\frac{1}{45}} = 41.0\,\text{mi/h}$$

Speed–Volume–Density Relationships

Many traffic flow models are based on relationships between the critical mobility variables—speed, volume, and density. Fundamental definitions of some of these parameters are given below.

The flow rate is the equivalent hourly rate at which vehicles pass a point on a highway during a time period less than 1 h. The flow rate V (vehicles per hour) is equal to the product of speed S (mi/h) and density D (vehicles per mile)

$$V = SD \tag{15.10}$$

The average spacing d_a (feet per vehicle) between corresponding points on successive vehicles is given by

$$d_a = \frac{5280}{D} \tag{15.11}$$

The average headway h_a (seconds per vehicle) between successive vehicles is given by

$$h_a = \frac{3600}{V} \tag{15.12}$$

The saturation headway—the average time gap between successive vehicles discharging at capacity of a roadway—is a concept often used for the design of signal timing at an intersection.

Greenshield's Linear Model

Greenshield's model is appropriate for both light and dense traffic situations. In this model, it is hypothesized that a linear relationship exists between space-mean speed S and density D. This is expressed as

$$S = S_f\left(1 - \frac{D}{D_j}\right) \tag{15.13}$$

where S = average travel speed (space mean)
S_f = free-flow speed
D = actual density
D_j = jam density, the density that (theoretically) causes speed and flow rate to become zero

According to this model, the travel speed varies linearly from the free-flow speed (S_f) at zero density to zero at the jam density D_j.

Flow rate V is given by the product of the space-mean speed and density

$$V = SD \tag{15.14}$$

Therefore, speed S and flow rate V are related by

$$S^2 = S_f S - \frac{S_f V}{D_j} \tag{15.15}$$

Flow rate V and density D are related by

$$V = S_f D - \frac{S_f}{D_j} D^2 \tag{15.16}$$

In the Greenshield model, maximum flow V_m occurs when the speed S is exactly half the FFS (S_f) and when the density D is exactly half the jam density D_j.

D_o is the critical or optimum density, which corresponds to the critical speed.
This maximum flow rate (capacity) V_m is given by

$$V_m = \frac{D_j S_f}{4} \tag{15.17}$$

Figure 15.1 shows these relationships graphically. The corresponding points on the graphs are marked as A, B, and C.

At point A, density $D \to 0$, speed $S \to$ FFS S_f, and flow rate $V \to 0$

At point C, density $D \to$ jam density D_j, speed $S \to 0$, and flow rate $V \to 0$

At point B, density D = optimum density D_o, speed S = optimum speed S_o, and flow rate V = maximum flow rate V_m

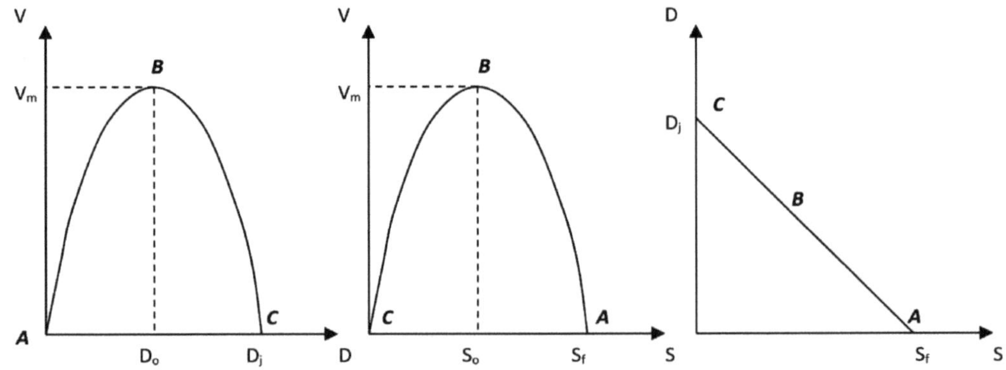

Figure 15.1 Speed–volume–density relationships (Greenshield).

Example 15.5

The average vehicular speed on a highway lane is 45 mi/h. If the peak flow rate in this lane is 1200 vehicles per hour, what is most nearly the density of vehicles (vehicles per mile) on this stretch of highway? Also, what is the average spacing between vehicles (ft)?

Solution

$$\text{Density } D = \frac{V}{S} = \frac{1200}{45} = 26.7 \text{ vehicles per month}$$

$$\text{Average spacing } d_a = \frac{5280}{D} = \frac{5280}{26.7} = 198 \text{ ft}$$

Example 15.6

Fifteen-minute traffic counts for a highway were conducted over a contiguous 2-h window. These are shown in the table below. What is the peak hour factor?

Solution The bidirectional flow rate is given by

$$v_p = \frac{V}{\text{PHF}} = \frac{242}{0.85} = 285$$

The directional flow rate = 0.6 × 285 = 171 bicycles per hour
If the average speed S = 14 mi/h, the corresponding density is

$$D = \frac{v_p}{S} = \frac{171}{14} = 12.2 \frac{\text{bicycles}}{\text{mile}}$$

This translates into an average spacing = 5280 ÷ 12.2 = 433 ft.

Level of Service (LOS) of Freeway Segments

According to the *Highway Capacity Manual*, the density of passenger car equivalents per mile per lane of a freeway segment (pc/mi/ln) is used to define the level of service. These LOS boundaries can also be expressed in terms of the following proxy variables—speed (mi/h), volume to capacity ratio, and service flow rate (pc/h/ln).

The density thresholds for LOS of freeway segments are

LOS	A	B	C	D	E	F
Density (pc/mi/ln)	≤11	>11–18	>18–26	>26–35	>35–45	>45 Demand exceeds capacity

Free-Flow Speed

For freeway segments, the *Highway Capacity Manual* gives the FFS according to Eq. (15.18), which makes speed deductions (penalties) from a base value of 75.4 mi/h. The penalties are based on lane width, lateral clearance, and total ramp density (TRD).

$$\text{FFS} = 75.4 - f_{\text{LW}} - f_{\text{LC}} - 3.22 \times \text{TRD}^{0.84} \tag{15.18}$$

Speed Adjustments to FFS

The parameters f_{LW} and f_{LC} are adjustments for lane width and right shoulder lateral clearance, respectively. These are tabulated on page 170 of the *FE Reference Handbook,* version 9.4. The TRD is the total number of ramps (on and off) per mile based on a 6-mi-length centered on the analysis segment.

Flow Rate

The peak flow rate v_p is calculated from the peak hourly volume V (vehicles per hour) using

$$v_p = \frac{V}{\text{PHF} \times N \times f_{\text{HV}} \times f_p} \tag{15.19}$$

where PHF = peak hour factor based on peak 15 min vehicular count
N = number of lanes
f_p = driver population factor
f_{HV} = heavy vehicle factor

Once the peak flow rate v_p has been calculated, the corresponding density can be calculated using

$$D = \frac{v_p}{S} \tag{15.20}$$

Peak Hour Factor

The peak hour factor is an indicator of the variability of flow. The greater the variability in the flow, the lower is the PHF. A lower value of PHF results in a higher value of peak flow rate v_p [see Eq. (15.19)].

Example 15.7

The two-way hourly volume on a bicycle path is 242 bicycles per hour. The peak hour factor is 0.85. The typical directional split is 60:40. If the average speed is 14 mi/h, what is the maximum density of bicycle in the design direction? What is the corresponding minimum spacing between bicycles?

Time (hh:mm)	Vehicles
09:00–09:15	234
09:15–09:30	256
09:30–09:45	245
09:45–10:00	231
10:00–10:15	237
10:15–10:30	266
10:30–10:45	245
10:45–11:00	210

Solution Using a 1-h sliding window, we can calculate five different hourly volumes (for the 9:00 to 10:00, 9:15 to 10:15, 9:30 to 10:30, 9:45 to 10:45, and 10:00 to 11:00 windows) as 966, 969, 979, 979, and 958, respectively.

The peak hourly volume is then $V = 979$ vehicles per hour and the peak 15 min flow is 266 vehicles/15 min. The PHF is calculated as

$$\text{PHF} = \frac{979}{4 \times 266} = 0.92$$

and the peak flow rate v_p is given by

$$v_p = \frac{979}{0.92} = 1064 \text{ vehicles per hour} \quad (\text{which is also the same as } 4v_{15} = 4 \times 266 = 1064)$$

Heavy Vehicle Factor

The heavy vehicle factor converts the mixed traffic (passenger cars, trucks, buses, and recreational vehicles) into an equivalent number of passenger cars. The factor is calculated in terms of equivalence factors E_T (trucks and buses) and E_R (RVs), which in turn are dependent primarily on terrain. The factor f_{HV} is given by

$$f_{\text{HV}} = \frac{1}{1 + P_T(E_T - 1) + P_R(E_R - 1)} \tag{15.21}$$

where P_T and P_R are the truck fraction and RV fraction, respectively, and equivalence factors E_T and E_R are given in Table 15.1.

Table 15.1 Passenger Car Equivalents on Extended Freeway Segments

	Type of terrain		
Equivalence factor	**Level**	**Rolling**	**Mountainous**
E_T (trucks and buses)	1.5	2.5	4.5
E_R (RVs)	1.2	2.0	4.0

Example 15.8

Determine the level of service for an extended stretch of a four-lane (two lanes in each direction) rural freeway in rolling terrain. Lanes are 11 ft wide with 4-ft shoulders. Assume a PHF = 0.92. The traffic consists of approximately 5% trucks and 2% RVs. The one-way peak hour volume is 2300 vehicles per hour. There are 12 access (on and off) ramps in the 6 mi vicinity of the segment being analyzed. Assume that the traffic primarily consists of commuters.

Solution Lane width = 11 ft. The table on page 170 of the *Handbook* gives f_{LW} = 1.9 mi/h. Right shoulder clearance = 4 ft. For two lanes in each direction, the table on page 170 gives f_{LC} = 1.2 mi/h.
Total ramp density, TRD = 12/6 = 2 ramps per mile

$$\text{FFS} = 75.4 - 1.9 - 1.2 - 3.22 \times 2^{0.84} = 66.1 \text{ mi/h (use 65 mi/h)}$$

Commuter traffic implies population factor f_p = 1.0
Heavy vehicle factor (for rolling terrain, E_T = 2.5 and E_R = 2.0)

$$f_{\text{HV}} = \frac{1}{1 + P_T(E_T - 1) + P_R(E_R - 1)} = \frac{1}{1 + 0.05(2.5 - 1) + 0.02(2.0 - 1)} = 0.913$$

V = 2300 vehicles per hour

$$v_p = \frac{V}{\text{PHF} \times N \times f_{\text{HV}} \times f_p} = \frac{2300}{0.92 \times 2 \times 0.913 \times 1.0} = 1369 \text{ pc/h/ln}$$

For an FFS = 65 mi/h, breakpoint = 1400 pc/h/ln (Page 170 of Handbook), and flow rate, v_p = 1369 pc/h/ln. Since v_p < breakpoint, speed = FFS = 65 mi/h.

$$\text{Density: } D = \frac{v_p}{S} = \frac{1369}{65} = 21.1 \text{ vehicles per mile}$$

Therefore, level of service is C (density range >18 to 26).

Constant Acceleration and Deceleration

For applications where motion of a vehicle may be assumed to be either at constant speed (zero acceleration) or constant acceleration (uniform acceleration from rest to maximum speed or uniform braking from initial travel speed to final speed), Newton's laws of motion adapted for uniform acceleration may be used. These are summarized in Eqs. (15.22) to (15.24):

$$V_f = V_i + at \tag{15.22}$$

$$s = s_o + V_i t + \frac{1}{2}at^2 \tag{15.23}$$

$$V_f^2 = V_i^2 + 2a(s - s_o) \tag{15.24}$$

where V_i = initial speed (at time t = 0)
V_f = final speed (at time t)
a = constant acceleration
s_o = initial position (at time t = 0)
s = final position (at time t)

Example 15.9

A commuter train has stations spaced 1 mi apart. The top speed of the train is 80 mi/h. The train accelerates at 5.5 ft/s² and decelerates at 4.5 ft/s². It dwells at each platform for 20 s to pick up and discharge passengers. What is the average running speed of the train?

Solution The motion of the train occurs in three phases: (1) acceleration from rest up to peak speed, (2) travel at peak speed, and (3) deceleration from peak speed to rest, as the train enters the destination station.
Peak speed = 80 mi/h = 80 × 1.47 = 117.6 ft/s

Phase 1: From rest to 117.6 ft/s @ 5.5 ft/s²

$$\text{Time to reach peak speed: } t_1 = \frac{V_f - V_i}{a} = \frac{117.6 - 0}{5.5} = 21.4 \text{ s}$$

$$\text{Distance traveled during this phase: } s_1 = \frac{at_1^2}{2} = \frac{5.5 \times 21.4^2}{2} = 1259 \text{ ft}$$

Phase 3: From 117.6 ft/s to rest @ 4.5 ft/s²

$$\text{Time to decelerate to rest: } t_3 = \frac{V_f - V_i}{a} = \frac{0 - 117.6}{-4.5} = 26.1 \text{ s}$$

$$\text{Distance traveled during this phase: } s_3 = V_i t_3 + \frac{at_3^2}{2} = 117.6 \times 26.1 - \frac{4.5 \times 26.1^2}{2} = 1536 \text{ ft}$$

Phase 2: Constant velocity at 117.6 ft/s

$$\text{Distance traveled during this phase: } s_2 = s - s_1 - s_3 = 5280 - 1259 - 1536 = 2485 \text{ ft}$$

$$\text{Time of travel at constant speed: } t_2 = \frac{s_2}{V_2} = \frac{2485}{117.6} = 21.1 \text{ s}$$

Therefore, total travel time between stations = $t_1 + t_2 + t_3 = 68.6$ s

Average running speed (not including dwell time) = 5280 ÷ 68.6 = 77 ft/s = 52.3 mi/h.

On the other hand, space mean speed is calculated by including the dwell time (train traverses 1 mi in 68.6 + 20 = 88.6 s).

Space mean speed = 5280 ÷ 88.6 = 59.6 ft/s = 40.5 mi/h

Example 15.10

A commuter train has a travel time of 70 s between uniformly spaced stations. During rush hour, the uniform headway between trains is 5 min. Trains have a dwell time of 15 s at each station. Each train has four coaches, each with a capacity of 240 passengers. What is the peak capacity of the line (passengers per hour)?

Solution Note that the headway (time gap between successive trains) is 5 min. This means that every 5 min, 960 passengers are being moved. No dwell time needs to be added to this. The difference between the travel time for one train (85 s) and the headway between trains (5 min) simply means that approximately 3½ min of every 5 min is "dead time" along the commuter line.

The capacity is therefore $\frac{960}{5} \times 60 = 11{,}520$ passengers per hour.

Example 15.11

A commuter train has a travel time of 70 s between uniformly spaced stations. During rush hour, the uniform headway between trains is 5 min. Trains have a dwell time of 15 s at each station. Each train has four coaches, each with a capacity of 240 passengers. What is the minimum allowable headway between trains?

Solution In order to avoid conflicts on the time-space diagram for successive trains, the minimum headway between trains is equal to the dwell time. If trains are spaced every 20 s, as soon as one train departs a station, the next one pulls in behind it.

The minimum headway is therefore 20 s.

Intersections

An intersection may be defined as an area shared by two or more roads, thus allowing drivers an option to change route direction. Intersections may be divided into three broad categories: grade separated without ramps (which simply allows the two routes to share the same space without making any provision for route changes), grade separated with ramps (more commonly known as an interchange), and at-grade intersections.

At-grade intersections may be divided into three categories—T or three-leg intersections, cross or four-leg intersections, and multileg intersections.

The two primary aspects of an at-grade intersection that impact design are the potential for conflict between different streams of traffic flow and the delay experienced by motorists using the intersection.

Figure 15.2 shows the conflict points between the various traffic streams at a four-approach unsignalized intersection. The 32 conflict points represent the maximum number possible at such an intersection, if all movements are permitted, that is, through left turn and right turn from all four approaches. If certain movements are prohibited, the number of conflict points will be less than 32.

Signalization of an intersection is a very costly design alternative and should be recommended in those locations where it is fully warranted. Occasionally, a very inexpensive option to improve the safety of an unsignalized intersection is to prohibit certain movements (e.g., no left turns from a particular approach). This eliminates a certain number of conflict points between traffic streams, thereby reducing or eliminating certain types of accidents. This recommendation can only be made if the prohibited movements can be accommodated at a nearby location.

Intersection Sight Distance

Unsignalized intersections can have all way stop control (AWSC), where the order of arrival determines the right of way. For such an intersection, a single driver has the right of way at any time, thereby eliminating conflicts. On the other hand, at an unsignalized intersection controlled by *stop* or *yield* signs, the driver on the minor road must yield to the traffic on

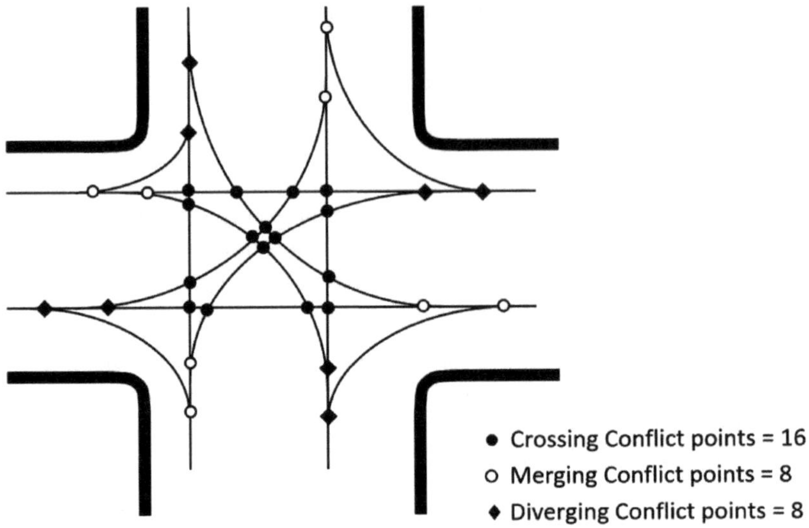

Figure 15.2 Points of conflict between traffic streams at a four-way intersection.

the major road. The AASHTO *Green Book* lists several cases of this type of intersection. The common element is that the driver on the minor roadway must have adequate sight distance (called the intersection sight distance) along the approach direction of the major road, in order to safely complete a particular maneuver. The safe time gap t_g accepted by the driver varies from case to case (such as left turn from minor, right turn from minor, etc.). The basic formula for intersection sight distance (ft) is

$$\text{ISD} = 1.47V_{\text{major}}t_g \tag{15.25}$$

where V_{major} is the design speed on major approach (mi/h) and t_g = acceptable time gap (s).

Change and Clearance Intervals at an Intersection

The change and clearance interval of a signal cycle is the sum of the yellow (amber) and all-red intervals that are provided between phases to allow pedestrian and vehicular traffic to clear the intersection before the release of conflicting movements. The all-red interval must provide enough time for a vehicle that has entered the intersection on yellow to clear the intersection.

Length of Yellow Phase

A poor choice of the yellow interval can create a dilemma zone—an area near the intersection where a vehicle can neither stop safely, nor clear the intersection without speeding before the signal turns red. According to the Institute of Transportation Engineers (ITE), the yellow interval may be calculated as

$$y = t_R + \frac{V}{2a \pm 64.4G} \tag{15.26}$$

where y = length of the yellow interval (s)
V = approach speed (ft/s). This may be an 85th percentile speed of approach
a = deceleration rate (50th percentile comfortable deceleration = 10 ft/s^2)
t_R = perception-reaction time (commonly assumed as 1.0 s)
G = grade (decimal)
W = width of the intersection (ft), measured curb to curb
L = length of the vehicle

The *Manual on Uniform Traffic Control Devices* states that a yellow change interval should be approximately 3 to 6 s, and the *Traffic Engineering Handbook* states that a maximum of 5 s is typical for the yellow change interval. The red clearance interval, if used, should not exceed 6 s (MUTCD), but 2 s or less is typical (*Traffic Engineering Handbook*). The traffic laws in each state may vary from these suggested practices. ITE recommends that the yellow interval not exceed 5 s, so as not to encourage driver disrespect for signals. If the calculated value of the yellow interval is greater than 5.0 s, an all-red phase is inserted following the yellow interval, so that the required clearance interval is still available.

Length of All-Red Phase

The all-red interval is determined according to Eq. (15.27) (ITE)

$$r = \frac{W + L}{V} \tag{15.27}$$

Example 15.12

The intersection shown below is signalized with a two-phase signal cycle. The green phase in each direction is to be followed by a yellow clearance interval. The design speed for both approaches is 45 mi/h. Main Street is on level grade, while 3rd Avenue has a longitudinal grade of 5%. If the design vehicle length = 48 ft, reaction time = 1 s, and

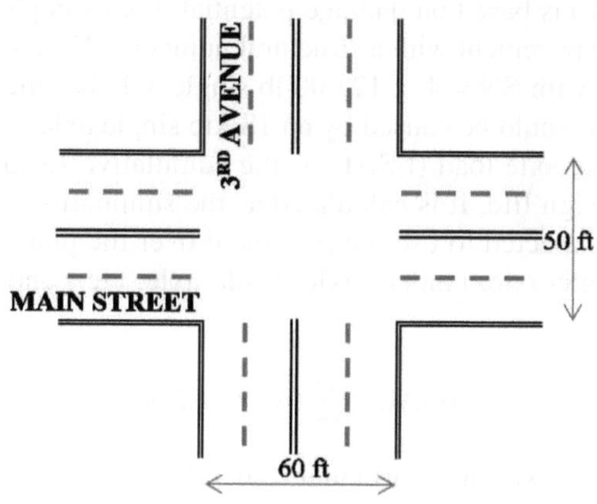

deceleration rate = 10 ft/s^2, what is the length of the yellow interval required following the north-south green phase?

A. 4 s

B. 5 s

C. 6 s

D. 7 s

Solution Design speed = 45 mi/h = 66 ft/s

The longer (and therefore, critical) value of the clearance interval will occur using the negative grade ($G = -5\%$).

The length of the yellow interval is given as:

$$y = t_R + \frac{V}{2a \pm 64.4G} = 1 + \frac{66}{2 \times 10 - 64.4 \times 0.05} = 4.93 \text{ s}$$

Yellow time of 5 s should be used.

Answer is B.

Pavement Design

Equivalent Single Axle Load

For design of highway pavements—rigid and flexible—one of the most significant inputs into the design process is the traffic volume and composition. Heavier vehicle axles have significantly higher potential for causing damage to pavements than lighter axles. Passenger cars and pickup trucks are excluded from the calculation of wheel load impact on pavements. In order to standardize the vehicular impact on the condition and life expectancy of a pavement, AASHTO *Guide for Design of Pavement Structures* converts all vehicle axles (single, double, and triple axles) of various axle loads to an equivalent number of 18-kip single axles. The load equivalence factor (LEF) is based on damage potential. For example, if the LEF for a 12-kip single axle on a flexible pavement with a structural number (SN) of 4 is 0.213, this means that on a flexible pavement with SN = 4, a 12,000 lb single axle has the potential to cause about 21% of the damage that would be caused by an 18-kip single axle.

The equivalent single axle load (ESAL) is the cumulative 18-kip equivalent for a pavement over its entire design life. It is calculated as the summation of the LEF values for the total number of axles expected to use the pavement over the plan duration. If the axles are classified by type (single-axle, tandem-axle, triple axle, etc.) and load, then the ESAL is calculated as

$$\text{ESAL} = \sum (N_i \times \text{LEF}_i) \tag{15.28}$$

where N_i is the number of axles in a particular category and LEF_i is the LEF for that category.

The truck factor (TF) is defined as

$$\text{TF} = \frac{\text{ESAL}}{\text{No. of trucks}} \tag{15.29}$$

Load Equivalence Factors

LEF values were generated based on AASHTO road tests conducted in Ottawa, Illinois. Test traffic consisted of thousands of single axles (ranging from 2 to 30 kips) and tandem axles (ranging from 24 to 48 kips) being driven over pavements composed of an asphalt surface course, a well-graded crushed limestone base course, and a uniformly graded sand-gravel subbase.

The serviceable life of a pavement is related to the difference in present serviceability index (PSI) between construction and end of life. Typical values used for PSI are

Postconstruction: 4.0 to 5.0 depending upon construction quality, smoothness, etc.

End-of-life (called "terminal serviceability" and designated "p_t"): 1.5 to 3.0 depending upon road use (e.g., interstate highway, urban arterial, residential). This chapter tabulates LEFs for terminal serviceability $p_t = 2.5$. Page 173 of the *Handbook* has a representative table of LEF values.

Example 15.13

Traffic data for a section of two-lane, bidirectional roadway show the following truck axle loads. The number of average daily trips (heavy vehicles only) is 8500 with a 60/40 directional split. What is the annual ESAL for the design lane? Assume all trucks are two-axle vehicles.

Axle type	Gross load (lb)	LEF	% ADT
Single	6000	0.017	67
	10,000	0.118	21
Tandem	14,000	0.042	6
	22,000	0.229	6

Solution Since the ADT is 8500, the number of axles = 17,000. Of these, 67% + 21% are single axles and 6% + 6% are tandem axles. The ESAL is calculated as

$$\text{ESAL} = \sum N_i \text{LEF}_i = 17{,}000 \times (0.017 \times 0.67 + 0.118 \times 0.21 + 0.042 \times 0.06 + 0.229 \times 0.06)$$

$$= 891/\text{day}$$

$$\text{Annual ESAL} = 891 \times 365 = 325{,}328 \text{ ESAL/year (bidirectional)}$$

For design lane: ESAL = 0.6 × 325,328 = 195,197 ESAL/year (directional)

Example 15.14

The annual ESAL_{18} for a highway is calculated to be 280,000 in 2005. Expecting a 4% growth in traffic per year over the next 10 years, what is the cumulative ESAL_{18} for this highway over the 10-year period?

Solution The situation described is a geometric series with the first term, $a = 280{,}000$ and the rate of increase, $r = 1.04$. The sum of the first 10 terms is given by

$$S_n = \frac{a(r^n - 1)}{r-1} = \frac{280{,}000 \times (1.04^{10} - 1)}{1.04 - 1} = 280{,}000 \times 12.0061 = 3{,}361{,}710$$

Note that the factor 12.0061 in the example above may be called a growth factor and it is numerically identical to the *F*/*A* factor in the engineering economics tables (Chap. 5) for $n = 10$ and $i = 4\%$.

Flexible Pavements

Flexible pavements are composed of a wearing surface, usually composed of bituminous materials, underlain by a layer of granular material (base course) and a layer of blended aggregates (subbase). This arrangement is underlain by a well-compacted subgrade that serves as the foundation of the pavement. Flexible pavements are also classified as high type, intermediate type, and low type. High-type pavements support the traffic loads without any visible distress and are not susceptible to weather conditions. Low-type pavements have wearing surfaces that range from untreated to loose materials to surface-treated earth. Intermediate-type pavements, as the name suggests, have qualities between those of high-type and low-type pavements.

The base course is that portion of the pavement structure which is immediately beneath the surface course. It lies either above the subbase course or if no subbase course is used, it may lie directly on the roadbed soil. It usually consists of aggregates such as crushed stone, crushed gravel, and sand which may be untreated or treated with stabilizing admixtures such as asphalt, lime, Portland cement, etc.

The subbase course is the portion of the flexible pavement structure between the roadbed soil and the base course. It usually consists of a compacted layer of granular material or of a layer of soil treated with an admixture. The subbase may be omitted from the pavement cross-section design if the underlying soil bed is of high quality. In addition to contributing to the overall structural strength of the pavement, the subbase course may have the following secondary functions—improve drainage characteristics, prevent intrusion of fines into the base course, and minimize frost damage.

Structural Number

Flexible pavements are characterized by the structural number (SN), which is calculated as

$$\mathrm{SN} = a_1 D_1 + a_2 m_2 D_2 + a_3 m_3 D_3 \tag{15.30}$$

where a_1, a_2, and a_3 are strength coefficients for layers 1, 2, and 3 and m_2 and m_3 are drainage coefficients for subsurface layers 2 and 3.

Default values for layer strength coefficients are

Asphalt concrete surface course	$a_1 = 0.44\ \text{in}^{-1}$
Crushed stone base course	$a_2 = 0.14\ \text{in}^{-1}$
Sandy gravel subbase	$a_3 = 0.11\ \text{in}^{-1}$

Coefficients m_2 and m_3 represent drainage coefficients of base course and subbase, respectively. Values of these coefficients range from 0.4 to 1.4. Values of m higher than 1.0 are assigned where these courses have very good drainage characteristics.

Example 15.15

The design SN of a flexible pavement is 5. The pavement cross-section consists of an asphalt surface course (minimum thickness = 4.0 in) underlain by a granular base course (maximum thickness 18 in). The following layer coefficients are given

Asphalt concrete surface course	$a_1 = 0.45\ \text{in}^{-1}$
Crushed stone base course	$a_2 = 0.15\ \text{in}^{-1}$

What is the minimum required thickness of the surface course (in)?

Solution In order for the surface course to have minimum thickness, we must use the maximum permissible thickness for the base course, D_2 = 18 in. Assuming drainage coefficient $m_2 = 1.0$, we can write

$$\text{SN} = a_1D_1 + a_2m_2D_2 \Rightarrow D_1 = \frac{\text{SN} - a_2m_2D_2}{a_1} = \frac{5 - 0.15 \times 18}{0.45} = 5.11 \text{ in}$$

Since this is greater than the minimum requirement of 4 in, the required thickness of the asphalt surface course is 5.5 in.

Flexible Pavement Design

The following criteria are important guideposts for the overall design of asphalt pavements:

- Sufficient asphalt to ensure a durable pavement
- Sufficient stability under traffic loads
- Sufficient air-voids—lower limit to allow room for initial densification due to traffic and upper limit to prevent excessive environmental damage
- Sufficient workability

Purposes of Compaction

- To prevent further compaction and settlement
- To increase shear strength
- To improve water tightness of mixture
- To prevent excessive oxidation of the asphalt binder

Asphalt

Hot mix asphalt (HMA) is a mixture of asphalt binder and well-graded, high quality aggregate heated and compacted. Asphalt is placed in multiple lifts (layers). Using deep lifts has the following advantages: (1) thicker layers hold heat longer and it is therefore easier to roll the layers to the required density, (2) deeper lifts can be placed in cooler weather, (3) one deep lift is more economical than multiple lifts, and (4) deep lifts suffer less distortion due to rolling than thin lifts.

Asphalt mix design may be performed using the Hveem, Marshall, or Superpave mix design methods.

Asphalt Grading In the past, asphalt cement (AC) was graded by either penetration resistance or viscosity.

Penetration Grading

Penetration graded asphalts were specified by a measurement by a standardized penetrometer needle (mass = 100 g) under a standard load at a standard temperature. Penetration graded asphalts were typically expressed as "Penetration Grade 85-100," meaning that the needle penetration was between 85 and 100 mm. Higher penetration signifies a softer AC. Five different penetration grades ranging from hard (40 to 50 mm penetration) to soft (200 to 300 mm penetration) are specified in this classification system.

Penetration grading describes only the consistency at an intermediate temperature (25°C). Low temperature properties are not directly measured by this grading system.

Viscosity Grading

Viscosity graded asphalts were specified by determining the viscosity of AC. A temperature of 60°C (140°F) was considered to be a typical summer pavement temperature, and at this temperature, the unit of viscosity used was the poise. Standard terminology referred to AC-10 and AC-20, meaning that the viscosity of the AC was 1000 or 2000 poise, respectively. AC-20 was thicker or harder than AC-10. A temperature of 135°C (275°F) was considered the mixing and handling control point. At this temperature, different laboratory equipment was used and the unit of viscosity used was the centistoke (cS).

Although viscosity is a fundamental measure of flow, it only provides information about high temperature viscous behavior, not about the low or intermediate temperature elastic behavior.

Performance Grading

In 1994, a new system of design for asphalt paving materials known as Superpave, which introduced a new concept called performance grading (PG), was introduced based on research done under the Strategic Highway Research Program (SHRP). The PG system of specifying binder is based on a complex series of performance-based tests.

The new system for specifying asphalt binders is based on performance at specified temperatures. Physical property requirements are the same, but the temperature at which the binder must attain the properties changes. For example, the high temperature, unaged binder stiffness ($G*/\sin \delta$) is required to be at least 1.0 kPa, but this must be achieved at higher temperatures if the binder is to be adequate in a hot climate.

Binder physical properties are measured using four devices:

1. *Dynamic shear rheometer:* The dynamic shear rheometer is used to characterize the viscoelastic properties of the binder. It measures the complex shear modulus G^* and phase angle δ. For totally elastic materials, there is no lag ($\delta = 0$) between the applied shear stress and the shear strain response of the sample. For totally viscous materials, $\delta = 90°$. The binder specification uses either $G^*/\sin\delta$ at higher temperatures ($T > 46°C$) or $G^*\sin\delta$ at intermediate temperatures ($7°C < T < 34°C$) as a means of controlling asphalt stiffness.
2. *Rotational viscometer:* This test characterizes the stiffness of the asphalt at 135°C, at which temperature it behaves almost entirely as a viscous fluid. The asphalt rotational viscometeris a rotational coaxial cylinder that measures viscosity by the torque required to rotate a spindle submerged in a sample of hot asphalt at a constant speed. The binder specification requires that binders have a viscosity of less than 3 Pa · s.
3. *Bending beam rheometer (BBR):* The BBR measures the creep stiffness S and the logarithmic creep rate m by measuring the response of a small binder beam specimen to a creep load at low temperatures. Binders with low creep stiffness and/or higher m values will not crack in cold weather.
4. *Direct tension tester:* High creep stiffness (at low temperatures) may be acceptable if a direct tension test shows that the binder is sufficiently ductile at low temperatures.

The new specification system no longer refers to AC, but rather to binder, which includes modified and unmodified asphalts. It specifies asphalt binders as PG followed by two numbers, for example PG 66-20. The first number is always higher and positive, while the second number is smaller and negative. The first number represents the *high pavement temperature* and is based on the 7-day average high air temperature of the surrounding area, while the second number represents the *low pavement temperature* and is based on the 1-day low air temperature of the surrounding area. Both numbers referred to are in degrees Celsius. PG asphalt binders are specified in 6°C increments. If the sum of the two numbers (absolute value) is greater than 90, then use of polymer-modified asphalt is indicated.

The Superpave software calculates high pavement temperature 20 mm below the pavement surface and low temperature at the pavement surface. Design pavement temperature calculations are based on HMA pavements subjected to fast moving traffic. Pavements subject to slow traffic, such as at intersections, toll booths, and bus stops should contain a stiffer asphalt binder than that which would be used for fast-moving traffic. Superpave allows the high temperature grade to be increased by one grade (6°C) for slow transient loads and by two grades (12°C) for stationary loads. Additionally, the high temperature grade should be increased by one grade for anticipated 20-year loading in excess of 30 million ESALs. For pavements with multiple conditions that require grade increases, only the largest grade increase should be used. For example, for a pavement intended to experience slow loads (one grade increase) and greater than 30 million ESALs (one grade increase), the asphalt binder high temperature grade should be increased by only one grade.

Example 15.16

What is the appropriate PG for an asphalt binder which is subjected to

Average 7-day maximum pavement temperature	55°C
Mimimum pavement temperature	–20°C

A. PG 52-10

B. PG 55-20

C. PG 58-20

D. PG 58-22

Solution PG 58 is appropriate if 52°C ≤ T_{max} < 58°C.
PG 58-22 is appropriate if minimum pavement temperature > –22°C.

Answer is D.

Example 15.17

The following results are given from conducting tests on an asphalt binder as part of the Superpave protocol.

Dynamic Shear Rheometer

Test temperature (°C)	Dynamic shear $G^*/\sin\delta$ (kPa)
58	1.92
64	1.08
70	0.81

Rolling Thin Film Oven (RTFO) residue

Test temperature (°C)	Dynamic shear $G^*/\sin\delta$ (kPa)
58	4.04
64	2.41
70	1.75

Pressure Aging Vessel (PAV)

Test temperature (°C)	Creep stiffness S (MPa)
–6	272
–12	304
–18	421

Based on these results, the recommended range of pavement temperatures where this asphalt can be used is

A. 55 to −18°C

B. 50 to −24°C

C. 65 to −20°C

D. 60 to −26°C

Solution From the dynamic shear rheometer, the minimum value of $G^*/\sin\,\delta$ should be 1.0 kPa. By interpolation, this is achieved at a temperature of 65.78°C.

From the RTFO residue, the minimum value of $G^*/\sin\,\delta$ should be 2.2 kPa. By interpolation, this is achieved at a temperature of 65.91°C.

Therefore, the maximum temperature is 65.78°C. Note that performance grades are specified in increments of 6°C (52°C, 58°C, 64°C, …).

For the low temperature, one looks at the results of the pressure aging vessel (PAV). The maximum value of the creep stiffness S should be 300 MPa. By interpolation, this is achieved at a temperature of −11.25°C.

The low temperature is set 10°C lower than this value. Therefore, the lowest paement temperature is −21.25°C.

Choices A and C are within these limits. However, C is the wider limit.

Answer is C.

NOTE Since the PG limits are discrete, this would actually be specified as PG 64-16.

Rigid Pavement Design

The design of a rigid pavement involves design of the thickness of the concrete slab, choice of reinforcement, and load transfer devices for joints. The basic materials in the pavement slab are—Portland cement concrete, reinforcement steel, either in the form of reinforcement bars or welded wire fabric, joint transfer devices, and joint sealing materials. There are four primary types of concrete pavement. They are (1) plain jointed concrete pavement (JCP), (2) jointed reinforced concrete pavement (JRCP), (3) continuous reinforced concrete pavement (CRCP), and (4) prestressed concrete pavement (PCP).

Highway Design: Horizontal Curves

The geometry of a horizontal circular curve is shown in Fig. 15.3. PC is the *point of curvature*, PT is the *point of tangent*, and PI is the *point of intersection* (of tangents). The *internal angle* subtended by the major chord (PC to PT) is I. This is also the *deflection angle between tangents* at PI. For the sense of the curve as shown in the figure, one would say that the deflection angle is “to the right.” E is called the *external distance* and M is the *middle ordinate distance*.

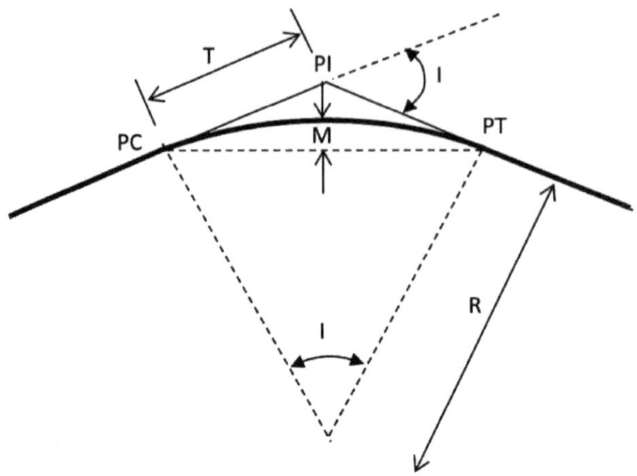

Figure 15.3 Geometry of a circular curve.

Specifying a circular curve as "degree of curve = 2°" means that a 100-ft *arc* along the curve subtends an angle of 2° at the center. The radius of the curve is given by

$$R = \frac{5729.578}{D} = \frac{LC}{2\sin\left(\frac{I}{2}\right)} \tag{15.31}$$

Where the length of the long chord (from PC to PT) = LC = $2R\sin\left(\frac{I}{2}\right)$

The length of curve is then given by

$$L = RI\frac{\pi}{180} = \frac{100I}{D} \tag{15.32}$$

The tangent length T is the distance from the PC to the PI (also the distance from the PI to the PT, by symmetry) and is given by

$$T = R\tan\left(\frac{I}{2}\right) = \frac{LC}{2\cos\left(\frac{I}{2}\right)} \tag{15.33}$$

The middle ordinate distance M is the radial distance from the midpoint of the major chord (which connects the PC to the PT) to the curve.

$$M = R\left[1 - \cos\left(\frac{I}{2}\right)\right] \tag{15.34}$$

The external distance E is the radial distance from the PI to the curve.

$$E = R\left[\sec\left(\frac{I}{2}\right) - 1\right] \tag{15.35}$$

Example 15.18

A circular curve is laid out by deflection angles. The degree of curve is 3°. The PC is at sta. 12 + 30.65 and the PI is at sta. 15 + 43.76. What is the deflection angle for station 13 + 00.00 on the curve?

Solution Tangent length T is the distance from the PC to the PI
$T = 1543.76 - 1230.65 = 313.11$ ft

Radius of curve is calculated as $R = \dfrac{5729.578}{D} = \dfrac{5729.578}{3} = 1909.86$

Using $T = R\tan\left(\dfrac{I}{2}\right)$, the deflection angle between tangents is calculated as $I = 18.62°$.

Length of curve $L = \dfrac{100I}{D} = \dfrac{100 \times 18.62}{3} = 620.67$ ft

The arc length from PC to sta. 13 + 00.00 = 69.35 ft.

The corresponding deflection angle is given by

$$\alpha = 69.35 \times \left(\frac{18.62}{2 \times 620.67}\right) = 1.0403^\circ = 1^\circ 02'25''$$

Example 15.19

Two tangents intersecting at an angle of 36° 42' are to be joined by a 6° 30' circular curve. The PI for the curve is at 9 + 31.00 and the degree of curve is based on a 100-ft arc. Determine the station of PC and PT.

Solution Radius of curve is calculated as $R = \dfrac{5729.578}{6.5} = 881.47\,\text{ft}$

$$T = R\tan\frac{I}{2} = 881.47 \times \tan 18°21' = 292.37\text{ft} = 2 + 92.37 \text{ sta.}$$

PI sta. = PC sta. + T = 9 + 31.00

Therefore, PC sta. = (9 + 31.00) – (2 + 92.37) = 6 + 38.63.

Length of curve L is the arc length corresponding to R = 881.47 ft and intersection angle = 36°42'.

$$L = 881.47 \times \frac{36.7}{57.29578} = 564.61 \text{ ft} = 5 + 64.61 \text{ sta.}$$

Therefore, PT sta. = (6 + 38.63) + (5 + 64.61) = 12 + 03.24.

Example 15.20

The PC of a 6° circular curve is at sta. 15 + 27.00. Determine the deflection angle from the tangent at the PC to each of the following points on the curve: (a) sta. 16 + 00.00, (b) sta. 17 + 00.00, (c) sta. 20 + 98.00.

Solution Degree of curve = 6

Using the equation sta. $A_{ft} = \text{sta. PC}_{ft} + \dfrac{200\alpha}{D}$

$$\alpha = \frac{D}{200} \times (\text{sta. } A - \text{sta. PC}) \text{ we obtain}$$

For station 16 + 00 $\alpha = 2.19° = 2°11'\,24''$
For station 17 + 00 $\alpha = 5.19° = 5°11'\,24''$
For station 20 + 98 $\alpha = 17.13° = 17°7'\,48''$

Example 15.21

The PC of a 14° circular curve, whose degree is based on an arc of 100 ft, is at sta. 49 + 27.36 and the angle of intersection of the tangents is 60°58". Calculate the deflection angle from the tangent at the PC to station 49 + 50.00 on the curve.

Solution

$$\text{Length of curve } L = \frac{100I}{D} = \frac{100 \times 60.967}{14} = 435.48$$

Therefore, PT is at station 53 + 62.84.

$$\text{Radius } R = \frac{5729.578}{14} = 409.26 \text{ ft}$$

For station 49 + 50.00, arc length = 22.64 ft, central angle = 0.0553 radians, deflection angle = 0.0277, radians = 1.585° = 1° 35'05".

Highway Design: Superelevation

Centrifugal forces on curves are proportional to the local curvature κ which is the inverse of the radius of curvature. The centrifugal force acts along the outward normal and tends to destabilize a vehicle by tipping it about the outside wheel line. One way to counter this destabilizing effect is to "bank" the roadway. Superelevation is the banking (rotation) of a highway to counter some of the lateral force. In Fig. 15.4, the components of the two body forces—the gravity force and the centrifugal force—are shown as solid lines. As shown in the figure, the banking causes a portion of the lateral acceleration to act normal to the pavement ($mV^2/R \sin \phi$). This is felt as a downward force by the vehicle occupants. The remaining portion of the lateral force may act one of three ways depending on the banking and speed of the vehicle.

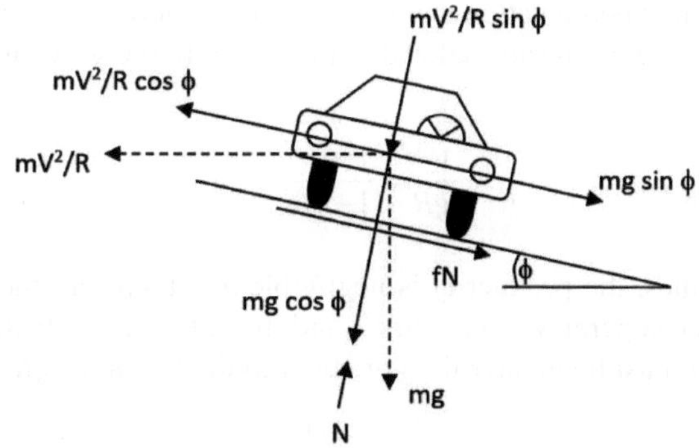

Figure 15.4 Forces on a vehicle on a superelevated roadway.

If the speed is balanced for the banking, the lateral force acting outward on the vehicle will be countered by the forces pushing the vehicle down the slope of the banking. This is a neutral equilibrium condition.

If the vehicle is traveling faster than the equilibrium speed, the resultant lateral force acts outward on the vehicle and occupants. At excessive speeds, the vehicle will skid or roll off the road.

If the speed is lower than the equilibrium speed, the vehicle and occupants are forced inward. Extreme banking can cause top heavy vehicles to rollover toward the inside of the curve. Additionally, icy conditions can cause the vehicle to slide down the pavement.

Coefficient of Side Friction

The wheel on wet pavement side friction factor is designated *f*. Note that this is different from the coefficient of friction (a combination of rolling and skidding friction) which is used to calculate stopping distance.

NOTE The current version of the AASHTO *Green Book* no longer uses a friction coefficient approach for computing stopping distance. Instead it uses a deceleration rate a, default value $a = 11.2$ ft/s².

The superelevation factor is given by $e = \tan\phi$, where ϕ is the banking angle (with the horizontal).

Solving the equilibrium equations for the vehicle traveling on a banked curve, to prevent the outward slide of the vehicle, the required superelevation factor e for a design speed V, curve radius R, and side friction coefficient f is given by

$$e = \tan\phi = \frac{V^2 - fgR}{fV^2 + gR} \tag{15.36}$$

On the other hand, if superelevation e and side friction coefficient f are known, the dimensionless centrifugal acceleration (normalized with respect to the gravitational acceleration) is given by

$$\frac{V^2}{gR} = \frac{e+f}{1-ef} \tag{15.37}$$

For most practical values, the product ef is negligible and therefore the centrifugal factor is given by the sum of the superelevation factor e and the side friction factor f. For stability, the centrifugal factor $e+f$ must be greater than or equal to the factor V^2/gR

$$e+f \geq \frac{V^2}{gR} \tag{15.38}$$

Given maximum superelevation e_{max} and design side friction factor f, the minimum safe radius R_{min} and maximum degree of curve D_{max} for a horizontal curve are given by

$$R_{min} = \frac{V^2}{15(e_{max}+f)} \tag{15.39}$$

$$D_{max} = \frac{5729.578}{R_{min}} \tag{15.40}$$

where V = design speed (mi/h)
e_{max} = superelevation (decimal)
f = recommended side friction factor (decimal)
R = curve radius (ft)
D = degree of curve (degrees)

Example 15.22

The exit ramp from a highway has 4% superelevation. The design speed for the ramp is 35 mi/h. If the side friction factor is 0.18, what is the maximum radius (ft) of the horizontal curve for the ramp?

Solution The minimum radius is calculated from

$$R_{min} = \frac{V^2}{15(0.01e_{max}+f_{design})} = \frac{35^2}{15(0.04+0.18)} = 371.2 \text{ ft}$$

Use a minimum radius of 375 ft for the ramp.

Example 15.23

The alignment of a roadway section is a circular curve with radius R = 765 ft with a deflection angle = 12°30' turning right. The cross section has 6% superelevation, rotated about the centerline. The design speed is 40 mi/h and the coefficient of side friction is 0.14. The factor of safety for lateral stability is most nearly

A. 0.71

B. 0.93

C. 1.08

D. 1.43

Solution For lateral stability, the superelevation factor $e + f$ must be greater than V^2/gR. Using speed V (mi/h), radius R (ft), and g (ft/s^2), this factor becomes

$$\frac{V_{\text{mph}}^2}{15R_{ft}} = \frac{40^2}{15 \times 765} = 0.139$$

$$e + f = 0.06 + 0.14 = 0.20$$

$$\text{FS} = 0.2/0.139 = 1.43$$

Highway Design: Spiral Curves

A spiral curve can be used to produce a gradual transition from tangents to circular curves. While the cross section is being rotated from the crowned to the fully superelevated section, the spiral may be used to gradually change the radius of curvature (from infinite on the tangent to the specified curve radius).

The recommended minimum length L_s of a spiral transition is given by

$$L_s \geq \frac{3.15V^3}{RC} \qquad (15.41)$$

where V = design speed (mi/h)
R = radius of circular curve (ft)
C = rate of change of lateral acceleration (recommended value 1 ft/s^3)

Highway Design: Vertical Curves

Vertical highway curves are parabolic curves connecting an approach tangent (gradient G_1) and an exit tangent (gradient G_2). If the x coordinate is measured horizontally (rather than along the curve), the equation of a simple parabolic vertical curve (see Fig. 15.5) is

$$y(x) = y_{\text{PVC}} + G_1 x + \frac{1}{2} R x^2 \qquad (15.42)$$

where R is the rate of gradient change is given by

$$R = \frac{G_2 - G_1}{L} \qquad (15.43)$$

The PVC is the beginning of the vertical curve, PVT is the end of the vertical curve, and PVI is the point of intersection of the two tangents. If the stationing of points (x coordinate) is expressed horizontally, the PVI is exactly midway between the PVC and the PVT.

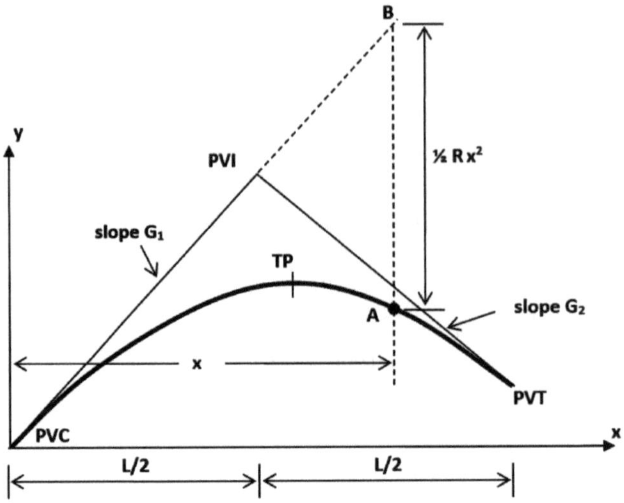

Figure 15.5 Geometry of a vertical curve.

In Eqs. (15.42) and (15.43), gradient G and distance x may be expressed in percent (%) and stations (sta.), respectively, only if 100 ft (or 100 m) stations are employed. The effect of the two factors of 100 implicit in such a choice of units will cancel each other and elevation y will have units of ft (or m). Thus, for 100 ft (m) stations, the rate of gradient change R will have units of %/sta. If the stationing interval is not 100, gradients G must be expressed in decimal (rather than percent) and distance x must be expressed in ft (or m) rather than stations. As a result, the units of R will be ft^{-1} or m^{-1}.

Note also that for crest curves, the rate of gradient change R is negative, resulting a negative (downward) offset distance, while for sag curves the offset is positive.

The offset from the tangent at any location grows as a parabolic function. This offset distance is the last term in the second-order equation for $y(x)$. Since this term is ½ Rx^2, we can say that if two locations x and x_1 are considered, the ratio of the tangent offsets at these two locations will be equal to the square of the ratio of the distances x and x_1

$$\frac{AB}{A'B'} = \left(\frac{x}{x_1}\right)^2 \tag{15.44}$$

The vertical offset at the midpoint of the curve is given by $H_{L/2} = \dfrac{(G_2 - G_1)L}{8}$ and at the PVT is given by $H_L = \dfrac{(G_2 - G_1)L}{2}$.

The location of the high or low point (turning point, marked TP in the figure) of the curve is given by

$$x_{\text{TP}} = -\frac{G_1}{R}. \tag{15.45}$$

For a sag curve, the low point would be the optimum point for locating a drainage outlet.

Example 15.24

A –4% grade meets a +5% grade at station 34 + 00. If a vertical curve of length 600 ft is used to connect the two tangents, find the station of the low point.

Solution

$$\text{Sta. PVC} = \text{sta. PVI} - \tfrac{1}{2} L = 34.00 - 6/2 = 31.00$$
$$R = (G_2 - G_1)/L = [5 - (-4)]/6 = 1.5\%/\text{sta.}$$
$$x_{TP} = -G_1/R = 4/1.5 = 2.67 \text{ stations}$$

Thus, the low point is at station (31 + 00) + (2 + 66.677) = 33 + 66.67.

Example 15.25

In the figure shown below, vertical curve no. 1 is followed by a tangent section, which is followed by vertical curve no. 2. What is the vertical clearance under the bridge structure at station 40 + 55.00?

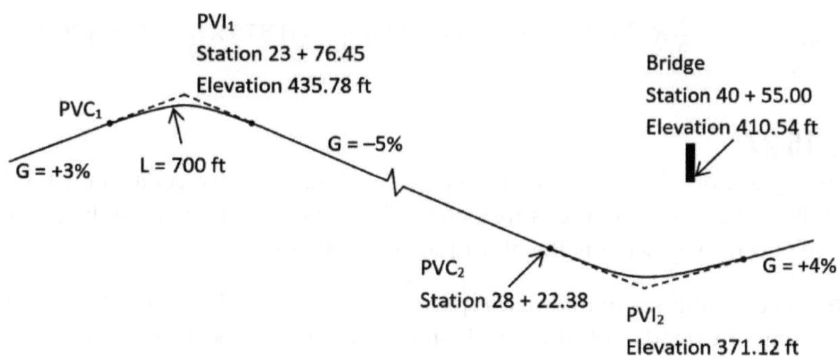

Solution The distance between PVI_1 and PVC_2 = 2822.38 – 2376.45 = 445.93 ft = 4.4593 sta.

The elevation at PVC_2 = 435.78 – 5 × 4.4593 = 413.48 ft

The elevation difference between PVC_2 and PVI_2 = 413.48 – 371.12 = 42.36 ft

The distance between PVC_2 and PVI_2 = 42.36 ÷ 5 = 8.472 sta.

Therefore, the length of curve no. 2 = 16.944 stations

The rate of gradient change for curve no. 2 = $R = \dfrac{4-(-5)}{16.944} = 0.5312$ %/sta.

The elevation of the point on curve no. 2 at sta. 40 + 55.00 is calculated as

$$y = 413.48 + (-5)\times 12.3262 + \frac{1}{2}\times 0.5312 \times 12.3262^2 = 392.20$$

Vertical clearance = 410.54 – 392.20 = 18.34 ft

Example 15.26

A vertical curve connects tangents of $G_1 = +2\%$ and $G_2 = -5\%$ that intersect at station 58 + 25.60 having elevation = 128.54 ft and the vertical curve is to be 800 ft long, what should be the elevation at station 59 + 00.00?

Solution The conditions in this case represent a crest vertical curve. Here, the PVI is at sta. 58 + 25.60 and, since the curve extends 400 ft on each side of that point, it starts at sta. 54 + 25.60 and ends at sta. 62 + 25.60

$$G_1 = +2,\ G_2 = -5,\ \mathrm{L} = 8 \text{ sta.}$$

Therefore, the rate of gradient change $R = \dfrac{-5-2}{8} = -0.875\%/\text{sta.}$

The elevation at PVC (station 54 + 25.60) is given by

$$y_{\text{PVC}} = y_{\text{PVI}} - G_1\frac{L}{2} = 128.54 - 2\times 4 = 120.54 \text{ ft}$$

The elevation at station 63 + 00.00 is given by

$$y = y_{\text{PVC}} + G_1x + \frac{1}{2}Rx^2 = 120.54 + 2\times 4.744 + \frac{1}{2}\times -0.875\times 4.744^2 = 120.18 \text{ ft}$$

Example 15.27

A slope with a grade of −3.2% and one with a grade of +2.6% intersect at sta. 34 + 28.00 at an elevation of 49.20 ft. If a sag curve is formed by the slopes and they are to be connected by a vertical curve 600 ft long, what is the elevation at sta. 35 + 0.00?

Solution The conditions in this case represent a sag curve. Here, the PVI is at sta. 34 + 28 and, since the curve extends 300 ft on each side of that point, the PVC is at sta. 31 + 28 and the PVT is at sta. 37 + 28.

$$G_1 = -3.2,\ G_2 = +2.6,\ L = 6 \text{ sta.}$$

Therefore, the rate of gradient change $R = \dfrac{2.6-(-3.2)}{6} = +0.967\%/\text{sta.}$

The elevation at PVC (station 31 + 28.00) is given by

$$y_{\text{PVC}} = y_{\text{PVI}} - G_1\frac{L}{2} = 49.20 - (-3.2)\times 3 = 58.8 \text{ ft}$$

The elevation at station 35 + 00.00 is given by

$$y = y_{\text{PVC}} + G_1x + \frac{1}{2}Rx^2 = 58.8 - 3.2\times 3.72 + \frac{1}{2}\times 0.967\times 3.72^2 = 53.58 \text{ ft}$$

Sight Distance

Stopping Sight Distance—AASHTO Recommendations

The stopping sight distance (SSD) is the sum of the distance traveled during perception-reaction time t_R and the distance traveled during braking to a stop from the design speed. The recommended value for t_R is 2.5 s. The default value of the deceleration rate in the *AASHTO Green Book* is 11.2 ft/s². The total distance traveled during the stopping maneuver, from the first sighting of the obstruction is given by

$$\text{SSD} = 1.47Vt_R + \frac{V^2}{30\left(\frac{a}{32.2} \pm G\right)} \qquad (15.46)$$

where V = design speed (mi/h)
t_R = perception-reaction time (s)
a = deceleration rate (ft/s²)
G = longitudinal grade (decimal)

Table 15.2 shows the stopping distance for various combinations of design speed (mi/h) and grade G (%). The default values of t_R = 2.5 s and a = 11.2 ft/s² have been used in generating the table.

Table 15.2 Stopping Sight Distance for Various Speeds and Grades

Design speed (mi/h)	Reaction distance (ft)	Stopping distance $S_R + S_B$ (ft) for various gradients G (%)						
		Downgrades			Level	Upgrades		
		−6	−4	−2	0	+2	+4	+6
15	55.0	81.1	79.4	77.9	76.6	75.4	74.4	73.4
20	73.3	119.8	116.7	114.1	111.7	109.7	107.8	106.1
25	91.7	164.2	159.5	155.3	151.7	148.4	145.5	142.9
30	110.0	214.4	207.7	201.7	196.4	191.7	187.5	183.7
35	128.3	270.5	261.3	253.1	246.0	239.6	233.8	228.7
40	146.7	332.3	320.3	309.7	300.3	292.0	284.5	277.7
45	165.0	400.0	384.7	371.3	359.5	348.9	339.4	330.9
50	183.3	473.5	454.6	438.1	423.4	410.4	398.7	388.1
55	201.7	552.7	529.9	509.9	492.2	476.4	462.2	449.4
60	220.0	637.8	610.6	586.8	565.7	546.9	530.1	514.9
65	238.3	728.6	696.8	668.8	644.1	622.0	602.2	584.4
70	256.7	825.3	788.4	755.9	727.2	701.6	678.7	658.0
75	275.0	927.8	885.4	848.1	815.2	785.8	759.5	735.7
80	293.3	1036.1	987.8	945.4	907.9	874.5	844.5	817.5

In Table 15.2, the braking distance may be obtained by subtracting the reaction distance from the total stopping distance. For example, for a vehicle traveling at 40 mph on a downgrade of 4%, the braking distance is equal to 320.3 – 146.7 = 173.6 ft.

The braking distance traveled to slow a vehicle from an initial speed V_i (mi/h) to a final speed V_f (mi/h) at a uniform rate of deceleration a (ft/s²) is given by

$$\text{SSD}_{\text{braking}} = \frac{(V_i^2 - V_f^2)}{30\left(\dfrac{a}{32.2} \pm G\right)} \tag{15.47}$$

Example 15.28

Calculate and verify the stopping distance on a rural highway with a downgrade of 4% and a design speed of 70 mi/h.

Solution Assuming a reaction time $t_R = 2.5$ s, $a = 11.2$ ft/s², and $G = -4$, we get
Reaction distance $d_R = 1.47 \times 2.5 \times 70 = 257.3$ ft

$$\text{Braking distance } d_B = \frac{1.075 \times 70^2}{11.2 - 0.32 \times 4} = 531 \text{ ft}$$

Therefore, total stopping distance = 788.3 ft (the Table 15.2 shows 788.4 ft).

Example 15.29

A car traveling at 55 mi/h is followed by another traveling at 65 mi/h. The lead car suddenly brakes to a stop within a distance of 270 ft. Assuming the following car to have a perception-reaction time of 2 s, what must be the minimum following distance in order to avoid a collision?

Solution The lead car travels a distance of 270 ft during the braking maneuver alone. (The braking to a stop is *not* in response to the sighting of an obstruction.) The key assumption in this case is that under identical road conditions, the same deceleration rate and grade should be used for both vehicles.

For the lead car, equating the braking distance to 270 ft, we have

$$270 = \frac{1.075 \times 55^2}{2(a \pm gG)} \Rightarrow a \pm gG = 6.022$$

Therefore, for the following car, the SSD is given by

$$S = 1.47 \times 65 \times 2 + \frac{1.075 \times 65^2}{2 \times 6.022} = 568.2 \text{ ft}$$

During the time that car 1 travels 270 ft, car 2 will travel 568 ft. Therefore, to avoid a collision, the minimum following distance = 568 – 270 = 298 ft.

Sight Distance on Vertical Curves

Stopping Sight Distance on Crest Curves

Simple (parabolic) vertical curves are either crest curves or sag curves. On crest curves, the vertical profile itself obstructs the line of sight, and the length of curve should provide adequate SSD based on height of driver's eye (h_1) and height of the obstruction being sighted (h_2).

AASHTO defines Eq. (15.48) for SSD on a crest vertical curve.

$$L = \frac{AS^2}{100\left(\sqrt{2h_1} + \sqrt{2h_2}\right)^2} \quad \text{for } S \le L \tag{15.48a}$$

$$L = 2S - \frac{200\left(\sqrt{h_1} + \sqrt{h_2}\right)^2}{A} \quad \text{for } S > L \tag{15.48b}$$

For stopping and passing sight distance calculations for passenger cars, the height of driver's eye above the roadway is assumed to be $h_1 = 3.5$ ft. For large trucks, the *AASHTO Green Book* recommends a value of $h_1 = 7.6$ ft.

For SSD calculations, the recommended value for height of object $h_2 = 2.0$ ft. For passing sight distance calculations, the recommended value of $h_2 = 3.5$ ft.

Thus, for SSD for a passenger car, adopting $h_1 = 3.5$ ft and $h_2 = 2.0$ ft yields Eq. (15.49)

$$L = \frac{AS^2}{2158} \quad \text{for } S \le L \tag{15.49a}$$

$$L = 2S - \frac{2158}{A} \quad \text{for } S > L \tag{15.49b}$$

where S = stopping sight distance (ft)
L = length of vertical curve (ft)
A = absolute grade difference $= |G_1 - G_2|$ (percent)

Example 15.30

The alignment of a roadway contains a parabolic curve which transitions from a grade of +5% to a grade of −3%. Based on the AASHTO guidelines for SSD, what is the minimum length of curve required for a design speed of 60 mi/h?

Solution For a design speed of 60 mi/h, the stopping sight distance $S = 566$ ft.
Using $A = 8$, Eq. (15.49a) gives

$$L = \frac{AS^2}{2158} = \frac{8 \times 566^2}{2158} = 1187.6 \text{ ft}$$

This satisfies the criterion $S < L$.
Therefore, the required length of curve = 1188 ft.

Stopping Sight Distance on Sag Curves

On sag curves, the vertical profile does not obstruct the line of sight. However, the length of curve should provide adequate SSD based on headlight illumination distance. AASHTO defines Eq. (15.50) for SSD on a sag vertical curve.

$$L = \frac{AS^2}{400 + 3.5S} \quad \text{for } S \le L \tag{15.50a}$$

$$L = 2S - \frac{400 + 3.5S}{A} \quad \text{for } S > L \tag{15.50b}$$

Comfort Criterion: Sag Curves

Another criterion for minimum length of curve is based on rider comfort. Physiologically, discomfort is produced by the perception of acceleration, rather than velocity. The vertical acceleration a_v is equal to the rate of change of vertical velocity (the product of the velocity V and the gradient G).

$$a_v = \frac{\Delta v_v}{\Delta t} = \frac{\Delta v_v}{\frac{L}{V}} = \frac{V\Delta v_v}{L} = \frac{V(VG_2 - VG_1)}{L} = \frac{V^2(G_2 - G_1)}{L} = \frac{V^2 A}{L} \tag{15.51}$$

If the upper limit of vertical acceleration is $a_{v,\max}$, the minimum length of curve may be given by

$$L = \frac{AV^2}{a_{v,\max}} \quad \text{(consistent units: e.g., } L \text{ ft, } A \text{ decimal, } V \text{ ft/s, } a_{\max} \text{ ft/s}^2\text{)} \tag{15.52}$$

The AASHTO default for $a_{v,\max} = 1$ ft/s^2

Thus, on sag vertical curves, the length of the vertical curve must be greater than

$$L_{\min} = \frac{AV^2}{46.5} \tag{15.53}$$

where $L_{\min}$ = minimum length of sag curve (ft)
A = absolute difference in grades $= |G_1 - G_2|$ (percent)
V = design speed (mi/h)

This formula may be generalized to Eq. (15.54) (for situations where the specified value of maximum vertical acceleration is different from 1 ft/s^2, such as for railroad alignments):

$$L_{\min} = \frac{AV^2}{46.5a_{v,\max}} \tag{15.54}$$

Limited Sight Distance under Obstructing Object over Sag Curve

On sag curves, drivers of high vehicles such as trucks may have limited sight distance due to the presence of obstructing structures (such as an overpass). This situation is illustrated in Fig. 15.6.

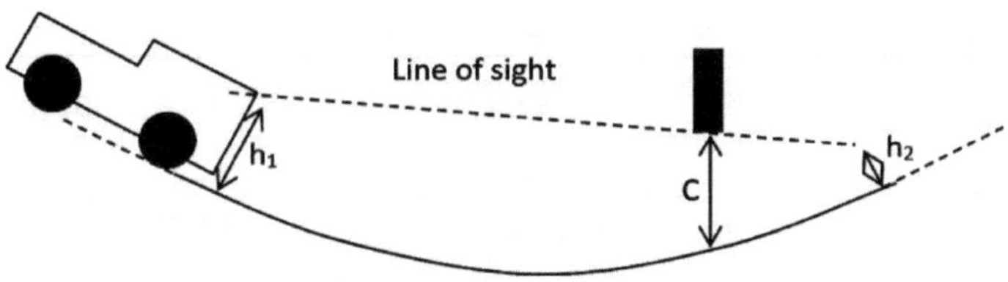

Figure 15.6 Sight distance on a vertical curve under an obstruction.

The minimum length of curve is given by

$$L = \frac{AS^2}{800\left(C - \frac{h_1 + h_2}{2}\right)} \quad \text{for } S \le L \tag{15.55a}$$

$$L = 2S - \frac{800}{A}\left(C - \frac{h_1 + h_2}{2}\right) \quad \text{for } S > L \tag{15.55b}$$

Example 15.31

A parabolic curve of length 800 ft is used to transition from a grade of –4% to a grade of +5%. The minimum vertical clearance under an overpass is 13 ft 8 in. What is the maximum safe speed for a truck (assume height of driver's eye = 7.6 ft and height of obstruction = 2.0 ft) in order to satisfy the AASHTO criterion for sight distance?

Solution The data for the problem are length of curve $L = 800$ ft, vertical clearance $C = 13.67$ ft, height of driver's eye, $h_1 = 7.6$ ft, height of obstruction $h_2 = 2.0$ ft, algebraic grade difference $A = 5 - (-4) = 9$. Using the second Eq. (15.55b)

$$2S = L + \frac{800}{A}\left(C - \frac{h_1 + h_2}{2}\right) = 800 + \frac{800}{9}\left(13.67 - \frac{7.6 + 2.0}{2}\right) \Rightarrow S = 794 \text{ ft}$$

This does not satisfy the criterion $S > L$. Therefore, we use Eq. (15.55a)

$$S = \sqrt{\frac{800\left(C - \frac{h_1 + h_2}{2}\right)L}{A}} = \sqrt{\frac{800\left(13.67 - \frac{7.6 + 2.0}{2}\right)800}{9}} = 794 \text{ ft}$$

This does satisfy the criterion $S \le L$

The design speed (for level roadway) corresponding to a sight distance $S = 794$ ft is 73.8 mi/h.

Table 15.3 shows a summary of the SSD criteria.

Stopping Sight Distance on Horizontal Curves

On horizontal curves, the obstruction to sight lines is due to proximity of adjacent structures (signposts, buildings, etc.). In Fig. 15.7, a two-lane roadway is shown. The dashed line shows

Table 15.3 Stopping Sight Distance on Vertical Curves (AASHTO)

	$S \le L$	$S > L$
Crest vertical curve general:	$L = \dfrac{AS^2}{100\left(\sqrt{2h_1} + \sqrt{2h_2}\right)^2}$	$L = 2S - \dfrac{200\left(\sqrt{h_1} + \sqrt{h_2}\right)^2}{A}$
Stopping sight distance crest vertical curve $h_1 = 3.5$ ft and $h_2 = 2.0$ ft	$L = \dfrac{AS^2}{2158}$	$L = 2S - \dfrac{2158}{A}$
Stopping sight distance sag vertical curve (headlight criterion)	$L = \dfrac{AS^2}{400 + 3.5S}$	$L = 2S - \dfrac{400 + 3.5S}{A}$
Sag vertical curve (comfort criterion): $a_{v,max} = 1$ ft/s²)	$L = \dfrac{AV^2}{46.5}$	
Sag vertical curve (limited vertical clearance under overhead structure) C = vertical clearance (ft)	$L = \dfrac{AS^2}{800\left(C - \dfrac{h_1 + h_2}{2}\right)}$	$L = 2S - \dfrac{800}{A}\left(C - \dfrac{h_1 + h_2}{2}\right)$
For $h_1 = 8$ ft and $h_2 = 2$ ft	$L = \dfrac{AS^2}{800(C - 5)}$	$L = 2S - \dfrac{800}{A}(C - 5)$

the centerline of the roadway (usually the baseline curve). The centerlines of the inner and outer lanes are also shown. A sight obstructing object is shown located adjacent to the inner edge of the roadway. The obstructing effect of this object is more pronounced on the driver traveling on the inside lane.

The lateral offset of the object from this curve (centerline of the inside lane) is shown as M in the figure. The design speed of the vehicle can be directly related to the required SSD S. This distance must be minimally available with the sight obstruction at distance M as shown.

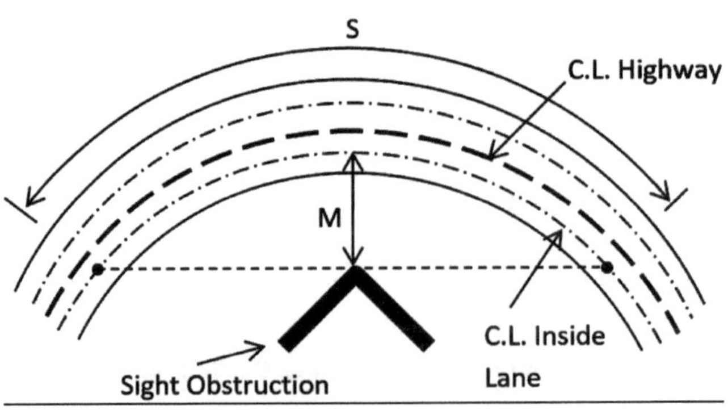

Figure 15.7 Sight distance on a horizontal curve around an obstruction.

For a horizontal circular curve of radius R, the middle ordinate distance M that corresponds to a particular arc length S is given by

$$M = R\left[1 - \cos\left(\frac{28.65S}{R}\right)\right] \tag{15.56}$$

This may also be written as

$$S = \frac{R}{28.65}\left[\cos^{-1}\left(1 - \frac{M}{R}\right)\right] \tag{15.56}$$

The middle ordinate distance M should be measured to the centerline of the traveled lane. In Eqs. (15.55) and (15.56), angles are expressed in degrees, not radians.

Example 15.32

The centerline of the local horizontal alignment of a two-lane highway is a circular curve with degree of curve = 3° (based on 100 ft arc). A sight obstructing object is located at a clear distance of 10 ft from the inside edge of the highway. What is the safe design speed (mi/h) on the curve? Assume standard lane width = 12 ft.

Solution The radius of curve (highway centerline) is given by

$$R = \frac{5729.578}{3} = 1909.86 \text{ ft}$$

The radius of the inside lane = 1909.86 − 6.0 = 1903.86 ft.

The middle ordinate distance is calculated as the offset between the obstruction and the centerline of the inside lane = 10.0 + 6.0 = 16 ft.

The corresponding SSD is calculated as

$$S = \frac{R}{28.65}\left[\cos^{-1}\left(1 - \frac{M}{R}\right)\right] = \frac{1906.86}{28.65}\left[\cos^{-1}\left(1 - \frac{16}{1906.86}\right)\right] = 494.35 \text{ ft}$$

From Table 15.2 (for level grade), the maximum speed, for which the available sight distance (494 ft) is adequate is 55 mi/h. Thus, the design speed is 55 mi/h.

Highway Safety

Accident Rates

Accident rates for a particular location are a critical parameter for decision making about possible improvements to the location or facility. Once the crash rate for the location (stretch of roadway or intersection) has been measured, it is then compared to the critical crash rate for the surrounding region (district, state, etc.).

The *equivalent number of accidents* for a facility is often calculated by adopting weighing factors for various types of accidents depending on their severity. These are not economic weighting factors, but are meant to represent the accident risk level for that facility.

Example 15.33

The example below outlines various ways for normalizing and reporting accident rates for a study jurisdiction. The table below shows some gross accident statistics for a relatively small urban jurisdiction in 2003.

Fatalities	75
Fatal accidents	60
Injury accidents	300
PDO accidents	2000
Total involvement	4000
Vehicle miles traveled	1.5×10^9
Registered vehicles	100,000
Licensed drivers	150,000
Population of area	300,000

The following "accident rates" may then be calculated and reported

1. Accidents per hundred million miles

$$\frac{2360}{1.5 \times 10^9} \times 10^8 = 157/\text{HMVM}$$

2. Deaths per 100,000 population

$$\frac{75 \times 10^5}{3 \times 10^5} = 25$$

3. Deaths per 10,000 registered vehicles

$$\frac{75 \times 10^4}{10^5} = 7.5$$

For Highway Segment (length L)
For a segment of highway of a finite length, the rate of accidents over a study period of N years is reported as the number of accidents per hundred million vehicle miles (number per HMVM) is defined as

$$\text{Accidents per hundred million vehicle miles} = \frac{\text{No. accidents} \times 10^8}{L \times \text{ADT} \times 365 \times N} \tag{15.57}$$

where ADT is the average daily traffic for the study period. If ADT values for these N years are reported individually, then the average ADT is calculated as

$$\text{ADT} = \frac{\sum_1^N \text{ADT}_i}{N} \tag{15.58}$$

For Intersections
For an intersection, the rate of accidents over a study period of N years is reported as the number of accidents per million entering vehicles (number per MEV) and is defined as

$$\text{Accidents/million entering vehicles} = \frac{\text{No. of accidents} \times 10^6}{\text{ADT} \times 365 \times N} \tag{15.59}$$

The number of entering vehicles is counted for all approaches into the intersection.

Accident Countermeasures

If accident rates at a particular site are unacceptable, then several countermeasures can be implemented. Some effective countermeasures are lane widening, sightline improvements, parking restrictions, speed control, etc. If empirical data are available about these countermeasures and their impact on accident rates, then a particular site can be evaluated for the overall impact (reduction of the number of accidents) of implementing several countermeasures.

The *Highway Safety Manual*, 2010, American Association of State Highway & Transportation Officials, Washington, DC includes several countermeasures grouped into categories and subcategories. Some examples of countermeasures and the category and subcategory they are nested under are listed below.

Category: Delineation

- Subcategory: Supplemental delineation
  - Countermeasure: Install post-mounted delineators
  - Countermeasure: Install snow-plowable, permanent raised pavement markers
- Subcategory: On-pavement markings
  - Countermeasure: Install edgelines, centerlines, and post-mounted delineators

Each countermeasure has a crash modification factor (CMF) which is a multiplicative factor used to compute the expected number of crashes after implementing a given countermeasure at a specific site. The crash reduction factor (CRF) is equal to 1 − CMF.

The overall crash reduction achieved by implementing a series of countermeasures ($1 \ldots n$) is given by Eq. (15.60):

$$\text{CR} = \text{CR}_1 + (1-\text{CR}_1)\text{CR}_2 + (1-\text{CR}_1)(1-\text{CR}_2)\text{CR}_3 + \cdots + (1-\text{CR}_1)\cdots(1-\text{CR}_{n-1})\text{CR}_n \tag{15.60}$$

Example 15.34

An urban intersection had 23 crashes during the planning year 2001 to 2002. The ADT (all entering vehicles) for this year was 6400. Two specific countermeasures were implemented. They are described below:

Countermeasure 1

Widening of lanes

Expected reduction of crashes = 12%

Countermeasure 2

Eliminating curbside parking

Expected reduction of crashes = 32%

If the ADT is expected to increase by 3% every year, what is the number of crashes expected during the year 2011 to 2012?

A. 12

B. 19

C. 26

D. 31

Solution Growing at the rate of 3% every year, ADT after interval of 10 years = $6400 \times 1.03^{10} = 8601$.

CRF for countermeasure 1: $CR_1 = 0.88$

CRF for countermeasure 2: $CR_2 = 0.68$

Overall CRF: $CR = CR_1 + (1 - CR_1)CR_2 = 0.12 + 0.88 \times 0.32 = 0.4016$

Therefore, the number of crashes prevented in 2011 to 2012:

$$N \times CR \times \frac{ADT_2}{ADT_1} = 23 \times 0.4016 \times \frac{8601}{6400} = 12.4.$$

It may be assumed that without the countermeasures, the number of crashes would increase in proportion to the ADT. This would mean $23 \times 8601/6400 = 30.9$

With the countermeasures, the expected number of crashes = 30.9 – 12.4 = 18.5

Answer is B.

CHAPTER 16

Environmental Engineering

Main U.S. Environmental Laws

Clean Air Act (1955, amended 1977)

The Clean Air Act (CAA) authorizes the U.S. Environmental Protection Agency (EPA) to establish National Ambient Air Quality Standards (NAAQS) to protect public health and the environment. The primary objective of the NAAQS is to protect human health, and the secondary objectives are nonhealth-related such as aesthetics, agriculture, etc.

The CAA divides the country into air quality regions. It also sets goals for the concentration of various pollutants in the ambient air. These goals are set so that the health risk from the various pollutants is essentially zero. The ambient air pollutants (also called "criteria pollutants") are carbon monoxide (CO), lead (Pb), nitrogen dioxide (NO_2), sulphur dioxide (SO_2), ozone (O_3), and particulates ($PM_{2.5}$ and PM_{10}). The CAA established technology-based emission standards for specific industry categories. These standards specify the technology and the emission limits that are allowed for pollutants discharged to the air.

Clean Water Act (1948, amended 1977)

The primary objective of the Clean Water Act (CWA) is to eliminate discharge of all pollutants, while the secondary objective is to restore and maintain the quality of the nation's waters so they are fishable and swimmable.

The current trend is toward regions setting watershed specific goals for principal pollutants (oxygen demand, nutrients, pathogens, suspended solids, salts, toxic metals, toxic organics, heat, and pH). The CWA established technology-based effluent standards for specific industry categories. These standards specify the technology and effluent limits that should be used to treat wastewater prior to disposal in a water body. The original focus of the law was on industrial point sources of pollution. However, the 1987 amendments added requirements for the control of nonpoint source pollution such as agricultural and urban runoff.

Under the auspices of the CWA, the National Pollutant Discharge Elimination System (NPDES) is a comprehensive national program for addressing the nonagricultural sources of stormwater discharges which adversely affect the quality of our nation's waters. The NPDES program uses the permitting mechanism to require the implementation of controls designed to prevent harmful pollutants from being carried by stormwater runoff into local water bodies.

The CWA prohibits anybody from discharging pollutants through a point source into a water of the United States unless they have an NPDES permit. The permit contains limits on what can be discharged, monitoring and reporting requirements, and other provisions to ensure that the discharge does not hurt water quality or people's health. The permit translates general requirements of the CWA into specific provisions tailored to the operations of each entity discharging pollutants.

NPDES permits make sure that state's mandatory standards for clean water and the federal minimums are being met. There are various methods used to monitor NPDES permit conditions. The permit will require the facility to sample its discharges and notify EPA and the state regulatory agency of these results. In addition, the permit will require the facility to notify EPA and the state regulatory agency when the facility determines it is not in compliance with the requirements of a permit. EPA and state regulatory agencies also will send inspectors to companies in order to determine if they are in compliance with the conditions imposed under their permits.

EPA and state regulatory agencies may issue administrative orders which require facilities to correct violations and that assess monetary penalties. They may also pursue civil and criminal actions such as mandatory injunctions or penalties, as well as jail sentences for persons found willfully violating requirements and endangering the health and welfare of the public or environment.

Members of the general public can also enforce permit conditions. If any member of the general public, upon reviewing the monitoring documents, finds that a facility is violating its NPDES permit, he/she can independently start a legal action, unless EPA or the state regulatory agency has taken an enforcement action.

The term *pollutant* includes any type of industrial, municipal, and agricultural waste discharged into water. Some examples of pollutants covered by the NPDES requirements are dredged soil, solid waste, incinerator residue, sewage, garbage, sewage sludge, munitions, chemical wastes, biological materials, radioactive materials, heat, wrecked or discarded equipment, rock, sand, and industrial, municipal, and agricultural waste.

The term *point source* means any discernible, confined, and discrete conveyance, such as a pipe, ditch, channel, tunnel, conduit, discrete fissure, or container. It also includes vessels or other floating craft from which pollutants may be discharged. By law, agricultural stormwater discharges and return flows from irrigated agriculture are not considered point sources.

For discharging pollutants into a municipal *sanitary* sewer system, an NPDES permit is not required. For discharging pollutants into a municipal *storm* sewer system, a permit may be required depending on the constituents.

The CWA limits the length of NPDES permits to 5 years. NPDES permits can be renewed (reissued) at any time after the permit holder applies. Federal laws provide EPA and authorized state regulatory agencies with various methods of taking enforcement actions against violators of NPDES permit requirements.

Safe Drinking Water Act

The primary objective of the Safe Drinking Water Act (SDWA) is to ensure potable water (safe for human health), while the secondary objective is to ensure palatable water (aesthetics such as taste, color, and odor). A tertiary objective is to protect the quality of underground sources of water.

The SDWA has four categories of standards that water suppliers must meet. The categories are physical, chemical, microbiological, and radiological. The physical standards include guidelines for total solids, suspended solids, dissolved solids, turbidity, color, tastes, odors, and temperature. These are nonenforceable standards aimed at improving palatability of the water supply.

The chemical standards set maximum contaminant levels (MCLs) for various chemical contaminants. These MCLs must be met and are enforceable. The SDWA also sets more stringent maximum contaminant level goals (MCLGs) that are not enforceable. Similarly, there are MCLs, MCLGs, and technology standards for microbiological and radiological contaminants.

Resource Conservation and Recovery Act (1976, amended 1986)

The primary objectives of Resource Conservation and Recovery Act (RCRA) are to manage the generation, storage, transport, treatment, and disposal of solid and hazardous wastes from operating facilities, and minimize waste disposal to land.

RCRA is divided up into various sections that regulate various activities. One section details how a waste can be characterized as a solid waste and as a hazardous waste. Another section details a tracking framework for hazardous waste that includes a paper trail from the time the waste is generated to when it is finally disposed. Another section details requirements for facilities that treat, store, or dispose of hazardous waste. These requirements must be met before a facility is granted a permit to operate. Another section details requirements for the design of solid waste landfill facilities. Another section specifies how the various hazardous wastes must be treated before disposal. Additional sections exist for special issues such as underground storage tanks.

Surface Water Treatment Rule

The Surface Water Treatment Rule (effective December 31, 1990) seeks to prevent waterborne diseases caused by viruses, *Legionella*, and *Giardia lamblia*. These disease-causing microbes are present at varying concentrations in most surface waters. The rule requires that water systems filter and disinfect water from surface water sources to reduce the occurrence of unsafe levels of these microbes. This rule governs water supplies whose source of drinking water is surface water, which it defines as "all water which is open to the atmosphere and subject to surface runoff." Surface water is particularly susceptible to microbial contamination from sewage treatment plant discharges and runoff from storm water and snow melt. These sources often contain high levels of fecal microbes that originate in livestock wastes or septic systems.

Municipal Separate Storm Sewer System

The regulatory definition of a municipal separate storm sewer system (MS4) is a conveyance or system of conveyances (including roads with drainage systems, municipal streets, catch basins, curbs, gutters, ditches, man-made channels, or storm drains) which are (1) owned or operated by a state, city, town, borough, county, parish, district, association, or other public body (created to or pursuant to state law) or a designated and approved management agency under section 208 of the CWA that discharges into waters of the United States, (2) designed or used for collecting or conveying stormwater, (3) which is not a combined sewer, and (4) which is not part of a publicly owned treatment works (POTW).

In practical terms, operators of MS4s can include municipalities and local sewer districts, state and federal departments of transportation, public universities, public hospitals, military bases, and correctional facilities. The Stormwater Phase II Rule added federal systems, such as military bases and correctional facilities by including them in the definition of small MS4s.

EPA's NPDES Stormwater Program regulates medium, large, and regulated small MS4s. A medium MS4 is a system that is located in an incorporated place or county with a population between 100,000 and 249,999. A large MS4 is a system that is located in an incorporated place or county with a population of 250,000 or more. A regulated small MS4 is any small MS4 located in an "urbanized area," as defined by the Bureau of the Census, or located outside of an urbanized area and brought into the program by the NPDES permitting authority. An NPDES permit must be obtained by the operator of an MS4 covered by the NPDES Stormwater Program.

Water Quality–Based Effluent Limits

Water quality goals for bodies of water within state boundaries are set by individual states. For ecologically important water bodies that abut several states, such as the Chesapeake Bay, these water quality standards are set by agreement between the affected states, with EPA having overall jurisdiction. Section 303(c) of the CWA requires every state to develop water quality standards at or above EPA standards.

Water quality standards have three components:

1. *Use classification:* The defining authority (the state) must make clear the intended use classification of the water body—water supply, recreation or propagation of fish and wildlife.
2. *Water quality criteria:* The defining authority must establish numerical and narrative criteria for all identified pollutants. These will be in the form of specific limits on the concentration of each pollutant.

   Water quality criteria consist of three components:

   a. *Magnitude:* This is a specific level (concentration) of the pollutant allowable in the water body.
   b. *Duration:* This the period over which the concentration is averaged. For example, the limiting concentration of pollutant A may be on the 1-h average, while that for pollutant B may be on the 8-h average.
   c. *Frequency:* This is how often the criteria may be exceeded.

3. *Anti-degradation policy:* The defining authority must adopt policies and identify methods used to implement these policies to prevent/remediate degradation of the water bodies below their defined minimum quality standards. There are three tiers of protection provided, depending on the importance of the water body, with Tier 3 protection being the most stringent.

Water Quality

Biochemical Oxygen Demand

Microorganisms in water digest organic material (substrate or food). During this process, they use oxygen and thus place an *oxygen demand* on the water. This oxygen demand is biochemical oxygen demand (BOD). The oxygen demand on a water body varies in accordance with the population dynamics of the microbial population. A typical model for such a population model includes a lag phase, a constant growth phase, a stationary phase, and a decay phase as shown in Fig. 16.1.

A mathematical model for the growth of BOD is the exponential model where the BOD grows exponentially and asymptotes to the so-called ultimate BOD of the water, which is its oxygen demand potential at $t = 0$. This model does not account for the lag or decay phases.

According to this model, $y(t)$, the BOD exerted at time t is given by the function shown below

$$y(t) = L(1 - e^{-kt}) \tag{16.1}$$

where L is the ultimate BOD = oxygen equivalent of organics at $t = 0$ and k = reaction rate constant (base e, day^{-1}). This is also called the deoxygenation rate constant or the endogenous decay constant (alternate symbol k_d).

Equation (16.1) employs the natural logarithmic base e (2.71828…). The same model could also be expressed as

$$y(t) = L(1 - 10^{-kt}) \tag{16.2}$$

where k is the reaction rate constant (base 10, day^{-1}).

Figure 16.1 Stages in life cycle of microorganisms.

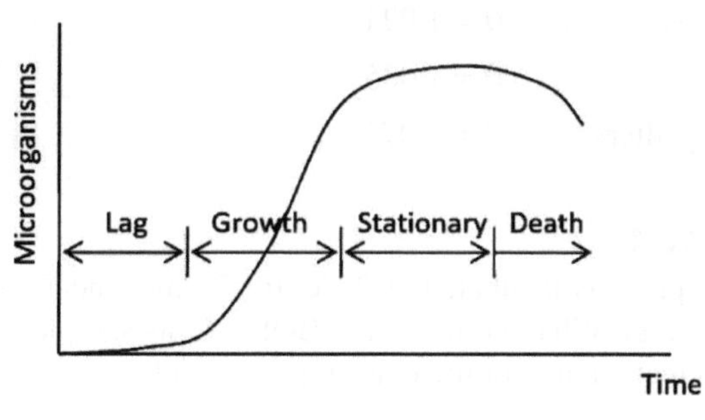

Figure 16.2 Biochemical oxygen demand.

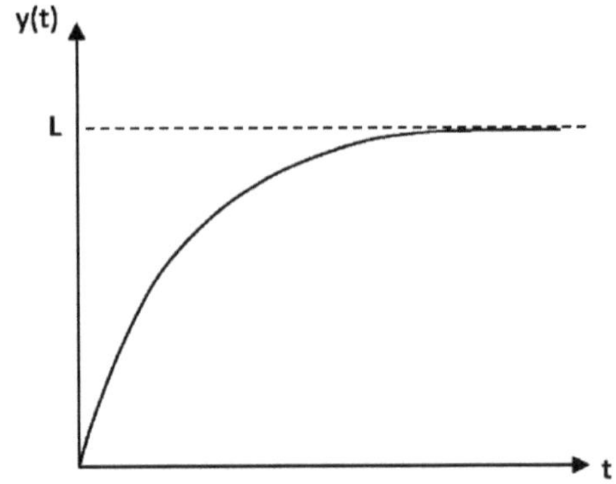

The growth constants in the two models are related by

$$k(\text{base} - e) = 2.303 \times k(\text{base} - 10)$$

A representative plot of BOD is shown in Fig. 16.2.

Temperature Dependence of BOD Rate Constant

The reaction rate constant is temperature dependent, due to the fact that metabolism progresses faster at higher temperatures. It should be noted, however, that the ultimate BOD (the oxygen demand potential at $t = 0$) remains the same while the rate of growth increases for higher temperatures.

Commonly, the base or reference temperature is taken as 20°C. The rate constant at other temperatures is given by

$$k_T = k_{20} \cdot \theta^{T-20} \tag{16.3}$$

where θ is the temperature coefficient. The *FE Reference Handbook,* version 9.4, recommends the following values for θ:

Activated sludge:	$\theta = 1.136$	$T > 20°C$
	$\theta = 1.056$	$T < 20°C$
Reaeration:	$\theta = 1.024$	
Biotowers:	$\theta = 1.035$	
Trickling filters:	$\theta = 1.072$	

Example 16.1

A BOD sample was incubated at 27°C for 7 days and found to be 100 mg/L. Given K_{20} = 0.23/day (base e). What would be the BOD of this sample recorded at 5 days, if incubated at 20°C? Use kinetic temperature constant $\theta = 1.047$.

Solution The reaction rate constant at 27°C is given by

$$k_T = k_{20}\theta^{T-20} \Rightarrow k_{27} = 0.23 \times 1.047^{27-20} = 0.32/\text{day}$$

Since BOD_7 at 27°C is 100 mg/L, the ultimate BOD is given by

$$L = \frac{BOD_7}{1-e^{-0.32\times 7}} = \frac{100}{1-e^{-0.32\times 7}} = 112 \text{ mg/L}$$

Based on this ultimate BOD, the BOD_5 at 20°C is given by

$$BOD_5 = L(1-e^{-0.23\times 5}) = 112 \times 0.683 = 76.5 \text{ mg/L}$$

Standard BOD Test

The standard BOD test uses a BOD incubation bottle of volume 300 mL. A wastewater sample is added to the bottle and diluted by filling the bottle with aerated dilution water. Untreated municipal wastewaters and unchlorinated effluents usually have adequate microbial populations and do not require addition of seed material. In a *seeded BOD test*, seed material—usually derived from aged wastewater—is added to industrial wastewaters and chlorinated effluents to accelerate biological activity. The dissolved oxygen of the diluted sample is measured at $t = 0$. The diluted sample is then incubated at a specified temperature for a specified time (commonly 5 days at 20°C). The dissolved oxygen of the diluted sample is measured at the end of the incubation period.

For the standard BOD test, the BOD is given by

$$BOD = \frac{DO_i - DO_f}{\dfrac{V_{\text{sample}}}{V_{\text{sample}} + V_{\text{dilution}}}} \tag{16.4}$$

where DO_i and DO_f are the measured values of dissolved oxygen at the beginning and the end of the incubation period, respectively. The denominator is the dilution fraction that extrapolates the oxygen depletion from the sample to full volume.

Example 16.2

Calculate the BOD of a wastewater sample, given the following data:

Sample volume = 10 mL

BOD incubation bottle volume = 300 mL

Initial DO of diluted sample = 9 mg/L

Final DO of diluted sample = 2 mg/L

Solution

Volume of sample = 10 mL

Sample volume fraction = 10/300 = 0.0333

Volume of dilution water = 290 mL

$$\text{BOD} = \frac{(9-2)}{0.0333} = 210 \text{ mg/L}$$

Dilution Purification of Wastewater Streams

When a stream of untreated or partially treated sewage is discharged into a large body of water, the wastewater is diluted. Assuming instantaneous mixing at the point of discharge, the parameters of the wastewater-river mixture can be calculated using a weighted average approach. This can be applied for calculating parameters such as temperature/dissolved oxygen/BOD/suspended solids, etc. If the resulting levels are within water quality standards for the stream, the wastewater does not require pretreatment prior to discharge.

Consider a wastewater flow rate Q_w with ultimate BOD L_w and dissolved oxygen DO_w mixing with a river with flow rate Q_r, ultimate BOD L_r, and dissolved oxygen DO_r.

The initial (immediately after mixing) ultimate BOD of the river-wastewater mix is

$$L_o = \frac{Q_w L_w + Q_r L_r}{Q_w + Q_r} \tag{16.5}$$

Initial dissolved oxygen immediately after mixing (DO_o) is given by

$$\text{DO}_o = \frac{Q_w \text{DO}_w + Q_r \text{DO}_r}{Q_w + Q_r} \tag{16.6}$$

Temperature immediately after mixing (T_o) is given by

$$T_o = \frac{Q_w T_w + Q_r T_r}{Q_w + Q_r} \tag{16.7}$$

Initial (at $t = 0$) oxygen deficit after mixing (D_o) is given by

$$D_o = \text{DO}_{\text{sat}} - \frac{Q_w \text{DO}_w + Q_r \text{DO}_r}{Q_w + Q_r} \tag{16.8}$$

where DO_{sat} is the saturation dissolved oxygen, which is a function of temperature and salinity.

Example 16.3

A wastewater flow of 1200 gal/min is discharged into a stream with a flow rate of 12 ft³/s. The ultimate BOD of the wastewater is 1800 lb/day. The stream has negligible BOD upstream of the mixing point. Other parameters are

Upstream temperature of river = 15°C

Upstream dissolved oxygen of river = 5.2 mg/L

Temperature of wastewater = 38°C

Dissolved oxygen of wastewater = 1.2 mg/L

Saturated dissolved oxygen downstream of mixing = 9.2 mg/L

(a) What is the ultimate BOD (mg/L) immediately downstream of the mixing point?

(b) What is the oxygen deficit (mg/L) immediately downstream of the mixing point?

Solution The wastewater flow rate is 1200 gal/min = 2.674 cfs = 1.728 million gallons per day (MGD).

The BOD load of the wastewater needs to be converted to a concentration (mg/L) using

$$\text{mg/L} \times \text{MGD} \times 8.3454 = \text{lb/day}$$

Therefore, the ultimate BOD of the wastewater can be calculated as

$$L_r = \frac{1800}{1.728 \times 8.3454} = 124.8 \text{ mg/L}$$

(a) The ultimate BOD of the stream-wastewater mix is given by the weighted average

$$L_o = \frac{2.674 \times 124.8 + 12 \times 0}{2.674 + 12} = 22.75 \text{ mg/L}$$

(b) The dissolved oxygen of the stream-wastewater mix is given by the weighted average

$$\text{DO}_o = \frac{2.674 \times 1.2 + 12 \times 5.2}{2.674 + 12} = 4.47 \text{ mg/L}$$

Saturated dissolved oxygen = 9.20 mg/L

Therefore, oxygen deficit immediately downstream of mixing point = 9.20 − 4.47 = 4.73 mg/L.

Example 16.4

A factory produces chromium at a concentration of 3000 μg/L in a wastewater stream of 500 gal/min. If the upstream chromium content of the receiving stream (flowrate 7.5 ft³/s) is 30 μg/L and the EPA limit on chromium is 50 μg/L, what must be the removal efficiency of chromium the factory must achieve before discharging into the stream?

Solution

Q_w = 500 gal/min = 500 × 0.002228 = 1.114 ft³/s

If the concentration of chromium discharged into the stream is C_w

The weighted average of the stream and wastewater is $\dfrac{C_w \times 1.114 + 30 \times 7.5}{1.114 + 7.5} \leq 50$

Therefore, $C_w < 184.65$ μg/L

Therefore, percent removal during pretreatment = $\dfrac{3000 - 184.65}{3000} \times 100\% = 93.85\%$.

Streeter–Phelps Equations

The Streeter–Phelps equations describe the dissolved oxygen "sag curve" obtained as a result of the simultaneous deoxygenation and reoxygenation that occurs when a wastewater is discharged

Figure 16.3 Dissolved oxygen sag curve (Streeter–Phelps).

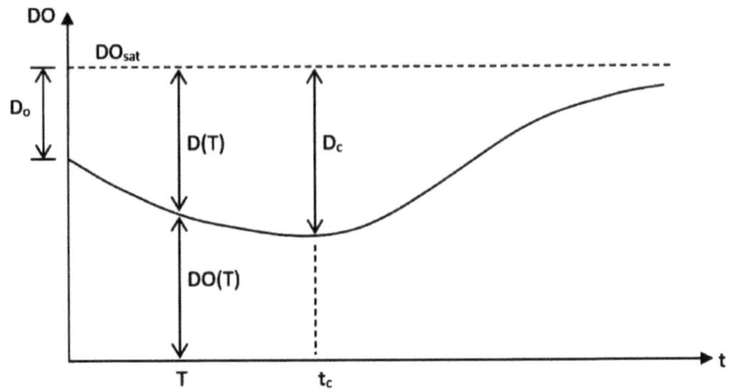

into a receiving stream. The organic load of the effluent (represented by the ultimate BOD or BOD potential) brings in microorganisms which continuously exert an oxygen demand on the receiving water, resulting in a lowering of the dissolved oxygen. The rate of deoxygenation is expressed in terms of the coefficient k_d. At the same time, depending on the extent of stream turbulence, surface aeration entraps atmospheric oxygen and tends to increase the dissolved oxygen. The rate of reoxygenation is expressed in terms of the coefficient k_r.

Under the influence of these two effects, the dissolved oxygen follows a sag curve, showing a minimum value at the critical time t_c. The dissolved oxygen sag curve is shown in Fig. 16.3. The maximum oxygen deficit at t_c is designated D_c. Far from the point of mixing (shown as $t = 0$ in the figure), the dissolved oxygen asymptotes to the saturation value of the dissolved oxygen corresponding to the stream temperature.

Oxygen Deficit

According to the Streeter–Phelps model, the oxygen deficit after time t ($t = 0$ at the instant of mixing) is given by

$$D_t = \frac{k_1 L_o}{k_2 - k_1}(e^{-k_1 t} - e^{-k_2 t}) + D_o e^{-k_2 t} \tag{16.9}$$

where L_o = ultimate BOD immediately after mixing (mg/L)
D_o = oxygen deficit immediately after mixing (mg/L)
k_1 = deoxygenation rate constant, base e (day^{-1})
k_2 = reaeration rate constant, base e (day^{-1})

Note that in these equations, dissolved oxygen concentration, oxygen deficit, and ultimate BOD should all be expressed in similar units (mg/L). However, the BOD L_o is often expressed as a daily load (as in lb/day). In that case, it must be converted to a concentration using

$$L_o(\text{mg/L}) = \frac{L_o(\text{lb/day})}{(Q_r + Q_w)(\text{MGD}) \times 8.3454} \tag{16.10}$$

The flow rate in the denominator ($Q_r + Q_w$) is the combined flow rate of the stream and wastewater.

Critical (Maximum) Oxygen Deficit (Minimum Dissolved Oxygen)

The time at which the dissolved oxygen is critical (minimum) is given by

$$t_c = \frac{1}{k_2 - k_1} \ln\left[\frac{k_2}{k_1}\left(1 - D_o \frac{k_2 - k_1}{k_1 L_o}\right)\right] \tag{16.11}$$

The distance (from the point of mixing) to the critical location is given by

$$x_c = vt_c \tag{16.12}$$

where v is the average stream velocity.

NOTE In the Streeter–Phelps model, the equations are written with $e = 2.81828 \ldots$ as the base for the exponential terms. This is consistent with the rate constants (k_1 and k_2) being defined with respect to the same base. It is just as legitimate for these equations to be written with base 10, as long as the rate constants are also defined similarly.

Hardness in Water

Hardness in water is quantified as the sum of the concentration of all polyvalent cations such as Ca^{++}, Mg^{++}, Fe^{++}, etc. Each of these species has a different potential for causing hardness and therefore cannot be combined unless expressed in equivalent terms such as milliequivalents per liter (mEq/L). For practical purposes, these concentrations are expressed in consistent units mg/L as $CaCO_3$. This is accomplished by converting concentrations of individual species by a multiplicative factor given by

$$\frac{\text{Eq. weight } CaCO_3}{\text{Eq. weight of species}} = \frac{50}{\text{Eq. weight of species}}$$

The table on page 196 of the *Handbook* shows the relevant data for the most dominant ions and compounds causing hardness in water. Of the total hardness, the carbonate hardness (or temporary hardness) is the lesser of alkalinity or total hardness. If the hardness is greater than alkalinity, the difference between the total hardness and the alkalinity is called the noncarbonate hardness (or permanent hardness).

Hardness Removal

The following chemical compounds are commonly used for removal of hardness from municipal water.

Lime (calcium oxide)	CaO
Slaked lime (hydrated lime)	$Ca(OH)_2$
Soda ash (sodium carbonate)	Na_2CO_3
Caustic soda (sodium hydroxide)	$NaOH$

The different stages in hardness removal are outlined below.

Meeting Carbon Dioxide Demand CO_2 demand is met first before any softening occurs. Removal of carbon dioxide raises the pH and makes hardness removal more efficient

$$CO_2 + Ca(OH)_2 \rightarrow CaCO_3 \downarrow + H_2O$$

Removal of Carbonate Hardness Next, lime or caustic soda reacts with carbonate hardness. This category of hardness in water is caused primarily due to calcium and magnesium bicarbonates. This can be accomplished by the addition of slaked lime or caustic soda. The chemical reactions shown below describe the removal of calcium and magnesium carbonate hardness using lime ($Ca(OH)_2$) or caustic soda (NaOH).

Calcium carbonate hardness removal with $Ca(OH)_2$

$$Ca(HCO_3)_2 + Ca(OH)_2 \rightarrow 2CaCO_3 \downarrow + 2H_2O$$

Calcium carbonate hardness removal with NaOH

$$Ca(HCO_3)_2 + 2NaOH \rightarrow CaCO_3 \downarrow + 2Na^+ + CO_3^{2-} + H_2O$$

Magnesium carbonate hardness removal with $Ca(OH)_2$

$$Mg(HCO_3)_2 + 2Ca(OH)_2 \rightarrow 2CaCO_3 \downarrow + Mg(OH)_2 \downarrow + 2H_2O$$

Magnesium carbonate hardness removal with NaOH

$$Mg(HCO_3)_2 + 4NaOH \rightarrow Mg(OH)_2 \downarrow + 4Na^+ + 2CO_3^{2-} + 2H_2O$$

Removal of noncarbonate hardness

To remove noncarbonate hardness, soda ash (Na_2CO_3) is needed—with or without lime.

Calcium noncarbonate hardness removal

$$CaSO_4 + Na_2CO_3 \rightarrow CaCO_3 \downarrow + Na_2SO_4$$

Magnesium noncarbonate hardness removal

$$MgSO_4 + Ca(OH)_2 + Na_2CO_3 \rightarrow CaCO_3 \downarrow + Mg(OH)_2 \downarrow + Na_2SO_4$$

Destruction of excess alkalinity

$$2HCO_3^- + Ca(OH)_2 \rightarrow CaCO_3 \downarrow + CO_3^{2-} + 2H_2O$$

Recarbonation: Recarbonation (injection with carbon dioxide) is then used to lower the pH

$$Ca^{2+} + 2OH^- + CO_2 \rightarrow CaCO_3 \downarrow + H_2O$$

pH and pOH

Potential hydrogen (pH) is a measure of the activity of hydrogen ions (H^+) in a solution. pH is a quantification of the acidity or alkalinity of a solution. In dilute solutions

$$pH = -\log_{10} [H^+] \quad (16.13)$$

$$pOH = -\log_{10} [OH^-] \quad (16.14)$$

where $[H^+]$ is the concentration of H^+ ions in mol/L.

For a neutral solution (H^+ ion activity equals OH^- ion activity), the pH is approximately 7. Aqueous solutions with pH values lower than 7 are considered acidic, while pH values higher than 7 are considered alkaline.

At temperatures other than 25°C, the pH of pure water will not be 7. Note also that pure water, when exposed to the atmosphere, will absorb carbon dioxide, some of which reacts with water to form carbonic acid, thereby lowering the pH to about 5.7.

Most substances have a pH in the range 0 to 14, although extremely acidic or basic substances may have pH less than 0 or greater than 14.

Alkalinity

Alkalinity is a measure of the ability of water to absorb hydrogen ions without significant change in pH. In municipal water supply, alkalinity is primarily due to hydroxyl (OH^-), carbonate (CO_3^{2-}), bicarbonate (HCO_3^-) ions, and traces of nitrate (NO_3^-). As with hardness computations, these species must be converted to a common basis (commonly "mg/L as $CaCO_3$").

Below a pH of 4.5, no alkalinity exists. Between pH = 4.5 and 8.3, the alkalinity is entirely due to bicarbonate, and above pH = 8.3, the bicarbonate ions are converted to carbonate. Hydroxyl ions appear above a pH = 9.5 and react with carbon dioxide to produce carbonates and bicarbonates. Alkalinity is measured by titrating a water sample with a standard acidic titrant (sulphuric acid). The volume of acid needed to lower the pH to 8.3 (phenolphthalein end point) and the volume needed to further lower the pH to 4.5 (bromocresol red/methyl red end point) are both recorded. Alkalinity is expressed in units mg/L as $CaCO_3$. V_P is the volume of acidic titrant needed to reach pH 8.3 and V_M is the volume of acidic titrant needed to reach pH 4.5.

Let P equal the phenolphthalein alkalinity and M the total alkalinity. These are calculated as

$$P = \frac{V_P \times N_{\text{titrant}}}{V_{\text{sample}}} \times 50 \times 1000 \quad (16.15)$$

$$M = \frac{V_M \times N_{\text{titrant}}}{V_{\text{sample}}} \times 50 \times 1000 \quad (16.16)$$

The initial pH determines the concentration of hydroxyl ions. If the initial pH is x, then the initial pOH is $14 - x$, and the hydroxyl ion concentration (converted from mol/L to mg/L as $CaCO_3$) is given by

$$[OH^-] = 10^{x-14} \times 50{,}000$$

As the titration progresses, the neutralization of the hydroxyl ions is followed by the neutralization of the carbonate ions. The phenolphthalein alkalinity (corresponding to a pH = 8.3) corresponds to the sum of the hydroxyl ions plus half the carbonate ions. The methyl red alkalinity (corresponding to a pH = 4.5) corresponds to the sum of the hydroxyl ions plus carbonate ions plus bicarbonate ions.

Example 16.5

A 200-mL water sample has an initial pH of 10.3 and is titrated using a 0.05 N sulphuric acid solution. A pH of 8.3 (phenolphthalein end point) is reached after 5.5 mL of acid is added and a pH of 4.5 (methyl red end point) is reached after addition of a further 10.5 mL of acid. What are the alkalinity species present and in what concentrations (mg/L as $CaCO_3$)?

Solution Initial pH = 10.3

Therefore, pOH = 14 – pH = 3.7 = $-\log_{10}$ [OH^-]

[OH^-] = $10^{-3.7}$ = 1.995 × 10^{-4} mol/L = 1.995 × 10^{-4} × 50,000 = 9.97 mg/L as $CaCO_3$ (say 10 mg/L as $CaCO_3$)

This can be directly calculated as: $[OH^-] = 50{,}000 \times 10^{10.3-14} = 9.98 \text{ mg/L}$

The phenolphthalein alkalinity is given by

$$P = \frac{V_P \times N_{\text{titrant}}}{V_{\text{sample}}} \times 50 \times 1000 = \frac{5.5 \times 0.05 \times 50{,}000}{200} = 68.75 \text{ mg/L as } CaCO_3$$

and the total alkalinity as (total acid added = 16 mL)

$$M = \frac{V_M \times N_{\text{titrant}}}{V_{\text{sample}}} \times 50 \times 1000 = \frac{16 \times 0.05 \times 50{,}000}{200} = 200 \text{ mg/L as } CaCO_3$$

In reaching pH of 8.3, all the hydroxide alkalinity and half of the carbonate alkalinity must be neutralized.

½ CO_3 + OH = 68.75

Since OH^- = 10 mg/L as $CaCO_3$

CO_3^{2-} = 2 × 58.75 = 117.5 mg/L as $CaCO_3$

HCO_3^- = 200 – 10 – 117.5 = 72.5 mg/L as $CaCO_3$

Wastewater Flow Rates from Various Sources

Approximately 70% to 80% of a community's domestic and industrial water supply returns as wastewater. Sanitary sewer sizing is commonly based upon a flow rate of 100 to 125 gallons per capita per day (gpcd). Hourly variation in this number is the most significant for designing sewer size.

According to the Ten States Standards (Recommended Standards for Sewage Works), new sanitary sewer systems should be designed to carry the average flow of 100 gpcd, but sewer pipes should be sized to carry the peak flow.

Page 166 of the *Handbook* shows several empirical curves for maximum and minimum sewage flow ratios based on population P. For example, curve G gives the peak factor for a community (P is the population in thousands of people) as

$$\frac{Q_{\text{peak}}}{Q_{\text{ave}}} = \frac{18+\sqrt{P}}{4+\sqrt{P}} \tag{16.17}$$

where P is the population in thousands.

Municipal Wastewater Treatment

In municipal wastewater treatment plants, the sewage first passes through a *screen*, where large floating objects such as rags and sticks are removed. The screened sewage then passes through a *grit chamber*, where the heavier particles sink to the bottom and are removed from the through flow. The through flow still contains organic and inorganic matters along with other suspended solids. Minute suspended solids can be removed in a *sedimentation tank* in the form of sludge. Primary treatment removes a significant fraction of the inorganic (fixed) solids but it reduces volatile solids to a much lesser extent.

In the United States, some of the most common secondary treatment processes are trickling filters, activated sludge process, and rotating biological contactors. Filter media for recently installed trickling filters comprise interlocking pieces of corrugated plastic or other synthetic materials to increase surface area for biological populations. These synthetic media filters can be significantly higher than rock media trickling filters and therefore have a much smaller footprint.

The activated sludge process is favored both for economy of space and time over trickling filters. The fundamental idea of the activated sludge process is to bring air and highly concentrated (activated) sludge into contact with sewage for several hours in an aeration tank. From the aeration tank, the partially treated sewage is sent to another sedimentation tank (secondary clarifier) to remove excess bacteria. The process usually involves recirculation of the activated sludge in order to maintain a design concentration of suspended solids in the aeration tank.

The EPA requires that secondary treatment should reduce BOD_5 below 30 mg/L and suspended solids below 30 mg/L. The secondary effluent is usually disinfected with chlorine, typically killing more than 99% of the harmful bacteria. Excess chlorine can lead to the formation of toxic byproducts such as trihalomethanes (THMs) and therefore should be removed by dechlorination. Ultraviolet radiation and ozone are alternatives to chlorine disinfection.

Depending on the pollutants found in the wastewater, advanced treatment processes that target specific pollutants (such as ammonia removal by air-stripping or phosphorus removal by precipitation) may be used following secondary treatment.

Reactors Used for Wastewater Treatment

The various unit operations and processes of wastewater treatment—physical, chemical, and biological—take place in the following types of reactors: batch reactor, complete-mix reactor, complete-mix reactors in series, plug-flow reactor, packed bed reactor, and fluidized bed reactor.

In a *batch reactor*, the wastewater enters the reactor, is treated and once this batch is discharged, it is replaced by a new batch of untreated wastewater. An example of an operation using a batch reactor is flocculation. In an ideal *complete-mix reactor*, it is assumed that complete mixing occurs instantaneously as the wastewater enters the reactor. An example of a complete-mix reactor is the aeration tank of the activated sludge process. A large number of complete-mix reactors in series approximate a plug-flow reactor. In the *plug-flow reactor*, fluid packets pass through the reactor with no mixing. Longitudinal dispersion is minimal. An example of a plug-flow reactor is a chlorination contact chamber. In a *packed-bed reactor*, the fluid moves either in upflow or downflow past a bed of packed media. An example of a packed-bed reactor is a trickling filter. A *fluidized-bed reactor* is similar to a packed-bed reactor except that the packed media is expanded by the upward movement of the fluid through the media. An example of a fluidized-bed reactor is air stripping.

Reaction Kinetics and Reactor Hydraulics

In complete mixing, the chemical (such as a coagulant) entering the reactor is immediately distributed throughout by impellers. If the mixing basin is small, it may be known as a rapid mixer. In plug-flow reactors, the water flows through a long chamber, the chemical is added at the inlet (as a "plug") and there is no mechanical agitation to aid mixing.

In the treatment of wastewater, operations in which some transformation is effected through the use of chemical reactions are termed chemical unit operations. The rate at which a chemical reaction progresses depends on the nature of the reaction. Broadly, chemical reactions may be classified as zero-order, first-order, second-order, etc. The basic characteristics of different types of reactions are given below.

Zero-Order Reactions

Zero-order reactions proceed at a rate which is independent of the concentration of the reactants. This may be expressed as

$$\frac{dC}{dt} = -k \qquad \text{leading to} \qquad C(t) = C(0) - kt \tag{16.18}$$

If concentration data are available at various times, a zero-order reaction would exhibit the following property: equal intervals of time would lead to equal changes in concentration. The plot of concentration $C(t)$ versus time is linear (slope k).

First-Order Reactions

A reaction is classified as a first-order reaction if the rate of reaction is directly proportional to the concentration of the reactant. This may be expressed as

$$\frac{dC}{dt} = -kC \qquad \text{leading to} \qquad C(t) = C(0)e^{-kt} \tag{16.19}$$

If concentration data are available at various times, a first-order reaction would exhibit the following property: equal intervals of time would lead to equal changes in ln C. The plot of ln C versus time is linear (slope k).

Second-Order Reactions

A reaction is classified as a second-order reaction if the rate of reaction is proportional to the square of the concentration of the reactant. This may be expressed as

$$\frac{dC}{dt} = -kC^2 \qquad \text{leading to} \qquad \frac{1}{C(t)} = \frac{1}{C(0)} + kt \tag{16.20}$$

If concentration data are available at various times, a second-order reaction would exhibit the following property: equal intervals of time would lead to equal changes in $1/C$. The plot of $1/C$ versus time is linear (slope k)

Saturation Reactions

A reaction is classified as a saturation reaction if the rate of reaction saturates as the reaction progresses. This may be expressed as

$$\frac{dC}{dt} = \frac{kC}{a+C} \qquad \text{leading to} \qquad a\ln\frac{C(0)}{C(t)} + C(0) - C(t) = kt \tag{16.21}$$

The plot of $1/t \ln(C_o/C_t)$ versus $(C_o - C_t)/t$ is linear (slope k).

Table 16.1 shows a summary of the required hydraulic detention time for complete-mix and plug-flow reactors for various reaction orders.

The variables in Table 16.1 have the following meaning

C_o = concentration in the influent

C_e = required concentration in the effluent

K = reaction coefficient

K_s = half saturation coefficient

Table 16.1 Required Hydraulic Detention Time for Various Orders of Reaction

Reaction order	Required hydraulic detention time	
	Complete-mix reactor	Plug flow reactor
Zero order	$\frac{C_o - C_e}{K}$	$\frac{C_o - C_e}{K}$
First order	$\frac{1}{K}\left(\frac{C_o}{C_e} - 1\right)$	$\frac{1}{K}\ln\left(\frac{C_o}{C_e}\right)$
Saturation	$\frac{(C_o - C_e)(K_s + C_e)}{KC_e}$	$\frac{\left(K_s \ln\left(\frac{C_o}{C_e}\right) + C_o - C_e\right)}{K}$

Physical Unit Operations in Wastewater Treatment

In the treatment of wastewater, operations in which some transformation is effected through the use of physical forces are termed physical unit operations. Some of the physical unit operations used for wastewater treatment are (1) screening, (2) mixing and flocculation, (3) grit removal, and (4) primary sedimentation.

Mixing and Flocculation

Mixing is employed in many parts of wastewater treatment to improve the efficiency of unit processes. Most mixing operations in wastewater treatment can be classified as continuous-rapid mixing (less than 30 s) or continuous mixing.

Power Requirements for Mixing

The energy imparted by impeller blades causes turbulence in the fluid, thereby inducing mixing. According to studies by Camp and Stein (1943), the power requirement of an impeller in a mixing operation is given by

$$P = G^2 \mu V \qquad (16.22)$$

where G = the desirable value of the velocity gradient (T^{-1})
μ = dynamic viscosity of the fluid (FTL^{-2})
V = volume of the tank (L^3)

For a turbulent flow impeller mixer, the power requirement is written as

$$P = K_T \rho n^3 D^5 \qquad (16.23)$$

where P = power requirement for mixer
K_T = impeller constant (also called power number) (dimensionless, provided by manufacturer; see table on page 196 of the *Handbook*)
ρ = fluid density
n = revolutions per second of impeller blades
D = diameter of impeller

The power requirement for a paddle system can be written as

$$P = \frac{1}{2} C_D A \rho v_P^3 \qquad (16.24)$$

where C_D = drag coefficient of the paddle moving perpendicular to the fluid
A = paddle area
ρ = fluid density
v_P = relative velocity of the paddles with respect to the fluid, typically assumed to be about 70% of the tip-speed of the paddles

Example 16.6

A rapid mixer has impellers that deliver an effective power = 25 hp to the fluid, whose dynamic viscosity is approximately the same as that of water at 15°C. If the chamber volume of the mixer = 5 m^3, what is the velocity gradient?

A. 1500 s^{-1}

B. 1650 s^{-1}

C. 2050 s^{-1}

D. 3100 s^{-1}

Solution Power $P = 25$ hp $= 25 \times 745.7 = 18{,}643$ Nm/s

Viscosity $\mu = 0.00139$ Pa·s (page 114 of the *Handbook*)

$$G = \sqrt{\frac{P}{\mu V}} = \sqrt{\frac{18643}{0.00139 \times 5}} = 1638\ \text{s}^{-1}$$

Answer is B.

Flocculation

Flocculation typically follows rapid mixing where chemicals (coagulants) have been added to destabilize the repulsive forces between particles. Flocculation may be divided into (1) microflocculation, for particles in the size range 0.001 to 1 μm, which is the aggregation of particles induced by random thermal motion of the fluid, also known as Brownian motion, and (2) macroflocculation, for particles larger than about 1 μm, which is induced by velocity gradients or differential settling. A velocity gradient imparted to the fluid causes faster moving particles to collide with slower moving particles, thus creating larger particles that are easier to remove. In differential settling, larger particles tend to settle faster than smaller particles and therefore collide and coalesce with the smaller particles, thereby forming larger particles.

Settling of Particles

Particle settling can be subdivided into four types: discrete particle settling, flocculent settling, hindered or zone settling, and compression settling.

Type I Settling—Discrete Particle Settling

Particles that settle discretely without any flocculation settle according to the laws of sedimentation by Newton and Stokes. For spherical articles, the terminal settling velocity is calculated by equating the gravitational force to the drag force. This yields Newton's law for Type I settling

$$V_t = \sqrt{\frac{4g(\rho_p - \rho_f)d}{3C_D\rho_w}} \tag{16.25}$$

where V_t = terminal settling velocity of the particle
d_p = particle diameter
ρ_p = density of the particle
ρ_f = density of the fluid
C_D = drag coefficient (dependent on Reynolds number)

Type I settling occurs commonly in presedimentation for sand removal prior to coagulation, settling of sand and other particle during cleaning of rapid sand filters and grit chambers in wastewater treatment plants.

Laminar Region For low Reynolds number (Re < 1.0), we have Stokes flow, for which the drag coefficient is given by $C_D \approx \frac{24}{\text{Re}}$ leading to

$$V_t = \frac{g(\rho_p - \rho_f)d^2}{18\mu} \tag{16.26}$$

Transition Region For intermediate values of Reynolds number ($1.0 < \text{Re} < 10^4$), the drag coefficient is given by

$$C_D = \frac{24}{\text{Re}} + \frac{3}{\sqrt{\text{Re}}} + 0.34 \tag{16.27}$$

Turbulent Region In the turbulent region ($\text{Re} > 10^4$), the coefficient of drag is often taken as 0.4, leading to

$$V_t = \sqrt{\frac{10g(\rho_p - \rho_f)d}{3\rho_f}} \tag{16.28}$$

Percent Removal of a Particular Size by Settling For discrete particle settling, all particles having settling velocity greater than the overflow rate $V^* = Q/A$ will be removed entirely. Finer particles will be partially removed according to

$$\%\text{removal} = \frac{V_s}{V^*} \times 100\% \leq 100\% \tag{16.29}$$

where V_s = settling velocity of chosen particle size
V^* = overflow rate = Q/A
Q = influent flow rate
A = surface area of the sedimentation tank

The hydraulic detention time t_d is given by

$$t_d = \frac{\text{tank volume}}{\text{flow rate}} \tag{16.30}$$

The design depth of the sedimentation tank is given by

$$D = V^* t_d \tag{16.31}$$

Example 16.7

What is the settling velocity for sand particles with nominal diameter = 0.10 mm. Assume specific gravity SG =2.65 and drag coefficient C_D = 10.0.

Solution

$$V_s = \sqrt{\frac{4gd}{3C_D}\left(\frac{\rho_p - \rho_f}{\rho_f}\right)} = \sqrt{\frac{4 \times 9.81 \times 10^{-4}}{3 \times 10} \times 1.65} = 0.0147 \text{ m/s}$$

Example 16.8

What is the percent removal of the 0.1 mm particle (in Example 16.7) if a flow rate 0.1 m^3/s was flowing through a grit chamber of diameter 2 m?

Solution The overflow rate is calculated as $V^* = \dfrac{0.1}{\frac{\pi}{4} \times 2^2} = 0.0318 \text{ m/s}$.

From the previous example, the settling velocity for the 0.1 mm size is 0.0147 m/s. Therefore, the percent removal for this particle size is given by

$$\%R = \frac{0.0147}{0.0318} = 0.46 = 46\%$$

Type II Settling (Flocculent Settling)

For particles which are too fine to settle by discrete settling, addition of a coagulant causes particles to flocculate as they settle. Particle size changes continuously due to flocculation. As a result, settling rates also change continuously. In waste water treatment plants, floc from ferric chloride or alum coagulation in primary sedimentation tanks and secondary settling tanks follows Type II settling.

Type III Settling (Hindered or Zone Settling)

Particles present in high concentration tend to settle as a mass, creating a clear zone above the sludge mass. In water treatment, sludge from lime softening exhibits Type III settling. In waste water treatment plants, Type III settling occurs in activated sludge and gravity thickeners.

Type IV Settling (Compression Settling)

As settling progresses, a compressed layer of particles forms on the bottom of the tank. There is continuous variation of solids concentration in the compression zone.

Primary Sedimentation Tanks

The objective of primary sedimentation is to remove approximately 50% to 70% of suspended solids and approximately 25% to 40% of the BOD. Primary sedimentation tanks are rectangular or circular concrete tanks that are mechanically cleaned. Typical dimensions of rectangular tanks are: 80 to 120 ft long, 15 to 30 ft wide, and 15 ft deep. Typical diameter of circular primary sedimentation tanks is between 40 and 150 ft. Typical design parameters for sedimentation tanks are summarized in the table on page 194 of the *Handbook*.

The overflow rate is defined as $V^* = \frac{Q}{A}$, where A is the surface area of the tank.

The settling time for a particular size particle is given by $t_s = H/V_s$, where V_s is the settling velocity and H is the depth of the tank. Particles whose settling velocity is greater than the overflow rate (or whose settling time is shorter than the detention time $t_d = V_{\text{tank}}/Q$) are presumed to be removed completely. Finer sizes are removed at a removal rate given by Eq. (16.29).

Example 16.9

A primary clarifier receives a wastewater flow rate of 2 MGD containing 350 mg/L of total suspended solids. The suspended solids removal efficiency of the clarifier is 65%. The primary sludge contains 3.5% solids. Calculate (1) the volumetric flow rate of primary sludge (in gal/day) and (2) the mass of suspended solids removed (lb/day).

Solution

Approach 1

If 65% of incoming solids are being removed, then the solids removed is $0.65 \times 350 = 227.5$ mg/L. The daily solids removed is given by

$$W_{\text{solids}} = 227.5 \times 2 \times 8.3454 = 3797 \text{ lb/day}$$

Since the solids represent 3.5% of the primary sludge, the weight of the sludge is

$$W_{\text{sludge}} = \frac{3797}{0.035} = 108490 \text{ lb/day}$$

At 3.5% solids, the sludge is mostly water, and therefore the unit weight is 8.34 lb/gal

$$V_{\text{sludge}} = \frac{108{,}490 \text{ lb/day}}{8.34 \text{ lb/gal}} = 13{,}008 \text{ gal/day}$$

Approach 2

If 65% of incoming solids are being removed, then the concentration of solids going through is $0.35 \times 350 = 122.5$ mg/L.

Solids fraction = 3.5% = 35 g/1000 g which is approximately equal to 35 g/L (since the sludge is mostly water) = 35,000 mg/L. One may then draw the following quantities flowing into and out of the clarifier.

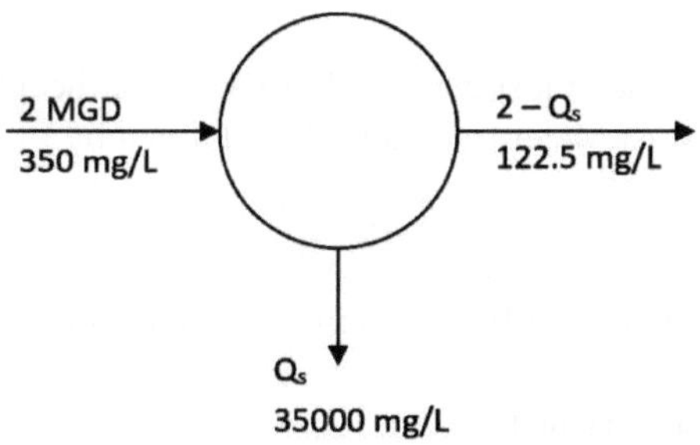

Balancing solids (entering versus leaving), we get

$$2 \times 350 = (2 - Q_s) \times 122.5 + Q_s \times 35{,}000 \Rightarrow Q_s = 0.01305\,\text{MGD} = 13{,}050 \text{ gal/day}$$

The two approaches have less than 0.3% disagreement.

Absorption versus Adsorption

Separation operations involve the transfer of material from one homogeneous phase to another, thereby reducing the concentration (of the substance to be removed) in one stream and increasing the concentration in another. Mass transfer of the substance relies on the inequilibrium between these streams and therefore ceases when equilibrium is reached. Two examples of such mass transfer processes are (1) absorption—the transfer of a substance across a gas-liquid interface as in air-stripping and (2) adsorption—the transfer of a substance across a liquid–solid interface as in adsorption onto granulated activated carbon (GAC).

Activated Carbon Adsorption

Adsorption, the binding of molecules or particles to a surface, is different from *absorption*, the filling of pores in a solid. The binding to the surface is usually weak and reversible. The most common industrial adsorbents are activated carbon, silica gel, and alumina, because they present very large surface areas per unit weight. Activated carbon is produced by roasting organic materials (such as wood, coconut shell) to decompose it to granules of carbon. Silica gel is a matrix of hydrated silicon dioxide. Alumina is mined or precipitated aluminum oxide and hydroxide. The black color of activated carbon adds a gray tinge if even trace amounts are left after treatment; which can be mitigated by using filters with fine pores.

Adsorption progresses in four distinct stages: (1) movement of the organic material to be adsorbed through the bulk liquid to the fixed film of liquid surrounding the adsorbent, (2) diffusion of the organic material through the film, (3) transport of the organic material through the pores of the adsorbent, and (4) the attachment of the material to an available adsorption site.

The quantity of adsorbate that can be adsorbed by an adsorbent is a function of the type and concentration of the adsorbate as well as the temperature and this function is called an adsorption isotherm. Two such functions—the Freundlich isotherm and the Langmuir isotherm—are discussed below.

Freundlich Isotherm

The Freundlich isotherm is used most commonly to describe the adsorption characteristics of activated carbon used for wastewater treatment. It is defined by

$$\frac{x}{m} = K_f C_e^{1/n} \tag{16.32}$$

where x/m = mass of adsorbate adsorbed per unit mass of adsorbent (mg adsorbent/g activated carbon)
K_f = Freundlich capacity factor
C_e = equilibrium concentration of adsorbate in solution after adsorption (mg/L)
$1/n$ = Freundlich intensity parameter

The ratio x/m is also related to the adsorbent dose by

$$\frac{x}{m} = \frac{V}{M}(C_o - C_e) \tag{16.33}$$

where V = volume of liquid in the reactor (L)
M = mass of adsorbent (g)
C_o = initial concentration of adsorbate (mg/L)
C_e = equilibrium concentration of adsorbate in solution after adsorption (mg/L)

Example 16.10

A chemical formed as a byproduct of wastewater treatment is to be removed/mitigated by GAC adsorption. The Freundlich isotherm coefficients are: K_f = 3.6 and $1/n$ = 0.70. If the original concentration of the chemical is 0.10 mg/L and the concentration in the effluent is to be 0.03 mg/L, how much GAC is required (lb/day) to treat 2.6 MGD of wastewater?

Solution $C_o = 0.10$ mg/L; $C_e = 0.03$ mg/L

$$\frac{x}{m} = K_f C_e^{1/n} = 3.6 \times 0.03^{0.7} = 0.309$$

$$\frac{x}{m} = \frac{V}{M}(C_o - C_e) = 0.309 \Rightarrow \frac{V}{M} = \frac{0.309}{0.10 - 0.03} = 4.414 \text{ L/g} \Rightarrow \frac{M}{V} = 0.227 \text{ g/L} = 227 \text{ mg/L}$$

Thus, the quantity of adsorbent (GAC) required = 2.6 MGD × 227 mg/L × 8.3454 = 4925 lb/day.

Langmuir Isotherm

The Langmuir isotherm, also used to describe the phenomenon of adsorption, is given by

$$\frac{x}{m} = \frac{abC_e}{1 + bC_e} \tag{16.34}$$

where x/m = mass of adsorbate adsorbed per unit mass of adsorbent (mg adsorbent/g activated carbon)
a, b = Langmuir empirical constants
C_e = equilibrium concentration of adsorbate in solution after adsorption (mg/L)

Coagulation

Untreated surface waters contain many inorganic and organic materials such as clay, minerals, bacteria, inert solids, microbiological organisms, oxidized metals, and other suspended solids. All of these particles can cause cloudiness (turbidity) of the water. Besides lowering the aesthetic quality of the water, turbidity is associated with the formation of cells that can contain pathogens and shield them from treatment processes. The purpose of coagulation is to remove these particles. Whereas larger particles may be removed physically by gravity settling within a reasonable length of time, colloidal particles ranging in size from 0.001 to 1 μm in diameter require long periods for complete settling. Since detention times in water treatment processes are typically less than 12 h, the rate of settling of these colloidal particles must be increased. This is accomplished in the coagulation process when tiny particles agglomerate into larger, denser "superparticles" which settle more quickly.

Coagulation involves three basic steps: charge destabilization, flocculation, and sedimentation. Colloidal particles in suspension have negative charges. These charges produce electrical repulsion forces that keep them in suspension. Addition of an appropriate cation (such as Al^{3+}) reduces the charge on these particles—causing formation of floc. This process is called flocculation.

The conventional choices for appropriate coagulants with appropriate characteristics are alum ($Al_2SO_4 \cdot 14H_2O$), ferric chloride ($FeCl_3$), or ferric sulphate ($Fe(SO_4)_3 \cdot H_2O$).

Properties of a Good Coagulant

A good coagulant must be nontoxic, be insoluble in water within the expected pH range and have a high-charge density. Some compounds that have these properties are aluminum sulphate $Al_2(SO_4)_3 \cdot xH_2O$ (effective for water containing organic matter), $FeSO_4 \cdot 7H_2O$, ferric chloride $FeCl_3$, and ferric sulphate $Fe_2(SO_4)_3 \cdot H_2O$.

Factors Influencing Coagulation

The pH range in which a coagulation process occurs is the single most important factor in satisfactory progress of the coagulation process. The presence of some anions such as sulphate (SO_4^{2-}) or phosphate (PO_4^{3-}) in a water supply typically causes the optimum pH range to shift lower. Thus, abrupt changes in the concentrations of these anions can disrupt the coagulation process. Generally, higher turbidity levels require higher coagulant dosages. On the other hand, water with low turbidity can be very difficult to coagulate due to the difficulty in inducing collision between the colloids. Low water temperatures can cause retarded rates of coagulation, due to lower chemical reaction rates and slower floc settling velocities at lower temperatures. Poor or inadequate mixing, resulting in an uneven dispersion of the coagulant, can affect overall coagulation effectiveness. The choice of the proper coagulant for the given conditions is of critical importance for efficient coagulation. The chemicals most commonly used in the coagulation process are aluminum sulphate, ferric chloride, ferric sulphate, and cationic polymers.

Table 16.2 shows chemicals commonly used for chemical coagulation and precipitation.

Table 16.2 Chemicals Commonly Used for Chemical Coagulation

Chemical	Formula	Molecular weight	Equivalent weight
Alum	$Al_2(SO_4)_3 \cdot 14H_20$	594.4	99.1
Aluminum chloride	$AlCl_3$	133.3	44.4
Calcium hydroxide	$Ca(OH)_2$	74.1	37
Ferric chloride	$FeCl_3$	162.2	54.1
Ferric sulphate	$Fe_2(SO_4)_3$	399.9	66.6
Ferrous sulphate	$FeSO_4 \cdot 7H_20$	278.0	139.0
Sodium aluminate	$Na_2Al_2O_4$	163.9	82.0

Alum (Aluminum Sulphate) as a Coagulant On a theoretical basis, 1.0 mg/L of dry alum will react with

- 0.50 mg/L of natural alkalinity as calcium carbonate
- 0.33 mg/L of 85% quicklime as calcium oxide
- 0.39 mg/L of 95% hydrated lime as calcium hydroxide
- 0.54 mg/L of soda ash as sodium carbonate

Alum can be effective in the pH range of 5.5 to 8.0, but seems to work best in a pH range of 6.8 to 7.5.

Alum is typically shipped and fed in dry granular form, although it is also available as a powder or liquid slurry. Aluminum sulphate reacts with natural alkalinity to form aluminum hydroxide as the precipitate (floc).

$$Al_2(SO_4)_3 \cdot 14.3H_2O + 3Ca(HCO_3)_2 \rightarrow 2Al(OH)_3 + 3CaSO_4 + 6CO_2 + 14.3H_2O$$

Polymers Synthetic polymers are high molecular weight organic compounds that have multiple electrical charges along a long chain of carbon atoms. Depending on whether the ionic groups have positive or negative charges, the polymer may be classified as a cationic or anionic polymer, respectively. In the absence of electrical charges, the polymer is called a nonionic polymer.

In water and wastewater treatment, anionic and nonionic polymers are effective aids to coagulation. Polymers promote the formation of larger and tougher floc by a mechanism known as particle bridging. If used in conjunction with alum, a polymer can significantly reduce the alum dosage. In order for the polymer to be effective, thorough mixing (without hindering particle bridging) is necessary. Polymers are available dry (powdered or granular), emulsion (in oil), solution (in water), or Mannich polymers (water soluble). If dry polymers are used, the solution must be aged before use, so that the long molecular chains of the polymer "uncoil" and become effective.

Chemical Unit Processes in Wastewater Treatment

Chemical Neutralization

Wastewater treatment has many operations and processes that may create either excess acidity or excess alkalinity in the water. All waters that are either too acidic (low pH) or too alkaline (high pH) require neutralization. Some of the most common chemicals used for raising the pH (i.e., for treatment of acidic waters) are calcium carbonate, lime (available as quicklime or slaked hydrated lime), sodium hydroxide (caustic soda), and sodium carbonate (soda ash).

Some chemicals used for lowering the pH (i.e., for treatment of alkaline waters) are carbon dioxide, carbonic acid, hydrochloric acid, and sulphuric acid.

Disinfection

Disinfection (the removal or destruction of bacteria) follows secondary treatment of wastewaters to bring it into compliance with safe discharge limits that depend on its intended use. Disinfection may be accomplished via chemical agents such as chlorine, calcium or sodium hypochlorite, and ozone, via physical agents such as light, heat, and sound waves, via mechanical processes such as screening and settling or by radiation such as ultraviolet (UV) light and gamma radiation.

Chlorine as a Disinfectant

Chlorine is universally favored as a disinfectant because it possesses many of the desired properties—it is available at low cost in large quantities, it is moderately soluble in water (7 g/L at 1 atm and 15°C) and highly toxic to microorganisms if used in adequate concentration. However, while it is transported as a liquid under high pressure, it vaporizes if unconfined and can pose a health risk if released. Some of the byproducts of chlorination can be harmful.

The principal chlorine compounds used for disinfection are chlorine (Cl_2), sodium hypochlorite (NaOCl), calcium hypochlorite [$Ca(OCl)_2$], and chlorine dioxide (ClO_2).

Stages in the Chlorination Process The various stages in chlorination can be generally described as follows:

1. The initial chlorine added is destroyed by reacting with reducing compounds.
2. After the demand exerted by the reducing compounds is met, the chlorine residual increases in the form of chloro-organics and chloramines.
3. Chlorine residual decreases as chloro-organics and chloramines are partly destroyed. This will continue until the breakpoint, when no ammonia remains.
4. Chlorine added after breakpoint will contribute to the free chlorine residuals (such as Cl_2, HOCl, OCl^-).

Chlorine Demand Added chlorine can be "used up" in many ways without producing significant disinfection. The term breakpoint chlorination is used for the process where enough chlorine is added to the water to react with all the oxidizable compounds so that any additional chlorine appears as free available chlorine.

The appearance of *free chlorine* is a signal that all chemical reactions (using chlorine) and disinfection are complete. Chlorine demand or dose is the amount of chlorine required to leave a desired chlorine residual (0.5 mg/L) 15 min after mixing. Using a very high chlorine dose can lead to the formation of THM.

Log Inactivation

Log inactivation levels are related to removal efficiency. For example, for 99% removal efficiency, the remaining fraction is 0.01. This corresponds to 2-log inactivation since

$$-\log_{10} 0.01 = 2$$

As listed on page 198 of the *Handbook*, the following log inactivation levels are mandated.

Microorganism	Required log inactivation	Removal efficiency
Giardia	3-log	99.9%
Viruses	4-log	99.99%
Cryptosporidium	2-log	99%

Also listed on that page are typical log inactivation credits that are issued following previous treatment stages such as (1) conventional treatment, (2) direct filtration, (3) slow sand filtration, and (4) diatomaceous earth filtration. The net difference between the overall inactivation requirement (3.0 for Giardia, 4.0 for viruses) and these credits will be required during the actual disinfection process.

Chlorination effectiveness depends on, and is often tabulated by the CT value, which is the product of the concentration of free available chlorine (C) and the contact time (T). Typical units are mg/L and minutes.

CT values for 3-log inactivation of Giardia cysts by free chlorine are tabulated on page 199 of the *Handbook*. The same page also lists some typical CT values for 4-log inactivation of viruses by free chlorine.

The time (T) that the water is effectively in contact with the disinfectant is calculated as the theoretical detention time (TDT) times the baffling factor (BF)

$$T = \text{TDT} \times \text{BF} \tag{16.35}$$

BFs (listed on page 197 of the *Handbook*) range from 0.1 (for unbaffled mixed flow) to 1.0 (plug flow). Also note that, as of the current printing, there is some ambiguity about the meaning of TDT on that page. The second definition of TDT equates it with the CT product, which is not true.

Example 16.11

A baffled chlorination chamber (BF = 0.8) has volume 1000 ft^3. The free chlorine concentration is 0.8 mg/L. If the flow rate into the chamber is 0.9 MGD, what is the contact time (minutes)?

A. 10 min

B. 12 min

C. 15 min

D. 18 min

Solution Flow rate $Q = 0.9$ MGD $= 0.9 \times 1.5472 = 1.3925$ ft³/s
The TDT is calculated as

$$\text{TDT} = \frac{V}{Q} = \frac{1000}{1.3925} = 718.1 \text{ s} = 11.97 \text{ min}$$

The actual (or effective) contact time:

$$T = \text{TDT} \times \text{BF} = 11.97 \times 0.8 = 9.58 \text{ min}$$

Answer is A.

This would make the actual CT value realized at this chamber = 0.8 × 9.58 = 8.0 min·mg/L

Example 16.12

What is the required CT value for 3-log inactivation of Giardia cysts using the following data:

Chlorine dose = 2.0 mg/L

Chlorine demand = 1.2 mg/L

Water temperature = 20°C

pH of water = 6.5

A. 42 mg/L·min

B. 46 mg/L·min

C. 48 mg/L·min

D. 52 mg/L·min

Solution Free chlorine = 2.0 − 1.2 = 0.8 mg/L
From page 199, for free chlorine concentration = 0.8 mg/L, pH = 6.5, and $T = 20$°C, the required CT value = 46 mg/L·min

Answer is B.

Biological Wastewater Treatment

Biological processes used for wastewater treatment can be broadly divided into two categories: suspended growth and attached growth processes. Examples of suspended growth processes are activated sludge, aerated lagoons, aerobic digestion, etc. Examples of attached growth processes are trickling filters, rotating biological contactors, packed bed reactors, etc.

In suspended growth processes, the microorganisms are maintained in liquid suspension at sufficient concentration to effect treatment. These processes can be either aerobic or anaerobic. The most common suspended growth process of the aerobic variety is the activated sludge process where the wastewater with the microbial suspension—known as the mixed liquor suspended solids (MLSS) or mixed liquor volatile suspended solids (MLVSS)—is aerated in an aeration tank. Mechanical equipment is used to facilitate mixing and air entrainment. The activated sludge process produces floc particles that can be removed by gravity settling in a secondary clarifier. Removal efficiency for the secondary clarifier is typically around 99%.

In attached growth processes, the microorganism population resides on the surface of an inert packing material. Packing materials include rock, gravel, sand to a wide range of plastic and other synthetic materials. The microbe layer is called the biofilm. The most common attached growth process is the trickling filter, in which the water is distributed (typically using a radial rotating arm) over the top of a bed of packing material. Air circulates in the void spaces in the bed and provides oxygen for the microorganisms to facilitate their metabolic processes and "consume" the organic material in the influent wastewater.

Biochemistry, Biology, and Bacteriology

Wastewater engineering involves designing processes and facilities that utilize microorganisms to destroy organic and inorganic substances. Metabolism is a term used to describe all chemical activities in the cell. Those processes that allow the bacterium to synthesize new cells from stored energy (ATP: adenosine triphosphate) are called anabolic. All processes where the cells convert substrate into useful energy and waste products are called catabolic.

To grow and reproduce, microorganisms need an energy source, a carbon source for the synthesis of new cellular material and inorganic nutrients such as nitrogen, phosphorus, potassium, calcium, magnesium, and organic nutrients (growth factors). Carbon is present in large quantities in wastewater. Treatment of wastewater involves converting this carbon into cellular material in microorganisms. This cellular material then forms a sludge that can be removed by settling. Thus, organisms that use organic material are encouraged.

Microorganisms derive their carbon from either organic matter (heterotrophs or organotrophs) or from carbon dioxide (autotrophs). Organisms that use inorganic material as a carbon source are also called lithotrophic. The energy needed for metabolic processes is derived from light (phototrophs) or from chemical oxidation–reduction reactions that transfer electrons from an electron donor to an electron acceptor (chemotrophs). In these reactions, the electron donor gets oxidized and the electron acceptor gets reduced. When oxygen is the electron acceptor, the reaction is termed aerobic. Chemical end products of aerobic decomposition are carbon dioxide, water, and cellular material. Organisms that can only exist in an environment that is devoid of oxygen generate energy by fermentation and are called obligate anaerobes. On the other hand, facultative anaerobes have the ability to grow in either the presence or absence of molecular oxygen. The term anoxic is used for organisms that use nitrite and nitrate ions as the electron acceptor. These ions get reduced to nitrogen gas, a process known as denitrification. Microaerophiles are organisms that may use oxygen, but only at low concentration. Microaerophiles carry out aerobic respiration, and some of them can also do anaerobic respiration. Aerotolerant organisms can survive in the presence of oxygen, but they are anaerobic because they cannot use it. Almost all animals, most fungi and several bacteria, are obligate aerobes. Most anaerobic organisms are bacteria. An example of a facultative aerobe is yeast.

Obligate (strict) anaerobes die in the presence of oxygen. Obligate anaerobes may use fermentation or anaerobic respiration. In the presence of oxygen, facultative anaerobes use aerobic respiration; without oxygen some of them ferment, some use anaerobic respiration. Aerotolerant organisms are strictly fermentative.

For dilute wastewater (BOD_5 < 500 mg/L), aerobic oxidation is preferred because decomposition is rapid and efficient. For strong (BOD_5 > 1000 mg/L) wastewater, aerobic decomposition requires large quantities of oxygen and sludge production is very high. For this reason, aerobic decomposition may not be the preferred mechanism for strong wastewaters. Conditions in a wastewater treatment plant are adjusted so that chemoheterotrophs (such as bacteria) predominate.

Effect of Temperature and pH

Temperature and pH have an important effect on the growth and survival of microorganisms. Psychrophiles or cryophiles grow in the temperature range of 10 to 30°C (optimum range 12 to 18°C), mesophiles grow in the temperature range of 20 to 50°C (optimum range 25 to 40°C), thermophiles grow in the temperature range of 35 to 75°C (optimum range 55 to 65°C) while ultrathermophiles grow in the temperature range of 60 to 80°C. For example, *Escherichia coli*, a mesophile, operates best in the range 20 to 50°C, but can continue to reproduce at temperatures around 0°C.

Most bacteria cannot tolerate pH values in excess of 9.5 or less than 4.0. Generally, the optimum pH for bacteria is in the range 6.5 to 7.5.

Fungi are aerobic, multicellular, nonphotosynthetic, heterotrophic eukaryotes. Most fungi are obligate or facultative aerobes. Molds (true fungi) produce a filamentous mass called the mycelium. Yeasts are unicellular fungi that cannot form a mycelium. Fungi are able to grow in low moisture, low nitrogen environments and are tolerant of low pH (acidic) environments. Fungi release CO_2 and N_2 during breakdown of organic material. Fungi use half the amount of nitrogen as bacteria for formation of cell material. (For this reason, they may replace bacteria as the dominant species in nitrogen-deficient wastewater.)

Bacteria are single-celled prokaryotes which reproduce primarily by binary fission. Their cytoplasm contains ribonucleic acid (RNA) which is responsible for protein synthesis and deoxyribonucleic acid (DNA) which contains all the information necessary for reproduction.

Algae are autotrophic, photosynthetic eukaryotes. They can be unicellular or multicellular. Algae and bacteria have a symbiotic relationship in aquatic systems, where algae produce the oxygen used by bacteria. In the presence of sunlight, photosynthetic production of oxygen is more than the amount used in respiration. At night, algae use up oxygen for respiration. Thus, long summer days result in a net production of oxygen. Excessive algal growth (algal blooms) can cause supersaturated oxygen conditions in daytime and anaerobic conditions at night. Algae produce turbidity in water. Some algae cause a smell and taste in natural water.

Protozoa are usually single-celled autotrophic eukaryotes. Most of protozoa are aerobic heterotrophs, while some are aerotolerant anaerobes and a few are anaerobic. They are desirable in wastewater effluent, as they act as polishers by consuming bacteria and particulate matter.

Viruses are parasitic organisms—pass through filters that retain bacteria—grow and reproduce inside a host but can survive outside. They are not cells, but a nucleic acid core (DNA or RNA) surrounded by a protein sheath. Viruses range in size from 10 to 25 nm (typical size found in water supply). Viruses are host specific. Bacteriophages are viruses that infect bacteria as the host.

Aerobic versus Anaerobic Biological Treatment

The primary advantages of anaerobic wastewater treatment (compared to aerobic processes) are low-energy requirement, less sludge production, smaller reactor volumes, and often a net production of energy. Also, an anaerobic process requires fewer nutrients than a comparable aerobic process and for this reason, it is often a better option for industrial wastewaters.

The primary disadvantages of anaerobic processes are as follows: need for addition of alkalinity, nitrogen and phosphorus removal is not possible, potential for production of odors, longer start-up times required to build up the necessary biomass, and the possibility of needing aerobic treatment to polish the effluent.

For anaerobic processes, the fundamental design parameter is the solids retention time (SRT). By definition, the SRT is the average time the solids in the activated sludge spend in the system. The SRT is calculated as the mass of solids in the system divided by the mass of solids removed per day.

In general, at temperatures around 30°C, SRT values are typically 20 days or more. Three types of anaerobic suspended growth treatment processes are (1) the complete-mix suspended growth anaerobic digester, (2) the anaerobic contact process, and (3) the anaerobic sequencing batch reactor. For the complete-mix process, the SRT and the hydraulic detention time t_d are equal, on the order of 15 to 30 days. This leads to very large reactor volumes compared to the other two types of processes.

Hydraulic Detention Time

The hydraulic detention time is purely a function of the flow rate being treated by the reactor and the volume of the reactor. It does not represent the organic load being imposed on the reactor. The hydraulic detention time is given by

$$t_d = \frac{V}{Q} \tag{16.36}$$

where V is the volume of the reactor and Q is the influent flow rate.

Example 16.13

What is the hydraulic detention time for a circular tank (diameter = 8 ft, height = 4 ft) that treats a flow Q = 15,000 gal/day. The hydraulic efficiency of the tank is 75%.

Solution Since the tank has a hydraulic efficiency of 75%, the detention time is less (by a factor of 0.75) of the ideal hydraulic detention time.

$$\text{Tank volume } V = \pi R^2 H = \pi \times 4^2 \times 4 = 201 \text{ ft}^3$$

Flow rate $Q = 15{,}000$ gal/day $= 0.0232$ ft^3/s

At 100% efficiency, detention time $t_d = \dfrac{201}{0.0232} = 8664\ \text{s} = 2.4\ \text{h}$

At 75% efficiency, detention time $= 0.75 \times 2.4 = 1.8$ h

Activated Sludge Process

The most fundamental of aerobic secondary treatment processes is the activated sludge process, which consists of the following basic components: (1) an aeration tank where a concentration of microorganisms is exposed to the influent wastewater and air, (2) a clarifier where solids (primarily organic solids) are settled, and (3) a recycle system to separate waste activated sludge from the return activated sludge, which is then returned to blend with the primary effluent. A schematic of the activated sludge process, showing its relative position in a wastewater treatment plant, is shown in Fig. 16.4.

Most activated sludge plants receive wastewaters that have been treated with primary sedimentation, where much of the settleable solids are removed. The secondary treatment is then responsible for removing soluble, colloidal, and suspended organic solids and nutrients such as nitrogen and phosphorus. The primary design elements for the activated sludge process are (1) the volume of the aeration tank, (2) the sludge production rate, and (3) oxygen requirements.

In the schematic of the activated sludge process shown in Fig. 16.4, the blending of the return activated sludge with the primary effluent produces the "mixed liquor," and the suspended solids concentration in it is called the "MLSS." The organic component of the MLSS is called the "MLVSS." In the process diagram, the MLVSS is designated as X.

X = volatile solids concentration in the aeration tank = MLVSS

V_a = volume of aeration tank

S_o = incoming (influent) BOD_5 (effluent from primary clarifier)

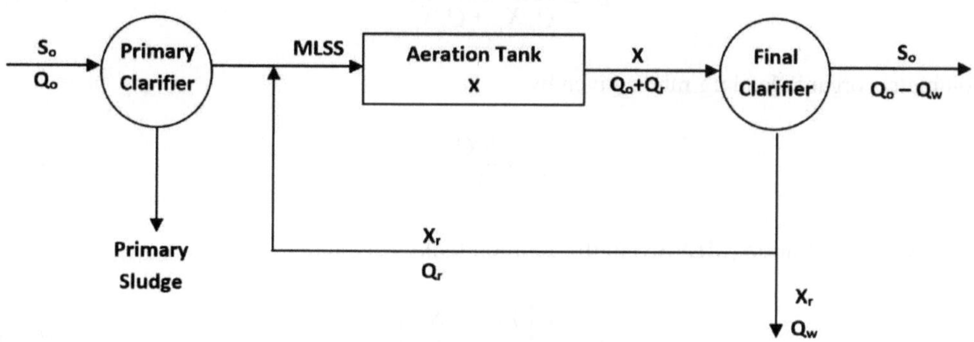

Figure 16.4 Schematic of the activated sludge process.

Q_o = incoming flow rate

MLSS = mixed liquor suspended solids

MLVSS = mixed liquor volatile suspended solids

Y = maximum cell yield ratio (mg VSS/mg BOD_5)

Y_{obs} = observed cell yield ratio (mg VSS/mg BOD_5)

f_d = fraction of biomass that remains as biomass (g · VSS/g · VSS)

The following critical parameters of the activated sludge process may be defined using the nomenclature shown above. Typical range of values of some of these parameters are listed in a table on page 191 of the *Handbook*.

Food arrival rate is given by

$$F = S_o Q_o \tag{16.37}$$

Mass of microorganisms is given by

$$M = V_a X \tag{16.38}$$

Food to microorganism ratio is given by

$$F : M = \frac{S_o Q_o}{V_a X} \tag{16.39}$$

Hydraulic detention time is given by

$$t_d = \frac{V_a}{Q_o} \tag{16.40}$$

Solids residence time is given by

$$\theta_c = \frac{V_a X}{Q_w X_w + Q_e X_e} \tag{16.41}$$

Volumetric organic loading rate is given by

$$L_{org} = \frac{S_o Q_o}{V_a} \tag{16.42}$$

Biomass concentration (MLVSS) in the aeration tank is given by

$$X = \frac{\theta_c}{\theta}\left[\frac{Y(S_o - S_e)}{1 + k_d \theta_c}\right] \tag{16.43}$$

Note that, as a result of the preceding definitions, one may also write the following relationship between the sludge production rate P and other key variables such as volume of aeration tank V_a, solids residence time SRT, and mixed liquor volatile suspended solids X

$$P \times \text{SRT} = V_a \times X \tag{16.44}$$

Typically, for BOD removal, the oxygen requirement varies from 0.9 to 1.3 kg O_2 per kg BOD removed for SRT between 5 and 20 days, respectively.

Mass Balance for a Closed System Boundary

Performing a mass balance around the aeration tank (boundary 1 in Fig. 16.5) yields

$$X_r Q_r + X_o Q_o = (Q_o + Q_r)X \tag{16.45}$$

Performing a mass balance around the secondary clarifier (boundary 2 in Fig. 16.5) yields

$$(Q_o + Q_r)X = Q_e X_e + Q_r X_r + Q_w X_w \tag{16.46}$$

Assuming the concentration X_e to be negligible, we have the following approximate relationships for the recycle ratio and the waste sludge

$$R = \frac{Q_r}{Q} = \frac{1 - \theta/\theta_c}{X_r/X - 1} \tag{16.47}$$

$$Q_w = \frac{V}{\text{SRT}} \tag{16.48}$$

Note from these equations that if sludge recycling is not used in a biological process ($R = 0$), the SRT θ_c and the hydraulic retention time θ are identical. This leads to a very large reactor volume for typical BOD removal goals. The purpose of using recycled sludge is to increase

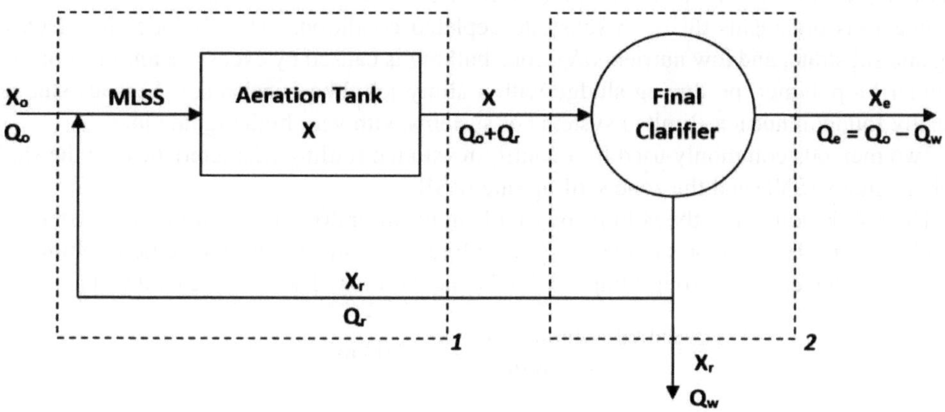

Figure 16.5 Mass balance for closed volumes.

the SRT to be significantly greater than the hydraulic retention time, thus keeping the reactor volume within practical limits.

Example 16.14

For an activated sludge process, the following data is given

Influent flow = 2 MGD

MLSS = 2500 mg/L

Sludge recycle ratio = 70%

Return activated sludge biosolids concentration = 6000 mg/L

Secondary clarifier—65 ft diameter × 12 ft deep

The solids loading rate to the secondary clarifier is most nearly:

A. 12 lb/day/ft^2

B. 17 lb/day/ft^2

C. 21 lb/day/ft^2

D. 25 lb/day/ft^2

Solution $Q_o = 2$ MGD; $Q_r = 0.7 \times 2 = 1.4$ MGD
Secondary clarifier surface area = $3.14 \times 32.5^2 = 3316.6$ ft^2
Solids load to the clarifier = 3.4 MGD × 2500 mg/L × 8.34 = 40,890 lb/day
Per unit surface area, solids load to the clarifier = 40,890 lb/day ÷ 3316.6 = 21.4 lb/day/ft^2

Answer is C.

Sludge Bulking When sludge shows poor settling characteristics in the clarifier, it may lead to poor system performance due to high levels of suspended solids in the effluent. With good settling characteristics, the sludge levels are typically 0.5 to 1.0 ft at the bottom of the clarifier. Two primary types of sludge bulking are filamentous bulking and viscous bulking. Filamentous organisms thrive in substrate depleted conditions. This includes low DO, low organic substrate, and low nutrients. Viscous bulking is caused by excessive amounts of extracellular biopolymer, producing sludge with a slimy jellylike consistency. Viscous sludge is usually found in nutrient-limited systems or systems with very high organic load.

Two methods commonly used for quantifying sludge settling characteristics are the sludge volume index (SVI) and the zone settling rate (ZSR).

The SVI is defined as the volume of 1 g of sludge after 30 min of settling. It is determined by placing a mixed liquor sample in a 1-L cylinder and measuring the settled volume after 30 min and the corresponding sample MLSS concentration. The SVI is calculated as

$$\text{SVI} = \frac{\text{settled volume (mL/L)}}{\text{MLSS (mg/L)}} \times 1000 \text{ mg/g} = \text{mL/g} \tag{16.49}$$

Typically an upper limit of 100 mL/g is considered desirable for good settling characteristics. SVI values above 150 mL/g are typically associated with filamentous bacteria.

Example 16.15

For a sludge, MLSS concentration is 3000 mg/L and the settled sludge volume after 30 min is 250 mL. What is the SVI?

A. 56 mL/g

B. 83 mL/g

C. 105 mL/g

D. 125 mL/g

Solution Sludge volume index: $\text{SVI} = \dfrac{\text{SSV}_{30} \times 1000}{\text{MLSS}} = \dfrac{250 \times 1000}{3000} = 83.3 \text{ mL/g}$

Answer is B.

Specific Gravity of Sludge

The SG of solid matter in sludge can be computed using

$$\frac{1}{S} = \frac{P_f}{S_f} + \frac{P_v}{S_v} \tag{16.50}$$

where S = specific gravity of sludge solids
S_f = specific gravity of fixed (inorganic) solids, usually assumed 2.6 (approx.)
S_v = specific gravity of volatile (organic) solids, usually assumed 1.0
P_f = fraction of fixed solids
P_v = fraction of volatile solids = $1 - P_f$

Example 16.16

A biological sludge contains 15% solids, with a volatile fraction of 60%. What is the SG of the sludge?

Solution The wet sludge may be considered 85% water (SG = 1.0), 6% (40% of 15%) fixed solids (SG = 2.6), and 9% (60% of 15%) volatile solids (SG = 1.0). Therefore, the SG of the sludge is

$$\frac{1}{S} = \frac{0.85}{1.0} + \frac{0.06}{2.6} + \frac{0.09}{1.0} = 0.963 \Rightarrow S = 1.038$$

Air Pollution

The concentration of a pollutant can be expressed either as parts per million, parts per billion, or as a concentration such as μg/m³.

$$\frac{\mu\text{g}}{\text{m}^3} = \text{ppb} \times \frac{P \times MW}{RT} \tag{16.51}$$

Figure 16.6 Migration of plume from stack.

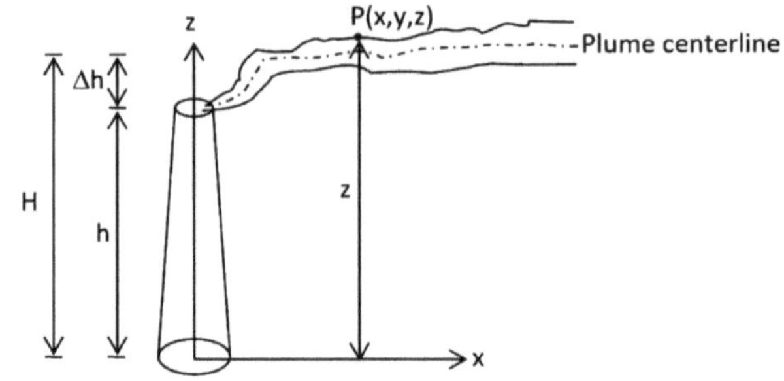

where P = pressure (atm)
R = ideal gas constant = 0.0821 L·atm/mol·K
T = absolute temperature (K)
MW = molecular weight (g/mol)

Pages 179 to 180 of the *Handbook* have a summary of the atmospheric dispersion model of a plume emerging from a stack of physical height h and effective height H, as shown in Fig. 16.6. Page 181 shows dispersion parameters σ_y and σ_z as a function of downwind distance x for six different stability classes (A to F). Page 182 shows the downwind distance x_{max} at which maximum concentration occurs as a function of effective stack height H (m) and stability class.

The maximum ground level concentration C_{max} is given by

$$C_{max} = \frac{Q}{\pi u \sigma_y \sigma_z} \exp\left(-\frac{1}{2}\frac{H^2}{\sigma_z^2}\right) \qquad (16.52)$$

where Q = emission rate (μg/s)
u = average wind speed at stack height (m/s)
σ_y = horizontal dispersion parameter (m)
σ_z = vertical dispersion parameter (m)
H = effective stack height = $h + \Delta h$ (m)

Example 16.17

A chimney stack of height 30 m emits a pollutant at the rate of 200 μg/s. The average wind speed at 30 m elevation is 10 m/s and the plume rise is 3 m. The atmospheric condition is overcast. Calculate the following:

A. The downwind distance at which the maximum concentration occurs (km)

B. The maximum downwind concentration (μg/s) at ground level

Solution For overcast conditions, use stability class D

Effective stack height = 30 + 3 = 33 m

From page 182 of the *Handbook*, $x_{max} = 0.7$ km = 700 m

From page 181, at a distance of 700 m, $\sigma_z = 25$ m and $\sigma_y = 55$ m

The maximum ground level concentration,

$$C_{max} = \frac{Q}{\pi u \sigma_y \sigma_z} \exp\left(-\frac{1}{2}\frac{H^2}{\sigma_z^2}\right) = \frac{200}{\pi \times 10 \times 55 \times 25} \exp\left(-\frac{1}{2}\frac{33^2}{25^2}\right) = 0.00194\ \mu\text{g/s}.$$

Cyclonic Separation

Cyclonic separation is a method of removing particulates from a gas or liquid stream through the mechanism of vortex separation. A high-speed rotating airflow is established within a cylindrical or conical container called a cyclone. Air flows in at the top of the cyclone, follows a helical path, and ends at the narrow bottom end before exiting the cyclone in a straight stream through the center of the cyclone. Larger (denser) particles in the rotating stream have too much inertia to follow the tight curve of the stream, and strike the outside wall, then fall to the bottom of the cyclone where they can be removed. In a conical system, as the rotating flow moves toward the narrow end of the cyclone, the rotational radius of the stream is reduced, thus separating smaller and smaller particles. The *cut point* of the cyclone is the particle size d_{pc} that will be removed from the stream with 50% efficiency. Particles larger than the cut point will be removed with a greater efficiency, while smaller particles with a lower efficiency. The cut point is defined by the geometry of the cyclone and the volumetric flow rate.

$$d_{pc} = \sqrt{\frac{9\mu W}{2\pi N_e V_i(\rho_p - \rho_g)}} \qquad (16.53)$$

where μ = dynamic viscosity of the gas

r_p = density of the particle

r_g = density of the gas

W = inlet width of cyclone

N_e = effective number of turns gas makes in the cyclone (function of cyclone geometry)

V_i = inlet velocity into the cyclone

The formula for efficiency of removal and its graphical representation are shown on page 183 of the *Handbook*.

Example 16.18

The percent removal of 30 μm particles in a cyclone precipitator is 50%. What is the percent removal of 20 μm particles?

A. 23%

B. 31%

C. 57%

D. 71%

Solution Efficiency: $\eta = \frac{1}{1+(d_{pc}/d_p)^2} = \frac{1}{1+(30/20)^2} = 0.31$

Answer is B.

Electrostatic Precipitator

In an electrostatic precipitator, fine particles such as dust and smoke are removed from a gas stream by inducing a charge to the particles and causing them to be collected through forces of electrostatic attraction. Compared to cyclones, electrostatic precipitators have significantly higher capital cost but much lower operating costs. They also have higher removal efficiency and are able to handle very large gas flow rates with low-pressure losses. Unlike the cyclone, which is limited to dry particles, the electrostatic precipitator can remove dry as well as wet particles. However, they do not work well on particles with high electrical resistivity.

The Deutsch-Anderson gives the efficiency of an electrostatic precipitator

$$\eta = 1 - \exp\left(-\frac{WA}{Q}\right) \tag{16.54}$$

where η = fractional collection efficiency
W = terminal drift velocity
A = total collection area
Q = volumetric flow rate

Example 16.19

The effective drift velocity for an electrostatic precipitator is 0.002 m/s. If the volumetric flow rate = 2000 m^3/min, what is the total required collection area (m^2) to achieve 99.9% removal efficiency?

Solution

$Q = 2000\ m^3/min = 33.33\ m^3/s$
$W = 0.002$ m/s

If $\eta = 0.999$, then $\exp\left(-\frac{WA}{Q}\right) = 0.001 \Rightarrow \frac{WA}{Q} = 6.91 \Rightarrow A = \frac{6.91 \times 33.33}{0.002} = 115{,}167\ m^2$.

Destruction and Removal Efficiency

For an incinerator, the destruction and removal efficiency (DRE) is a percentage that represents the number of molecules of a principal organic hazardous compound (POHC) removed or destroyed relative to the number of molecules that entered the system. (For example, a DRE of 99.99% means that 9999 molecules are destroyed for every 10,000 that enter.) The 99.99% is commonly referred to as "four nines efficiency."

$$\text{DRE} = \frac{W_{in} - W_{out}}{W_{in}} \times 100\% \tag{16.55}$$

Reverse Osmosis

When a semipermeable membrane separates two regions of a solvent with different solute concentrations, the process of osmosis causes the spontaneous movement of solvent across the membrane from the low to the high concentration side. This tends to equalize the solute concentration across the membrane, thereby establishing equilibrium.

Osmotic pressure refers to the external pressure required to be applied so that there is no solvent migration across the membrane. Osmotic pressure depends on the molar concentration of the solute.

Reverse osmosis is a filtration process often used to purify water. In reverse osmosis, an applied pressure is used to overcome osmotic pressure. It can be used to remove many types of dissolved and suspended matter from water. The result is that the solute is retained on the pressurized side and pure solvent is allowed to pass to the other side.

The osmotic pressure (Π) is given by

$$\Pi = \phi \nu \frac{n}{V} RT \tag{16.56}$$

where P = osmotic pressure (Pa)
ϕ = osmotic coefficient
ν = number of ions formed from one molecule of electrolyte
n = number of moles of electrolyte
V = specific volume of solvent m^3/kmol
R = universal gas constant = 8314 Pa · m^3/(kmol · K)
T = absolute temperature (K)

Solid Waste Management

A landfill site is a site for the disposal of waste materials by burial. Soil is then used to cover the waste in order to control odor and destabilization by vectors such as wind and pests such as rodents. In addition to disposal, some landfills can also be used for waste management activities such as waste storage and/or sorting, treatment, and recycling.

Waste received at the landfill is only slightly compacted (typical specific weight 1000 lb/yd^3), and must be compacted further (typically to around 2000 lb/yd^3) so that it occupies as small a volume as possible. After the waste is inspected, accepted, and deposited, bulldozers are used to spread and compact the waste. The working face of the landfill cell is covered with soil or alternative materials, typically on a daily basis. The final density of the waste is affected by factors such as waste compressibility, layer thickness, number of passes made by the compactor, etc.

Leachate

Landfill cells are usually lined, in an effort to capture and treat leachate before it reaches groundwater. A liner cannot prevent the leaking of leachate, but a well-engineered liner can slow down the migration considerably. The *Handbook* gives the following formula for the penetration time of leachate through a clay liner.

$$t = \frac{d^2 n}{K(d + h)} \tag{16.57}$$

where t = breakthrough time (year)
d = clay liner thickness (ft)
h = hydraulic head (ft)
n = porosity (typical value between 0.1 and 0.3)
K = coefficient of permeability (typical value 0.01 to 1.00 ft/year)

CHAPTER 17

Construction

Stages of Construction

The construction process is composed of six distinct stages, as outlined below:

Stage 1—Concept

Once a particular need has been identified, a preliminary conceptual design of the proposed facility must be created. At this stage, the broad requirements of the facility are defined, but not translated into specific design requirements. This preliminary design produces a list of solutions, alternatives, feasibility studies, and cost estimates. These are followed by preparation of schematic plans.

Stage 2—Bid Documents and Working Drawings

In order to solicit bids, the owner must provide a set of working drawings as well as proposed specifications. These drawings show the architect's design which specifies key dimensions, materials, etc. Engineers design specific elements of the concept according to their specialty area—structural, electrical, heating, ventilation, and air conditioning (HVAC).

The second part of the bid documents stage is the preparation of project specifications. The most common system for arranging construction specifications is the CSI (Construction Specification Institute) Master format, which requires four types of information:

1. Bidding requirements (invitation to bid, instructions to bidders, information available to bidders, bid forms and attachments, bid security forms).
2. Contract forms—the agreement, performance and payment bonds and certificates.
3. Contract conditions (general and supplementary).
4. Technical specifications of all items that will be installed in the project.

CSI format consists of the following 16 "Divisions of the Work":

Division 1—General requirements

Division 2—Site work

Division 3—Concrete

Division 4—Masonry

Division 5—Metals

Division 6—Woods and plastics

Division 7—Thermal and moisture

Division 8—Doors and windows

Division 9—Finishes

Division 10—Specialties

Division 11—Equipment

Division 12—Furnishings

Division 13—Special construction

Division 14—Conveying systems

Division 15—Mechanical

Division 16—Electrical

Each CSI division is further subdivided into three subparts: general, products, and execution.

Construction Documents

Construction documents set the parameters for the building process, once the design has been completed. They must clearly and completely describe all the components of a project that need to be built. At the same time, they must prescribe rights and responsibilities of all parties involved—typically the owner, engineer, and contractor.

Construction documents typically include the following types of information:

- Legal and contractual information
- Invitation to bid
- Instructions to bidders
- Bid forms contract forms
- Contract modifications (changes or additions after the contract is signed)
- Project information—surveys, borings, drawings of existing conditions
- Contract conditions (rights, responsibilities and duties of owner, engineer and contractor)
- Drawings, including definition of symbols, abbreviations, all necessary dimensions
- Specifications and standards (quality requirements to be met during construction)
- Addenda (additions or changes issued by the owner during the bidding or contract negotiation process)

Stage 3—Bidding

Once the owner is satisfied that the proposed project is feasible and affordable, a bid solicitation is put out publicly. Interested contractors obtain a copy of the plans and specifications and plans for the process of how to build the project. The primary contractor then solicits bids from any subcontractors that may be required to complete the project. The contractor then submits a bid to the owner. Once the bid deadline has passed, the owner formally opens the qualified bids and awards the contract. The contract documents will contain general information such as the start and completion dates, payment procedures, required insurance, etc.

Stage 4—Construction

Prior to engaging in actual construction, the contractor and/or the client must ensure that all necessary permits have been issued. The contractor and subcontractor(s) must also translate working drawings into more specific shop drawings after the final design phase but before actual construction begins.

Submittals

During construction, the contractor makes submittals to the owner and owner's representatives that outline specific design and construction materials and methods. Submittals are supposed to be approved by the owner's representative (engineer) within a specified length of time. Submittals sometimes go through multiple iterations (revise and resubmit) until they are approved.

Request for Information

Any clarifications needed about missing or incorrect information on the working drawings must be submitted by the contractor in the form of an RFI (request for information). RFIs are supposed to receive a formal response by the owner's representative (engineer) within a specified length of time.

Stage 5—Payments

The budget for the project is broken down into bid items. The contractor's bid consists of dollar figures—either unit rate or lump sum—for each of the bid items. At the end of every payment period (typically monthly), the contractor submits a payment estimate to the resident engineer for the project. The amount requested for each bid item is based upon the incremental progress for that item during the period of the estimate. Once the resident engineer (as the client's representative) and the contractor agree, this payment estimate forms the basis for a certified payment authorization. The contractor is responsible for staging work in order to maintain an adequate cash flow for the project.

Stage 6—Project Closure

At the conclusion of all project activities, the contractor "delivers" the project to the client. To ensure that the final product is to the client's satisfaction, certain tasks need to be performed by the resident engineer, acting on behalf of the client. This brings closure to the project and releases any funds held back in the form of "retainage." The following activities must take place in order to close out a project.

As Built Drawings

As-built plans must be updated to reflect conditions that were different than that shown on the project drawings. These could be revised location of utilities "as found" or changes in configuration of certain components, such as slight changes in orientation of a sewer pipe or exact location of a proposed bend.

Notice of Substantial Completion

Substantial completion is generally defined as the point at which the project is suitable for occupancy or use for its intended purpose. Once the notice of substantial completion is issued, the owner can no longer assess liquidated damages for late completion, nor withhold any payment, except costs associated with punch list items. Substantial completion is typically also the start of the warranty period.

Punch List

A punch list is a document prepared near the end of a construction project listing work that does not conform to the contract specifications and that the contractor must complete prior to full and final payment.

Notice of Completion

The notice of completion is a written notice issued by the owner of the project to notify all concerned parties that all work has been completed. This notice will also typically set out the time frame in which entities can file a lien against the owner. If the owner files the document within the required time limit (typically 15 days) from the date of completion, then the time allowed for the contractor to file a lien drops to as short as 30 days. Not recording the notice within the deadline extends the amount of time a contractor has to file a lien—prime contractor 90 days, subcontractors 60 days.

Procurement Methods

The procuring department is responsible for acquiring goods and services for a business. This may involve shopping for goods at competitive prices, handling all legal procedures associated with obtaining a contract, budgeting costs for the goods and studying financial trends to ensure that the owner's money is being spent wisely.

Generally speaking, there are six procurement methods used by the procurement team in a company. The type of procurement method chosen is highly relative to the conditions of the procurement effort and the type of goods or services being acquired. All procurement methods follow tight legal frameworks to ensure all standards are being met and quality in the selection process exists.

The six types of procurement are open tendering, restricted tendering, request for proposal (RFP), two-stage tendering, request for quotations (RFQ), and single-source (SS) procurement.

Open Tendering

Open tendering is basically another term for competitive bidding. It allows companies to bid on goods in an open competition. For this procurement method, the procurer must have clear technical specifications and objective evaluation measures. The competition must be openly advertised and be open to all qualified bidders. The award is typically made to the lowest bidder. The open tendering method of procurement encourages effective competition to obtain goods with an emphasis on the value for money. However, this method is not very suitable when the requirements are very complex. There may be complications in defining the exact needs of the requirement by the procuring company. The time required to procure goods and services via this method also can be excessive.

Restricted Tendering

Like open tendering, restricted tendering is considered a competitive procurement method. However, the competition is limited to agencies that are invited by the procuring team. The procuring entity should establish a set of guidelines to use when selecting the suppliers and service providers that will be on the invitation list. This method is selective to find the best-suited and most qualified agencies to procure goods and services from. Since the bids are submitted by a selected few qualified bidders, this method also saves the procurement team time and money due to a shorter selection process.

Request for Proposal

Request for proposal is a method used when suppliers or service providers are proposing their goods or services to a procurement team for review. Procurement teams are often searching for the best valued, most marketable items to bring into circulation. A well-qualified proposer submits a proposal to outline their qualifications and value for services. Usually the technical proposal is separated from the cost proposal so that the former can be evaluated solely based on merit.

Two-Stage Tendering

This procurement method has a two-stage process. This can be disadvantageous for some procurement teams if there is a time limit on securing a contract. On the other hand, this method allows for discussion between the two parties to meet mutual needs. Bidders are required to submit a technical proposal that highlights their solutions to fulfilling the requirements as specified by the procuring department.

These proposals are scored according to the relevance of the solution to the needs of the procurer. The highest scored proposal is invited for further discussion in an attempt to reach an agreement. After the final agreement for the technical proposal is reached, the bidder is invited to submit their financial proposal and then further discussions ensue to negotiate a contract.

Request for Quotations

This procurement method is used for goods or services of small value. Request for quotation is the least complex procurement method available. This method ensures a fast procurement process and not a lot of paperwork. There is no formal proposal drafted from either party in this method. Essentially, the procurement entity selects a minimum of three suppliers or service providers that they wish to get quotes from. The submitted quotes are analyzed and the best quotation (based on cost and requirement compliance) is chosen.

Single Source

Single source procurement occurs when the procuring entity intends to acquire goods or services from a sole provider. This method should undergo a strict approval process from management before being used. It is a noncompetitive method that should only be used under specific circumstances, such as (a) emergencies, (b) if only a single provider is available and qualified to fulfill the requirements, (c) continuation of previous work that cannot be reproduced by another provider.

Project Delivery Methods

A project delivery system is, simply, the contractual structure (exclusive of the financial arrangements) for how the final project is produced and delivered to the owner. The various components of the project are planning, design, permitting, constructing, testing, and accepting the final product.

In construction projects, there are three primary project delivery methods: (1) design-bid-build (DBB), (2) construction management at risk (CMAR), and (3) design-build (DB).

Design-Bid-Build

This traditional project delivery method involves three sequential stages: the design phase which requires the services of a designer who designs the project, the bidding phase where a contractor is chosen and a construction phase, where the chosen contractor builds the project.

The engineer chosen by the owner prepares the design, including drawings and specifications. These become components of the bid package. Once completed, the bid package is presented to interested contractors, who submit bids within a declared deadline. Usually, the lowest responsive and responsible bid is chosen.

The advantages of DBB are that it is widely applicable, well understood with clearly defined roles for all parties. The owner has a significant amount of control over the end product because all requirements and specifications are clearly defined before construction begins.

The timeline for design-bid-build projects may be longer because of being sequential, that is, design must be completed before bids are solicited. For long-duration projects, certain methods and technologies may become obsolete mid-project and require changes. In this structure, once the owner accepts the design from the designer, the owner also faces more exposure to change orders and claims over design and constructability issues.

Construction Management at Risk

This project delivery method utilizes the services of a construction manager at risk (CMAR), who serves as a general contractor during construction. The owner and the CMAR negotiate a guaranteed maximum price (GMP) and the CMAR holds the risk of construction performance and guarantees completion for this price. Any costs exceeding the GMP that are not change orders are the financial liability of the CMAR. This project delivery method offers the opportunity to start construction before design is completed.

The biggest disadvantage for this method is that once the GMP is fixed, the CMAR switches from an advisory role of construction manager to the contractual role of general contractor, possibly creating adversarial relationships over matters of construction quality.

Design-Build

In the design-build (DB) project delivery method, the owner enters into a single contract with the DB firm, which is responsible for both design and construction. The advantage to the owner is that there is a single entity responsible for the design and construction of the project. At some point early in the process, the DB team establishes a fixed price to complete the design and construction. In this method of project delivery, construction can start significantly before the design is complete. The kinds of conflicts that can and do occur in the other methods become internal management issues for the DB team.

There are two kinds of design-build projects: progressive DB and lump sum DB.

Progressive DB (negotiated approach)—selection is based primarily on qualifications: In the first phase, DB works with the owner to develop project scope, preliminary design, initial permits, schedule, and GMP for owner approval. In the second phase, DB proceeds to complete design, permitting, construction, testing, acceptance, and closeout.

Lump sum DB (best value basis)—the request for qualifications (RFQ) is the shortlist of best qualified (typically 3) proposers. A technical package is developed by the owner (with a consultant) including project scope, design and performance criteria, and quality specs. The request for proposal (RFP) provides the technical package and invites bids from a select few qualified firms. The DB firm selection is based on a weighted combined score of technical approach and price (best value).

Construction Operations and Methods

Machine Production

When using construction equipment, concepts such as capacity, cycle time, efficiency, production rate are used to calculate the overall rate of production of that equipment. When calculating earth excavation operations, adjustments must be made for volume changes due to shrinkage and swell.

Example 17.1

A dump truck having a capacity of 15 yd^3 is used at a job site to haul excavated material. The distance from the excavation pit to the dump site is 5 mi and the average speed of the truck is 25 mi/h. The total time to load and unload a single load off the dump truck is 10 min. The overall efficiency factor for the dump truck is 0.85. The swell factor for the excavated soil is 1.25.

(a) What are the number of trips the dump truck needs to make to move material from a pit with bank volume = 120,000 ft^3.

(b) How many 8-h working days are needed if using a single truck?

(c) If a fleet of trucks is used to transport the material, how many trucks are needed to finish the haul operation in two days?

Solution

(a) Bank volume = 120,000 ft^3 = 120,000 ÷ 27 = 4444 yd^3

Hauled volume = 1.25 × 4444 = 5555 yd^3

At 15 yd^3 hauled per trip, the number of trips needed at 100% efficiency = 370.4 trips

Operating at 85% efficiency, number of trips = 370.4 ÷ 0.85 = 435.7 (436 trips)

(b) One way trip time = 5 miles ÷ 25 mi/h = 0.2 h = 12 min

Each trip requires 10 min (load unload) + 24 min (two-way travel) = 34 min = 0.567 h

In an 8-h day, the number of daily trips = 8 ÷ 0.567 = 14.1 trips (14 complete trips).

Therefore, a single truck (operating at 85% efficiency) needs 436 ÷ 14 = 31.1 days.

(c) To finish hauling 5555 yd^3 in 2 days, daily haul = 2778 yd^3 (loose)

Each truck, operating at 85% efficiency, makes 14 trips per day = 210 yd^3
Therefore, number of trucks needed = 2778 ÷ 210 = 13.23 trucks (use 14 trucks)

Production Cycle Time

Every piece of construction equipment has an ideal capacity, an overall operational efficiency and a cycle time. The number of work cycles in a working day is obtained by dividing the length of the working day by the cycle length. If the capacity of the equipment is then multiplied by the number of daily work cycles, we obtain the daily production rate of the equipment. For example, for an excavator with capacity C (typical units yd^3), cycle time T (typical units min), operating at an overall efficiency E, the production P (typical units yd^3/h) is given by

$$P = E \times \frac{C}{T} \times 60 \tag{17.1}$$

Example 17.2

An excavator with a capacity of 3 yd^3 has a cycle time of 45 s. The overall efficiency of the operation is 0.84. What is the soil export (yd^3) from the jobsite in an 8-h day?

Solution

Number of daily cycles = 8 h ÷ 45 s = 640
Daily production (ideal) = 640 × 3 = 1920 yd^3
Daily production (actual) = 0.84 × 1920 = 1613 yd^3

Equipment Balancing

When several pieces of equipment need to work in tandem for accomplishing a task, the overall productivity will depend on the equipment with the longer cycle time, while the other equipment will have idle time. Consider the following example.

Example 17.3

A front end loader has an output of 220 yd^3/h (bank measure). It is used to load a fleet of seven trucks, each of which has a capacity of 15 yd^3 (loose). The truck cycle time, which includes load time, travel time an unload time is 18 min. The excavated soil has 20% swell and 10% shrinkage. The project requires 25,000 yd^3 of compacted soil. How many hours are needed to complete the job?

Solution Overall requirement = 25,000 yd^3 (compacted). This represents 90% of the bank measure.

Bank measure = $25{,}000 \div 0.9 = 27{,}778$ yd^3

Due to 20% swell, this soil will occupy $1.2 \times 27{,}778 = 33{,}333$ yd^3 in the loose condition

Loader productivity = 220 yd^3/h (bank)

Each truck makes 3.33 trips per hour (cycle time = 18 min). Therefore, the fleet of seven trucks makes 23.33 trips per hour, transporting $23.33 \times 15 = 350$ yd^3 loose soil, which is equivalent to $350 \div 1.2 = 291.67$ yd^3 bank soil per hour. This is higher than the loader's capacity and so the loader's capacity will determine overall productivity.

Therefore, the time required = $27{,}778 \div 220 = 126.3$ h

Under these conditions, the fleet of seven trucks has some idle time.

Example 17.4

For example 17.3, what is the size of the largest truck fleet for which trucks have no idle time?

Solution As before, we compute loader productivity = 220 yd^3/h (bank)

Each truck has productivity = $3.33 \times 15 = 50$ yd^3 loose soil per hour, which is equivalent to $50 \div 1.2 = 41.67$ yd^3 bank soil per hour.

For the productivity of the trucks to equal that of the loader, minimum number of trucks needed = $220 \div 41.67 = 5.28$.

Therefore, if five trucks are used, they will have no idle time.

Scheduling

A schedule is essential for managing any kind of project. A schedule shows relationships between project tasks and can be used to obtain a snapshot of project progress. Methods such as the critical path method (CPM) can be used to understand which tasks are "critical" and the extent of "float" in the noncritical tasks. Because conditions in the field hardly ever remain as initially conceived, a project schedule becomes a "living" document that must be updated periodically.

Gantt Charts

A Gantt chart is a bar chart that is often used to track the scheduling of activities in a project. It can also be used to track budget, resources and equipment usage. A Gantt chart, when updated regularly, can give an accurate snapshot of the status (percent completion) of activities, as well as incurred expenditures and equipment allocations. The major advantage of the Gantt chart is that time relationships between tasks are immediately apparent. It is also quite simple to tweak the information displayed on a Gantt chart to ask "what-if" questions.

A major disadvantage is that precedence relationships are difficult to display, and are therefore typically not shown on a Gantt chart. This disadvantage can be overcome by drawing a time-scaled network.

An example of a time-scaled network is shown in Fig. 17.1.

Critical Path Analysis

The benefit of using critical path analysis (CPA) over Gantt charts is that CPA formally identifies tasks which *must* be completed on time for the whole project to be completed on time, and also identifies which tasks can be delayed for a while if resources need to be reallocated to catch up on missed tasks. A further benefit of CPA is that it helps calculate the minimum length of time needed to complete a project. The disadvantage of CPA is that the relation of tasks to time is not as immediately obvious as with Gantt charts.

The following conventions and abbreviations are used in this section.

ES	Early start
EF	Early finish
LS	Late start
LF	Late finish
TF	Total float
FF	Free float
D	Duration

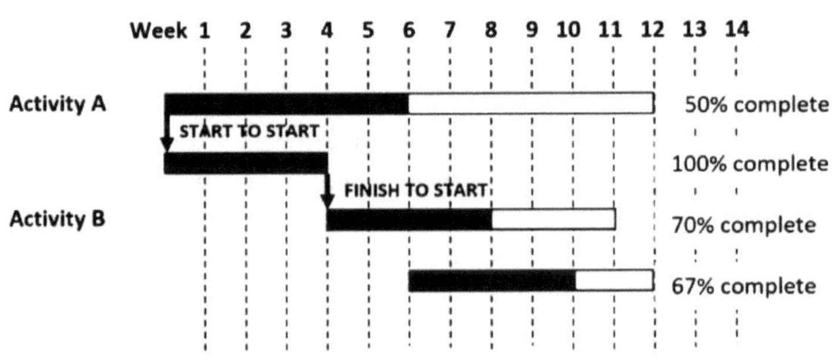

Figure 17.1 Time-scaled network.

The start date for the project is designated as zero. The first activity therefore has $ES = 0$. The finish date for an activity is calculated as

$$EF = ES + D \tag{17.2}$$

$$LF = LS + D \tag{17.3}$$

The early start date for any following activity is taken to be equal to the latest finish date of all predecessor activities.

Total Float

Float or slack is the amount of time that a task in a project network can be delayed without causing a delay to subsequent tasks (free float) or project completion date (total float).

The total float (*TF*) of an activity is calculated as the difference between late start and early start dates, or as the difference between late finish and early finish dates

$$TF = LS - ES = LF - EF \tag{17.4}$$

An activity that has a total float equal to zero is said to be a "critical activity," which means that a delay in the finish time of this activity will cause the entire project to be delayed by the same amount of time. A critical activity typically has free float equal to zero, but an activity that has zero free float may not necessarily be on the critical path. For example, see activities C, G and I in Example 17.5.

Free Float

The free float (*FF*) of an activity is the maximum time that it can be delayed without delaying the start of its successors. It is calculated as the latest time the activity could start but still finish before the early start of the next activity minus the earliest start time for the activity. In other words, the *FF* of an activity can be expressed as

$$FF_i = \min(ES_{i+1}) - EF_i \tag{17.5}$$

where $\min(ES_{i+1})$ is the smallest of early start times of all successors of activity *i* and EF_i is the early finish time of activity *i*.

NOTE For a terminal activity in the network, there are no successor activities. For such activities, the $\min(ES_{i+1})$ is taken as the *project duration.*

Activity on Node and Activity on Arrow Networks

An activity network may be represented using either activity-on-node (AON) or activity-on-arrow (AOA) diagrams. In an AON diagram, the nodes represent the activities and the arrows indicate the predecessor–successor relationships between these activities.

In an AOA diagram, the arrows represent the activities and the nodes represent instants in time indicating the start and end of each activity. In the AOA notation, arrows are scaled to represent activity durations. An activity connecting nodes i and j is represented by duration t_{ij} and scaled accordingly. A major disadvantage of the AOA notation is that spurious relationships between activities may arise. The use of dummy arrows is common with AOA networks to avoid such misinformation.

Example 17.5

Construct the flow diagram for the project described in the table. Identify the critical path and parameters such as early start, late start, early finish, late finish for each task.

Task	Predecessors	Successors	Duration
A	–	B, C	2
B	A	D, E	3
C	A	G	1
D	B	F	3
E	B	F	4
F	D, E	J	1
G	C	H, I	2
H	G	J	1
I	G	K	4
J	F, H	L	3
K	I	L	2
L	J, K	–	2

The activities networks for the project are shown in Figs. 17.2 and 17.3.
Executing the *forward pass* through the network, we obtain the following:

For activity A: $ES_A = 0, EF_A = 2$

For activity B: $ES_B = 2, EF_B = 5$

For activity C: $ES_C = 2, EF_C = 3$

For activity G: $ES_G = 3, EF_G = 5$

For activity H: $ES_H = 5, EF_H = 6$

For activity I: $ES_I = 5, EF_I = 9$

For activity K: $ES_K = 9, EF_K = 11$

For activity D: $ES_D = 5, EF_D = 8$

For activity E: $ES_E = 5, EF_E = 9$

For activity F: $ES_F = 9, EF_F = 10$

For activity J: $ES_J = 10, EF_J = 13$

For activity L: $ES_L = 13, EF_L = 15$

The project duration is therefore 15 days.

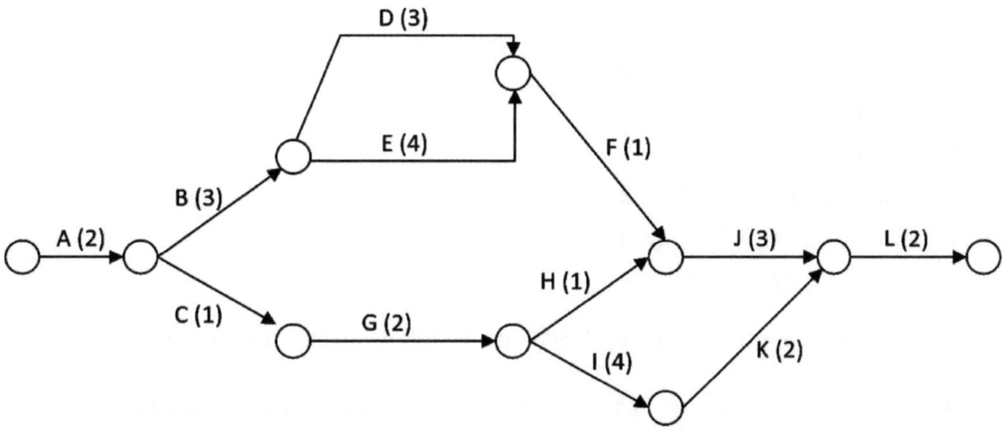

Figure 17.2 Activity on arrow representation of the example network.

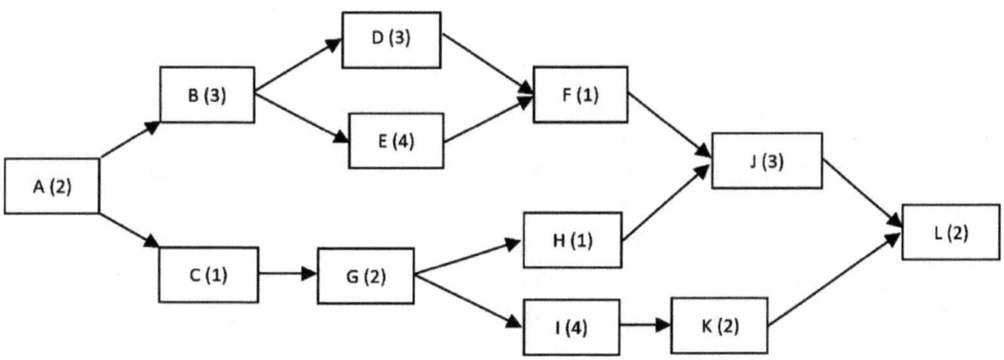

Figure 17.3 Activity on node representation of the example network.

Activity L is the terminal task of the project and its early finish date is the same as the project duration. If there were multiple ends to parallel paths, the activity with the latest finish date (same as the project duration) would be on the critical path and therefore have late finish date equal to the early finish date. *All terminal tasks would then have their late finish date set equal to the project duration.* Therefore, the late finish date for L is 15 and the late start date for L is equal to 15 – $\text{duration}_L = 13$.

For activity L: $EF_L = 15$, $ES_L = 13$

For activity J: $EF_J = 13$, $ES_J = 10$

For activity K: $EF_K = 13$, $ES_K = 11$

For activity F: $EF_F = 10$, $EF_F = 9$

For activity H: $EF_H = 10$, $ES_H = 9$

For activity I: $EF_I = 11, ES_I = 7$

For activity G: $EF_G = 7, ES_G = 5$

For activity D: $EF_D = 9, ES_D = 6$

For activity E: $EF_E = 9, ES_E = 5$

For activity B: $EF_B = 5, ES_B = 2$

For activity C: $EF_C = 5, ES_C = 4$

For activity A: $EF_A = 2, ES_A = 0$

If all the details are filled in for each activity, the following table may be constructed.

Task	D_i	ES_i	EF_i	LS_i	LF_i	TF_i	Successors	min(ES_{i+1})	FF_i
A	2	0	2	0	2	**0**	B, C	2	0
B	3	2	5	2	5	**0**	D, E	5	0
C	1	2	3	4	5	2	G	3	0
D	3	5	8	6	9	1	F	9	1
E	4	5	9	5	9	**0**	F	9	0
F	1	9	10	9	10	**0**	J	10	0
G	2	3	5	5	7	2	H, I	5	0
H	1	5	6	9	10	4	J	10	4
I	4	5	9	7	11	2	K	9	0
J	3	10	13	10	13	**0**	L	13	0
K	2	9	11	11	13	2	L	13	2
L	2	13	15	13	15	**0**	—	15	0

In the table above, free float for each activity has been calculated using Eq. (17.5). The critical path for this project is therefore A-B-E-F-J-L, which has a total duration of 15 days. This is the longest path through the project and is therefore the minimum time in which the project can be completed. Note that every activity on the critical path has zero total float and zero free float. However, there are noncritical activities (C, G, I) that have zero free float. This simply means that delaying these activities will delay the start time for their successors, but will not necessarily delay the entire project. The critical path is highlighted in Fig. 17.4.

NOTE If the available time for the entire project is the same as the duration of the longest path, then the float time or slack on the critical path will be zero. In the table, note that the late finish time for the last activity L is set to be the same as the early finish time for that activity. The backward sweep through the network starts from this value.

However, if the time available for the project were different (say 17 days), then that value would be the start value for the backward sweep and would change all the values in

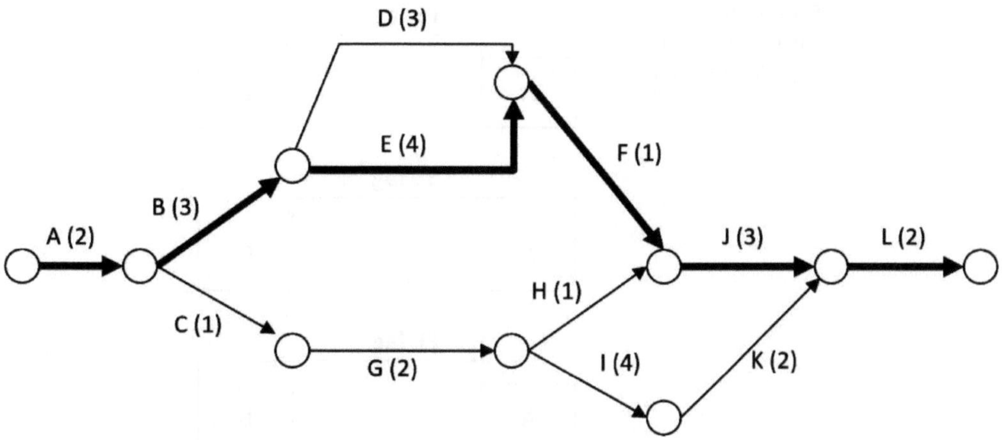

Figure 17.4 Critical path for the example network.

the *late finish* and *late start* columns. The float time along the critical path would then no longer be *zero*, but equal to the difference between the time available and the duration of the longest path.

NOTE If we look at the original task diagram, we see that for this relatively simple project, there are only four task-chains from project start to project finish. These are

ABDFJL—duration 14 days

ABEFJL—duration 15 days

ACGHJL—duration 11 days

ACGIKL—duration 13 days

Of these, ABEFJL is the longest and therefore is the critical path. All tasks on this path are critical and therefore must have total float and free float equal to zero. Depending on the question asked, such a quick approach may be adequate.

Lag Information

The previous discussion has assumed that there is no specified lag between activities. All predecessor–successor relationships have been assumed to have zero FS (finish-to-start) lag. In other words, as soon as a predecessor activity finishes, the successor activity can start. Activities that share the same start time are designated as SS and activities that have the same finish time are designated FF.

In many projects, two tasks are related but with a specified lag. A simple example of FS lag may be the time that must elapse between the end of a concrete pour and the construction of the next stage (thus allowing the concrete to reach a specified minimum strength). The FS lag can also be negative, which is referred to as *finish-to-start lead*.

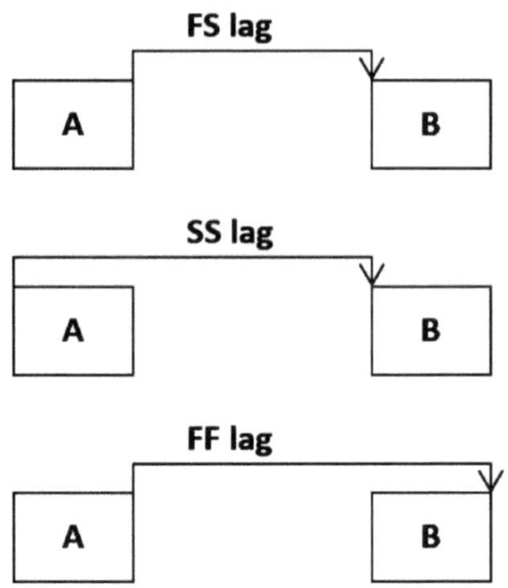

Figure 17.5 Unconventional lag relationships.

The lag between two activities can also be a SS (start-to-start) lag. In such a case, the start time of one activity lags the start time for another. If, on the other hand, the lag between the two activities is specified as a FF (finish-to-finish) lag, then there is a specified lag between the finish times of the two activities (Fig. 17.5). Symbolically, these relationships may be shown in the following ways.

Program Evaluation and Review Technique[1]

Program evaluation and review technique (PERT) is similar to critical path analysis (CPA). However, whereas CPA uses a single deterministic value for each activity duration, PERT uses a probabilistic approach that uses three time estimates for the duration of each activity since accurate estimates of task duration require good historical data.

Optimistic time a: Time an activity will take if everything goes perfectly.

Most likely time m: Most realistic time estimate to complete the activity.

Pessimistic time b: Time an activity takes if everything goes wrong.

These three time estimates can be used to compute the expected time μ and the standard deviation σ for each activity.

$$\text{Expected time } \mu = \frac{(a+4m+b)}{6} \tag{17.6}$$

$$\text{Standard deviation } \sigma = \sqrt{\text{variance}} = \left[\frac{(b-a)}{6}\right] \tag{17.7}$$

[1]See the Industrial Engineering section of the *FE Reference Handbook*, version 9.4.

Thus, if a path within the project comprises the tasks A, E, H, and K, the expected time for the path may be given by

$$\mu = t_A + t_E + t_H + t_K \tag{17.8}$$

where t_A, t_E, t_H, and t_K are mean time estimate for activities A, E, H, and K, respectively, and the variance for the path is given by

$$\sigma^2 = \sigma_A^2 + \sigma_E^2 + \sigma_H^2 + \sigma_K^2 \tag{17.9}$$

The project variance is the sum of the variances of each of the tasks on the critical path. The square root of this is the standard deviation of the path duration.

From the normal distribution equation:

$$z = \frac{(\text{due date} - \text{expected date of project completion})}{\text{project standard deviation}}$$

Looking this value up in the normal distribution tables gives the probability of the project being completed on our due date.

Example 17.6

Consider the project described in the following table:

Activity	Predecessor	*a*	*m*	*b*	Mean μ	Variance σ^2
A	—	1	2	4	2.17	0.25
B	—	5	6	7	6.00	0.11
C	—	2	4	5	3.83	0.25
D	A	1	3	4	2.83	0.25
E	C	4	5	7	5.17	0.25
F	A	3	4	5	4.00	0.11
G	B, D, E	1	2	3	2.00	0.11

What is the probability that the project completion time will be between 10 and 13 days?

Solution The mean and variance of the activity durations are calculated and tabulated in the last two columns of the table

$$\mu = \frac{(a + 4m + b)}{6}$$

$$\sigma = \left[\frac{(b-a)}{6}\right]$$

The activity on node network for the project is shown below. Start and Finish nodes represent "zero duration" activities.

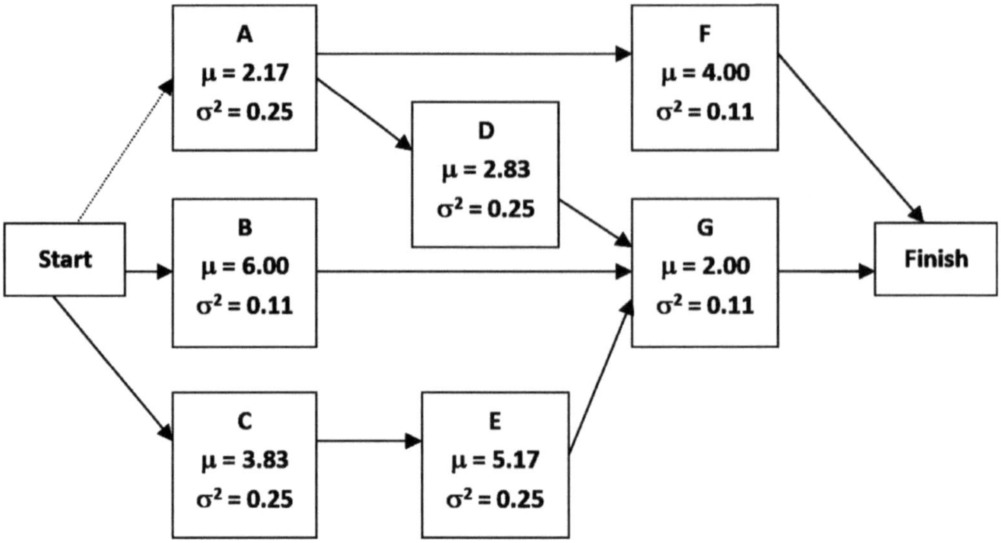

The network is simple enough that the critical path is seen as Start-C-E-G-Finish (mean duration 11 days).

$$\mu = \mu_C + \mu_E + \mu_G = 3.83 + 5.17 + 2.00 = 11.0$$

The variance of this path is calculated as

$$\sigma^2 = \sigma_C^2 + \sigma_E^2 + \sigma_G^2 = 0.25 + 0.25 + 0.11 = 0.61$$

$$\sigma = 0.782$$

The probability that the project duration is between 10 and 13 days is given by

$$P(10 \le T \le 13) = P\left(z \le \frac{13-11}{0.782}\right) - P\left(z \le \frac{10-11}{0.782}\right)$$

$$= P(z \le 2.558) - P(z \le -1.279) = P(z \le 2.558) - [1 - P(z \le 1.279)]$$

$$= 0.9948 - (1 - 0.8997) = 0.8945$$

Earned Value Management

In project management, *budgeted cost of work performed* (BCWP) or *earned value* (EV) is the budgeted cost of work that has actually been performed. On the other hand, the *budgeted cost of work scheduled* (BCWS) is the approved budget that has been allocated to complete a scheduled task during a specific time period. Other names for BCWS are *planned value* (PV) and *budget at completion* (BAC). *Actual cost of work performed* (ACWP) is the actual cost that has been spent, rather than the budgeted cost.

As an illustration of these terms, assume that a schedule contains a task that is budgeted to cost \$10,000 to perform, and is expected to begin on January 1st and complete at the end of January 10th.

At the end of January 5th, the work is scheduled to be 50% complete (5 days of the scheduled 10 days). So, at the end of January 5th, the BCWS is \$10,000 (the budgeted cost) times 50% (the scheduled completion percentage), or \$5000.

Suppose that by the end of January 5th, the work is actually only 30% complete. In this case, the BCWP would be \$10,000 (budgeted cost) times 30% (the actual completion percentage), or \$3000.

Also, suppose that to reach the 30% complete level at the end of January 5, \$2700 was actually spent. Then, the ACWP would be \$2700.

The cost variance (CV) is defined as the difference between BCWP and ACWP. Thus, in this example, CV = BCWP – ACWP = 3000 – 2700 = \$300 (positive value indicates under-budget).

The schedule variance (SV) is defined as the difference between BCWP and BCWS. Thus, in this example, SV = BCWP – BCWS = 3000 – 5000 = –\$2000 (negative value indicates behind schedule).

Cost performance index (CPI) is defined as the ratio of the BCWP to the ACWP. CPI greater than 1.0 means "under budget." CPI less than 1 means that the cost of completing the work is higher than planned.

In this example, CPI = BCWP/ACWP = 3000/2700 = 1.11.

Schedule performance index (SPI) is defined as the ratio of the BCWP to the BCWS. SPI greater than 1.0 means ahead of schedule.

In this example, SPI = BCWP/BCWS = 3000/5000 = 0.6 (indicating behind schedule)

The *estimate to complete* (ETC) is the estimate (*additional funds*) to complete the remaining part of the project. It is calculated as

$$\mathrm{ETC} = \frac{\mathrm{BCWS} - \mathrm{BCWP}}{\mathrm{CPI}} \tag{17.10}$$

In this example:

$$\mathrm{ETC} = \frac{5000 - 3000}{1.11} = 1800$$

The estimate at complete (EAC) represents the *total funds to complete* the project. It is calculated as

$$\mathrm{EAC} = \mathrm{ACWP} + \mathrm{ETC} \tag{17.11}$$

In this example, EAC = 2700 + 1800 = \$4500

Example 17.7

During exercising project controls at a certain time for a project, the project manager computes the following:

BCWS = \$435,000

ACWP = \$510,000

BCWP = \$488,000

Which of the following is true?

A. The project is behind schedule and over budget.

B. The project is ahead of schedule and under budget.

C. The project is behind schedule but under budget.

D. The project is ahead of schedule but over budget.

Solution The cost performance index is calculated as

$$\text{CPI} = \frac{\text{BCWP}}{\text{ACWP}} = \frac{488{,}000}{510{,}000} < 1$$

Therefore, the project is currently over budget

The schedule performance index is calculated as

$$\text{SPI} = \frac{\text{BCWP}}{\text{BCWS}} = \frac{488{,}000}{435{,}000} > 1$$

Therefore, the project is currently ahead of schedule

Answer is D.

OSHA Regulations for Construction Projects

Provisions in this section are based on the Part 1926 of the Code of Federal Regulations CFR29, which is compliant with section 107 of the Contract Work Hours and Safety Standards Act, commonly known as the Construction Safety Act.

Rules of Construction

By contracting for full performance of a contract, the prime contractor assumes all obligations prescribed as employer obligations, whether or not he subcontracts any part of the work. The prime contractor assumes the entire responsibility under the contract and the subcontractor assumes responsibility with respect to his portion of the work. With respect to subcontracted work, the prime contractor and any subcontractor or subcontractors shall be deemed to have joint responsibility.

All employees required to enter into confined or enclosed spaces shall be instructed as to the nature of the hazards involved, the necessary precautions to be taken, and in the use of protective and emergency equipment required. The employer shall comply with any specific regulations that apply to work in dangerous or potentially dangerous areas.

Confined or enclosed space means any space having a limited means of egress, which is subject to the accumulation of toxic or flammable contaminants or has an oxygen deficient atmosphere. Confined or enclosed spaces include, but are not limited to, storage tanks, process vessels, bins, boilers, ventilation or exhaust ducts, sewers, underground utility vaults, tunnels, pipelines, and open top spaces more than 4 ft in depth such as pits, tubs, vaults, and vessels.

Hazardous substance means a substance which, due to being explosive, flammable, poisonous, corrosive, oxidizing, irritating, or otherwise harmful, is likely to cause death or injury.

Site Layout

The controlling contractor shall ensure that the following is provided and maintained:

1. Adequate access roads into and through the site for the safe delivery and movement of cranes, trucks, other necessary equipment, and the material to be erected and means and methods for pedestrian and vehicular control.
2. A firm, properly graded, drained area, readily accessible to the work with adequate space for the safe storage of materials and the safe operation of the erector's equipment.

Emergency Response

The employer shall establish an emergency action plan which should list the types of evacuation to be used in emergency situations. Before implementing the emergency action plan, the employer shall designate and train a sufficient number of persons to assist in the safe and orderly emergency evacuation of employees.

The employer shall review the plan with each employee covered by the plan at the following times: initially when the plan is developed, whenever the employee's responsibilities or designated actions under the plan change, and whenever the plan is changed.

The employer shall make provisions for prompt medical attention in case of serious injury. In addition, the employer shall provide proper equipment for prompt transportation of the injured person to a physician or hospital, or a communication system for contacting necessary ambulance service. In areas where 911 service is not available, the telephone numbers of the physicians, hospitals, or ambulances shall be conspicuously posted.

First aid supplies shall be easily accessible when required. The contents of the first aid kit shall be placed in a weatherproof container with individual sealed packages for each type of item, and shall be checked by the employer before being sent out on each job and at least weekly on each job to ensure that the expended items are replaced.

Where the eyes or body of any person may be exposed to injurious corrosive materials, suitable facilities for quick drenching or flushing of the eyes and body shall be provided within the work.

If it is reasonably anticipated employees will be exposed to blood or other potentially infectious materials while using first aid supplies, employers should provide personal protective equipment (PPE). Appropriate PPE includes gloves, gowns, face shields, masks, and eye protection.

Every company must measure it safety performance, using experience modification rate (EMR), OSHA safety statistics, etc., and communicate these results to its employees. The EMR indicates whether the company is at, below or above average (EMR <1.0 better than average, EMR >1.0 worse than average) for number of injuries for its particular industry. OSHA uses EMR <1.0 as a criteria for enrollment in its voluntary protection program (VPP). In addition, OSHA uses EMR as enrollment criteria in some of its partnership programs.

The incident rate is an OSHA term for the number of recordable injuries and illnesses for a facility in a year, normalized to 100 employees working 2000 h/year. Thus, the incident rate is calculated as

$$\text{Incident rate} = \frac{\text{number of recordable injuries}}{\text{total annual hours worked by all employees}} \times 200{,}000 \qquad (17.12)$$

29 CFR 1926 provides guidelines about permissible limits on noise exposure (decibels), light exposure (W/cm^2), minimum illumination levels, minimum facilities (based on size of construction crew).

Exposure to impulsive or impact noise should not exceed 140 dB peak sound pressure level. When employees are subjected to permitted sound levels, feasible administrative or engineering controls shall be utilized. If such controls fail to reduce sound levels within the levels of the table, personal protective equipment shall be provided and used to reduce sound levels within the permissible levels.

Signs, Signals, and Barricades

Signs and symbols (danger signs which are predominantly red and caution signs which are predominantly yellow) shall be visible at all times when work is being performed, and shall be removed or covered promptly when the hazards no longer exist. In addition, exit signs and directional signs shall be provided wherever necessary.

Traffic Signs

All traffic control signs or devices used for protection of construction workers shall conform to Part VI of the *Manual on Uniform Traffic Control Devices*, Millennium Edition.

Fall Protection

The critical fall height (requiring fall protection) at a construction site is 6 ft. Any potential fall hazard shall be protected by passive fall protection devices such as guardrails, or employees shall be provided with fall arrest harnesses or other safety devices.

Safety Nets

Safety nets shall be installed as close as practicable under the walking/working surface on which employees are working, but in no case more than 30 ft (9.1 m) below such level. Each safety net (or section of it) shall have a border rope for webbing with a minimum breaking strength of 5000 lb (22.2 kN). The maximum size of each safety net mesh opening shall not exceed 36 in^2 nor be longer than 6 in on any side, and the opening, measured center to center of mesh ropes or webbing, shall not be longer than 6 in. All mesh crossings shall be secured to prevent enlargement of the mesh opening.

When nets are used on bridges, the potential fall area from the walking/working surface to the net shall be unobstructed. Safety nets shall extend outward from the outermost projection of the work surface according to provisions in 29CFR.

Lanyards

Lanyards and vertical lifelines shall have a minimum breaking strength of 5000 lb (22.2 kN). Self-retracting lifelines and lanyards which automatically limit free fall distance to 2 ft (0.61 m) or less shall be capable of sustaining a minimum tensile load of 3000 lb (13.3 kN) applied to the device with the lifeline or lanyard in the fully extended position.

Self-retracting lifelines and lanyards which do not limit free fall distance to 2 ft (0.61 m) or less, ripstitch lanyards, and tearing and deforming lanyards shall be capable of sustaining a minimum tensile load of 5000 lb (22.2 kN) applied to the device with the lifeline or lanyard in the fully extended position.

Quantity Estimating

A very important step in the process of preparing a bid is preparing an estimate of quantities (materials, labor, and equipment) involved in the proposed solution. Two significant components of this estimate are (1) estimating quantities accurately (quantity take-off) and (2) estimating unit prices as accurately as possible. The bill of quantities (BOQ) is an itemized list of materials, parts and labor and associated costs required to construct, maintain, or repair a structure.

There are various types of entities that require estimating. The unit costs for these entities are consistent with the entity type. For example, excavation quantities are estimated as a volume (yd^3 typical) and the unit cost must be specified in $\$/yd^3$. Some of the most commonly encountered building components are listed in Table 17.1

For example, in the cost estimating of the construction of a free-standing concrete wall, the following categories must be considered:

1. Material cost of concrete (including waste)
   a. Calculate volume of wall in yd^3.
   b. Obtain the unit cost of concrete per yd^3 from a source such as RS means.

Table 17.1 Units of Measure of Various Activities

Unit measure	Typical units	Examples
Each	—	Doors, windows
Length	ft	Pipes
Area	ft^2 sheet	Forms, painting, plastering Plywood is sold in 4 × 8 ft sheets
Volume	yd^3 board-foot	Excavation, backfilling, concrete Lumber (equivalent to a 12 × 12 × 1-in volume)
Weight	Ton	Reinforcement steel
Effort	LH (labor hour)	Equipment, labor

Table 17.2 Standard (ASTM) Reinforcing Bar Sizes

US size #	Weight (lb/ft)	Diameter (in)	Area (in²)
3	0.376	0.375	0.11
4	0.668	0.500	0.20
5	1.043	0.625	0.31
6	1.502	0.750	0.44
7	2.044	0.875	0.60
8	2.670	1.000	0.79
9	3.400	1.128	1.00
10	4.303	1.270	1.27
11	5.313	1.410	1.56
14	7.650	1.693	2.25
18	13.600	2.257	4.00

2. Material cost of reinforcement
   a. Estimate weight of reinforcement (ton) from structural drawings.
   b. Obtain unit cost of steel reinforcement ($ per ton) from a source such as RS means.
3. Labor cost of placing concrete
   a. Estimate productivity (crew hours per 100 yd^3).
   b. Estimate total crew hours for the total volume of concrete to be placed.
   c. Calculate labor cost using hourly labor cost data.
4. Formwork material cost. If forms are reused, the cost of initial form erection ($ per ft^2) will be different from the cost of reuse of forms (the latter is considerably smaller).
   a. Estimate total wall surface area to be formed. Estimate how much area will apply to initial erection and how much area will apply to form reuse.
   b. Apply unit material costs to calculate total form material costs.
5. Formwork labor cost. The operations of form erection and stripping usually have different values of productivity associated with them, so must be calculated separately.

Table 17.2 lists standard (ASTM) reinforcing bar sizes.

Example 17.8

The (outer) plan dimensions of a wall footing for a building are 54 × 72 ft as shown below. The 18-in-thick footing is reinforced concrete and the wall is masonry (12-in blocks). The width of the footing is 3 ft and the bottom of the footing is located 30 in below grade. The wall is 1 ft thick and 14 ft high (measured from the top of the footing).

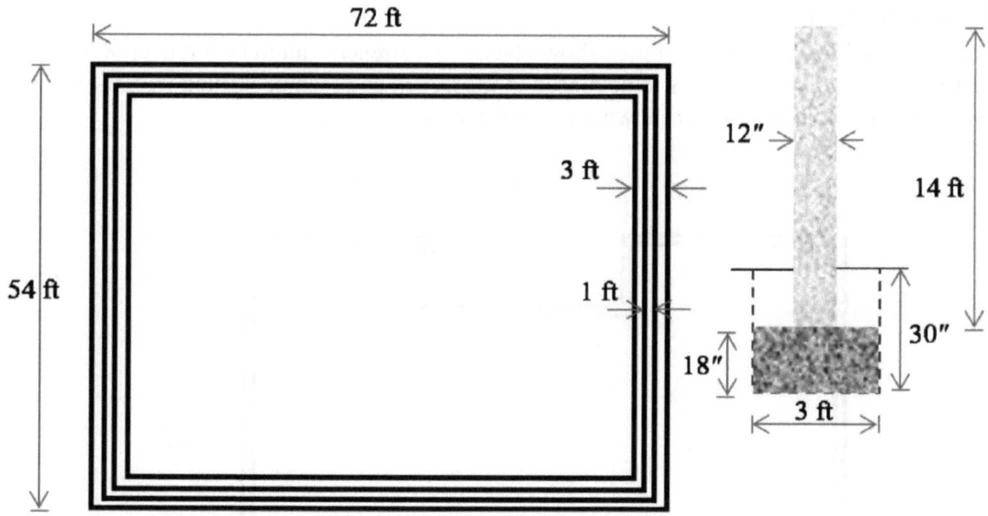

Estimate the following quantities: (1) total volume (yd^3) of concrete used for the footings, (2) the total number of 12-in masonry blocks (12D × 8H × 16L) used to construct the wall (extending 14 ft above the footing), and (3) the total excavation volume (yd^3).

Solution Dimensions of building (along footing centerline) = 69 × 51 ft
Perimeter (along footing centerline) = 2 × (69 + 51) = 240 ft

1. Cross section of footing = $1.5 \times 3 = 4.5\ ft^2$

   Volume of concrete in footing = $240 \times 4.5 = 1080\ ft^3 = 40\ yd^3$
2. Total wall "surface" area (centerline) = perimeter × height = $240 \times 14 = 3360\ ft^2$

   Each 12-in-wide masonry block is 8 in high and 16 in long.

   Wall surface area covered by one block = $8 \times 16 = 128\ in^2 = 0.89\ ft^2$

   Number of masonry blocks required = $3360 \div 0.89 = 3780$
3. Cross section of trench = $2.5 \times 3\ ft = 7.5\ ft^2$

   Volume of excavation = $7.5 \times 240 = 1800\ ft^3 = 66.7\ yd^3$

Quantity Estimation for Excavations

For excavations in soil, we must account for the fact that once the soil is excavated and released from the confining pressures in situ, it expands to a larger volume. This expansion is quantified in terms of the bulking factor (or swell) and typically expressed either as a percentage or a decimal. A bank volume V expands to a loose volume = $V(1+\text{bulking})$ upon being excavated. When this soil is placed and compacted into place, it usually occupies a lesser volume than the original bank volume. The shrinkage factor is also expressed as a decimal or percentage. The final compacted volume = $V(1 - \text{shrinkage})$.

Example 17.9

A contractor is excavating the trench shown below. The trench should be 5 ft deep × 3 ft wide. A 12-in-diameter water pipe is to be placed in the trench and then backfilled with the soil that was removed. All dimensions shown are on centerline.

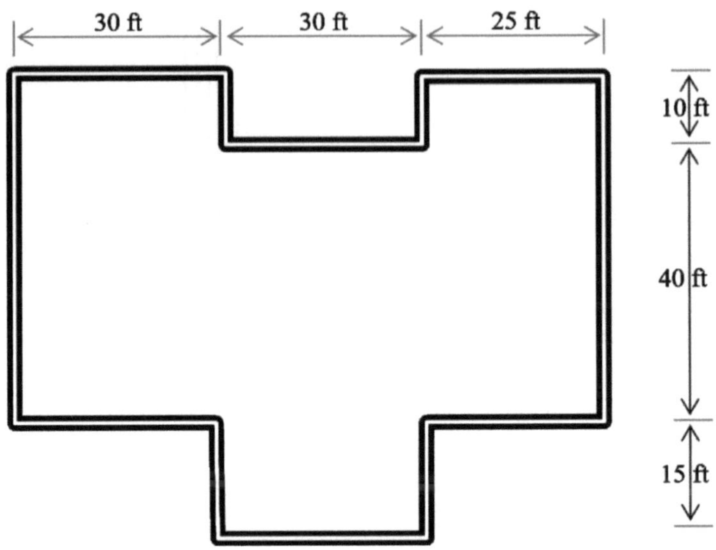

Does the contractor have enough soil to backfill the trench, or will some have to be borrowed? If borrow soil needs to be brought in, how much (LCY)?

Assume swell = 15% and shrinkage = 12%

Solution

Perimeter = $30 + 10 + 30 + 10 + 25 + 50 + 25 + 15 + 30 + 15 + 30 + 50 = 320$ ft
Cross section of trench = $3 \times 5 = 15$ ft^2
Volume of soil originally in the trench = $15 \times 320 = 4800$ ft^3
After compaction, excavated soil will fill a volume = $0.88 \times 4800 = 4224$ ft^3
Area of pipe = $\pi/4(1)^2 = 0.785$ ft^2
Area to be filled with soil = $15 - 0.785 = 14.215$ ft^2
Required soil (compacted) volume = $14.215 \times 320 = 4549$ ft^3
Deficit = $4549 - 4225 = 324$ ft^3 (compacted volume)

This soil is equivalent to $324 \div 0.88$ ft^3 = 368.2 ft^3 (bank volume) = $368.2 \times 1.15 = 423.4$ ft^3 (loose volume) = 15.68 LCY (loose cubic yards).

Estimating Mortar Quantity between Bricks or Masonry Blocks

A brick having dimensions $L \times W \times H$ is shown below. The front view of the wall is shown on the right, with the mortar joints between the brick and its adjacent units shaded in gray.

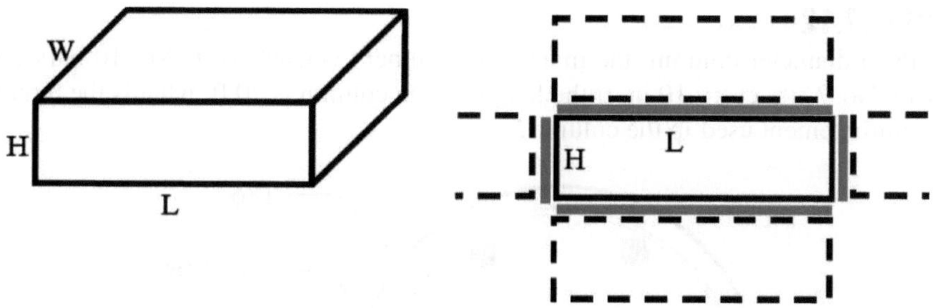

The top and bottom ($L \times W$) and left and right ($H \times W$) faces have mortar adjacent to them. If the mortar joint thickness is assumed to be t, then half that thickness is assumed to be associated with each brick. The total surface area (of four faces) covered by a thickness $t/2$ is then given by

$$S = 2W\left(L + \frac{1}{2}t + H + \frac{1}{2}t\right) = 2W(L + H + t)$$

When multiplied by the mortar thickness of $t/2$, the mortar volume is calculated as

$$V = 2W(L + H + t)\frac{t}{2} = W(L + H + t)t$$

Example 17.10

If mortar joint thickness in Example 17.8 is ½ in, what is the volume of mortar required for the entire wall?

Solution Thus, for the masonry blocks in Example 17.8, $L = 16$ in, $W = 12$ in, $H = 8$ in. If we assume mortar thickness = 0.5 in, the volume of mortar associated with each brick is

$$V = W(L + H + t)t = 12 \times (16 + 8 + 0.5) \times 0.5 = 147 \text{ in}^3$$

For Example 17.8, total mortar volume required (for 3780 bricks) = 555,660 in^3 = 321.6 ft^3 = 11.91 yd^3.

Example 17.11

The wall in Example 17.8 is reinforced with vertical reinforcement in the form of a No. 5 bar every 16 in. The wall height is 14 ft and the perimeter measured along the centerline is 240 ft. What is the total weight of reinforcement steel needed for the wall?

Solution

Wall perimeter = 240 ft
Number of rebars = 240 × 12 ÷ 16 = 180
Since each bar is 14 ft long, total length of reinforcement = 180 × 14 = 2520 ft
No. 5 bars have weight = 1.043 lb/ft (Table 17.2)
Total weight of reinforcement = 2520 × 1.043 = 2628 lb

Example 17.12

For the 18-in-diameter column, the main reinforcement consists of 6 No. 10 bars confined laterally by No. 3 ties every 10 in. If the length of the column is 20 ft, what is the total weight of steel reinforcement used in the column?

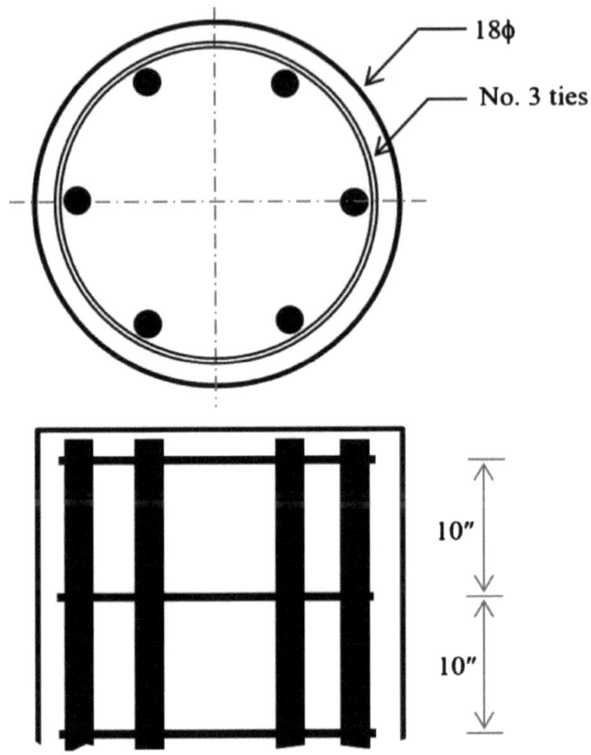

Solution Using the ACI recommended clear cover of 1.5 in, the diameter of the ties = $18 - 2 \times 1.5 - 0.375 = 14.625$ in

Circumference of each tie = $p \times 14.625 = 45.95$ in = 3.83 ft

Number of ties in a 20-ft-long column (using a spacing of 10 in) = $240/10 + 1 = 25$

Total length of ties (No. 3) = $25 \times 3.83 = 96$ ft

Total length of main reinforcement (No. 10) = $20 \times 6 = 120$ ft

Total weight of reinforcement (using values from Table 17.2) = $0.376 \times 96 + 4.303 \times 120 = 552.5$ lb

Example 17.13

A reinforcing steel quantity take-off needs to be done for a 32-ft-long wall footing. The cross section shown below shows all steel groups as designed. The footing thickness is 30 in and the width is 8 ft. Compute the total weight of steel (lb) required.

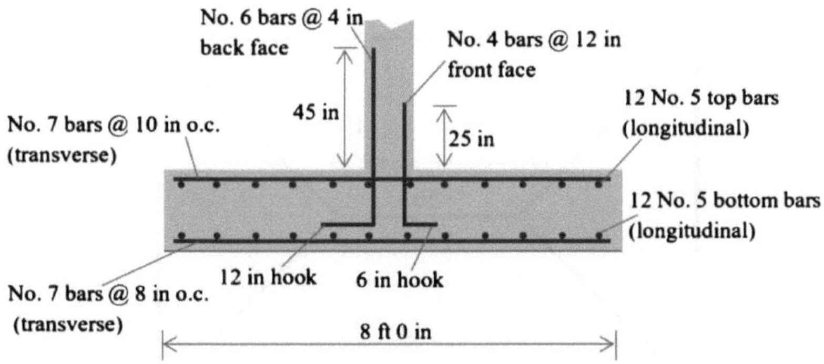

Solution The table below summarizes weight of reinforcement in each of the six categories.

Category	Size/Spacing	Number	Length (ft)	Weight (lb/ft)	Total weight (lb)
Top transverse	No. 7 @ 10 in	384/10 ≈ 39	39 × 8′ = 312	2.044	637.7
Bottom transverse	No. 7 @ 8 in	384/8 ≈ 48	48 × 8′ = 384	2.044	784.9
Top longitudinal	12 No. 5	12	12 × 32′ = 384	1.043	400.5
Bottom longitudinal	12 No. 5	12	12 × 32′ = 384	1.043	400.5
Front face	No. 4 @ 12 in	384/12 ≈ 32	32 × 57″ = 152	0.668	101.5
Back face	No. 6 @ 4 in	384/4 ≈ 96	96 × 83″ = 664	1.502	997.3
					3322.4

Therefore, total weight of reinforcement required for the wall = 3322.4 lb.

Note that length estimates for the various bars can be more precise than what is shown in the table. For example, the transverse bars that run the entire footing width will be a little shorter than 8 ft, because of clear cover requirements in the ACI specifications. Similarly, the top transverse bars (@ 10 in spacing o.c.) will span a little less than the 32 ft (384 in) length of the wall. This will lead to 38 spaces (380 ÷ 10 in) and therefore 39 bars. This is the same as that obtained by dividing 384 in by the 10 in spacing and rounding it up to 39.

Example 17.14

A building has horizontal dimensions 100 × 80 ft. It has a hipped roof with a pitch of 6 in/ft, and it has a 3-ft overhang.

How many sheets of plywood will be needed to cover the roof?

Solution If the eaves are 3 ft horizontally out from each wall, the horizontal dimensions of the roof are 106 × 86 ft. The plan and elevation views are shown below.

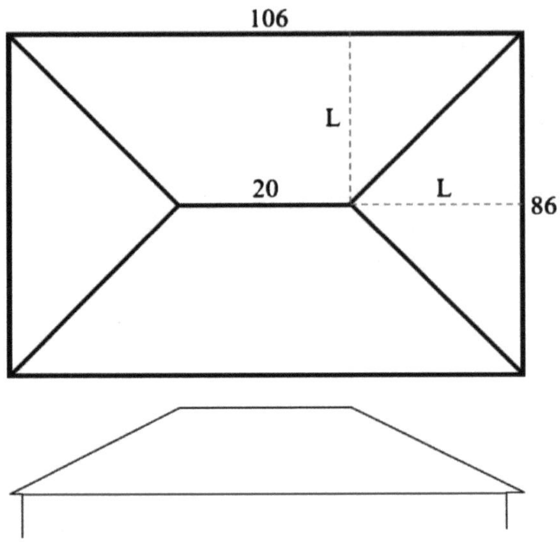

In the north-south direction, the horizontal projection from the eaves to the ridge = 43 ft. At a 1:2 pitch the height of the roof is 0.5 × 43 = 21.5 ft. The inclined length of the lines shown as dashed lines on the figure is given by the Pythagorean formula (using horizontal projection = 43 ft and vertical projection = 21.5 ft)

$$L = \sqrt{43^2 + 21.5^2} = 48.08 \text{ ft}$$

For both the triangular and the trapezoidal roof sections, L is the distance from the base to apex. The total area of the roof is equal to two each of these shapes.

Area of the triangle: $A_1 = \frac{1}{2} \times 86 \times 48.08 = 2067.4 \text{ ft}^2$

Area of the trapezoid: $A_2 = \frac{1}{2} \times (106 + 20) \times 48.08 = 2029.0 \text{ ft}^2$

The total roof area is twice the sum of these, since the roof consists of two of each type of section.

Total area of roof: $A = 2(A_1 + A_2) = 10{,}193 \text{ ft}^2$

Assuming 10% waste and each standard sheet of plywood being 4 × 8 ft in size, the number of sheets required is calculated as

$$N = \frac{10{,}193 \times 1.1}{4 \times 8} = 350$$

Suggested shortcut: A quick estimate can be made by noting that because of the pitch of the roof, a plan dimension of 43 ft corresponds to an inclined dimension of 48.08 ft. We can use this ratio (48.08/43) to estimate the roof area from the corresponding plan area as follows:

$$A = 106 \times 86 \times \frac{48.08}{43} = 10{,}193 \text{ ft}^2$$

CHAPTER 18

Surveying

Glossary of Terms

Backsight (BS): Elevation (rod reading) obtained by sighting back station. Also called plus sight (+S).

Benchmark: A fixed reference point or object, the elevation of which is known.

Contour: An imaginary line of constant elevation on the ground surface.

Deflection angle: An angle measured to a line from the extension of the preceding line.

Departure: The orthographic projection of a line on the east-west axis of the survey. East departures are considered positive.

Foresight (FS): Elevation (rod reading) obtained by sighting forward station. Also called minus sight (–*S*).

Height of instrument (HI): The elevation of the line of sight of the telescope above the survey station or control point.

Height of target: The elevation of the target or prism above the survey station or control point.

Latitude (astronomical): Angle measured along a meridian north (positive) and south (negative) from the equator. Latitude varies from 0° to 90°.

Latitude (traverse): The orthographic projection of a line on the north-south axis of the survey. North latitudes are considered positive.

Level surface: A curved surface, every element of which is normal to a plumb line.

Longitude: Angle measured at the pole, east, or west from the prime meridian. Longitude varies from 0° to 180° east or 180° west.

Meridian (astronomical): An imaginary line on the earth's surface having the same astronomical longitude at every point.

Zenith: The zenith is the point on the celestial sphere where the gravity vector, extended upward, intersects it.

Zenith angle: An angle formed between two intersecting lines in a vertical plane where one of these lines is directed toward the zenith.

Coordinate Systems

State Plane Coordinate System

Standardized in 1983 (NOAA Manual NOS NGS 5), the State Plane Coordinate system is a map projection system based on the North American Datum of 1983 (NAD 83). Given the geodetic coordinates (latitude and longitude) of a point, the equations in SPCS can be used to convert these coordinates to state plane coordinates (northings and eastings).

Global Positioning System

The global positioning system is based on a constellation of nongeosynchronous satellites, whose position at any time is known to a very high degree of precision. The Global Positioning System (GPS) was developed by the U.S. Department of Defense in 1978 and is officially named NAVSTAR. GPS receivers on the ground calculate the distance to these known reference points by measuring a time lag (phase shift) between a coded signal generated by the satellite and an identical signal generated by the receiver. When distances from four satellites are measured simultaneously, the intersection of the four imaginary spheres reveals the location of the receiver. Knowledge of the elevation of the point on the ground can reduce the number of necessary satellites to three. GPS data are susceptible to errors due to atmospheric and ionospheric effects, ephemeris and atomic clock errors, numerical errors, and errors due to multipath effects. Moving receivers, such as those in vehicles, are less susceptible to multipath errors.

Differential GPS may be used to achieve subcentimeter accuracy. The sum total of all errors inherent in measurements is estimated using a second receiver located at a known position. This error estimate is then used to adjust the computed position of other receivers in the same general locale.

Types of Surveys

Plane surveying methods consider the surface of the earth as a plane. Curvature of the earth surface is neglected. This type of survey is appropriate for small geographical areas.

Geodetic surveying takes into account the true (near-spherical or spheroidal) shape of the earth. When the survey covers a large geographical area, neglecting the curvature of the earth introduces significant errors.

Stationing

The concept of stations is commonly used for linear projects, such as in transportation engineering. In the United States, a common convention is to utilize 100-ft stations. Thus, a distance of 320.76 ft may be expressed as 3 + 20.76 stations. Earthwork (haulage) costs are often expressed in typical units of $\$/yd^3 \cdot sta$. This means the cost of hauling a soil volume of 1 yd^3 a distance of 1 station (100 ft).

Chaining Techniques

Distances used in surveying are generally horizontal distances, not sloped distances, for example, property lines on deeds, stationing, layout of building corners, etc. Even though chaining, as a distance measurement method, has been rendered obsolete by modern optical instruments, the history of surveying has involved the use of chains and tapes (typically steel) for measuring distance.

The results obtained by chaining need to be corrected for extension errors (created by the application of a tension on the tape to prevent excessive sag), thermal elongation errors (created by the difference between the average ambient temperature and the tape's standardization temperature), and sag errors (created by the self-weight induced sag of a tape which is held between two elevated supports).

Tension Correction

When a metal tape is held extended under tension P, greater than the tension at which it was standardized P_o, it experiences an elongation given by

$$e_P = \frac{(P - P_o)L}{AE} \tag{18.1}$$

where P = applied tension
P_o = tension at which tape or chain was standardized
L = length of tape or chain
A = cross-sectional area of tape or chain
E = modulus of elasticity of tape or chain (steel has E = 29,000 kips/in^2 or 200 GPa)

NOTE Applied tension greater than the standard tension makes the tape longer than its standard length, thereby recording a lesser tape reading (negative error), and the correction should be added (positive correction).

Temperature Correction

When a metal tape is used at a field temperature T different than the temperature at which it was standardized T_o, it experiences a change in length given by

$$e_T = L\alpha(T - T_o) \tag{18.2}$$

where α = coefficient of thermal expansion of chain or tape (steel has $\alpha = 6.5 \times 10^{-6}$/°F = 11.6×10^{-6}/°C)
L = measured length
T = ambient temperature
T_o = standardization temperature

NOTE Ambient temperature greater than the standard temperature makes the tape longer than its standard length, thereby recording a lesser tape reading (negative error), and the correction should be added (positive correction).

Sag Correction (Not in the *FE Reference Handbook*)

Surveying tape is best used if held elevated above ground, rather than laying on the ground, in which case it would be affected by the roughness of the ground. In spite of the fact that the tape must be held under tension, it does sag in the form of a catenary, thus increasing the recorded length. The correction due to sag is given by

$$e_S = \frac{w^2 L^3}{24P^2} \tag{18.3}$$

where w = weight of tape per unit length
L = horizontal distance between level supports
P = applied tension

NOTE Tape sag makes recorded length greater than actual (positive error), and the correction should always be subtracted (negative correction).

Angles and Distances

Azimuth

The azimuth of a line is the horizontal angle measured to the line from a specific meridian (usually north) in a particular direction (usually clockwise). Figure 18.1 shows the azimuths of lines AB and BA. The back azimuth of a line is the azimuth of the line running in the reverse direction. When the azimuth is less than 180°, the back azimuth equals the azimuth plus 180° and when the azimuth is greater than 180°, the back azimuth equals the azimuth minus 180°.

Bearings

Bearings of lines are directional (horizontal) angles with respect to a meridian (north or south) measured at the originating point on the line. For example, the bearing of a line headed in the north-east direction can be written as N45°E.

The angle between two lines whose bearings are given may be calculated as the difference between their azimuths. For example, if the two lines are N45°E and S22°W, then the

Figure 18.1 Azimuth and back azimuth of a line.

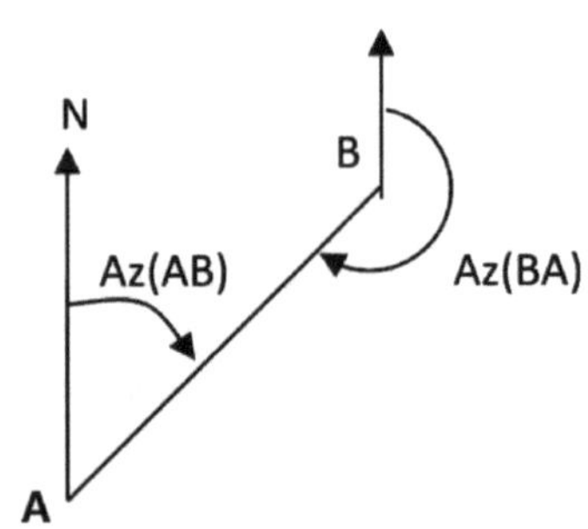

Figure 18.2
Bearings of lines and angle between lines.

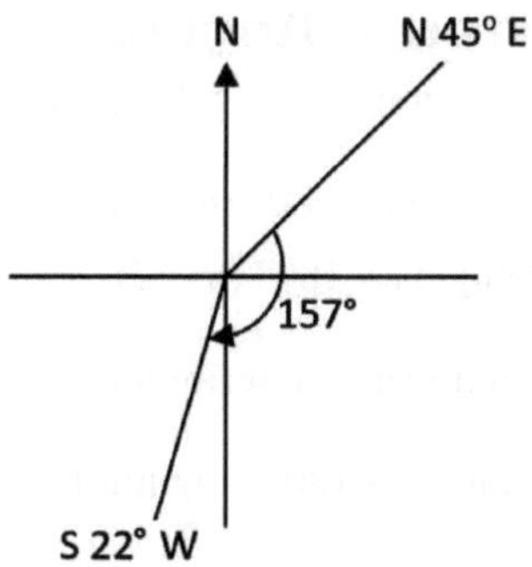

first step will be to describe them as azimuths. It is helpful to visualize the angles: S22°W is in the third quadrant; 22° west of south, as shown in Fig. 18.2. The azimuth of N45°E is 45° and the azimuth of S22°W is 22° + 180° = 202°. Therefore, the angle between the two lines is 202° – 45° = 157°.

Latitude and Departure

In the rectangular coordinate system, where the north-south meridian serves as the y axis and the east-west line serves as the x axis, the projections of a line are termed *latitude* (y projection) and *departure* (x projection), as shown in Fig. 18.3. The departure is considered positive to the east (i.e., the line shows an *increase in easting*) and the latitude is considered positive to the north (i.e., the line shows an *increase in northing*).

If the bearing of the line shown in Fig. 18.3 is NθE, where θ is the angle between the north meridian and the line (also known as the *azimuthal angle*), then the latitude is given by $L\cos\theta$ and the departure by $L\sin\theta$, where L is the length of line AB.

Northings and Eastings

Whereas latitudes and departures are north-south and east-west projections, respectively, of a line segment, northings and eastings are coordinates of points in a traverse. Thus, for line segment AB, the latitude is the difference of the northings of points A and B and the departure is the difference of the eastings.

Figure 18.3
Latitude and departure.

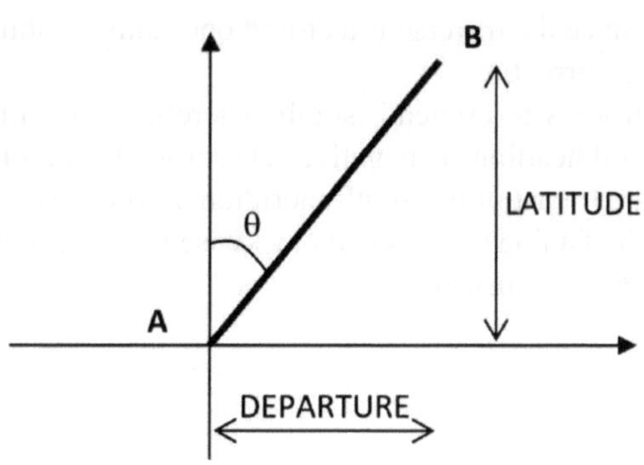

Interior and Exterior Angles

For a closed traverse (polygon with n sides), the following relations must hold

$$\text{Sum of all exterior angles} = 180^\circ\,(n + 2)$$

$$\text{Sum of all interior angles} = 180^\circ\,(n - 2)$$

For successive lines in a traverse, interior angles may be calculated from azimuths

$$\text{Interior angle} = \text{azimuth}_1 + 180^\circ - \text{azimuth}_2 = \text{back azimuth}_1 - \text{forward azimuth}_2$$

Example 18.1

A closed traverse ABCD is shown below. Which of the following statements is incorrect?

A. The bearing of line AD is S 23°E.

B. The azimuth of line BC is 73°.

C. The latitude of line AD is negative.

D. The sum of the latitudes (for all four lines) is equal to the sum of the departures.

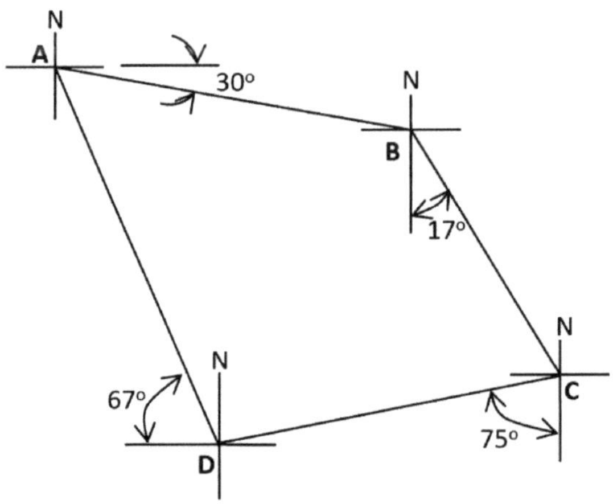

Solution Since the traverse is a closed one, sum of latitudes = 0 *and* sum of departures = 0. Therefore, D is correct.

Line AD (from A to D) heads south. Therefore, the latitude (which is considered positive with a northward heading) is negative. Therefore, C is correct.

Line AD is 23° east of the south meridian. A is correct.

The azimuth of a line is measured clockwise from the north meridian. For line BC, azimuth = 180 – 17 = 163°. B is incorrect.

Answer is B.

Example 18.2

The back tangent to a horizontal circular curve has a bearing of N 34°44'35"W. The deflection angle between tangents is 67° to the right. If the PC is located at coordinates 4453.51^m N, 643.29^m W, and the tangent length is 850.32^m, what are the coordinates of the PT?

A. 5657.19 N, 888.18 W

B. 5850.95 N, 643.29 W

C. 5871.32 N, 674.06 W

D. 4184.70 N, 147.33 E

Solution

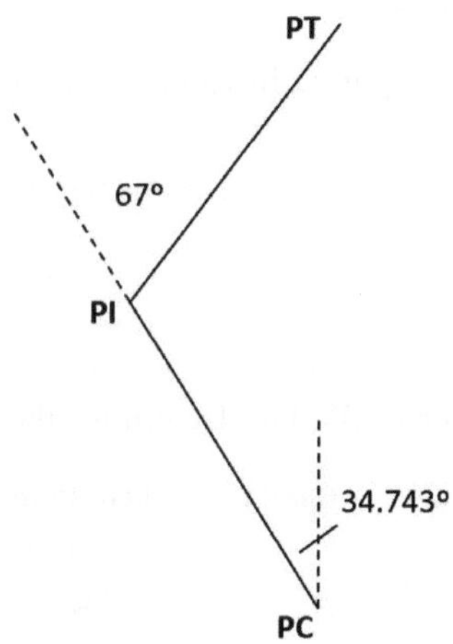

Bearing angle for forward tangent = 67° – 34°44'35" = N32°15'25"E (32.257°)

Coordinates of PI:

Northing = 4453.51 + (850.32)cos(34.743°) = +5152.23

Easting = –643.29 – (850.32)sin(34.743°) = –1127.88

Coordinates of PI: 5152.23 N, 1127.88 W

Coordinates of PT:

Northing = 5152.23 + (850.32)cos(32.357°) = +5871.32

Easting = –1127.88 + (850.32)sin(32.357°) = –674.06

Coordinates of PT: 5871.032 N, 674.06 W

Answer is C.

Example 18.3

Electronic distance measurement (EDM) instrument is used to sight a prism located at station B. The distance is measured as 234.78 m and the zenith angle is 73°32'30". If the height of instrument (H.I.) is 565.23 m above sea level, what is the elevation of the prism at B? (Neglect curvature and refraction effects.)

A. 776.85 m above sea level

B. 790.39 m above sea level

C. 394.92 m above sea level

D. 631.75 m above sea level

Solution

Zenith angle = 73°32'30" (73.542°)

Elevation difference between sighting instrument and prism = (234.78)cos(73.542°) = +66.52 m

Elevation of prism = elevation of sighting instrument + 66.52 = 565.23 + 66.52 = 631.75 m

Answer is D.

Example 18.4

The table below describes a traverse ABCDE. Determine the interior angle at C.

Line	Azimuth angle	Bearing	Length (m)	Deflection
AB	132°23'45"	–	271.12	–
BC	–	–	289.18	52°34' left
CD	137°46'32"	–	362.15	–
DE	–	–	213.32	96°56'22" left

A. 130°11'08"

B. 122°03'13"

C. 121°45'17"

D. 120°39'23"

Solution The azimuthal angles for all lines may be calculated starting with AB.

Az(AB) = 132°23'45" (given)

Az(BC) = 132°23'45" – 52°34'00" = 79°49'45" (deflection angle is subtracted because line deflects left)

Az(CD) = 137°46'32" (given)

Interior angle at C = azimuth(BC) – azimuth(CD) + 180° = 79°49'45" – 137°46'32" + 180° = 122°03'13"

Answer is B.

Example 18.5

The table below shows data for a closed traverse ABCD. All necessary closure corrections have been applied. What is the bearing of line AD?

Segment	Length (ft)	Bearing
AB	561.87	S 30°11'08" W
BC	1023.65	N 28°41'38" W
CD	653.67	N 73°16'22" E
DA		

A. N 76°09'16" W

B. N 13°50'44" W

C. S 13°50'44" E

D. S 76°09'16" E

Solution In the table below, the north-south and east-west projections of each line segment are calculated and listed (easting and northing positive).

Segment	Length (ft)	Bearing	Northing	Easting
AB	561.87	S 30°11'08" W	–485.68	–282.51
BC	1023.65	N 28°41'38" W	897.94	–491.49
CD	653.67	N 73°16'22" E	188.14	626.01

In order for the traverse to close perfectly, line DA must have northing = –600.40 and easting = +147.98.

Therefore, line AD has northing = +600.40 and easting = –147.98.

$$\tan\theta = \frac{147.98}{600.40} \Rightarrow \theta = 13.85°$$

Bearing of line AD = N 13.85 W = N 13°50'44" W.

Answer is B.

Example 18.6

A distance is measured as 65.123 m using steel tape on a day when the ambient temperature is 28°C. The tape is held "taut" by using a tension of 250 N. Use the following data:

Standardization temperature = 20°C

Thermal expansion coefficient, $\alpha = 11.6 \times 10^{-6}/°C$

Modulus of elasticity of steel = 205 GPa

Tape cross-sectional area = 3 × 0.5 mm

The tape length is standardized while supported on the ground with tension = 30 N

What is the distance, corrected for temperature and axial tension?

A. 65.132 m

B. 65.129 m

C. 65.117 m

D. 65.114 m

Solution

Temperature correction: $\alpha L\Delta(T - T_o) = 11.6 \times 10^{-6} \times 65.123 \times (28 - 20) = +0.00604$ m

Tension correction: $\dfrac{(P - P_o)L}{\text{AE}} = \dfrac{(250 - 30) \times 65.123}{(0.003 \times 0.0005) \times 205 \times 10^9} = +0.00283$ m

Corrected length = 65.123 + 0.00604 + 0.00283 = 65.132 m

Answer is A.

Example 18.7

A surveyor is using an instrument that can only measure angles (no distance measurement). She is using this instrument to sight the top of a building, as shown below. From position A, she measures the zenith angle to be 77°45'. She then moves the instrument to position B, which is 100 ft closer to the building and measures the zenith angle as 73°30'. What is the height of the building?

A. 61 ft

B. 67 ft

C. 81 ft

D. 86 ft

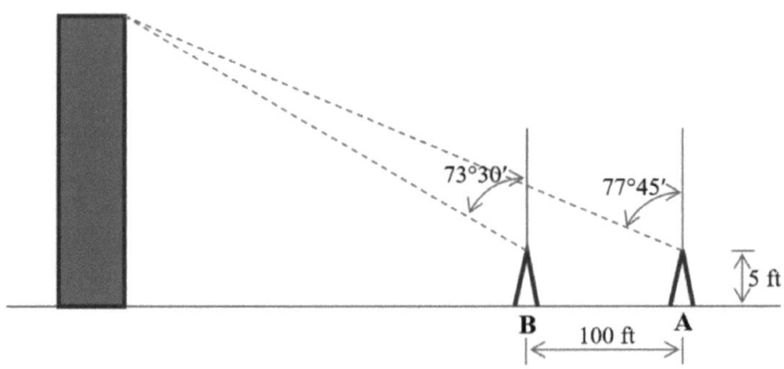

Solution Assuming the distance from B to the front of the building to be "X," we have the following relations:

$$\tan 16.5 = \frac{Y}{X} \Rightarrow Y = X \tan 16.5 = 0.2962X$$

$$\tan 12.25 = \frac{Y}{X + 100} \Rightarrow Y = X \tan 12.25 + 21.71 = 0.2171X + 21.71$$

Equating the two expressions for Y, we have $0.2171X + 21.71 = 0.2962X \Rightarrow 0.0791X = 21.71 \Rightarrow X = 274.5$ ft.

Therefore, $Y = 0.2962X = 81.3$ ft

Height of building = 86.3 ft

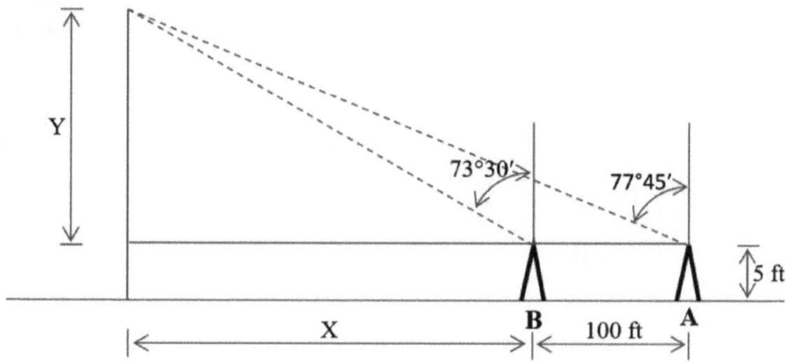

Answer is D.

Example 18.8

Points A, B, and C are keypoints on a traverse. The distance from A to B is 387.65 ft and the bearing of line AB is S 75°45' E. The distance from B to C is 712.12 ft and the bearing of line BC is N 25°30' E. What is the length of line AC?

A. 723.5 ft

B. 784.6 ft

C. 843.4 ft

D. 911.3 ft

Solution Since B is the vertex point (common to both rays), we can say that line BA (opposite to AB) has bearing N 75°45' W (azimuth 284.25) and line BC has azimuth 25.5.

Thus, the included angle at B is $360 + 25.5 - 284.25 = 101.25$.

Length AC (by the cosine rule) is

$$L_{AC} = \sqrt{387.65^2 + 712.12^2 - 2 \times 387.65 \times 712.12 \times \cos 101.25} = 843.35 \text{ ft.}$$

Answer is C.

Example 18.9

A 36-in-diameter concrete pipe (ID = 36 in, OD = 44 in) slopes at a longitudinal slope 0.5% as shown. The elevation of the pipe invert at A (sta. 34 + 45.30) is 238.15 ft above sea level. What is the soil cover over the pipe at location B (sta. 47 + 12.80) where the surface elevation is 239.23 ft?

A. 3.7 ft

B. 4.1 ft

C. 4.4 ft

D. 4.7 ft

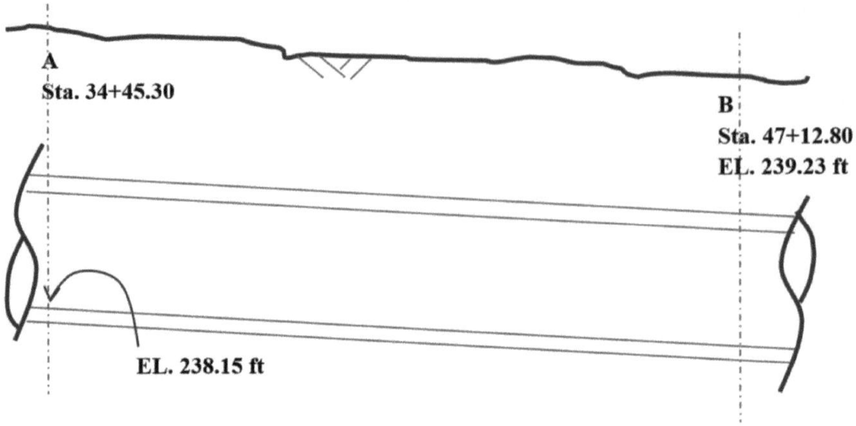

Solution

Distance from A to B = 4712.80 − 3445.30 = 1267.5 ft

Change in elevation = 0.005 × 1267.5 = 6.34 ft

Invert elevation at B = 238.15 − 6.34 = 231.81 ft

Pipe thickness = (44 − 36)/2 = 4 in. Therefore, top of pipe is 40 in above invert.

Therefore, elevation of top of pipe at B = 231.81 + 40/12 = 235.14 ft.

Therefore, cover at location B = 239.23 − 235.14 = 4.09 ft.

Answer is C.

Area of a Traverse by Coordinates

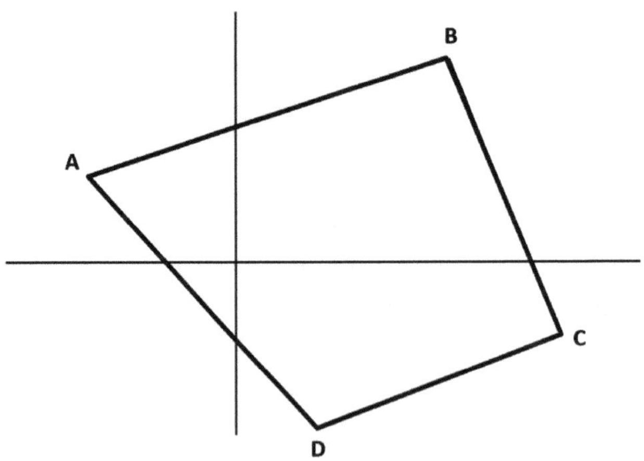

To calculate the area of any closed polygon by using the coordinates of its vertices, the formula is

$$A = \frac{1}{2}\left|\sum y_i \left(x_{i-1} - x_{i+1}\right)\right| \tag{18.4}$$

where x_i and y_i are orthogonal coordinates of the vertices. These are often given as eastings and northings, respectively.

Example 18.10

A closed traverse ABCD is described below in terms of coordinates (northings and eastings of stations A, B, C, and D). What is the area of the polygon ABCD?

A. 39.8 acres

B. 14.5 acres

C. 29.0 acres

D. 79.6 acres

	Northing (ft)	Easting (ft)
A	+520	−350
B	+850	+350
C	−450	+950
D	−700	−1150

Solution The area of a closed polygon is given by

$$A = \frac{1}{2}\left|\sum y_i \left(x_{i-1} - x_{i+1}\right)\right|$$

These are calculated in the table below:

	Northing (ft)	Easting (ft)	$X_{i-1} - X_{i+1}$	$Y_i(X_{i-1} - X_{i+1})$
A	+520	−350	−1150 − 350	−780,000
B	+850	+350	−350 − 950	−1,105,000
C	−450	+950	+350 − (−1150)	−675,000
D	−700	−1150	+950 − (−350)	−910,000
				−3,470,000

Therefore, the area of the polygon = 0.5 × 3,470,000 = 1,735,000 ft^2 = 39.83 acres.

Answer is A.

The calculation above can be quickly written by writing out all the coordinates (gray shaded cells in table below), repeating the first set of coordinates (+520, −350) at the end. The diagonal products are then grouped. One group yields the sum +1,752,000 and the other group yields the sum −1,718,000.

	+520	−350	
−297,500	+850	+350	182,000
−157,500	−450	+950	807,500
−665,000	−700	+1150	517,500
−598,000	+520	−350	245,000
−1,718,000			1,752,000

Area is given by $A = \frac{1}{2}[1,752,000 - (-1,718,000)] = 1,735,000 \text{ ft}^2$.

Area under an Irregular Curve

When calculating area under a curve (with known vertical ordinates), or volume of earthwork (with known end areas), several numerical schemes are available. Of these, the trapezoidal rule and Simpson's rule are commonly used. A requirement of both formulas is that the stations (marked as $i = 1, 2$, etc., in Fig. 18.4) be spaced equal distance apart (shown as Δ in the following equations). Also, for Simpson's rule, the number of stations (n) must be odd.

Using *linear* approximation between (regularly spaced) nodes we have the *trapezoidal rule*:

$$A = \frac{\Delta}{2}\left[y_1 + y_n + 2\sum_{k=2}^{n-1} y_k\right] \tag{18.5}$$

Using quadratic approximation between (regularly spaced) nodes, we have *Simpson's rule* (n must be odd):

$$A = \frac{\Delta}{3}\left[y_1 + y_n + 2\sum_{k=3,5,7,\ldots}^{n-2} y_k + 4\sum_{k=2,4,6,\ldots}^{n-1} y_k\right] \tag{18.6}$$

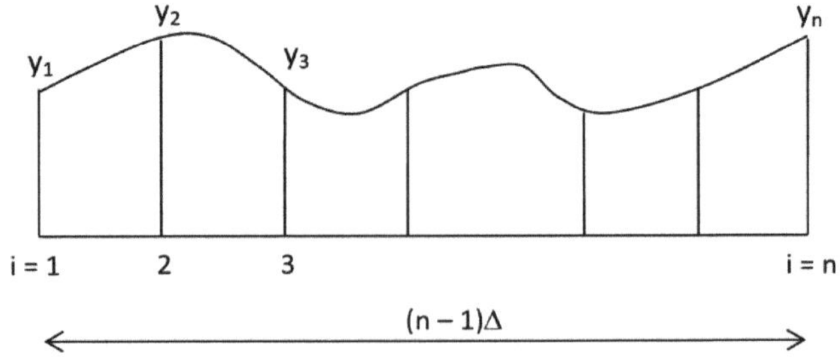

Figure 18.4 Calculating area under an irregular curve.

Example 18.11

The table below shows ordinates (y) at regularly spaced points (x). Estimate the area under the curve using (a) trapezoidal rule and (b) Simpson's rule.

	1	2	3	4	5	6	7
X_i (m)	0.0	0.5	1.0	1.5	2.0	2.5	3.0
Y_i (m)	2.6	3.5	2.4	4.2	4.1	2.0	3.0

Solution For the data shown above, we apply the trapezoidal rule as follows:

$$A = \frac{0.5}{2}[2.6 + 3.0 + 2(3.5 + 2.4 + 4.2 + 4.1 + 2.0)] = 9.5 \text{ m}^2$$

For the same data, applying Simpson's rule, we have

$$A = \frac{0.5}{3}[2.6 + 3.0 + 2(2.4 + 4.1) + 4(3.5 + 4.2 + 2.0)] = 9.57 \text{ m}^2$$

Example 18.12

Points A, B, and C are keypoints on a traverse. The distance from A to B is 387.65 ft and the bearing of line AB is S 75°45' E. The distance from B to C is 712.12 ft and the bearing of line BC is N 25°30' E. What is the length of line AC?

A. 723.5 ft

B. 784.6 ft

C. 843.4 ft

D. 911.3 ft

Solution Since B is the vertex point (common to both rays), we can say that line BA (opposite to AB) has bearing N 75°45' W (azimuth 284.25) and line BC has azimuth 25.5.

Thus, the included angle at B is 360 + 25.5 − 284.25 = 101.25.

Length AC (by the cosine rule) is

$$L_{AC} = \sqrt{387.65^2 + 712.12^2 - 2 \times 387.65 \times 712.12 \times \cos 101.25} = 843.35 \text{ ft}.$$

Answer is C.

Example 18.13

An excavation plan is outlined in the figure below. Each grid square is 50 × 50 ft. The numbers are depth of cut (ft) at indicated locations. The total volume of cut (yd^3) is most nearly

A. 1250

B. 2500

C. 3400

D. 4600

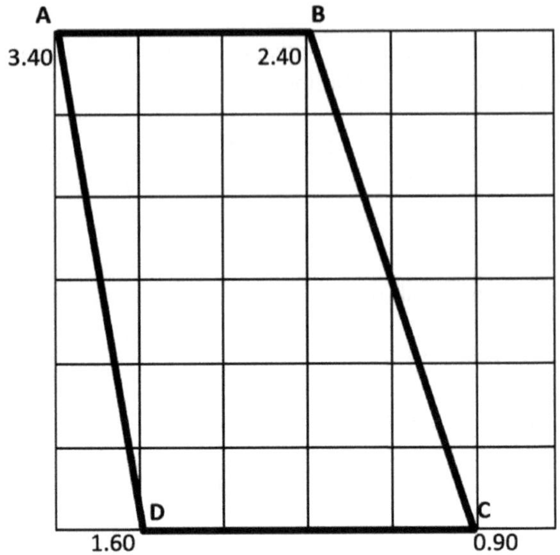

Solution Drawing the diagonal BD allows us to calculate the areas ABD and BCD as 9 squares and 12 squares, respectively (½ base times height).

$$A_{ABD} = ½bh = 0.5 \times 150 \times 300 = 22{,}500 \text{ ft}^2$$
$$d_{ave} = ⅓(3.4 + 2.4 + 1.6) = 2.43 \text{ ft}$$
$$V = 22{,}500 \times 2.43 = 54{,}750 \text{ ft}^3$$

$$A_{BCD} = ½bh = 0.5 \times 200 \times 300 = 30{,}000 \text{ ft}^2$$
$$d_{ave} = ⅓(2.4 + 0.9 + 1.6) = 1.63 \text{ ft}$$
$$V = 30{,}000 \times 1.63 = 49{,}000 \text{ ft}^3$$

Total volume = 103,750 ft^3 = 3842.6 yd^3

Example 18.14

Earthwork estimates for a highway project are summarized in the following table of end areas at stations 100 ft apart. The total earthwork volume (yd^3) between stations 20 + 0.00 and 24 + 0.00 is most nearly

A. 6500

B. 7000

C. 7500

D. 8000

Station	Area (ft^2)
20 + 00	405
21 + 00	576
22 + 00	432
23 + 00	378
24 + 00	630

Solution Using the average end area method (trapezoidal rule), earthwork volume is calculated as

$$V = \frac{1}{2} \times 100 \times [405 + 630 + 2 \times (576 + 432 + 378)] = 190.350 \text{ ft}^3 = 7050 \text{ yd}^3$$

Answer is B.

Example 18.15

For a particular site, the existing and the proposed ground profiles have been drawn to establish the following cross-sectional areas at stations spaced at 25 ft intervals. What is the volume of earthwork between stations 204 + 0.00 and 205 + 50.00 using Simpson's rule?

A. 2662

B. 2460

C. 1662

D. 1775

Station	End area (ft^2)
204 + 00.00	214.56
204 + 25.00	345.54
204 + 50.00	412.45
204 + 75.00	334.78
205 + 00.00	325.45
205 + 25.00	276.45
205 + 50.00	233.23

Solution Using Simpson's rule, earthwork volume is calculated as

$$V = \frac{1}{3} \times 25 \times [214.56 + 233.23 + 4 \times (345.54 + 334.78 + 276.45) + 2 \times (412.45 + 325.45)]$$

$$= 47{,}922.25 \text{ ft}^3 = 1774.9 \text{ yd}^3$$

Answer is D.

Example 18.16

Cut and fill areas are shown at three stations 50 ft apart. What is the total earthwork volume between station 12 + 0.00 and 13 + 0.00?

A. Cut 320 yd^3

B. Fill 320 yd^3

C. Cut 480 yd^3

D. Fill 480 yd^3

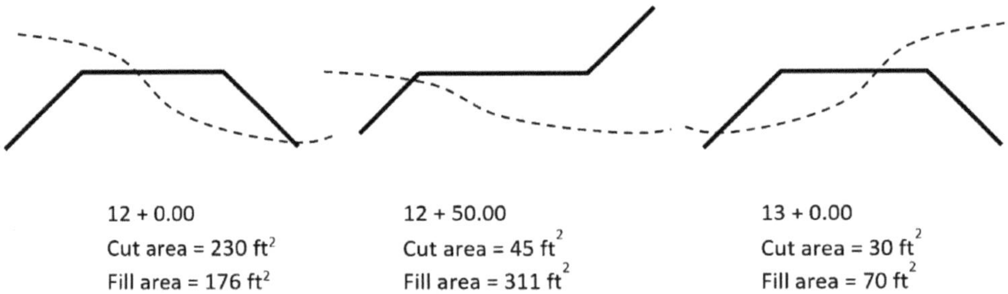

Solution

$$\text{Cut volume (positive earthwork): } V = \frac{50}{2} \times (230 + 30 + 2 \times 45) = +8750 \text{ ft}^3$$

$$\text{Fill volume (negative earthwork): } V = \frac{50}{2} \times (176 + 70 + 2 \times 311) = -21{,}700 \text{ ft}^3$$

Net earthwork volume (FILL) = 21,700 − 8750 = 12,950 ft^3 = 479.6 yd^3.

Answer is D.

Traverse Closure

A closed traverse must, in a perfect world, be closed, that is, it must have no closure error. However, in starting a traverse from a particular station and following several paths to finally return to the originating station, there are many length and angle measurements, each of which could have some degree of error. Errors can be instrument error or human error. The purpose of traverse closure methods is to distribute the closure error to all parts of the traverse, thereby attempting to approximate the true orientation of the lines and angles in the traverse.

Whereas greater accuracy in this correction may be achieved by more involved methods such as least-squares techniques, these are likely to be beyond the scope of the FE exam. Some less computationally demanding techniques using coordinate adjustment are described next.

Transit Rule

In this method, the coordinate error is distributed in proportion to the amount that various coordinates change between points. Thus, latitudes and departures of lines are adjusted, rather than their coordinates. This method is appropriate when accuracy of angular measurements is much better than accuracy of distance measurements.

$$\text{Latitude adjustment} = \frac{|\text{LAT}_i|}{\sum|\text{LAT}_i|} \times \delta y \tag{18.7}$$

$$\text{Departure adjustment} = \frac{|\text{DEP}_i|}{\sum|\text{DEP}_i|} \times \delta x \tag{18.8}$$

where LAT_i is the latitude (y projection) of the line being adjusted and δy is the northing closure error of the traverse and DEP_i is the departure (x projection) of the line being adjusted and δx is the easting closure error of the traverse.

Example 18.17

Transit Rule

The table below shows data for a traverse ABCD, with columns 2 and 3 giving the length and azimuth of each segment, respectively. Latitude and departure (columns 3 and 4) for each line segment are calculated using

$$\text{Latitude} = \text{length} \times \cos(\text{azimuth})$$

$$\text{Departure} = \text{length} \times \sin(\text{azimuth})$$

	Length (ft)	Azimuth	Latitude (ft)	Departure (ft)
AB	890.32	51°50'40"	+550.038	700.091
BC	1392.85	158°56'40"	−1299.853	500.413
CD	1079.35	256°36'00"	−250.137	−1049.966
DA	1011.20	351°28'00"	+1000.005	−150.047
			+0.054	+0.492

First, we need to calculate the closure error for latitude and departure. This is found by summing the latitude and departure values of all the lines. In a perfect closed traverse, these values would sum to zero. For the given data, the latitude and departure closure errors are 0.054 and 0.492 ft, respectively.

The negative of the closure error is the *closure correction*, and it must be distributed in proportion to each segment. So, the next step is to determine the appropriate proportions. The proportion to use is the ratio of the absolute latitude (or departure) of each segment to the sum of these values for all lines in the traverse [as in Eqs. (18.8) and (18.9)]. The results of these calculations are shown in the table below.

	Length (ft)	Azimuth	Latitude (ft)	Departure (ft)	ΔLat	ΔDep
AB	890.32	51°50'40"	+550.038	700.091	−0.010	−0.144
BC	1392.85	158°56'40"	−1299.853	500.413	−0.023	−0.103
CD	1079.35	256°36'00"	−250.137	−1049.966	−0.004	−0.215
DA	1011.20	351°28'00"	+1000.005	−150.047	−0.017	−0.031
			+0.054	+0.492	−0.054	−0.492

The modified latitude and departure values are obtained by adding the closure correction to the original values.

	Original latitude (ft)	Original departure (ft)	Latitude correction (ft)	Departure correction (ft)	Corrected latitude (ft)	Corrected departure (ft)
AB	+550.038	700.091	−0.010	−0.144	550.029	699.947
BC	−1299.853	500.413	−0.023	−0.103	−1299.876	500.311
CD	−250.137	−1049.966	−0.004	−0.215	−250.141	−1050.181
DA	+1000.005	−150.047	−0.017	−0.031	999.988	−150.077
	+0.054	+0.492	−0.054	−0.492	0.000	0.000

Compass Rule

This method for traverse closure is appropriate when accuracy of angular measurements is about the same as accuracy of distance measurements. The closure error has two components in Cartesian coordinates: the northing error (δy) and the easting error (δx). In this method, the coordinate error is distributed in proportion to the length of traverse lines. The assumption is that the greatest error will come from the longest shots.

Thus, every line in the traverse has its northing and easting adjusted according to

$$\text{Northing adjustment} = \frac{L_i}{\sum L_i} \times \delta y \tag{18.9}$$

where L_i is the length of the line being adjusted and δy is the northing closure error.

Easting and elevation values may also be adjusted using the same concept.

Example 18.18

Compass Rule

Solving the same problem using the compass rule, calculations are shown in the table below. The only difference between the two methods is the way in which the ratio is calculated for distributing the correction. For the compass rule, this ratio is calculated from the *lengths of the lines* AB, BC, CD, etc., rather than in terms of the latitudes and departures as in the transit rule.

	Length (ft)	Azimuth	Latitude (ft)	Departure (ft)	ΔLat	ΔDep
AB	890.32	51°50'40"	+550.038	700.091	−0.011	−0.100
BC	1392.85	158°56'40"	−1299.853	500.413	−0.017	−0.157
CD	1079.35	256°36'00"	−250.137	−1049.966	−0.013	−0.121
DA	1011.20	351°28'00"	+1000.005	−150.047	−0.013	−0.114
			+0.054	+0.492	−0.054	−0.492

The modified latitude and departure values are obtained by adding the closure correction to the original values.

	Original latitude (ft)	Original departure (ft)	Latitude correction (ft)	Departure correction (ft)	Corrected latitude (ft)	Corrected departure (ft)
AB	+550.038	700.091	−0.011	−0.100	550.027	699.991
BC	−1299.853	500.413	−0.017	−0.157	−1299.870	500.257
CD	−250.137	−1049.966	−0.013	−0.121	−250.150	−1050.087
DA	+1000.005	−150.047	−0.013	−0.114	999.993	−150.160
	+0.054	+0.492	−0.054	−0.492	0.000	0.000

As may be seen, the final adjusted values are close but not exactly the same for the transit rule and the compass rule.

Example 18.19

A triangular site ABC is surveyed. The table below shows latitudes and departures recorded in the field.

Segment	Latitude (ft)	Departure (ft)
AB	+250	+400
BC	−520	+350
CA	+272	−755

The transit rule for traverse closure says that corrections to latitudes and departures for each segment are to be done according to

$$\text{Latitude adjustment} = \frac{|\text{LAT}_i|}{\sum|\text{LAT}_i|} \times \Delta_y$$

$$\text{Departure adjustment} = \frac{|\text{DEP}_i|}{\sum|\text{DEP}_i|} \times \Delta_x$$

What are the corrected latitudes and departures of segment BC?

A. −521.00, +349.84

B. −519.00, +351.16

C. −521.00, +351.16

D. −519.00, +349.84

Solution As may be seen, the closure errors are $e_{LAT} = +2.0$ and $e_{DEP} = -5.0$. Therefore, the corrections must be negative (for latitude) and positive (for departure). Absolute values of latitudes and departures are shown in the table.

Segment	|LAT|	|DEP|
AB	250	400
BC	520	350
CA	272	755
Sum	1042	1505

Therefore, for segment, BC, the corrections are

$$\text{Latitude adjustment} = \frac{520}{1042} \times -2.0 = -0.998 \text{ ft}$$

$$\text{Departure adjustment} = \frac{350}{1505} \times +5.0 = +1.163 \text{ ft}$$

The corrected values (for BC) are: latitude = 519.002; departure = 351.163

Answer is B.

Example 18.20

A closed traverse ABCD has the recorded latitudes and departures as summarized in the table below.

Segment	Latitude (ft)	Departure (ft)
AB	−328.45	+702.29
BC	+1301.48	+589.38
CD	−238.93	−2076.91
DA	+1210.89	+786.23

What is the closure correction?

A. $\Delta X = +0.99$; $\Delta Y = +1.07$

B. $\Delta X = +1.07$; $\Delta Y = +0.99$

C. $\Delta X = -0.99$; $\Delta Y = +1.07$

D. $\Delta X = -1.07$; $\Delta Y = -0.99$

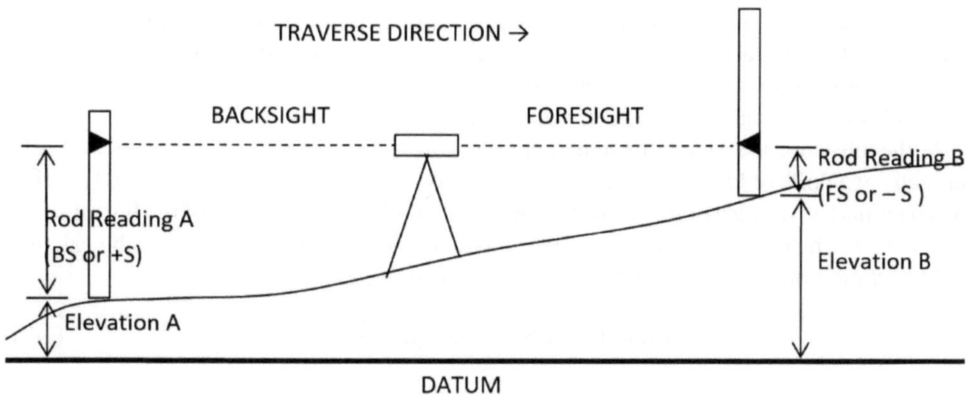

Figure 18.5 Differential leveling.

Solution As may be seen, the closure errors are $e_{\text{LAT}} = -1.07$ and $e_{\text{DEP}} = +0.99$.
Therefore, the corrections must be positive (for latitude, Y) and negative (for departure, X).

Answer is C.

Differential Leveling

Differential leveling is the process of measuring the difference in elevation between points as illustrated in Fig. 18.5. A transit level is used in combination with a rod or staff to determine elevations of points by obtaining differential vertical measurements. All elevations are referenced to a known elevation or *benchmark*. Table 18.1 shows typical field data for a series of staff readings obtained at successive stations A, B, C, … in a traverse. In this example, the

Table 18.1 Differential Leveling Example

Station	BS	HI	FS	Elevation	Notes
BM-A	6.35			1000.00	Benchmark—marker no. ES-12
		1006.35			
B	9.29		2.45	1003.90	
		1013.19			
C	6.15		3.78	1009.41	
		1015.56			
BM-D			5.56	1010.00	Benchmark—marker no. ES-13
Σ =	**21.79**	Σ =	**11.79**		**Check** ΣBS − ΣFS = 21.79 − 11.79 = +10.00 Ending elevation = 1000.00 +10.00 = 1010.00 OK

traverse is started at station A (known elevation 1000.00 ft and ended at station D (elevation 1010.00 ft). In each case, the height of instrument HI is equal to the elevation of the back station plus the backsight (BS) (e.g., 1000.00 + 6.35 = 1006.35) or the elevation of the forward station plus the foresight (FS) (e.g., 1003.90 + 2.45 = 1006.35).

As the telescope is leveled, the line of sight is horizontal (or level). A graduated rod is held vertically at a point of known elevation. The elevation of the benchmark *plus* the rod reading at the benchmark gives the height of instrument (HI), which is the elevation of the line of sight. If the rod is now held vertically at some other location, its unknown elevation can be calculated as the elevation of the line of sight *minus* the rod reading. Thus,

$$\text{Elevation}_A + \text{BS}_A = \text{Elevation}_B + \text{FS}_B \tag{18.10}$$

Example 18.21

The table below shows differential leveling data using a transit level. The starting station (A) is of known elevation. Find the elevation of station D.

Station	BS (m)	FS (m)	Elevation (m)	Notes
A	3.95	–	500.00	Benchmark
B	2.47	6.34		
C	3.81	5.51		
D	–	6.78		

A. 508.40 m

B. 489.77 m

C. 510.23 m

D. 491.60 m

Solution

Sum of all BS values = 10.23

Sum of all FS values = 18.63

Elevation of D = elevation of A + sum of BS − sum of FS = 491.60 m

Answer is D.

Example 18.22

A benchmark station (A) is located at coordinates 135,765.23 ft N and 349,231.67 ft W within the state coordinate system. The elevation at the top of a peg at this station is 453.94 ft above sea level. Station B is sighted from instrument located at station A. Height of instrument is 4.76 ft. The elevation of the top of a building located at coordinates 136,052.44 ft N and

348,535.29 ft W is 595.21 ft above sea level. What is the zenith angle of the line of sight from the instrument to the top of the building?

A. 89°21'01"

B. 86°37'23"

C. 82°34'52"

D. 79°43'42"

Solution Elevation of instrument = 453.94 + 4.76 = 458.70 ft

Using north as "Y, east as "X," and elevation as "Z" coordinate:

Coordinates of the instrument at A are (−349,231.67, +135,765.23, +458.70)

Coordinates of the top of building at B are (−348,535.29, +136,052.44, +595.21)

Elevation difference: $Y = 595.21 - 458.70 = 136.51$ ft

Horizontal distance: $X = \sqrt{(349231.67 - 348535.29)^2 + (135765.23 - 136052.44)^2}$

$= 753.28$ ft

Vertical angle: $\alpha = \tan^{-1}\left(\frac{Y}{X}\right) = \tan^{-1}\left(\frac{136.51}{753.28}\right) = 10.272$

Azimuth angle = 90 − 10.272 = 79°43'42"

Answer is D.

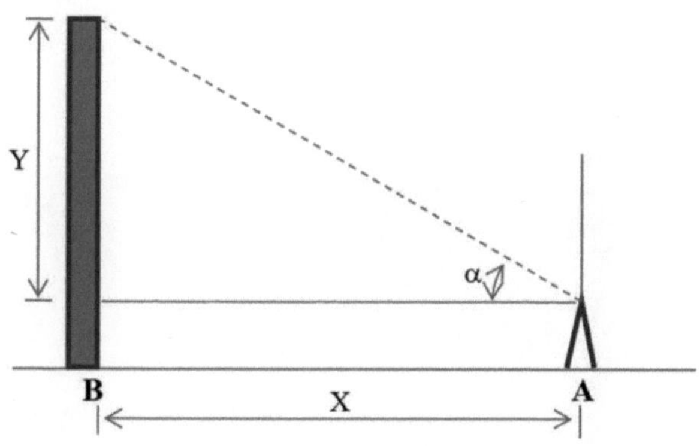

INDEX

B

E

F

G

H

I

M

Q

R

S

T

U

V

W

Y

Z

9 781260 011340

Printed by Libri Plureos GmbH in Hamburg,
Germany